P. E. Potter
J. B. Maynard
P. J. Depetris
Mud and Mudstones

The cover photograph taken by James Dixon from a helicopter in the Canadian Arctic is our entry for the most informative outcrop of mudstones in all the Americas. Located along Fish river in the northern Richardson Mountains of the Northwest Territories, this 150 m high outcrop has two contrasting mudstones separated by an unconformity above which is a large, detached slump fold.

The upper, light gray mudrock is Albion in age and rests discordantly on the unconformity at its base whereas the underlying dark gray to black, pyritic shale is lower Cenomanian in age, forms a steep cliff and is essentially flat lying. The sequence of events is: deposition of mud in an anoxic rift basin → uplift and erosion → post rift deposition of mud in a shallow and oxic foreland basin in front of the rising ancestral Rocky Mountains.

This spectacular outcrop invites many questions. Which mudstone would make the best base for a landfill? The best seal for a petroleum reservoir? What would study of the size, sorting, and abrasion of plant and animal microfossils tell about their environments of deposition? What is the significance of their contrasting styles of stratification? Did the shale and the mudrock have the same or different sources? And, how deeply were they buried? What is the significance of the detached asymmetrical fold? Is this unconformity related to a global event such as the opening of the nearby Arctic Ocean? These are but a few of the many fascinating questions that can be asked about mudstones of all ages everywhere (for a start on more questions-read our Challenges).

References

Dixon J (1993) Cretaceous tectonics and sedimentation in northern Canada. In: Caldwell WGE, Kauffman EG, eds., Evolution of the Western Interior Basin: Geol. Assoc. Canada Special Paper 39:119-130

Lane LS, Dietrich JR (1995) Tertiary structural evolution of the Beaufort Sea-Mackenzie Delta region, Arctic Canada. Bull Canadian Petrol Geol 43:293-314

P. E. Potter
J. B. Maynard
P. J. Depetris

Mud and Mudstones

Introduction and Overview

With 261 Figures and 48 Tables

Paul E. Potter
University of Cincinnati
Department of Geology
Cincinnati
45221-0013 Ohio
USA

J. Barry Maynard
University of Cincinnati
Department of Geology
Cincinnati
45221-0013 Ohio
USA

Pedro J. Depetris
Centro de Investigaciones Geoquímicas
y de Procesos de la Superficie (CIGeS),
FCEFyN, Universidad Nacional de Córdoba,
X5016GCA Córdoba
Argentina

ISBN 3-540-22157-3 Springer Berlin Heidelberg New York

Library of Congress Control Number: 2004106315

Springer is a part of Springer Science+Business Media
springeronline.com

Printed in Germany

Cover design: E. Kirchner, Heidelberg
Production: Almas Schimmel
Typesetting: LE-TeX Jelonek, Schmidt & Vöckler GbR, Leipzig
Printing and Binding: Stürtz AG, Würzburg
Printed on acid-free paper 32/3141/as 5 4 3 2 1 0

Preface

We wrote Mud and Mudstone for a broad audience whose work and interests include the wide spectrum of fine-grained argillaceous sediments and rocks. Thus our subtitle, Introduction and Overview.

Mud and Mudstone can be thought of as our answers to eight sequential questions,

- How are mud and silt produced?
- How are they transported and deposited?
- What is the role of oxygen at the site of deposition?
- What is known about mud in modern environments?
- What are the controlling processes and changes that occur with burial?
- How do we determine the provenance of mud and mudstone?
- How do we study ancient mudstone-rich basins?
- What are the practical aspects of muds and mudstones?

With this strategy in mind, Mud and Mudstone is the logical choice for both a one-semester course for geologists and engineers and for the self-taught. We especially hope it will be useful to professionals who need to quickly learn about selected aspects of the world's most abundant sedimentary rock and most widespread surface deposit.

To make Mud and Mudstone more readable for a wider audience, we only sparingly cited references in the text, but at the end of each chapter, we direct the reader to important sources in Digging Deeper. We note that GEOREF lists 61,219 references to clay, 34,841 to shale, 8908 to mudstone, 6355 to mud, 3802 to bentonite, 118 to underclay ... and 14 to mudrocks. Clearly there is an enormous literature related to fine-grained sediments; we have not attempted to cover this ground comprehensively, rather we provide the reader with a roadmap to follow into this territory that gives both highlight plus provides directions for individual exploration.

In our effort to make the book accessible to the widest audience, we have emphasized main themes, and avoided excessive jargon (and where we fail, we provide you with a glossary). We also use many tables to organize and summarize special aspects of a chapter, and use separate boxes to provide background for technical details. Most important, we have created a book rich in both number and quality of its illustrations. Finally, we provide a summary of essential methodology.

At the end of each chapter, we have set forth challenges, areas that we identify as having unresolved problems. We think of these as challenges to ourselves as much as to you? problems that we have long wondered about or research that we would like to do had we infinite time and resources.

We are most indebted to the many who have helped us over the last 10 years: typing (Sandi Cannell in Cincinnati, and Guilherme Mahlmann in Porto Alegre); library resources, especially interlibrary loans (Angela Gooden, Sarah Harper and Richard Spohn of the University of Cincinnati and Veleida Blank of the Universidade Federal do Rio Grande do Sul), and photography (Jay Yocis); and Evelyn Pence and Tim Phillips, our talented in-house illustrators, whose high skills and high standards are fully apparent in all the illustrations.

We especially thank all those who read the text. Foremost are Drs. Kenneth and Linda Fulton of Aberdeen, Ohio, who read all the text and greatly helped with its style and scope. Others include Kevin Bohacs of Exxon-Mobil, Michael Lewan and Charles W. Spencer of the U.S. Geological Survey, Scott McLennan of Stoney Brook, Linda Hinnov of The Johns Hopkins University, and Carl Brett and Warren Huff here in Cincinnati. Many geologists also sent us photographs and artwork that added much to the breadth of Mud and Mudstone. These include: Kevin Bohacs (Exxon-Mobil Upstream, Houston, TX), Arnt Bronger (Institut von Geographie, Kiel), James M. Coleman (Coastal Studies Institute, Baton Rouge, LA), John Dunham (Unocal Indonesia Balikpapon, Houston, TX), A. El Albani (Laboratoire de Géobiologie, Poitiers), Nicholas Eyles (University of Toronto, Scarborough, CN), T.D. Fouch (Geological Society of America, Boulder, CO), Warren D. Huff and Thomas Algeo (University of Cincinnati, OH), D.J. Miller (Virginia Polytechnic Institute, Blacksburg, VA), Neal R. O'Brien (SUNY, Potsdam, NY), Paul Olsen (Lamont-Dougherty Earth Observatory, Palisades, NY), H.H. Roberts and J.M. Coleman (Coastal Studies Institute, Baton Rouge, LA), S.-Y. Wu (Consultant, Houston, TX), Grant M. Young (University of Western Ontario, London), Scott M. McLennan (SUNY, Stony Brook), and Brandon C. Nuttall (Kentucky Geological Survey, Lexington). R.N. Ginsburg of Miami helped us with Figure 2.1.

We are most indebted for the sustained support we have received from the Department of Geology of the University of Cincinnati, the Department of Geociencias of the Universidade Federal do Rio Grande do Sul in Brazil, and the Consejo Nacional de Investigaciones Cientificas y Técnicos (CONICET) of Argentina. And kudos to Springer Verlag for patience, performance, and price.

Paul Edwin Potter,
J. Barry Maynard,
Pedro J. Depetris
Cincinnati, Ohio, USA and Córdoba, Argentina

Contents

CHAPTER 1

Overview

We need to know about mud and mudstones – you can't escape them

Detrital fine-grained sediments, all those composed of 50 percent or more of particles smaller than 62 micrometers, include terrigenous mud and mudstone, carbonate mud (micrite) and micstone and silt and siltstone plus deep sea ooze and sapropel. All of these may locally grade into one another and into carbonates and sandstones. Here we consider primarily the terrigenous mud-rich deposits and to a lesser degree their commonly associated silts and siltstones, which are nearly always present. We use the term mud for an unconsolidated deposit and the term mudstone for its lithified equivalent (see Box 1.1).

Mudstones provide us with many insights into major processes of sedimentation and thus effectively complement studies of both sandstones and carbonates (Table 1.1). Mudstones, for example, help us identify sea level changes, provide key marker beds for both local correlation in reservoirs and, because of their great lateral continuity, for regional correlations over hundreds of kilometers. And it is the mudstones that, better than any other lithology, permit us to identify the paleo oxygen levels of ancient basins and explore their low-energy transport processes. In addition, most Mesozoic and Cenozoic shales are rich in pelagic faunas, which yield refined global age determinations. The presence and abundance of mudstones is a key factor for limiting carbonate and chemical deposition. In addition, mud and mudstone are linked to all the other fine-grained rocks. Muds and mudstones are commonly interbedded with silts and siltstones and grade into carbonate mud, marl and also sandstone, limestone and dolostone. Fine-grained volcanic debris is also common along active margins. Thus the study of mudstones

Box 1.1. Usage

Although the terms clay, mud and shale are widely recognized, their technical definitions and usage have long been troublesome and are not fully agreed upon. There are at least two reasons for this – the term "*clay*" is used both as a size and a mineral term, plus many clays, muds and shales are rich in silt-sized particles and thus span the clay-silt boundary.

First consider clay as a size term. The upper limit of clay has been set at 2, 4 and even 20 µm so there is not a universal limit with respect to size. As a sediment, clay has been defined as an unconsolidated deposit that has 50% or more clay minerals by weight and is plastic when wet. These clay minerals are also informally referred to as "clays". So here are three untidy problems. A related but broader term is *mud*, a field term for a fine-grained deposit of any composition. It can consist dominantly of clay minerals, carbonate, volcanic ash, or contain much fine silt or even diatoms – as long as any of these form 50% or more by weight of a deposit, *which is plastic when wet*. Strictly speaking, in such a deposit this 50% should be less than 4 µm in size. In field use, however, this is not always true, because silt, 4 to 62 µm in size, is nearly always present and may rival or exceed in abundance the less than 4 µm fraction, so that many muds are in fact borderline silts, although clearly identified as clay or mud by the field engineer or geologist because they are plastic when wet.

Here we emphasize the terrigenous components of this broad spectrum and use the term mud as defined above for unconsolidated deposits and the term *mudstone* for their lithified equivalents. As used here mudstone, like mud, is a broad generic field term that includes many diverse subtypes. This usage permits us to explore a very wide spectrum of the compositions, process and products of low energy fine-grained sedimentation past and present. See Appendix A.2 for the classification we recommend and why we chose mudstone for the all-inclusive term.

Table 1.1. Why study muds and mudstones

General Sedimentary Geology	Most common sedimentary lithology Key to "quiet water", shallow or deep basin fill and its paleoxygen levels First order barriers to cross flow in sedimentary basins Paleoecology, paleontology and broad geologic history
Energy/Stratiform Ores	Maturation of contained organic matter to oil and gas Mudstones closely associated with coals Prediction of pore pressure Vertical migration of petroleum Organic-rich shales sources of heavy metals for stratiform ores
Waste Containment	Liners for landfills plus hazardous and nuclear waste Bounds to contaminant transport
Engineering	Landslide prediction and control Foundation design Subsidence/compaction Swelling/shrinkage Integrity of trapping seals
Industrial Uses (over 200)	Heavy and light clay industry (bricks/ceramics) Fillers/extenders Bleaching agents Pigments Sealants Suspending agents (drilling and transport) Absorbents

connects us to most sedimentary rocks and has rewards totally different from the terrigenous sandstones, which provide us with "big picture tectonics" and the distribution of high-energy environments in ancient basins.

The study of mudstones is clearly not only the natural complement to the study of sandstones and carbonates but, in addition, it is the mudstones of a basin that are its principal source rocks for petroleum and for the mineralized fluids that source many sedimentary ores. And because of their low permeability, muds and mudstones inhibit the vertical flow of petroleum, mineralized fluids and fresh water. Thus the distribution and continuity of mudstones in a reservoir, be it petroleum, water or an ore deposit, is all important for its development, because mudstones segregate the reservoir into separate, independent or semi independent parts. The clay seals of landfills. play exactly this role by dividing it into compartments and isolating it from underlying aquifers. And finally, abundance also shows the importance of the study of muds and mudstones: they are by far the dominant sedimentary rock. Rare, fine-grained rocks that either grade into or are closely associated with mudstones include phosphorites and some ironstones and ores as well as the more common carbonate muds, marls and volcanic ashes.

Important controls on mud include its supply, which relates to the climate and relief of the source region; low turbulence in the water column, especially near or at the bottom; the oxygen level at the bottom, which governs total organics, color, lamination and faunal activity; and the rate of sedimentation. The slow settling velocity of fine particles ensures far-ranging dispersion by even weak currents and explains the wide extent of many mudstones, even the thinnest of beds.

Changes in relative sea level are a major control on the deposition of marine mud and depend on global eustatic changes in sea level, subsidence rates of basins and the influx of mud. The sum of these three independent factors determines the space available for sediment accumulation, called *accommodation*. The greater this space, the deeper the water and the more likely it is that mud will accumulate. On continental shelves and cratons, supply permitting, changes in relative sea level shift mud deposition from far inshore during high stands to beyond the shelf edge during low stands of the sea. In this scenario, the low energy environments of mud deposition shift back and forth from shallow to deepwater, from well to poorly oxygenated water. They are deposited with widely different sedimentation rates and thus have different colors, textures, structures, and faunal and facies associations. On cratons lacking shelf breaks, a relative rise of sea level tends to produce vast, thin sheets of shale, whereas on a steep shelf or ramp there is little migration, geometry is wedge like, and resedimentation downdip is important.

Thick sections of mudstones accumulate in rapidly subsiding basins supplied either by large rivers draining to passive margins or by many small rivers closely coupled to actively subsiding foreland, forearc, backarc, and pull-apart basins. On cratons and on passive margins lacking large rivers, thinner but widely traceable shales commonly are linked to distal deltas, as are large submarine fans offshore.

In all of these basins, oxygen levels at the sediment-water interface and in the mud control not only bottom fauna, but also the color, pyrite abundance, and organic content of the mud plus its degree of bioturbation – a key factor for the preservation of lamination. Thus black shales typically are well laminated, have a sparse bottom fauna with low diversity and high total organic carbon and pyrite, whereas the common, greenish-gray mudstone has little pyrite and is likely to be bioturbated and poorly laminated. A secondary control on many of the above characteristics is the rate of sedimentation – how long the mud remains at or near the sediment-water interface.

Petrographically, most muds and mudstones consist of a mixture of metastable clay minerals and much less reactive silt-sized quartz and feldspar debris. The latter occurs either as millimetric wisps and laminations or as scattered silt grains plus fine-grained organic debris of many different kinds – a good example of hydraulic equivalence depositing many unlike particles together. In many mudstones, the proportion of terrigenous silt may equal or exceed the proportion of clays. The proportions of silt and organic material greatly influence the physical properties of mud and mudstone, but the dominant control on the strength of a mud is still its water content.

Mud deforms easily both on land and underwater and flows at exceedingly low angles. And after burial, rapidly deposited mud can become over pressured and form spectacular diapirs rising hundreds of meters into and through a sedimentary pile. Low angle slides and mud lumps are especially common in low energy, rapidly deposited deltas rich in mud. Mudstones, because of their low shear strength in comparison to sandstones and carbonates, deform and fail easily. These commonly form the soles of low angle overthrusts and, at the surface, are likely to fail, where they dip toward a valley. Artificial cuts in muds and shales are notably unstable and require special engineering attention even where dips are very shallow.

The expulsion of water from mud as it is compacted is central to its role in ancient basins, because it concentrates petroleum in traps and leads to the deposition of many sedimentary ores. With burial, the density of shale increases and rising temperatures and changing pore water chemistry produce many mineralogical changes. Transformations include glass to zeolites, clays to feldspar, smectite to illite, kaolinite to chlorite, opal A to opal CT and quartz, and complex changes in the organic matter (kerogen). New minerals include pyrite, marcasite and cements of iron, carbonate and phosphate.

The organic content of muds and mudstones has been intensively studied from both the viewpoint of micro- and body fossils, trace fossils and of organic geochemistry.

Oxygen permitting, muddy sediments may have a rich infauna so they commonly have many trace fossils and also accumulate calcareous pelagic microfossils, where deposition was not too deep. Jurassic and Cretaceous mudstones are notable for their great richness of pelagic micro fauna, which provide biozones of about 1,000,000 years duration. Bottom fauna can be present in mudstones, but are commonly not abundant, because muddy waters choke most filter feeders. Oxygen also is the principal control on the preservation of organic matter on the sea bottom.

Because both clay and organic matter need quiet water to accumulate on the bottom, mudstones contain most of the world's organic carbon. The total amount of organic matter in a mudstone, its types, and how much it has been heated are key to the study of petroleum source rocks. The type of organic matter also helps distinguish between marine and terrestrial sources of organic carbon while chemical fossils called biomarkers provide fingerprints for depositional environments, for identifying oils, and tracing them back to their sources.

The uses of mudstones are many. Major mineral deposits that occur in mudstones include lead and zinc, manganese, barite, and copper. Metamorphosed shales are also hosts for emeralds and gold. Industrial uses of shales surpass 200 – from toothpaste to high temperature ceramics to landfills to pigments. Thus, from the first manufacture of pottery and mud bricks to the high technology uses of today, mudstones have been important to mankind (see Table 1.1).

In sum, mud and mudstones are the most widespread and abundant deposits on the Earth's surface both today and in most of the past and deserve full attention.

Milestones

Above we have explored the many facets of mudstones and seen their wide significance for the earth sciences and society. But when did mudstone studies start and how did they develop? In Table 1.2 we identify, in first appraisal, our view of the milestones to our present knowledge. It all began, of course, with brick and pottery making followed by early mining. But from a scientific viewpoint we start with Hooson's 1747 definition of shale in the Miner's Dictionary (Fig. 1-2). Since then special concepts have been developed, societies formed and books written about muds and mudstones. And from Sorby's use of the microscope in 1853 into the twenty first century technology has played a most important role.

Supporting References for Table 1.2

Bates, T.F., 1949, The electron microscope applied to geological research: N.Y. Academy Science, Series 2, v. 11, p. 100–107.
Heezen, B.C., Hollister, C.D., and Ruddiman, W.F., 1966, Shaping of the continental rise by deep geostrophic contour currents: Science, v. 152, p. 502–508.
Hoosan, William, 1747, The Miner's Dictionary, T. Payne, Wrexham, unpaged.
Hower, J., Fairbairn, H. W., Hurley, P. M., Pinson, W. H. , 1963, The dependence of K-Ar age on the mineralogy of various particle size range in a shale: Geochimica et Cosmochimica Acta, v. 27, p. 405–410.
Kuenen, P.H., and Migliorini, C.I., 1950, Turbidity currents as a cause of graded bedding: Journal Geology, v. 58, p. 91–127.
Millot, G., 1964, Géologie des Argiles: Masson et Cie, Paris, 499 p.
Murray, J., and Renard, A.F., 1891, Report on deep-sea deposits based on specimens collected during the voyage of HMS Challenger in the years 1873–1876 *in* Challenger Reports, HMSO Edinburgh, 525 p.
Pauling, L., 1930, The structure of the micas and related minerals: Proceedings National Academy Sciences, USA, v. 16, p. 123–129.
Payton, C.E., Ed., 1977, Seismic Stratigraphy – Applications to Hydrocarbon Exploration: American Association Petroleum Geology, Memoir 26, 516 p.
Rhoads, D.C., and Morse, I.W., 1971, Evolutionary and ecologic significance of oxygen-deficient marine basins: Lethaia, v. 4, p. 423–428.
Rinne, F., 1924, Crystals and the Fine Structures of Matter: E. P. Dutton & Co., New York, 195 p.
Schieber, J., Zimmerle, W., and Sethi, P.S., Eds., 1998, Shales and Mudstones: E. Schweizerbart'sche Verlagsbuchhandlung, Stuttgart, v. 1, 384 p.; v. 2, 286 p.
Sorby, H.C., 1853, On the origin of slaty-cleavage: Edinburgh New Philosophical Journal, v. 60, p. 137–150.
Sorby, H.C., 1880, On the structure and origin of non-calcareous stratified rocks: Quarterly Journal Geological Society London, v. 36, p. 33–92.
Teichmüller, M., 1958, Metamorphisme du charbon et prospection du pétrole: Revue industriel minerale, Numero Espécial (Paris), p. 1–15.
Thode, H.G., MacNamara, J., and Fleming, W.H., 1953, Sulphur isotope fractionation in nature and geological and biological time scales: Geochimica et Cosmochimica Acta, v. 3. p. 235-243.
Treibs, A., 1936, Chlorophyll und Häminderivate in organischen Mineralstoffen: Angewandte Chemie, Jg. 49, p. 682-686.
Vinogradov, A.P., Chupakhin, M.S., and Grinneko, V.A., 1956, Isotopic ratios S^{32}/S^{34} in sulfides: Geokhimiya, v. 4, p. 3-9.
Wilgus, C., Hastings, B., Ross, C., Posamentier, H., Van Wagoner, J., and Kendal, C.G.S.C., Eds., 1988, Sea Level Changes: an Integrated Approach: Society Economic Paleontologists Mineralogists Special Publication 42, 407 p.
Woolnough, W.G., 1937, Sedimentation in barred basins, and sources of oil: American Association Petroleum Geologists Bulletin, v. 21, p. 1101–1157.

Books on Mudstones

The literature of mudstones is scattered far and wide in many different journals and across fields as diverse as highway and foundation engineering, sedimentary geology, clay mineralogy and technology, geomorphology, sedimentary geochemistry, petroleum geology and ore deposits. Books devoted to mudstones, however, are few in number even though they are the most common of all sedimentary rocks. We hope the annotations below will guide you to those that are most helpful.

References

Aplin AC, Fleet AJ, MacQuaher JHS (eds) (1999) Muds and Mudstones: Physical and Fluid-Flow Properties. Geological Society Special Publication 63, 190 p.

Twelve technical articles arranged in three parts plus an introduction. Advanced reading for the specialist. Many tables.

Bennett RH, Bryant WR, Hulbert MH (1991) Microstructures in Fine-Grained Sediments: From Mud to Shale. Springer, Berlin Heidelberg New York, 582 p.

Many short articles on mudstone fabrics – advanced technical reading covers almost all aspects.

Chamley H (1989) Clay Sedimentology. Springer, Berlin Heidelberg New York, 623 p.

Well-illustrated presentation of mostly recent through Cretaceous deposits arranged chiefly by environments.

Churchman GJ, Fitzpatrick RW, Eggleston RA (eds) (1995) Clays Controlling the Environment: Proceedings of the 10th Clay Conference, Adelaide, 18 to 23 July, 1993. CSIRO Publishing, Melbourne, 526 p.

This proceedings volume shows well the great range of uses, both practical and academic, of muds and mudstones. Seven parts and over 80 short papers

Clauer N, Chaudhuri S (1995) Clays in Crustal Environments. Springer, Berlin Heidelberg New York, 359 p.

A technical monograph on isotopic dating and tracing for specialists – advanced background required

Meunier A (2003) Argiles. Gordon and Breach, London, 433 p.

Introduction to the fundamentals of the crystallography and chemistry of clay minerals. With this physical-chemical background, one can explore the formation of clay minerals in soils and weathered rocks and sediments, burial diagenesis, and hydrothermal systems with full confidence. See also the forthcoming English edition

Table 1.2. Milestones. *Inspired by Dorrik Stow*

Year	*Event*	*Significance*
1998	*Shales and Mudstones by Schieber et al.*	*Large collection of papers on mudstones, many of which use sequence stratigraphic interpretations*
1988	*Sequence stratigraphy from Exxon and Wilgus et al.*	*Predictive stratigraphy leaps forward (especially so for mudstones), because sea level oscillations are analyzed in terms of supply, basin subsidence, and changes in world sea level*
1980	*Sedimentology of Shale by Potter et al.*	*First book-length treatment dealing with sedimentology of mudstones*
1977	*Seismic stratigraphy from Exxon and Wilgus et al.*	*Stratigraphy becomes an equal partner with structure when studying seismic sections*
1971	*Models of oxygen-dependent biofacies by Rhoads & Morse*	*Degree of bioturbation, body fossils, and organic content of mudstones linked to oxygen in bottom water*
1968	*Reports of the Deep Sea Drilling Program*	*Huge volume of new sediment, largely muds, available for study*
1966	*Contourites described by Heezen et al.*	*Taught us that there are many more currents in deep water than we had thought*
1964	*Geologie des Argiles by Millot*	*First book devoted to mudstones*
1963	*K-Ar analysis applied to mudstones by Hower et al.*	*Demonstrated considerable post-depositional mobility of elements*
1961	*Van Krevelen Diagrams of organic maturation*	*Quick method for evaluating sources and thermal history of organics in mudstones*
1961	*Plate tectonics*	*Essential dynamic framework to understand formation of sedimentary basins*
1958	*Vitrinite reflectance applied to sedimentary rocks by Teichmüller*	*Extended standard method of coal rank evaluation to mudstones, greatly extended understanding of thermal history of basins*
1956	*Isotopes of sulfur analyzed in sedimentary pyrite by Vinogradov et al.*	*Helps trace microbial pathways in diagenesis; reconstruction of seawater history*
1953	*Proceedings National Conference on Clays and Clay Minerals*	*With successor journal, Clays and Clay Minerals, provided outlet, for information about clays*
1950	*Turbidite concept articulated by Kuenen and Migliorini*	*Explained a major process that carries sand, silt and mud far into deep water*
1949	*Scanning electron microscopy introduced to geology by Bates*	*Method of visualizing clay particles in three dimensions*
1937	*"Euxinic" proposed for sulphur-rich anoxic muds*	*Van der Gracht's letter to Woolnough*
1935	*Biomarkers in petroleum from Treibs*	*Showed presence of plant-derived molecules in petroleum, the start of oil-to-source rock correlations*
1931	*Journal of Sedimentary Petrology*	*Emergence of sedimentology as a major subdiscipline in geology*
1930	*Structure of mica group by Pauling*	*Proposed the sheet structure of micas, Prototype for most clay minerals*
1927	*Schlumberger develops down-hole geophysical logging in France*	*Tool for long-distance correlation of mudstones in the subsurface and much, much more*
1927	*Reflection seismology in petroleum industry begins*	*Enables study of structure of rocks in the subsurface*
1924	*X-ray study of clay minerals by Rinne*	*Permits identification of the many types of clay minerals*
1881	*Pelagic samples from the Challenger dredges described by Murray and Renard*	*Showed how sediment types are distributed throughout the world ocean; their picture largely unchanged by later work*
1880	*Sorby's presidential address to Geol Soc*	*First systematic discussion of clastic sedimentology*
1853	*Thin-section study of (metamorphosed) mudstones by Sorby*	*First use of what became the main tool of sedimentary petrology*
1827	*Al isolated as an element; first wet chemical measurement of mineral compositions*	*Beginnings of geochemistry*
1747	*"Shale" defined by Hoosen*	*Recognition of mudstones as a separate class of rocks*

Millot G (1970) Geology of Clays. Springer, Berlin Heidelberg New York, 429 p.

Translated from the French edition of 1964, this book represents the summation of a lifetime of study of clays, shales, and clay minerals in France and Africa by Professor Millot and his many students. This first mudstone book has much emphasis on clay mineral transformations and their response to climate and burial.

O'Brien NR, Slatt RM (1990) Argillaceous Rock Atlas. Springer, Berlin Heidelberg New York, 141 p.

A most helpful atlas with 242 illustrations, 46 in color. Each stratigraphic unit illustrated has a summary of its geology, composition, and two to five illustrations (SEM, photomicrographs and drawings). Well worth your attention!

Paquet H, Clauer N (eds) (1997) Soils and Sediments. Springer, Berlin Heidelberg New York, 369 p.

Fifteen chapters range widely from weathering to laterite to nickel-bearing deposits to continental silicifications to burial diagenesis. A monograph directed to a wide audience.

Potter PE, Maynard JB, Pryor WA (1980) Sedimentology of Shale. Springer, Berlin Heidelberg New York, 306 p.

The second book devoted totally to shale is short and elementary – the first 74 pages are an overview followed by a short chapter on methods and a long illustrated, annotated bibliography.

Schieber J, Zimmerle W, Sethi PS (eds) (1998) Shales and Mudstones, vol 1 and 2. E Schweizerbart'sche Verlagsbuchhandlung, Stuttgart, 384 p. and 296 p.

Seven chapters and 29 articles by experts provide insights to mudstones for the advanced student. Forwards prepared by Schieber are useful overviews and introductions to each of the major sections of both volumes.

Scott ED, Bouma AH, Bryant WR (eds) Siltstones, Mudstones and Shales: Depositional Processes and Characteristics: Society for Sedimentary Geology, Tulsa, OK, and The Gulf Coast Association of Geological Societies, 125 p. CD-ROM.

Eleven articles arranged in five groups (Introduction, Oil and Gas Exploration, Transport and Depositional Processes, Characteristics of Deposits, and Post-Depositional Processes and Properties) emphasize the processes and environments of fine-grained deposits and their significance for the petroleum industry. Many valuable color illustrations. Power-point presentations related to most of the chapters are also included. Important step forward.

Stow DAV, Piper DJW (eds) (1984) Fine-Grained Sediments; Deep-Water Processes and Facies: Geological Society Special Publication 15, 657 p.

All time classic has seven parts and an extended introduction and summary, which deserves your full attention. Thirty three papers. Read especially the Introduction, which covers history of study, controlling processes, methods, classification and a thoughtful question set, much of which we are still working on…

Tyson RV (1995) Sedimentary Organic Matter. Chapman and Hall, London, 615 p.

Shows how the "sedimentology" of the palynomorphs (pollen, spores, acritarchs, etc), which are common in mudstones, provide insights to their provenance and depositional environment. Twenty five chapters. We learned much from this book.

Velde B (1992) Introduction to Clay Minerals: Chemistry, Origins, Uses and Environmental Significance. Chapman and Hall, London, 198 p.

Insightful, clearly written chapters with excellent illustrations. Recommended as a good starting point, but some background needed.

Velde B (ed) (1995) Origin and Mineralogy of Clays. Springer, Berlin Heidelberg New York, 334 p.

Subtitled, "Clays and the Environment," this book has seven chapters, 182 figures and 7 tables and provides a somewhat more advanced reference than Velde (1992). Broad in scope with simple, clear illustrations

Weaver CE (1989) Clays, Muds and Shales. Elsevier, New York, 819 p.

The longest of all books on shales has 10 chapters, 78 tables, 325 figures and over 1,600 references and was written by a widely experienced clay mineralogist as a summation of his life's work.

Wignall PB (1994) Black Shales. Clarendon Press, Oxford, 127 p.

Almost everything one needs to know before attending a specialist's conference on black shales. Short, well referenced and well illustrated monograph helpful to all who study black shales.

CHAPTER 2

Production of Mud and Silt

High, wet mountains produce a lot of mud

2.1. Introduction

Terrigenous clay-sized material, less than 4 microns, comes mostly from the chemical weathering of rocks at the Earth's surface, plus some contribution from volcanic ash and glacial rock flour. The origin of the silt component of mudstones is more controversial. Terrigenous silt, 4 to 64 microns in size, has been thought by many to be largely the product of physical processes – fracture or chipping in transport, freezing and thawing, thermal expansion, exfoliation, release of confining pressure – all processes that favor size reduction. Some terrigenous silt may also be "born and not made", when mudstones are deeply buried or become low-grade metamorphic rocks. Silt may also be formed biologically by the action of plants or animals to break up larger grains or to precipitate new silt-sized grains. Mudstones, especially from the Mesozoic and younger, contain clay- and silt-sized carbonate and fine siliceous debris of biogenic origin. These physical and biological processes are enhanced by the chemical transformation of parent materials, which releases both mineral particles and solutes from a rock. The flow diagram of Fig. 2.1 incorporates many of these possibilities.

The chemical processes that reduce mineral grains in size and transform primary minerals formed at depth into stable clay minerals are better understood both qualitatively and quantitatively than are the diverse physical and biological processes that produce silt-sized terrigenous debris. This difference in understanding reflects the great fundamental difference in the two processes – clay minerals form from the slow but continuous transformations of the primary minerals to new, lower P-T phases at the Earth's surface, whereas breakage and grinding, whatever their causes, are instantaneous, random events and thus much harder to inventory and understand.

Sources of terrigenous mud and silt include soils; erosion of unconsolidated clays and silts by slope wash on interfluves; gully and stream bank collapse; volcanic ash, especially on convergent margins; the deflation of arid and semi-arid regions by wind, which deposits loess on land or silt and clay directly into a lake or ocean; and glaciations, which produces outwash and till on land and, at tidewater, marine fine-grained glacial deposits with dropstones. In addition, submarine weathering (Fe and Mg-rich saponite and celadonite) and the formation of glauconite directly on the sea floor provide a small source.

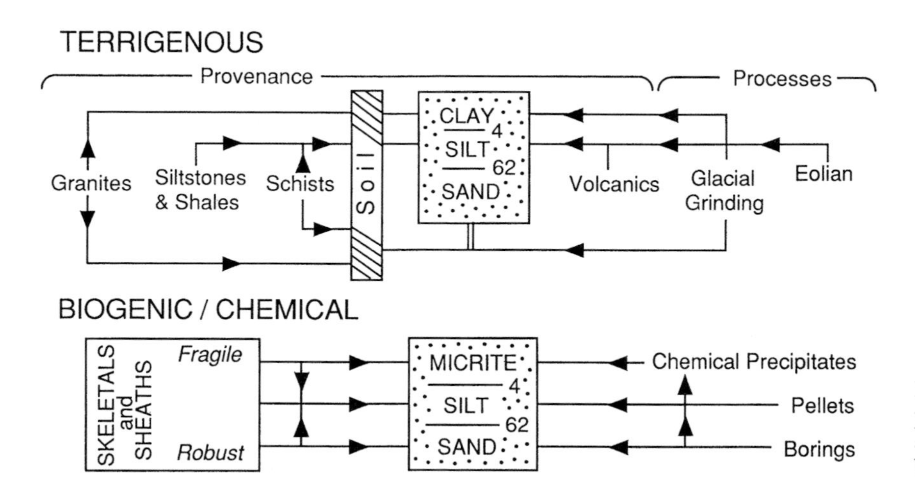

Fig. 2.1. Flow chart for the production of clay- and silt-sized terrigenous and biogenic/chemical debris

Of all of these, clay and silt-sized debris produced in soils, fines from volcanism, and glacial abrasion are the best understood and are the most important primary sources. To this add the weathering and erosion of ancient mudstones, which are abundant and disintegrate readily.

Above is a list of all the immediate sources. But where do these materials ultimately come from? Chiefly from the weathering of the two most common igneous rocks, granite and basalt, and from volcanic ejecta. From these materials, feldspar, amphiboles, pyroxenes and volcanic glass are the principal donors of clay minerals while the chief sources of silt-sized debris are quartz and feldspar, both ultimately derived from granites and gneisses. Thus silt is derived only from felsic rocks, whereas clay is generated from both mafic and felsic sources. The rock cycle (Fig. 2.2) is the ultimate control here – mud is transported to a basin, becomes indurated into mudstone with deep burial, and with advanced metamorphism, is transformed into a gneiss or granite. When uplifted and weathered once again, these crystalline rocks yield mud, silt, sand, and gravel and a new cycle starts. A smaller cycle exists at sea – the formation of saponite and celadonite from basalt on newly formed mid ocean ridges and their later emplacement into accretionary wedges and incorporation into magma along active margins.

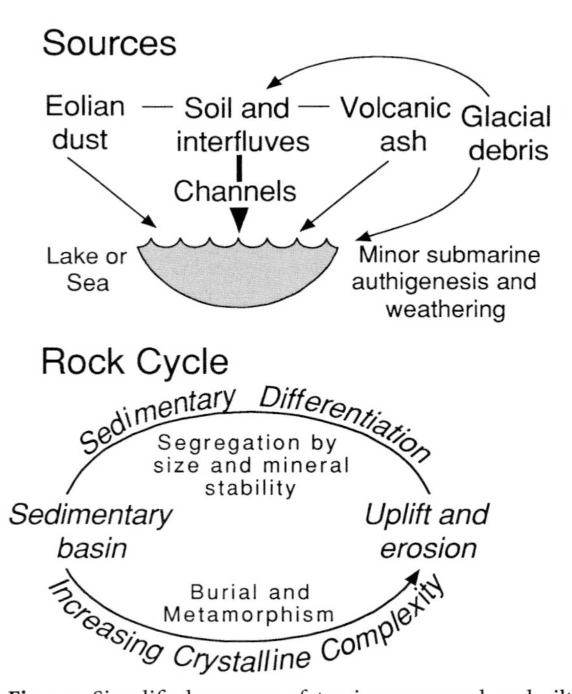

Fig. 2.2. Simplified sources of terrigenous mud and silt (*above*) and rock cycle (*below*)

2.2. Sedimentary Differentiation

The conversion of granite, gneiss, basalt, or an arkose into mud, silt, sand and gravel of variable composition is the result of *sedimentary differentiation* at the Earth's surface. Sedimentary differentiation is broadly defined as all those changes during weathering and transport that segregate detrital minerals by size and sorting – and thus by chemical composition.

The concept of sedimentary differentiation is an old one, but in recent years we have come to understand much better what controls both its efficiency and the great contrasts in surficial mineralogy that occur at the Earth's surface. Sedimentary differentiation starts in the profile of weathering in humid climates (more than 50 cm of rainfall per year) and follows the sequence below (Chamley 1989, p 22)

Parent rock + Cation deficient rainwater
→ Secondary clay minerals plus quartz, Fe and Mn oxides
+ Export of solutions rich in cations and dissolved silica

The processes represented by this equation transform large primary minerals formed at high P-T conditions into fine-grained secondary clay minerals stable at low P-T conditions (see Box 2.1 and Appendix A.2) so that finally insoluble minerals such as quartz, kaolinite, and aluminum and iron oxides accumulate at the Earth's surface. Virtually all these transformations occur in soils (Appendix A.3) or in alluvium as it sits in flood plains in transit to the sea. Although not explicit in the above equation, the *residence time* that the primary mineral spends in the zone of weathering is all-important – the longer this time, the greater the likelihood of mineralogical transformation between a mountain range and a distant basin.

There are also two other outcomes – weathering in semi-arid (10 to 50 cm rainfall) and arid (less than 10 cm) climates, where the above pattern does not apply. In semi-arid regions Na^+ and K^+ are both mainly in solution, but Ca^{++}, Mg^{++}, and H_4SiO_4 only partially. Consequently, mineralogical transformation is less complete. In arid regions or regions with excess irrigation, many soils do not follow the above depletion equation because evaporation exceeds precipitation and calcite and evaporitic minerals are precipitated to form crusts such as calcretes and gypcretes. Smectites and mixed layer clay minerals are also formed. This reverse process also occurs in low lying, poorly drained areas where cations and sil-

Box 2.1. Clay Minerals and Weathering

Clay minerals (Appendix A.3) chiefly form via the weathering of primary minerals in soils in the following way

$$H^+ + \text{primary mineral} \rightarrow \text{intermediate clay mineral} + \text{solutions} \rightarrow \text{gibbsite} + \text{solutions}\,.$$

Three examples of the above general equation are

$$H^+ + \text{K-feldspar} \rightarrow \text{illite} \rightarrow \text{smectite} \rightarrow \text{kaolinite} \rightarrow \text{gibbsite}\,,$$
$$H^+ + \text{muscovite} \rightarrow \text{illite} \rightarrow \text{smectite} \rightarrow \text{kaolinite} \rightarrow \text{gibbsite}\,,$$
$$H^+ + \text{glass} \rightarrow \text{gels (allophanes)} \rightarrow \text{smectite} \rightarrow \text{halloysite} \rightarrow \text{kaolinite} \rightarrow \text{gibbsite}\,.$$

The needed H^+ comes via release from water, which is facilitated by excess CO_2

$$H_2O \Leftrightarrow H^+ + OH^-\,,$$
$$CO_2 + H_2O \Leftrightarrow \underset{\text{carbonic acid}}{H_2CO_3} \Leftrightarrow \underset{\text{bicarbonate ion}}{H^+ + HCO_3^-}\,.$$

Thus the more CO_2 dissolved in the water (supplied by bacterial respiration), the faster the weathering process. Another factor, and probably more important, is the total flux of water through the soil system; the larger the flux, the greater the tendency for these reactions to move to the right, a process that always converts complex crystal structures into simpler ones. Conversely, with minimal H^+ all these reactions are sluggish or stall. This H^+ or its proxy, rainfall, is a key underlying driving force in weathering and the production of clay minerals.

Two other factors are time – with enough time even slow reactions go to completion – and, of course, starting materials in the source rocks. Hence, clay mineral compositions depend on residence time, rainfall, and source rocks. The above reactions tend to be reversed during burial, for example illite and quartz form at the expense of smectite (Chap. 6).

ica leached from uplands are added back to the clay mineral lattices.

2.2.1. Residence Time, Relief, and Rainfall

The interrelationships of residence time, relief and rainfall are well displayed in a simple matrix (Fig. 2.3). The tectonic stability of the site of weathering is the key factor that controls residence time. There are two end members – stable cratons and passive continental margins form one pole, while convergent margins form the other. Cratons and passive margins typically have low relief and relatively gentle slopes – except for some residual plateau-type mountains and some marginal, rift-related, escarpments along coasts like Brazil. Convergent margins, by contrast, have high relief and high summits plus steep unstable slopes and deep valleys. Compound-

Wet highlands	**Wet lowlands**
Major donor – large volumes of silt and mud with expandable and layered clays in suspension plus large volume of diverse chemical species in solution	Minor donor – small volumes of mud with gibbsite and kaolinite in suspension plus restricted chemical species in solution
Dry highlands	**Dry lowlands**
Minor to moderate donor – small to moderate volumes of mud and silt plus suspended and chemical loads of diverse composition. Minor export of eolian clay and silt	Negligible donor except for eolian export of clay and silt of diverse compositions. Appreciable export of eolian clay and silt

Fig. 2.3. Simplified 2 × 2 matrix of role of highlands and lowlands (relief) versus rainfall (wet/dry) on sediment production. Although other factors exist, these are the most important

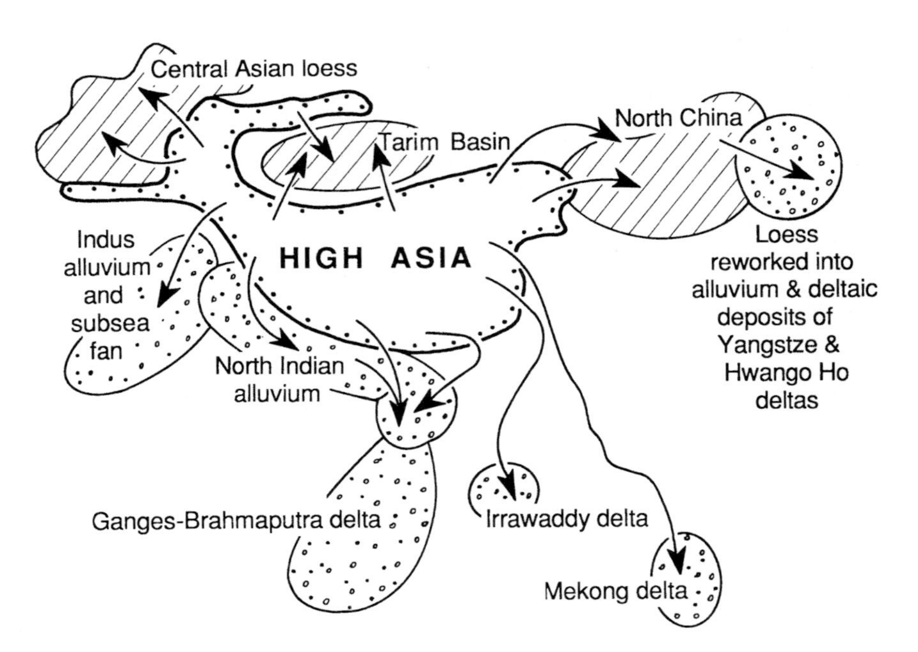

Fig. 2.4. Schematic of riverine silt and mud (coarse stipple) and loess derived from High Asia (After Assallay et al. 1998, Fig. 5). Published by permission of the authors and Elsevier Science

ing this effect is a high proportion of mechanically unstable volcanic ash and hydrothermally altered volcanic rock. Continent-continent collisions produce the greatest volume of mud. High Asia, the product of the India-Asia collision, is a superb example of a great source of detritus, especially fines (Fig. 2.4).

Convergent margins are of three types: ocean-to-ocean (island arcs), continent-to-ocean (Andean) and continent-to-continent (Himalayan) margins. In all three, uplift (sustained by isostatic rebalancing) is rapid and occurs chiefly by faulting. Consequently, erosion is rapid and large volumes of detritus are generated over long time spans, where climates are wet. Under these conditions, bedrock materials in the zone of weathering have short residence times and detritus is composed mostly of mineralogicaly immature debris. The great contribution, about 70%, of southeast Asia to the world's total erosion output (Milliman and Meade 1983, Fig. 4) is explained by its high relief and steep slopes combined with high monsoonal rainfall. Here large rivers such as the Ganges, Brahmaputra, Mekong, and Irrawaddy play important roles, but it is also easy to visualize a simple Andean margin with high monsoonal rainfall and many short rivers contributing large volumes of mud directly to a sea or ocean. An example would be the Cascade and Olympic Mountains of the Puget Sound area. Mt. Rainier is a prodigious producer of fine-grained detritus. Not only do its eruptions generate large volumes of ash, but hydrothermal activity on the flanks of the mountain produces a severe alteration of the original coarse-grained volcanic rocks into unstable, clay-rich deposits that are a major contributor to the giant mudflows the area has experienced, some of which have reached the sea (John et al. 2003).

On the other hand, on stable cratons and passive margins, uplift is mostly either epeirogenic or sea level controlled. Except for a few areas of active rifting in narrow belts along continental margins or within the craton, erosion rates are slow, residence times are long, and non-glacial soils are well developed and millions of years old. Here sediment output is low and, in tropical climates, such landscapes will yield mostly quartz, kaolinite, gibbsite and Fe and Mn oxides (Fig. 2.5) – the most stable end products of weathering (Edmond et al. 1996). Thus in the absence of glaciation on cratons and passive margins, a *chemical landscape* will prevail as long as rainfall exceeds evaporation and there is ample water to flush through the soil system. Consequently, such peneplained, non-glaciated, low-relief cratons and passive margins are insignificant sediment donors, except where they are crossed by a large river sourced in far distant, bordering mountains. This is well illustrated in Brazil at the junction of the Solimões and Rio Negro Rivers at Manaus, Amazonas, where they join to form the Amazon River (Fig. 2.6). The Solimões, sourced on the steep, high-rainfall, eastern slopes of the Andes, is always brown and turbid because of its large suspended load whereas the Rio

Fig. 2.5. Thin laterite developed over leached quartz sand in Bahia, Brazil

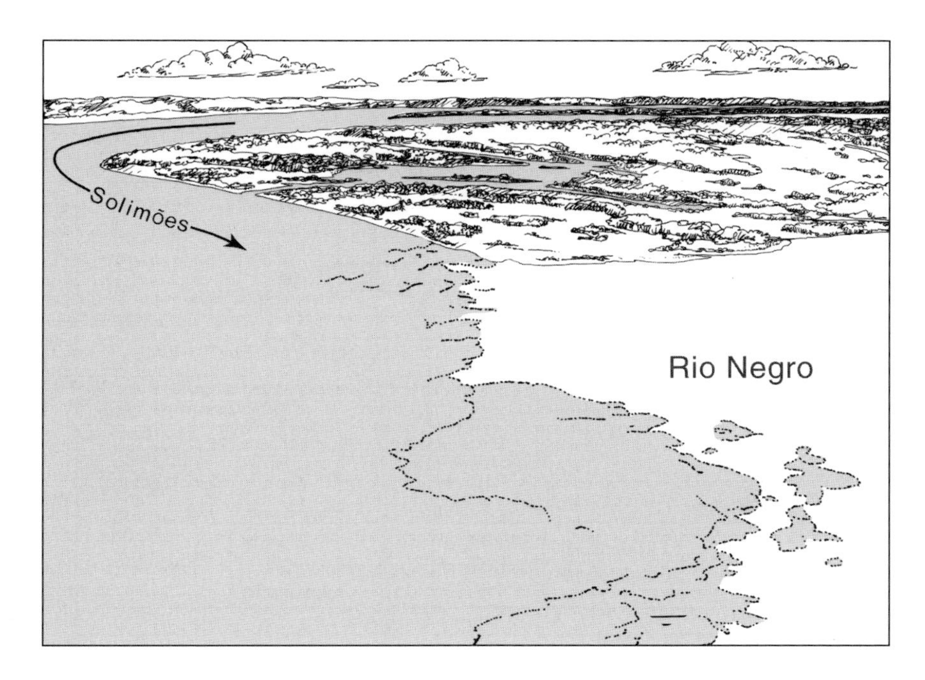

Fig. 2.6. "Meeting of the waters" at Manaus, Amazonas, Brazil, where the mud- and silt-charged Solimões River draining the Andes meets the acidic, mud- and silt-free waters of the Rio Negro draining a well-leached, tropical, peneplained landscape (after Potter 1998, Fig. 3). The brownish-black water of the Rio Negro are sufficiently acid that mosquito larvae cannot survive in it. Published by permission of E. Schweizerbart'sche Verlagsbuchhandlung

Negro, sourced on a low-lying, tropical peneplain, has minimal suspended load, its sands are almost all quartz and its waters have a dark tea-like color, because of abundant organic acids.

Even uplifted rift margins in humid climates are dominated by chemical weathering. The famous peaks that surround Rio de Janeiro in Brazil are granitic. The sands on nearby beaches, however, are composed almost entirely of quartz (Savage et al. 1988), and the soils of the area are dominated by kaolinite and iron oxide. Contrast this situation with the opposite margin in Namibia, one of the world's driest coastlines, where the rivers and the adjacent shelf have about 85% illite+smectite and only 15% kaolinite (Bremner and Willis 1993).

What happens to soil development, weathering products and sediment volumes when rainfall is minimal on a desert, either at mid-to-low latitude or the poles? Here the role of rainfall and temperature is very clear (Fig. 2.7 and Table 2.1). High relief still produces high potential energy, but low rainfall leads to minimal chemical weathering, soil development and erosion. Therefore basins proximal to such areas have low accumulation rates and contain deposits that closely reflect source area compositions. Because sediment input is low along most arid coasts, carbonates and even chemical sediments may accumulate in marginal basins and directly onlap a desert landscape. The carbonates along the Red Sea rift system are a Cenozoic example. An ancient example is the Middle Ordovician strata of southeastern Ontario, where tidal-flat carbonates directly overlie the Precambrian shield along the flank of the Frontenac Arch. This observation prompted the idea that deflation, in the absence of much vegetation, was responsible for the deficiency of mudstones in much of the Cambrian and Ordovician of the North American craton except along its margins (Dalrymple et al. 1985). Another example is in Australia where Devonian carbonates directly overlie and onlap a Precambrian shield (Playford 1980, p 815). *Thus it seems clear that rainfall is the intensity factor and residence time in the zone of weathering is ultimately governed by tectonic stability.*

Table 2.1. Clay mineralogy and surficial weathering (after Chamley 1989, p. 49)

High latitudes	Strong inherited signature with illite-chlorite dominant except along convergent margins
Temperate	Incompletely altered clays such as expanded illite-chlorite, variable mixed-layer clays, vermiculite and degraded smectites; the most varied clay mineral products with much recycled clay
Semiarid to arid	Well crystallized Fe-smectites and some fibrous clays, but usually only minor volumes
Hot and Wet	Weak source rock signature with almost complete transformation into kaolinite + goethite + gibbsite plus some smectite, but minor volumes where source is low lying; strong paleoclimatic signal

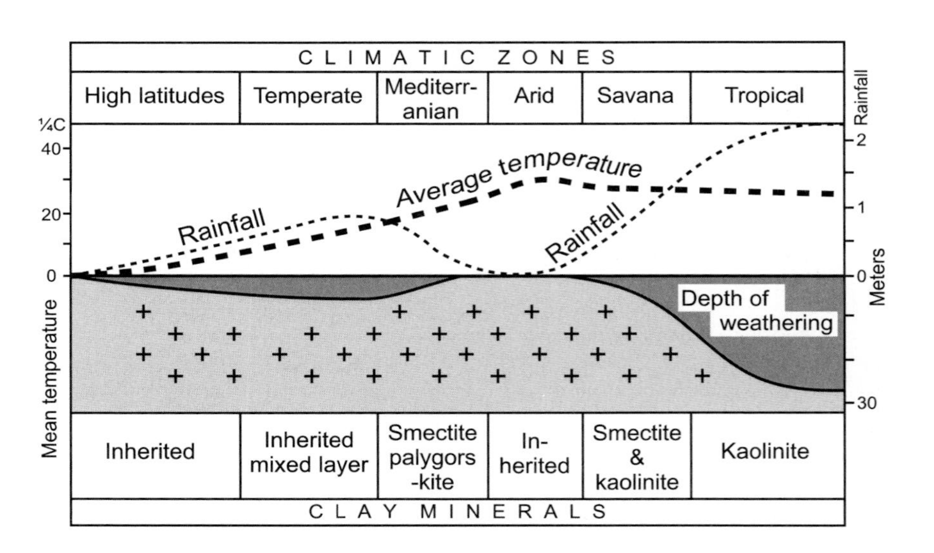

Fig. 2.7. Schematic of north-south depth of weathering and principal clay minerals (After Thiry 1999, Fig. 3). Note role of rainfall. Published by permission of the author and Elsevier Science

2.2.2. Source Rock Control

In the above discussion source rock control on clay mineral composition is not explicitly mentioned, but does it have a role? When and where weathering is incomplete, source rock composition controls the composition of the resultant detritus. The distribution of the bottom sediments of the South Atlantic Ocean shows this well (Petschick et al. 1996). In the southern part of the South Atlantic, illite from Antarctica and chlorite from the southern part of South America predominate, whereas kaolinite is typical of the northern part of the South Atlantic, because it is the end product of tropical weathering in Brazil and Africa (no source rock control). Totally opposite from tropical weathering is glacial flour, the finely divided, pulverized product of glacial abrasion with virtually no weathering. Here the clay size fraction and most of the silt fraction consist of primary minerals such as feldspar and amphiboles as well as quartz. It is estimated that presently about 20% of all the detritus entering the sea is from glaciers. Table 2.1 presents a simplified summary of clay minerals developed on selected different parent materials and shows two major points: at intermediate stages of weathering great differences exist, but that these all tend to converge to kaolinite and gibbsite with either sufficient time or rainfall. In addition to mineral composition, several chemical ratios are used to determine the degree of weathering of modern muds and paleosols (Maynard 1992). The most popular is the CIA index (Box 2.2), which is further discussed in Chap. 7.

Volcanism contributes significantly to mudstones in active margins both as a primary and secondary source (Table 2.2). The principal direct contributor is wind-transported volcanic ash produced by explosive, silica-rich acid magmas (Fig. 2.8). This is carried 10's to 100's of km downwind after an explosion. Such ashes, or *tephras*, form widespread, thin beds that blanket topography and become useful stratigraphic markers (Fig. 2.9). Where such ash falls accumulated in coal swamps, they formed thin, widely traceable kaolinite-rich bands in coal beds called *tonsteins;* where they accumulated on a sea bottom, they formed smectite-rich beds called *bentonites* (see Burial, Chap. 6).

Table 2.2. Sources of volcanic material in mudstones

Primary	Secondary
Wind-blown ash (tephra)	*On Land*
Nuées ardentes	Debris flows
Glowing avalanches	Alluvial deposits
Lahars	*At Sea*
Lava flows	Submarine slumps
Floating pumice	Turbidity currents
	Submarine weathering of basalt

The other three common products of volcanism, lava flows, ignimbrites formed by *nuées ardentes* and by ash flows, are secondary rather than primary sources of mud, because these cool rapidly from high temperatures to form rock-like masses. Hence their contribution to mudstones is indirect via later soil development. Other, and probably much more important sources of volcanic fines for mudstones are the sheet wash and stream erosion of newly deposited unconsolidated ash plus secondary debris

Box 2.2. The Chemical Index of Atlteration, CIA

Because of weathering reactions, the ratio of aluminum to alkalis and alkaline earths in secondary clay minerals is typically higher than in the parent materials. Nesbitt and Young (1982) have derived the Chemical Index of Alteration (CIA) to quantitatively express this relationship. In molecular proportions, the CIA index is

$$CIA = 100[Al_2O_3/(Al_2O_3 + CaO^* + Na_2O + K_2O)]$$

where CaO^* represents the calcium in the silicate fraction only, and moles are used rather than weight % to emphasize mineralogical changes.

The average upper crust has a CIA value of about 47 (McLennan 1993). Hence, CIA values of 45–55 indicate essentially no weathering. At the other extreme kaolinite, chlorite, gibbsite, and boehmite all have average CIA values of about 100, and smectite and illite have values in the 70–80 range. It follows that the proportions of clay minerals and primary minerals in a bulk sample will introduce substantial variation in the resulting CIA value.

Fig. 2.8. Steam and tephra plume from Rabaul stratovolcano on 19 September 1994 at Lat. 4.2 S, Long. 152.2 E, New Britain Island, Papua New Guinea (NASA Photo SSI 64 from space shuttle). Shifting winds cause the wide dispersion of the plume

flows or lahars that accompany eruption. Lahars and ignimbrites can extend down valley 10 to 100 km or more. Repeated reworking of all these deposits will disperse volcanic debris widely, so that lacustrine and marine mudstones with a volcanic signature of zeolites, smectites, opal, etc. are common along active margins. At sea, porous, low-density pumice composed of small watertight vesicles may also float hundreds of kilometers before sinking into a muddy bottom, where it is likely to form reverse graded bedding.

Very little is known about how much volcanic material is (or was) present in Paleozoic and older mudstones, although we know that some Mesozoic mudstones contain appreciable amounts (Zimmerle 1998). Long-term burial diagenesis converts any volcanogenic clay to illite-chlorite, so older mudstones are hard to evaluate. The presence of many well-defined bentonites in the Paleozoic and volcaniclastic debris in many older sandstones suggests that there is a larger volcanic component in mudstones than we think.

2.3. Silt

Silt-sized particles – those 4 to 62 microns in size – come from many different sources, both terrigenous and biological (Table 2.3). And like the sandstones, the dominant mineral of non-volcanic silts is quartz, in spite of the more rapid cleavage-induced abrasion of coarse feldspar by both air and water transport that should produce abundant feldspathic silt. We can think of these variables in terms of a *silt factory*:

$$\sum \text{silt} = \text{abrasion} + \text{weathering} + \text{biologic production} + \text{metamorphism} + \text{recycling}\,.$$

Fig. 2.9. Volcanic contributions to mud production, (*top*) Lahar from the 1991 eruption of Mount Pinatubo on Luzon destroyed homes along tributary of the Abacan River (Major 1996, Fig. 9). (*bottom*) Young ignimbrites in La Puna, JuJuy Province, Argentina. Ignimbrites can travel many kilometers downslope, are both subaerial and submarine, and, if climate is wet, weather readily to become, major donors of mud. Ignimbrites are commonly interbedded with ash fall deposits, which extend much farther from the volcano. Published by permission of the author and University of Washington Press

Table 2.3. Origins of silt-sized particles

By abrasion	
In streams	Fracture of polycrystalline quartz grains and metamorphic rock fragments; chipping of unit quartz; fracture along cleavages in feldspar
By wind	As in streams but over a much wider area and with longer residence times in the abrading system
Beneath glaciers	Crushing and grinding in and at base of ice
By weathering	
Weathering of older mudstones	Recycling of silt from older fine-grained rocks; greatest source?
Weathering of acidic volcanics	Silt-sized quartz rapidly released by breakdown of glass and fine-grained matrix
Weathering of greenschist metamorphics	Silt-sized quartz grown through authigenesis and metamorphism at temperatures 100 to 300 °C; released from surrounding micas and feldspars in soil by weathering
Frost shattering	Cyclic thermal expansion and contraction of quartz and feldspar (frost wedging)
By organisms	
In soils	Silt-sized biogenic opal accumulates in A-horizon of poorly-drained soils; easily eroded and transported to lakes and oceans where burial converts to quartz
In water column	Diatoms, radiolaria, and sponges secrete silt-sized amorphous silica particles that recrystallize to quartz during shallow burial

Little is known of the relative quantities of silt generated by these processes, and there is much controversy over the effectiveness of these mechanisms, particularly concerning the relative importance of abrasion in rivers and abrasion by wind in deserts. In part this controversy stems from inconsistencies between experiments (which evaluate the rates of silt production in water and air at constant volume) and field observations (which include the area and time over which abrasion processes act). Thus a desert may have a lower abrasion rate than the headwaters of a stream, but the effective area over which abrasion occurs is vastly greater, as is the residence time of grains in the zone of active abrasion:

$$\text{Silt volume} = (\text{abrasion rate}) \times (\text{area}) \times (\text{residence time}) .$$

But other factors beyond grinding may be equally important for silt generation. Consider what happens to mudstones with deep burial in the clay cycle of Fig. 2.4. At depths of about 2,000 m and temperatures of 80 to 100 °C (Loucks et al. 1984; Bjørlykke and Egeberg 1993) quartz overgrowths develop and presumably fine authigenic quartz too. So with deep burial and high temperatures, quartz silt is created by precipitation from pore fluids, most of which are supersaturated with respect to quartz at depths greater than 2–3,000 meters (Milliken 2003, Fig. 11). Support for this conclusion is provided by Blatt (1987, p 377), who argued, using oxygen isotopic evidence, that much of the fine quartz in mudstones was derived from slates and phyllites of the greenschist facies.

Biogenic silica is a large and under appreciated source of quartz silt, either directly through deposition of diatoms in lakes and the deep sea, or indirectly through erosion of soil particles. Clarke (2003) has estimated that almost all soils contain detectable opaline silica of biogenic origin, and that amounts can exceed 5% by volume in many cases. Poor soil drainage encourages proliferation of silica-secreting organisms, which include sponges as well as diatoms. This biogenic opal accumulates in the A-horizon, and so is readily eroded and redeposited by wind or water to depositional basins where it converts to microquartz during diagenesis. Most of this biogenic debris is found where terrigenous silt and mud is minimal – in basins bordering arid lands whose rivers have low suspended loads, in basins with recently flooded estuaries that trap most terrigenous detritus inshore (recent sea level rise), or beyond the shelf edge and reach of most terrigenous mud carried in suspension (hemipelagic mud) and in some lakes.

A biogenic origin of quartz silt in ancient rocks has been demonstrated by Schieber et al. (2000), using backscatter electron images (BSEI), cathodoluminescence, and oxygen isotopes. They found that up to 100% of the fine silt of a Devonian black shale was derived from radiolaria and diatoms. These results show the benefits of applying a variety of techniques and instrumentation in the study of the components of mudstones. A significant research problem is to determine how widespread this biogenic silica is in other mudstones, especially those that are not organic rich.

As with clays, quartz-feldspar silt production is maximized by high local relief combined with high rainfall in temperate to tropical climates. For example, weathering of low grade, fine-grained metamorphic rocks (mostly phylites) and rainfall of 5 to 10 m in the Southern Himalayas produces soils

Fig. 2.10. Loess section at Chashmanigar, Tajikistan (Photos courtesy of Professor Arnt Bronger). In this outcrop the loess is about 130 m thick. Note the calcareous nodules marking paleosols and the low angle discordance (*lower left*) at the base of the section. Just above this discordance is the contact with the Brunhes/Matuyama magnetic reversal interval, about 780 ka bp

with high silt contents, 20–30% by weight (Gardner 1994). When eroded, these contribute vast quantities of silt to the streams draining to the Indian Ocean. Other sources of silt are active margins with explosive andesitic volcanism. Here large quantities of silt-sized ash consisting of glass, amphiboles, plagioclase, etc. occur. The finest fraction of such ash, say less than 10 microns, is transported long distances in the high atmosphere.

Loess (Fig. 2.10) is a distinctive deposit consisting dominantly of wind-blown silt in the 20–50 micron size range. It accumulates by deflation downwind from glacial outwash valley trains (periglacial loess) or downwind from continental deserts (lee desert loess) as in Central Asia, North China, North Africa and the Arabian Peninsula (Fig. 2.11). In both glacial and desert environments, loess is formed by eolian size segregation from a source that readily generates silt-sized abrasion products. Where deposited on flat uplands, loess is little eroded, whereas on slopes it is easily eroded and quickly redeposited in alluvium and transported seaward. Deforestation of a loess-

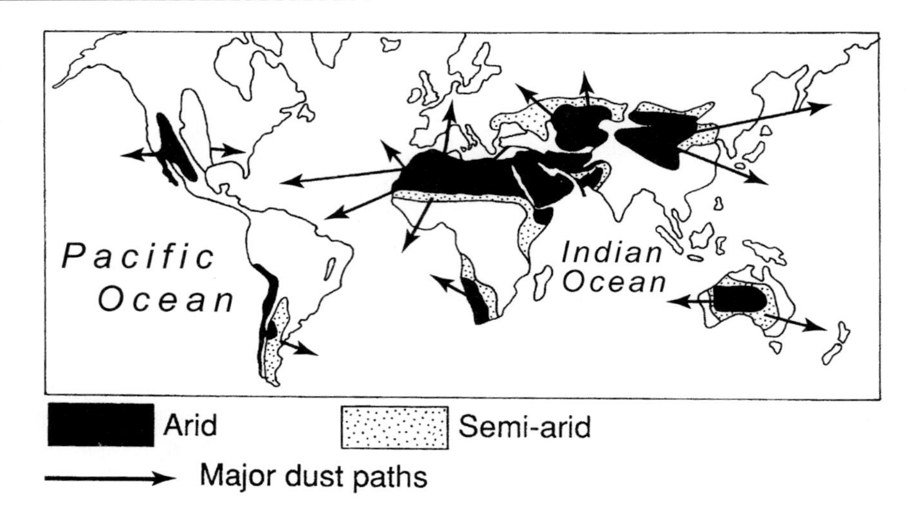

Fig. 2.11. Schematic of export of dust from world's arid and semiarid regions. After Pye and Tsoar (1987, Fig. 2). Published by permission of the authors and the Geological Society of London

covered upland quickly releases vast volumes of silt downstream as is well documented in the northern China loess area and also downstream from areas of thick ash fall deposits. See Assallay et al. (1998) in Digging Deeper for a detailed review. The dust source from the Sahara has been thoroughly documented, and is estimated to comprise between 200 and 700 million tones of dust each year (Goudie and Middleton 2001). Much of this load is transferred to the Atlantic, and the red soils of the Bahamas show evidences of Saharan derivation (Muhs et al. 1990).

Lee desert loess appears to be far greater in aerial extent and volume than periglacial loess. In Central Asia and China thicknesses of 100 to more than 200 m are common over vast areas. There are both thick valley fills of loess and also loess plateaus downwind from large deserts and semi arid steppes. In these thick deposits there are multiple buried soils defined by color, leaching, or large calcareous nodules and concretions (Fig. 2.10). These paleosols are valuable for their paleo climatic value (Bronger et al. 1998). Long-term aridity seems to be a prerequisite for such deposits (Miocene uplift of the Himalayas with consequent desertification.)

Patagonia, in far southern South America, is also an end member landscape. Patagonia is a dry tableland (molassic and glacial outwash plain plus loess) between the Andes and South Atlantic Ocean. Strong and persistent westerly winds and low humidity give Patagonia its character. Sediment supply to the ocean off Patagonia is about equal parts by coastal erosion and by dust, whereas the few and small streams only supply about 3 percent (Gaiero et al. 2003). Another important sediment source, however, is volcanism in the Andean Arc. This volcanic contribution is a key factor in the biological productivity of the South Atlantic Ocean, because it supplies available iron.

Many mudstones are calcareous, containing carbonate that is virtually all produced in lakes and oceans by organisms. Such carbonate-sized silt is produced either directly from the skeletals of small intact organisms (small pelagic foraminifers, for example) or as skeletal fragments of ostracodes, fish bones, etc. Carbonate particles are also produced indirectly via algal binding (mucous-like sheaths in shallow warm water) and by mixtures of micrite and silt in fecal pellets from a vast array of deposit feeders living in carbonate muds. Fine-grained carbonate in mudstones is particularly common in Mesozoic and younger mudstones following the great expansion of pelagic shelled organisms in the Cretaceous.

2.4. Pre-Devonian Landscapes

Residence time in the soil seems to be the overriding control on both mud and silt production, and above we emphasized slope and thus tectonics as the chief control on residence time. But there is yet another factor to consider – changes in plant cover with time, which in turn is linked to biological evolution and atmospheric oxygen (Fig. 2.12). Little is known of how past landscapes that lacked extensive plant cover affected mud production, but it must have been most significant. Judging from modern experience in tropical Africa (Goudie 1993, p. 160–170), the conversion of forest to bare soil increases erosion rates by two to three orders of magnitude or more depending on slope (Fig. 2.13). Geologically speaking, this increase may be a brief transient spike before a new steady state is established, or it may be a permanent condition, but one conclusion seems certain: much more of

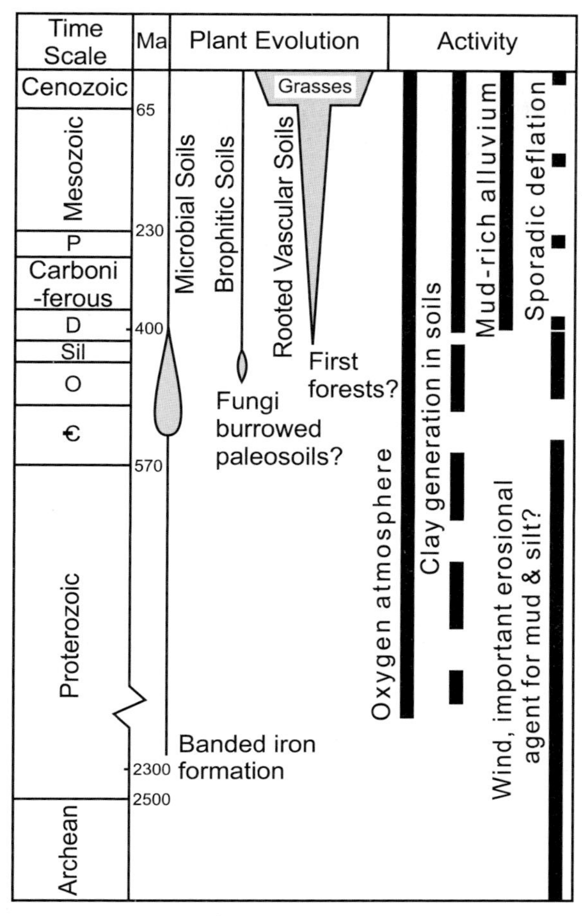

Fig. 2.12. Evolution of soils through time (adapted in part from Wright 1990, Fig. 1)

the Earth's surface would contain weathering-limited slopes than today. Under arid or extremely cold climate conditions, the rate of weathering is slower than the rate at which erosion removes the weathered debris. Therefore it is weathering that limits the rate of degradation of the landscape, and the slopes are said to be *weathering-limited*. Where there is thick plant cover, as is universal in warm wet climates today, weathering is faster than erosion, a thick mantle of weathered residuum accumulates, and the slopes are said to be *transport-limited*. Before plants invaded the uplands in the Devonian, most landscapes must have been weathering-limited.

Among the many possible consequences of this change in geomorphologic conditions, it is easy to think of at least three. Less clay would be produced in comparison to silt, more clay- and silt-sized mineral grains would be inherited from bedrock rather than modified in soils and, finally, wind-blown fines should be much more common in older marine and lacustrine mudstones than in post-Devonian mudstones. Another impact would have been on river channel form. Schumm (1967) speculated that the rivers of pre-Devonian landscapes must have been mostly braided (flash floods and less clay) rather than the mix of braided and meandering that we see today. A chemical effect has been hypothesized by Algeo and Sheckler (1998), who argued that the abundant black shales of the Upper Devonian were a response to the appearance of large trees with deeply-penetrating root systems. This change to the landscape produced deeper weathering and a spike in the flux of nutrients to the oceans.

Fig. 2.13. Spectacular gulley near Salvador, Bahia, Brazil. Remove the scant vegetation from this landscape and could it not be a model for a Precambrian landscape?

2.5. Summary

Sedimentary volumes eroded from the continents are maximized where convergent margins have high rainfall. In addition, large, tropical, high relief, wet islands and plateaus also provide large volumes. Typically, the large rivers draining convergent continental margins flow long distances to passive margins and mostly bring mud, silt, and fine sand to the sea. Where climates are humid, weathering can appreciably simplify the composition of the detritus of a large river as it is transported episodically to the sea. However, most stable cratons and passive margins are small donors of sediment volumes unless they have been glaciated or are bordered by distant marginal mountains. Unglaciated, they have soils that are millions of years old and, where climate is tropical, yield principally quartz, kaolinite, and gibbsite in contrast to the mineralogicaly more complex clays and silts of convergent margins. On ocean-to-ocean and ocean-to-continent active margins, volcanism is a major contributor of fine debris to mudstones and locally can be the dominant one for active margin mudstones. A special search should always be made for volcanic debris, because it can so radically alter the physical properties of a mudstone. Everywhere sedimentary differentiation controls the composition and texture of detritus at the Earth's surface, but its efficiency depends on both *residence time* (ultimately on tectonics) to respond to weathering and reworking and on the *intensity of weathering*, which is mostly controlled by the amount of rainfall that moves through the soil system. Thus time and intensity of weathering emerge as the two fundamental underlying controls on sedimentary differentiation, while both high rainfall and relief are needed for large volumes of sediment to be produced.

Where continents happen to lie with respect to the world's climatic belts also has much to do with both the volumes and kinds of fine debris that they supply to the sea. Where monsoonal rains meet high collisional mountains, great volumes of mud, silt, and sand erode rapidly (Fig. 2.4); on the other hand, in the rain shadows behind such mountains, much less is eroded. In both polar and low to mid-latitude deserts, erosion is minimal. In addition, tropical cratons with deep stable regolith contribute low amounts of mineralogicaly stable debris, mostly quartz and kaolinite. Thus the location and orientation of a continent with respect to the world's climatic belts is all important for the volumes and compositions of fine-grained detritus eroded from it.

Clay and silt of carbonate composition is common in many marine and lacustrine mudstones as a minor component and many such mudstones grade into fine-grained carbonates. Fine grained carbonate, micrite, typically is formed from the fragmentation of weak skeletal grains and micro fossils, by passing through the intestines of organisms (pelletization), by algae, by chemical precipitation in hypersaline environments and even some from boring organisms. Diatoms are the main sources of siliceous biogenic debris.

2.6. Challenges

Quartz silt is a significant fraction of the grains in mudstones, and it has diverse sources: glaciers, deserts, streams, older rocks, volcanoes and biota.

So far, sedimentologists have not come to agreement on the relative importance of these different sources, either for an individual mudstone or globally. Because of the large contribution of silt to mudstones, we suggest that this is the biggest unsolved problem in mud production.

Although there is a clear relationship between climate and relief on the one hand and mud production on the other, less is known about the role of rock types and tectonic settings.

Intuitively, a stratovolcano in the wet tropics should be by far the greatest producer of mud per km², but is this true? How would other settings compare?

Oxygen isotopes are widely used to study the origin of quartz in sandstones and in plutonic igneous rocks.

How can this technology be efficiently applied to the study of quartz in mudstones? A rapid and inexpensive way to do this analysis would yield enormous benefits.

Was the production of mud and silt significantly different before the advent of trees than it is today? Did the later appearance of grasses have an effect?

As interesting as these questions are, the real question is, "What can the study of mudstones tell us about this problem?"

We have a good estimate of the amount of dust brought into the oceans by winds in the modern.

How can eolian contributions to ancient marine or lacustrine mudstones be identified and quantified? Could "pathfinder minerals" like zircon be useful if the quartz grains themselves are indistinguishable?

The Archean atmosphere lacked appreciable oxygen. How did this affect rates of weathering and the nature of soil formation?

What characteristics would one look for in Archean mudstones to measure this effect? Is simply the ratio of sandstone to mudstone significant?

References

Algeo TJ, Sheckler SE (1998) Terrestrial-marine teleconnections in the Devonian, links between the evolution of land plants, weathering processes, and marine anoxic events: Philosophical Transactions Royal Society of London. Biol Sci 353:113–130

Assallay AM, Rogers CDF, Smalley IJ, Jefferson IF (1998) Silt: 2–62 μm, 9–4 Φ. Earth Science Reviews 45:61–88

Bjørlykke K, Egeberg PK (1993) Quartz cementation in sedimentary basins. Am Assoc Petrol Geol Bull 77:1538–1548

Blatt H (1987) Oxygen isotopes and the origin of quartz. J Sediment Petrol 57:373–377

Bremner JM, Willis JP (1993) Mineralogy and geochemistry of the clay fraction of sediments from the Namibian continental margin and the adjacent hinterland. Marine Geol 115:85–116

Bronger A, Winter R, Heinkele T (1998) Pleistocene climatic history of East and Central Asia based on paleopedological indicators in loess-paleosol sequences. CATENA 34:1–17

Chamley H (1989) Clay Sedimentology. Springer, Berlin Heidelberg New York, 623 p

Clarke J (2003) The occurrence and significance of biogenic opal in the regolith. Earth-Sci Rev 60:175–194

Dalrymple RW, Narbonne GM, Smith L (1985) Eolian action and the distribution of Cambrian shales in North America. Geology 13:607–610

Edmond JM, Palmer MR, Measures CI, Brown ET, Huh Y (1996) Fluvial geochemistry of the eastern slope of the northeastern Andes and its foredeep in the drainage of the Orinoco in Colombia and Venezuela. Geochim Cosmochim Acta 60:2949–2976

Gaiero DM, Probst J-L, Depetris PJ, Bidart SM, Lelyter L (2003) Iron and other transition metals in Patagonian river borne and windborne materials: geochemical control and transport to the southern Atlantic Ocean. Geochim Cosmochim Acta 67:3603–3623

Gardner RAM (1994) Silt production from weathering of metamorphic rocks in the southern Himalaya. In: Robinson DA, Williams RBC (eds) Rock Weathering and Landform Evolution. Wiley, New York, pp 487–503

Goudie A (1993) The Human Impact on the Natural Environment, 4th edn. Blackwell, Oxford, 454 p

Goudie AS, Middleton NJ (2001) Saharan dust storms: nature and consequences. Earth Sci Rev 56:179–204

John DA, Rytuba JJ, Ashley RP, Blakely RJ, Vallance JW, Newport GR, Heinemeyer GR (2003) Field Guide to Hydrothermal Alteration in the White River Altered Area and in the Osceola Mudflow. US Geol Survey, Bull, Washington 2217:52

Loucks RG, Dodge MM, Galloway WE (1984) Regional controls on diagenesis and reservoir quality in lower Tertiary sandstones along the Gulf Coast. In: McDonald DA, Surdam RC (eds) Clastic Diagenesis. Am Assoc Petrol Geol Memoir 37:15–46

Major JJ, Jarda RJ, Daag AS (1996) Watershed disturbance and lahars on the east side of Mount Pinatubo during the mid-June 1991 eruptions. In: Newhall CG, Punongbayan RS (eds) Fire and Mud. Eruptions and Lahars of Mount Pinatubo: Philippine Institute Volcanology and Seismology, Quezon City and University of Washington Press, Seattle, pp 895–919

Maynard JB (1992) Chemistry of modern soils as a guide to interpreting Precambrian paleosoils. J Geol 100:279–289

McLennan SM (1993) Weathering and global denudation. J Geology 10:295–303

Milliken KL (2003) Late diagenesis and mass transfer in sandstone-shale sequences. In: MacKenzie (ed) Treatise on Geochemistry. Elsevier, Amsterdam, vol. 7, pp 159–190

Milliman JD, Meade RH (1983) World-wide delivery of river sediment to the oceans. J Geology 91:1–22

Muhs DR, Bush CA, Stewart KC, Rowland TR, Crittenden RC (1990) Geochemical evidence of Saharan dust parent material for soils developed on Quaternary limestones of Caribbean and western Atlantic islands. Quaternary Res 33:157–177

Nesbitt HW, Young GM (1989) Formation and diagenesis of weathering profiles. J Geology 97:77–92

Petschick R, Ginglete FX, Kuhn G (1996) Clay mineral distribution in surface sediments of the South Atlantic: sources, transport and relation to oceanography. Marine Geol 130:203–229

Playford PE (1980) Devonian "Great Barrier Reef" of Canning basin, Western Australia. Am Assoc Petrol Geol Bull 64:814–840

Potter PE (1998) Shale-rich basins: controls and origin. In: Schieber J, Zimmerle W, Sethi PS (eds) Shales and Mudstones. E Schweizerbart'sche, Stuttgart, pp 21–32

Pye K, Tsoar H (1987) The mechanics and geological implications of dust transport and deposition of deserts with particular reference to loess formation and sand dune diagenesis in the Northern Negev, Israel. In: Frostick LE, Reed I (eds) Desert Sediments: Ancient and Modern. Geol Soc London (Special Paper) 35:139–158

Savage KM, Potter PE, deCesero P (1988) Mineral maturity of modern sand along a high-energy tropical coasts: Baixada de Jacarepagua, Rio de Janeiro, Brazil. J South Am Earth Sci 1:317–328

Schieber J, Krinsley D, Riciputi L (2000) Diagenetic origin of quartz silt in mudstones and implications for silica cycling. Nature 406:981–985

Schumm SA (1967) Speculations concerning paleohydraulic controls on terrestrial sedimentation. Geol Soc Am Bull 79:1573–1588

Thiry M (1999) Paleoclimatic interpretation of clay minerals in marine deposits: an outlook from the continental origin. Earth-Sci Rev 49:201–221

Wright VT (1990) T 8 Terrestrialization. In: Briggs DEG, Crowther PR (eds) Paleobiology a Synthesis. Blackwell Scientific, Cambridge, pp 57–59

Zimmerle W (1998) Petrography of the Boom Clay from the Rupelian type locality, Northern Belgium. In: Schieber J, Zimmerle W, Sethi PS (eds) Shales and Mudstones, vol II. E. Schweizerbart'sche, Stuttgart, pp 13–33

Digging Deeper

Assallay AM, Rogers CDF, Smalley IJ, Jefferson IF (1998) Silt: 2–62 μm, 9–4 Φ. Earth Sci Rev 45:61–88.

Excellent discussion of the distribution of silt in the modern. Includes a valuable diagram showing the size distribution of major loess deposits of the world, and useful information on the cohesiveness of granular deposits of different grain sizes and mineralogies.

Griffin DW, Kellogg CA, Garrison VH, Shinn EA (2002) The global transport of dust. Am Scientist 90:228–235.

We found this short summary to be most informative. For example, possibly as much as two billion tons of dust composed of fine rock and clay particles as well as bacteria, fungi, viruses and herbicides enter the atmosphere each year. In addition to stressing the significance of microorganisms in dust and aerial pollution, reading the bibliography was most enlightening.

Hiller S (1995) Erosion, sedimentation and sedimentary origin of clays. In: Velde B (ed) Origin and Mineralogy of Clays. Springer, Berlin Heidelberg New York, pp 162–219.

Well written easy-to-understand comprehensive overview of clay mineral production, transport of deposition including the formation of authigenic clays and their distribution in the present oceans. Excellent.

Milliman JD, Syvitski JPM (1992) Geomorphic/tectonic control of sediment discharge to the ocean: the importance of small mountainous rivers. J Geol 100:525–544.

Includes the most complete data set published so far on the sediment transport of world rivers – over 280 – and considers the role of runoff, basin area and relief, arguing that the two latter factors have order-of-magnitude control over sediment discharge of most rivers.

Nahon S, Trompette R (1982) Origin of siltstones: glacial grinding vs. Paleozoic weathering. Sedimentology 29:25–35.

Arguing from Paleozoic field experience in northwest Africa and from the literature, they suggest that much periglacial loess is derived not from glacial grinding, but from erosion of deep, pre-glacial residual soils.

Ollier C, Pain C (1996) Regolith, Soils, Landscape. Wiley, New York, 316 p.

Regolith in the geosystem is the theme of this well written, easy-to-read book, which provides many insights to what most of the ancient non-glaciated world must have been like.

Pye K (1995) The nature, origin and accumulation of loess. Quaternary Sci Rev 14:653–667.

A thoughtful, well referenced review. Figures 10 and 11 summarize well overall process.

Sigurdsson H, Houghton BF, McNutt SR, Rymer H, Stix J (eds) (2000) Encyclopedia of Volcanoes. Academic Press, London, p 1417.

This impressive volume has over 75 articles and nine parts, two of which have special value for the study of mudstones – effusive volcanism with 9 articles and explosive volcanism with 18 articles. If starting, begin here.

Wright J (2001) Making loess-sized quartz silt: data from laboratory simulations and implications for sediment transport pathways and the formation of 'desert' loess deposits associated with the Sahara. Quaternary Int 76/77:7–19.

Thorough review of previous experimental and field evidence on making silt by fracturing quartz. Concludes that abrasion in high-gradient streams gives the highest rate of silt formation, followed by glacial grinding and eolian abrasion, which are about an order of magnitude slower. This result is counter to some earlier experiments, which indicates considerable uncertainty in designing these experiments.

CHAPTER 3

Transport and Deposition

Mud goes a long way

3.1. Introduction

Mudstones consist dominantly of clays transported in suspension (wash load) and silt-sized grains transported both as bed-load and partial suspension. Energy is supplied mostly by flowing water in streams, waves, tidal currents, turbidity currents in lakes and oceans and oceanic currents, including contour currents in the deep ocean (Fig. 3.1). Rivers and deltas bring most of these fines to lakes and oceans, although exceptionally in geologic history, glaciers contribute significantly. Mass movements along the sides of streams, on delta fronts, and at the shelf edge and beyond reintroduce temporarily deposited fines to the transport process. Underwater, slides and slumps are the sources of most turbidity currents, which transport mud and silt far into deep waters. Additionally, fines are carried into deep waters by wind, which also deposits vast blankets of mostly silt-sized loess on land. Many of these processes are closely linked to distinct styles of stratification and sedimentary structures (Table 3.1), which permit us to reconstruct the details of basin filling and its paleocurrent system. Sedimentary structures also help predict lateral and vertical facies changes, and perhaps most importantly, help us recognize cycles and sequences in otherwise massive-looking mudstone sections. This chapter provides insight to how clays and fine silts are transported and deposited and the origin of their sedimentary structures. In Chapter 5 we use these results in our study of muddy environments. We start with mass movements, because on slopes both on land and under water, they initiate much transport.

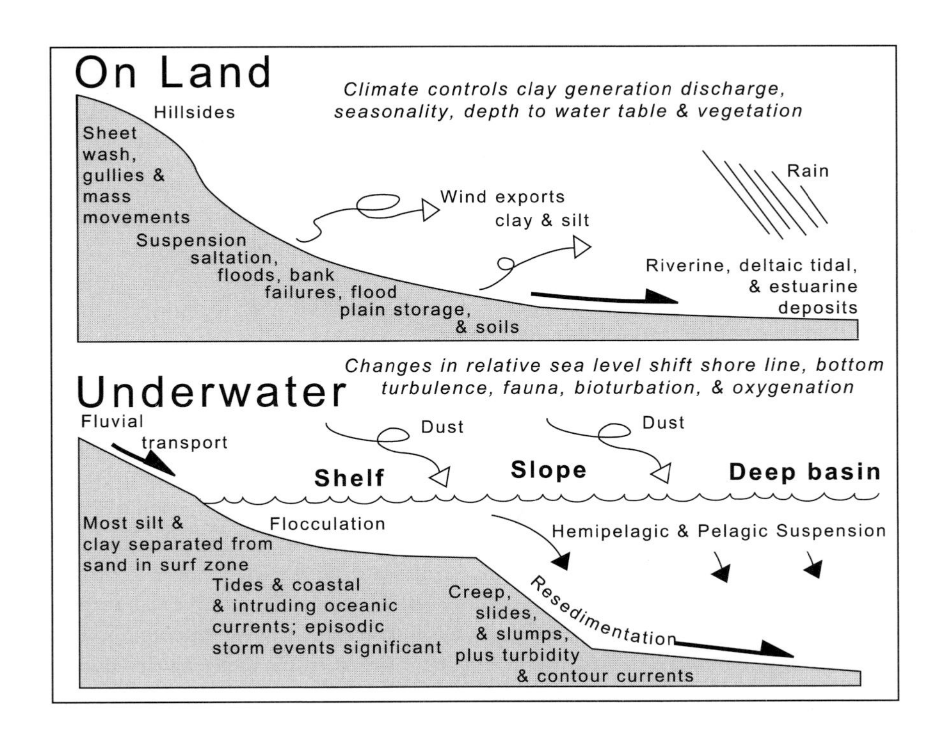

Fig. 3.1. Overview of major processes

Table 3.1. Comparative hydraulics of clay silt and sand

Differences in Transport	
Clay	Individual flakes and grains transported in suspension by weak currents travel long distances. Aggregates of clay and silt linked together by organic sheaths and chemical bonds form coarser floccules that settle more rapidly. More muds may be deposited as aggregates rather than separate flakes. Aggregation depends in part on water chemistry – fresh or saline – and pelletization by organisms. Bottom turbulence – not water depth – controls deposition. Once deposited, a clay-sized lamination or bed is sticky and cohesive and requires stronger currents for erosion than coarser silts or fine sands. Probably much unrecognised clay of eolian origin is present.
Silt	Intermediate in size, silt is transported both in suspension and by sliding, rolling and saltation. Lamination, either parallel or inclined, is typical and has many forms. Fifteen μm has been suggested as a lower limit for silt-sized ripples. Much silt transported by wind as well as water
Sand	Transport mostly by traction and rolling in streams, saltation and suspension in floods, turbidity currents, the surf zone, and by desert winds
Commonalities	
Grain size and thickness of stratification	A broad correlation exists between grain size and bed thickness, with the finer sizes forming thinner laminations and coarser sizes thicker laminations and beds, except when coarser sizes not available
Winnowing	The separation of fines from a poorly sorted mixture when a wave "feels the bottom" (breaking surf), when wind entrains mud and silt from a sand-rich mixture, or when currents entrain fines, leaving behind a lag of sand, pebbles, or shells
Episodic transport and deposition	In virtually all environments, except the deep sea, random floods, storms, and tsunamis create event beds of muddy sediment. Volcanic explosions and slope failures create event beds in all environments.

3.2. Mass Movements

Failure, liquefaction and creep dominate the processes that are important to mass movements. *Failure* is the separation or rupture of a rock or sediment mass into two or more parts; *liquefaction* is the conversion of unconsolidated sediment into a liquid state causing it to flow; and *creep* is the slow, imperceptible movement of material down slope. All three occur both on land and underwater and all three contribute significantly to the transport of fine-grained sediment. Creep and failure move colluvium down slope on hillsides to headwater streams or rivers, where bank failures reintroduce the material to the hydraulic part of the transport process. Creep also occurs on very gently dipping, "soupy" coastal mudflats, on some low-lying muddy savannas, and on submarine slopes. Failure and liquefaction are common on delta fronts, the sides of submarine canyons and on the continental slope beyond the shelf break. And finally, and most important, failure and liquefaction underwater are the source of most turbidity currents, which transport mud and silt on gentle slopes far into deep water.

The failure equation is fundamental to understanding mass movement (Box 3.1). Failure, liquefaction, and creep on slopes are favored by

- decrease in the effective pressure, N–P,
- increase in the slope angle, either by overloading a slope at its top, or undercutting its base (by stream or bulldozer),
- tilting toward a topographic low,
- cyclic shocks from either seismic (earthquake induced) or water waves (*cyclic dynamic loading*),
- and a change in cohesion caused by swelling of smectites or flushing of ions by freshwater.

Consider first liquefaction, because, whether complete or partial, it goes far to explain many features of unconsolidated muds and silts.

3.2.1. Liquefaction

Both mud and silt can be liquefied to produce both small-scale flow structures (see soft-sediment deformation) and larger much more dangerous slope failures. If underwater, these pass down-dip first into debris flows and then into turbidity currents. The universal prerequisite for liquefaction is that the sediment must be fully saturated with water that cannot readily escape. These conditions may exist in surface deposits or in shallowly buried unconsolidated sediment. Liquefaction is more likely to develop in silt than in either fine sand or mud, because silt is

Box 3.1. Failure Equation

The shear strength, σ_s, of a bed or deposit such as colluvium is defined as

$$\sigma_s = C + (N - P)\tan\phi$$

where C is the cohesion of the material, N is its pressure perpendicular to the basal contact of the bed, P is its internal pore water pressure, and ϕ is the internal angle of fraction of the material, a constant for a given material. The *effective pressure* $N - P$, varies with P so that as P increases after a heavy rain $N - P$ decreases, causing the strength of the deposit to decrease and failure to be more likely. Expressed differently, as P increases, the framework is more and more supported by water pressure so that when P equals N, the bed is liquefied and has no strength whatsoever. Thus it is changes in the effective water pressure within a bed that determine its strength. On a hillside, effective pressure is closely tied to rainfall, but in a sedimentary basin it can change with rapid change in water level by cyclic loading from storm waves. Other factors are cementation and increase in the salinity of the pore waters of the clays, both of which increase cohesion.

less permeable than fine sand so that its pore water escapes less readily (the undrained condition of soil mechanics) and because silt, unlike clay, is *non-cohesive*.

Liquefaction occurs when effective pressure goes to zero and the strength of the bed just equals the submerged weight of the overlying saturated sediment – when the framework is supported only by pore water pressure and not by particle-to-particle contact. This situation is also informally known as the *quick* condition. Consider for example a deposit of interstratified mud, silt, and sand with their pore water pressures all in equilibrium with that of a stable lake level. A sudden lowering of the lake (equivalent to the lowering of the water table) reduces hydrostatic pressure in the permeable sands (well drained condition), but not in the less permeable silts or muds (poorly drained condition), which are now over-pressured (effective pressure now decreases). As effective pressure decreases, creep may start and, when effective pressure reaches zero, liquefaction occurs. If the now liquefied, overpressured bed breaks out to the surface, a mud or silt volcano is formed. If the bed is confined below the surface and inclined even slightly, all the overlying beds will slide down-dip, slowly or rapidly, with serious consequences for any structures built on the surface. If this happens underwater, a submarine slump results with potential for a turbidity current.

The term *fluidization* is used when the fluid that supports the clasts of the debris is air (or volcanic gas), as for example, a debris flow resulting from the collapse of a high bluff of loess. Here the material is dry and *dispersive pressure* supports the particles.

A debris flow may move fairly slowly or may move catastrophically at many meters per second. Where under water, debris flows are diluted and transformed into turbidity currents (Fig. 3.2).

Another cause of liquefaction is *dynamic cyclic loading* either by seismically induced ground shaking (seismic shocks), or by the passage of water wave fronts. As successive wave fronts pass, bottom pressures oscillate in response to passing crests and troughs. This pulsing, which in storms extends well below fair weather wave base, compacts loose sediment, separates water from it, and because unit volume is now reduced, forces water upward through channels to form mud or silt volcanoes at the bottom. Another consequence of this cyclic pulsing by waves is periodic, high-frequency reduction of effective pressure below each trough, which may cause liquefaction (the same principle as used in vibrocoring). Should this pore water already contain dissolved methane gas from the microbial decomposition of organic matter, exsolution of the gas also reduces effective pressure and thus reduces shear strength. This extra reduction increases the likelihood of creep and failure on slopes. Where the sediment is permeable and fully drained, pore water readily escapes, but where permeability is low and the bed poorly drained, pore water cannot escape readily

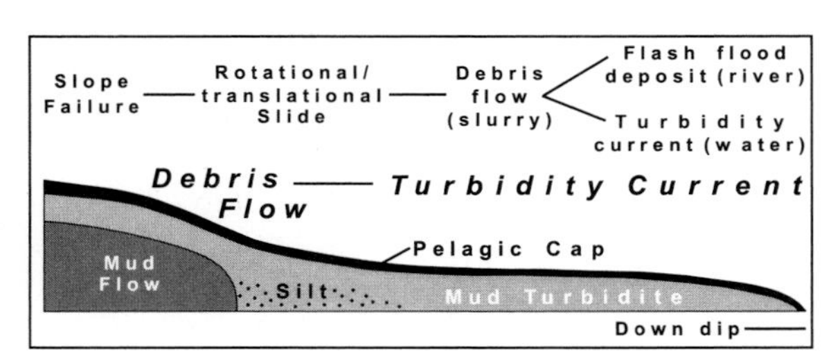

Fig. 3.2. Initial failure leads to slump, slide or debris flow, and to final distal turbidity current

and liquefaction may occur. The product would be a seismically or a wave deformed bed, called a *seismite* or a *quake sheet*, and usually composed of silt or fine sand.

Liquefaction can also exceptionally occur where there is a change in cohesion of the clay minerals. This happens onshore when the original saline pore water of glacial-marine muds is invaded and diluted by fresh water entering along fractures or from prolonged slow leaching. The initial open flocculated fabric of the marine clay (the result of high Na^+ content of seawater) is disrupted by the removal of ions by fresh water flushing through the clay. Consequently the cohesion of the clay is greatly reduced and the strength of the deposit is severely weakened.

3.2.2. Creep

The slow imperceptible plastic movement of a mass downslide is called *creep*, which is produced by the steady constant force of gravity. On land, creep tilts fence posts and poles and breaks walls and pipelines (Fig. 3.3). It is most rapid when internal pore water pressure is highest. Creep is also common in plastic, saturated muds on very low submarine slopes and in some inland swampy areas is the first step to failure.

3.2.3. Failure

Slides, slumps and debris flows are the direct result of failure. Slides are coherent masses that remain essentially intact and can be classed as either rotational or translational. Rotational slides (Fig. 3.4a) grow headward up slope and, if underwater, form a gully that funnels turbidity currents down slope. Rotational slides occur in thick to moderately thick unconsolidated soils and alluvium along streams, on the sides of channels and canyons on continental shelves, and on slope aprons. Such slides have a well-defined curved (listric), basal fault surface, a head scarp, and moraine-like ridges at their toe. Rotational slides are common on many delta fronts and steep aprons and they start resedimentation of near shore muds into deeper water. Translational slides (Fig. 3.4b) on the other hand, are largely confined to steep hillsides on land. The basal surface is planar rather than circular, and the slide mass tends to be thinner than in the rotational case. Such slides are important agents of erosion, carrying blocks or sheets of colluvium down slope over indurated bedrock to streams. Spectacular translational slides occur along sea cliffs in California where smectite-rich volcanic ash beds dip seaward and provide easy glide surfaces when water is added, expanding the clays and reducing their cohesion.

In contrast to slides, slumps are plastic or semi-plastic masses that contain remolded clays, deformed beds of many types and slump folds. An unconsolidated mass may start as a slide, but become plastic with movement pass into a slump. A slump in turn may totally disintegrate and finally turn into a debris flow or a mudflow or even into a turbidity current if under water (Fig. 3.2).

With complete liquefaction, slumps pass into cohesive debris flows with clasts suspended in a muddy matrix (*pebbly mudstone*). These are common on

Fig. 3.3. Broken old stone wall on colluvium-covered hillside shows long-term influence of creep, Walker Street, Cincinnati, Ohio

Fig. 3.4A,B. Rotational and translational slides: (**A**) rotational slides produced by riverbank collapse along Ohio Brush Creek, Adams County, Ohio and (**B**) debris scar formed by a translational slide at colluvium-basalt contact following a cloudburst, Planalto Plateau, Rio Grande do Sul, Brazil. Back-tilted trees (**A**) show rotation, whereas all vegetation is removed (**B**) by translational slide

Fig. 3.5. Purple, well-indurated mudstones (distal turbidites?) of the Neoproterozoic Upper Tindir Group, Tatanduk River of the Alaska-Yukon border has inter-bedded diamictite. This mass-flow deposit has a wide range of clast types, some of which are striated. Note the differential compaction both above and below the two largest clasts (from Young 1982, Fig. 16). Published by permission of the author and the Geological Society of America

Fig. 3.6. Trilobite, *Flexicalymene retrosa*, "buried alive" in mud by an Ordovician storm, Cincinnati Arch, USA

many slope aprons (Fig. 3.5) and in some canyons. On muddy shelves subject to storms, local mud flows with sharp bases can be recognized even on gentle slopes by their transported fossils and an irregular fissility rather than lamination. Such mudflows often bury fauna resting on the bottom (Fig. 3.6).

3.2.4. Stability of Underwater Slopes

The stability of a muddy subaqueous bottom depends on many factors. These include: water content (which is high when deposition is rapid because of less compaction), slope angle and presence of swelling clays, which also apply to subaerial failures. In addition underwater slopes respond to percentage of carbonate minerals, to content of organic matter and to degree of bioturbation. Mineralogically, the more expandable clays in a mud, the less stable it will be, whereas the more carbonate minerals it contains, the more stable. The effect of more carbonate in a mud is to raise its shear strength (Kenter and Schlager 1989) and thus to increase the steepness of the stable slope

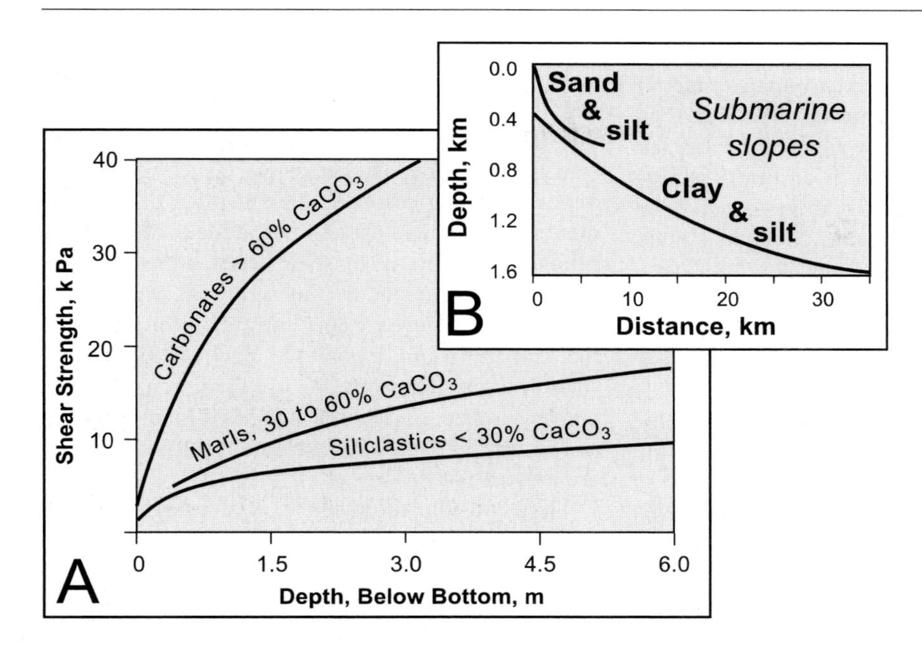

Fig. 3.7A,B. Subaqueous muds and silts on slopes: (**A**) Shear strength of fine-grained sediment increases with carbonate content (after Kenter and Schlager, 1989, Figs. 1, 2, and 3), and (**B**) slopes underlain by terrigenous muds with carbonate and by carbonate muds are steeper than those underlain by sand, silt, and noncalcareous muds (after Adams et al. 1998, Fig. 4). Published by permission of the authors and Elsevier Science

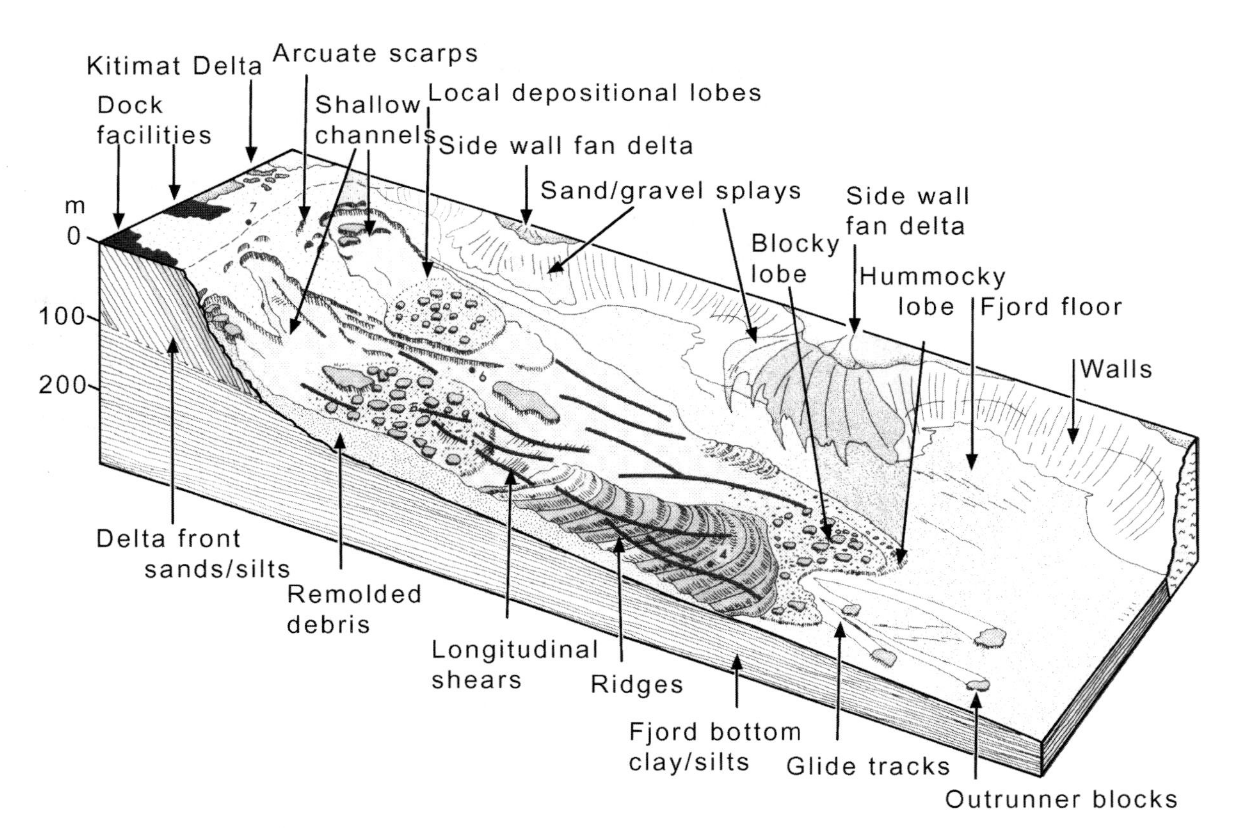

Fig. 3.8. Slope failure in the muds and silts of the Kitimat delta of British Columbia illustrate most of the features found in submarine slides (after Prior et al. 1984, Fig. 2). Length of block diagram is about 8 km. Published by permission of the author and the University of Chicago Press

angle (Fig. 3.7a and b). Early cementation (dissolution and reprecipitation of fine carbonate mud) is chiefly responsible for this. Consequently, slopes of limestone or carbonate muds or silts will be steeper than sandy, silty, or muddy slopes. In addition, sandy-silty submarine slopes are noticeably steeper than silty-muddy ones. Organic matter, if algal slimes rather than woody in origin, can make a mud "slippery" and will lower its shear strength. Bioturbation raises shear strength by reducing the parallelism of the clay flakes in the mud and by mixing sand and silt into the clay.

Because freshly-deposited mud can be so easily remobilized by waves, tides and seismic shocks even on the lowest of slopes, their subaqueous stability is of great interest for marine engineers, and is an especially important consideration for the construction of offshore oil platforms. The geological consequences of the submarine failure of muddy slopes are also many and include headward channeling by regressive rotational slides on upper slopes, lobes created by debris flows, and remolded (and thus weaker) clays and silts (Fig. 3.8). Although submarine failures occur on very low slopes, the steeper the slope and greater the water content of its fines, the more likely is failure to occur. Rapid sedimentation in front of a delta bordering deep water favors both.

3.3. Settling, Suspension, and Entrainment

Gravitational settling, flocculation and pelletization are the three chief processes that control the transport and deposition of the clay fraction. The depositional processes that form most of the sedimentary structures of the interbedded silts and fine sands are dominated by shear stress – the hydraulic force of the current at the sediment-water interface (Box 3.2). Here we concentrate on the processes that transport and deposit the clay fraction and silt smaller than 15 microns. We use 15 microns, because above this size, silt has the same bed form sequence as the coarser sands (Mantz 1978, pp 85–90), which are well known (Allen 1984).

First, consider the suspension of a single particle. A particle of any kind stays in suspension when the upward components of turbulence in the fluid equal or exceed the fall velocity, w, of the particle, where

$$w = \frac{1d^2g}{18\mu}\Delta\varrho = \frac{2r^2g}{9\mu}\Delta\varrho \, .$$

d and r are the diameter and radius of the particle, g is gravity, μ is the dynamic viscosity and $\Delta\varrho$ is the density difference between particle and fluid. This is called *Stokes' law* and best applies to particles smaller than 0.18 mm (180 micrometers) settling in

Box 3.2. Shear Stress and Criteria for Suspension

Shear stress, τ, is a force applied parallel to a contact – an eraser or block sliding down an inclined plane, a fluid moving along its boundary or between two fluids of different densities. The same force applied perpendicular to a boundary is called *pressure* and both have dimensions of MLT^{-2}.

The shear stress of a fluid on its boundary is given by

$$\tau_0 = \gamma RS$$

where γ is the specific weight of the fluid, R is the hydraulic radius of the channel (a measure of its shape and "efficiency"), and S is the slope of the channel. Increase in channel slope, channel efficiency and specific weight (colder water is denser than warm water) all increase the shear force of the fluid on its boundary and thus promote erosion and transport.

Shear stress, although but one of several measures of *flow intensity* in either water or air, is also used, along with friction velocity, as a criterion for suspension. *Friction velocity*, v_*, is defined as $v_* = \sqrt{\tau_0/\varrho}$ where ϱ is the density of the fluid. By dividing the shear stress by the density of the fluid and taking its square root, a velocity is obtained which can be directly related to fall velocity so that the criterion for suspension is $v_* > w$ or $\tau_0 > \varrho w^2$ (Middleton 1976, p. 409).

Another criterion uses the *Reynolds number*, which directly measures the turbulence of a flow, and combines it with depth of the flow, d, and several empirical constants to obtain

$$v_z = 0.17(Rd)^{-0.46}$$

where v_z is the upward component of velocity of the turbulence, which can be then directly compared to the fall velocity of a given size fraction (Graf 1971, p. 163).

Table 3.2. Fall veocities calculated from Stokes law

Diameter	Time to fall one meter			Fall velocity
(µm)	Days	Hours	Minutes	(cm/s)
60	0	0	5	0.223
30	0	0	30	0.0558
16	0	2	0	0.0139
8	0	7	48	0.00349
4	1	6	0	0.00087
2	5	6	0	0.000217
1	21	10	0	0.000054
0.5	89	0	0	0.000013

isothermal, quiet water. Table 3.2 gives transit times for particles of different sizes to settle one meter as calculated by Stokes' law.

Simple as it seems, Stokes' law goes far to explain one of the most important characteristics of mudstones – their great lateral continuity (Box 3.3). Using Stokes' law as a rough approximation to a more complex reality of irregular particle shape, we can calculate that a grain of fine sand 100 microns in diameter will fall through quiet water a unit distance about nine times faster than a silt-sized grain 30 microns in diameter, nearly 140 times faster than a fine silt grain 8 microns in diameter, and about 6,600 times faster than a particle 1 micron in diameter. The upward component of turbulent flow, even in weak currents, is large compared to the settling velocity of a clay particle, but the same upward component becomes insignificant when compared to the fall velocity of larger silt or fine sand grains. In other words, for a clay particle in a weak current, the upward and downward forces are closely balanced, whereas for larger particles the net force is downward. Thus contrasts in fall velocity are the key hydraulic control in down-current grain size decrease, fining (or coarsening) upward sequences, and the lateral transport of mud and fine silt in suspension that is responsible for the great lateral continuity of many mudstones (see Box 3.3). Sand is mostly transported on or near the bottom during high riverine discharge, during storms in shallow waters, and on or near the bottom in deeper water by turbidity currents. In contrast, mud and fine silt, if available, are transported even at low flows and carried farther downstream or spread more widely in suspension in lakes or marine basins. Muddy salt marshes behind shallow sand flats of a lagoon illustrate the same idea. In the salt marsh, reeds act as baffles and reduce currents and consequently fines settle to the bottom, where they are trapped in the root systems of the reeds (In carbonate rocks we have the term *bafflestone* applied to rocks formed by the trapping of carbonate mud by frame builders such as bryozoans and corals).

In addition to size, two other factors greatly affect the fall velocities of mud – its concentration in the water column and salinity of the water (Fig. 3.9). When concentrations exceed some 20 g/l, *hindered* settling occurs and a dilute, semi-liquid results. This is called *fluid mud*, and is fairly common in some estuaries and along coastlines with abundant mud supply. The presence of fluid mud along a coast dampens incoming waves and thus reduces the erosive power of the waves, effectively converting a high-energy coastline into a low-energy coastline; i.e., one that accumulates mud. The term *retarded settling* is used where high salinities increase the density of the fluid and thus retard settling.

But are most clays in nature deposited as single particles? Increasingly, studies indicate that most modern mud is deposited as aggregates – as flocs

Box 3.3. Why Mud "Goes Farther"

Let two particles, one a small sand grain and other a flake of clay, fall though a water column D meters deep, with a horizontal current, v. The sand grain has a fall velocity, w_s, many times greater than that of the clay particle, w_c, say 1,000 to 1 (see Stokes' Law and Table 3.2). Transit times, t_s and t_c, will then be

$$D/w_s = t_s \quad \text{and} \quad D/w_t = t_c \,.$$

Consequently, given any horizontal current, v, the horizontal distance traveled while the two particles fall D meters to the bottom will be vt_s and vt_c. But because the fall velocity of the clay particle is only 1/1,000th or less than that of fine sand grain, it will stay in suspension 1,000 or more times longer – and thus move laterally 1,000 or more times farther than a grain of fine sand or silt.

In sum, in settling basins large or small, small fall velocities of suspended particles equate to long transit times and long lateral transport before reaching the bottom, basins with no currents excepted.

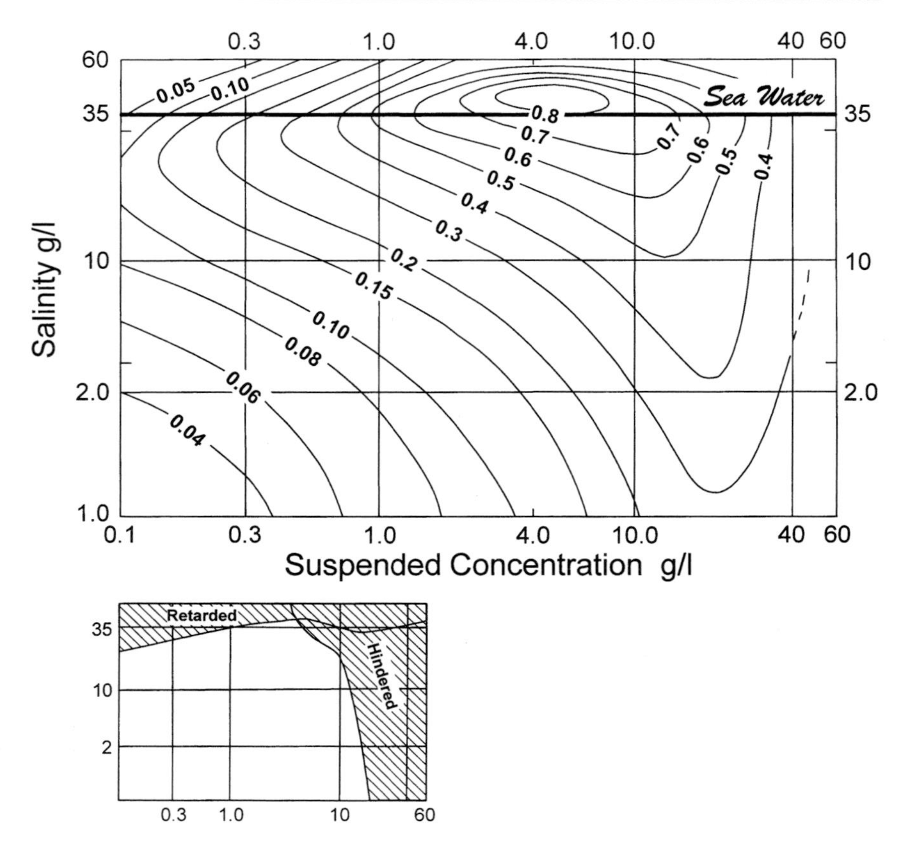

Fig. 3.9. Effect of changing salinity and concentration on fall velocities of fines. Redrawn from Owen (1970, Fig. 5)

and as fecal pellets (Figs. 3.10 and 3.11). Flocculation results from both physical and biological processes. The physical processes were recognized first and depend on the natural attraction of small platy particles through Van der Waals attractions. The excess negative charge present on the surface of clay particles, however, tends to keep clays too far apart for aggregation to occur unless large concentrations of charged ions are present in the solutions, as in seawater. These ions counteract the natural electrostatic repulsion of the clays, allowing them to contact one another and flocculate. The result is that clays are largely separate particles in river water, but are largely in aggregates

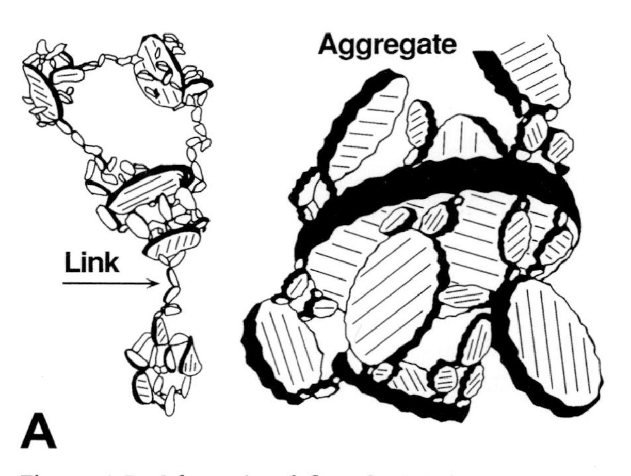

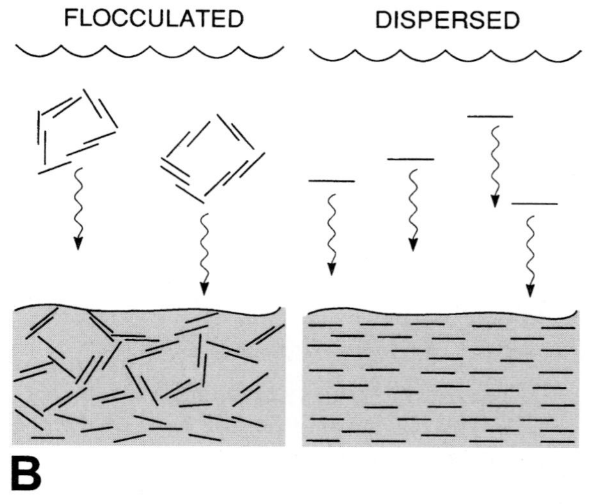

Fig. 3.10A,B. Schematic of floccules:(A) loose irregular chains of clay minerals, silt-sized quartz feldspar, fecal matter and fossil debris (after Pusch 1962 and 1970) and (B) of flocculated versus dispersed clay deposition (after O'Brien and Slatt 1990, p. 24). Published by permission of the authors and the National Research Council of Canada

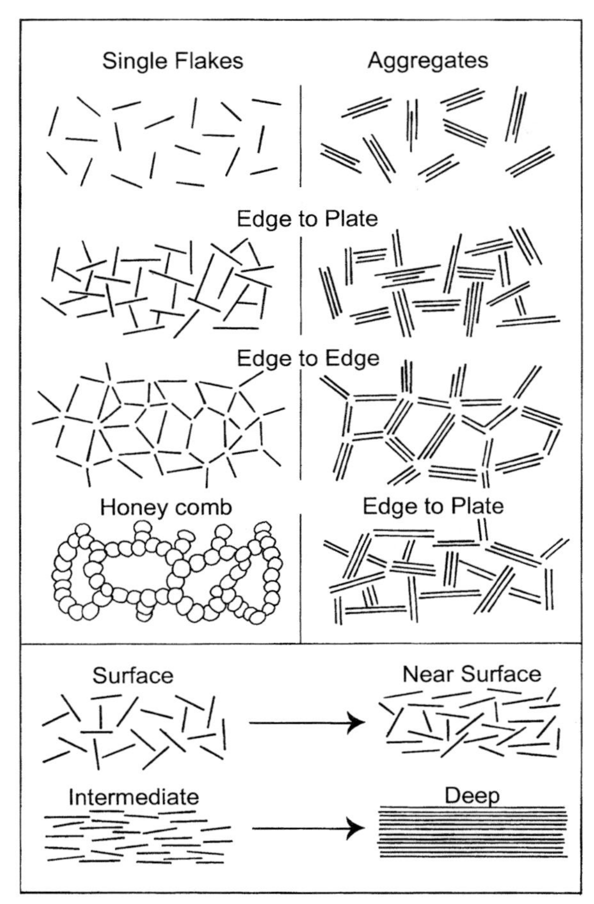

Fig. 3.11. Different forms of flocculation all have open fabrics. Deep burial, however, can obscure much of the initial fabric (Allen 1985, Fig. 8.5). Published by permission of Elsevier Science

in seawater. In fact public water systems that rely on river water employ flocculation by aluminum sulfate to raise the size of clay aggregates sufficiently for filtration to be effective in cleaning the water for consumption. Both high sediment concentrations and strong turbulence also produce more collisions and thus more flocs, although even stronger turbulence and shear destroy them.

Biologically induced flocculation, seemingly the more important process, is the result of sheaths and strands of mucus secreted by bacteria and algae that trap and combine clays, silts and even occasional larger skeletal grains and fine sand into open, unsorted, aggregates with weights and sizes many times greater than their individual components. Mucus from bacteria also plays a key role as a binder of particles at the sediment-water interface (Noffke et al. 2001). Because of the role of bacteria and fine organic matter, including that contributed by man, flocculation occurs in streams, harbors, estuaries and the ocean and is not just limited to the interface of fresh and saline waters as was originally thought. In essence, flocculation combines whatever was available in the water column into loose, irregular unsorted larger aggregates that settle faster than their individual components. In contrast to the above biogenetically formed aggregates, *pedogenic* silt and sand-sized aggregates are formed in expandable clays by alternate wetting and drying. These occur in soils and subsequently become bed load in streams (Rust and Nanson 1989) and also form wind-deposited clay dunes.

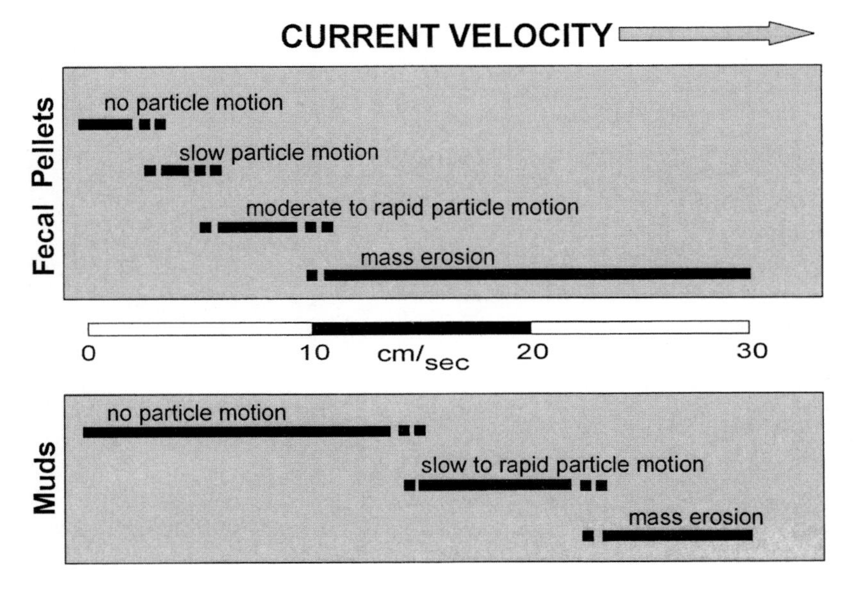

Fig. 3.12. Fecal pellets are entrained at much lower current velocities than equivalent-sized quartz silt and sand (after Minoura and Osaka 1992, Fig. 11)

The practical consequences of flocculated or non-flocculated clays in the water column are many. If flocculated, the initial fabric of the deposit is open and contains much water and is therefore unstable. If non-flocculated, the initial deposit has a more stable subparallel structure with less water. Consequently, its density and shear strength will be higher. Entrainment by flowing water will also be reduced, making a non-flocculated deposit more difficult to erode.

Given the small fall velocities of mud (Table 3.2), the predominance of turbulent over laminar flow in nature, and the great abundance of mudstones in the geologic column, Pryor (1975, p 1244) concluded that most mud must be deposited as aggregates rather than as single particles. This inference is additionally supported by the abundance of fecal pellets and related aggregates reported in modern marine and lacustrine mud. Such ovate pellets with circular cross sections are produced by both pelagic and benthic organisms and typically have sizes of 0.3 to 2 mm, but because of their low density, can be moved easily by weak bottom currents (Fig. 3.12). On some tidal flats in the southeastern United States, as much as 40% of the mud is thought to consist of pellets and in one Japanese bay almost 100% (Minoura and Osaka 1992). Thus pelletization by organisms probably ranks along with flocculation as the major process in the deposition of mud. A consequence of this is that more mud may be transported as "grains" by weak currents and deposited in low amplitude bed forms than has been recognized.

The main conclusion from the above analysis is that clay-sized particles settle to the bottom wherever bottom turbulence and currents are minimal regardless of water depth – the basin could be a lagoon only one meter deep or an offshore basin more than 1,000 m deep. Another important observation is that where organic material is available, it will concentrate in quiet, low-energy sinks along with mud, because its low density imparts a low fall velocity (even weak currents can keep it in suspension). Thus there is a strong inverse relationship between organic content and grain size – the more silt and fine sand, the less organic material. This explains why the protected topographic lows of almost every environment contain its darkest, most organic-rich, silt-free mud.

And what about the erosion of muddy bottoms? Although some studies have been made (Einsele et al. 1974; Young and Southard 1978), entrainment velocities suggested for muddy bottoms range from as little as 10 up to 150 cm/sec. Complicating factors include the variable water content of muddy bottoms, bioturbation, carbonate content (the more carbonate, the greater the shear strength), the effects of bacterial slimes on erodibility and possibly even clay mineralogy (think of the fibrous shapes of small smectitic clays in contrast to the large, platy shapes of kaolinites). At one extreme are water-rich soupy, flocculated, low-shear strength muds rapidly deposited from concentrated suspensions and at the other are slowly deposited, non-flocculated, illitic clays from which most water has been expelled. The weakest of currents stir and erode the soupy muds whereas the cohesive, compact illitic clays require much stronger currents (Fig. 3.13).

Bioturbation and pelletization illustrate well the need to consider all of the above factors. Bioturbation makes mud erode easier when it makes mud less dense, but decreases erodibility when it makes the mud more cohesive (organic binding). In addition, mud pellets are likely to be transported at lower velocities than non-pelletal mud, because of their lower densities (Fig. 3.12). Pellets of silt, organically bound, have also been reported. And, finally, pedogenic clay

Fig. 3.13. Large starved ripples in marine mudstone of Upper Ordovician Kope Formation along Middle Creek, western Boone County, Kentucky. Although the starved ripples, composed of coarse bioclastic shell debris, needed strong currents to form, the underlying cohesive muds at their base were already sufficiently stiff not to be eroded. Also note the differential compaction of the mud above the ripples

aggregates are also fairly common in some alluvial and eolian deposits (clay dunes). Because of the above factors, a universal, well-defined entrainment curve for muddy bottoms has yet to be established.

3.4. Dispersion of Fines

Rivers, waves, turbidity and contour currents in water and wind disperse clay- and silt-sized fines. Below each of these processes is briefly summarized.

3.4.1. Rivers

The governing equation for suspension transport of fines in a river is

$$-Q\frac{dc}{dy} = cw$$

where Q is a diffusion or transfer coefficient, c is the solid concentration in the water column and dc/dy is its vertical change. Q is assumed to be proportional to the upward components of turbulence (thus its negative sign) so that the faster the flow, the greater is Q and the more and coarser are the particles carried in suspension (and along the bottom as bed load). In a river, turbulence is generated by roughness of its boundary, in contrast to standing water where turbulence is also generated by waves and currents and by shear between different water masses. In rivers, the concentration of particles decreases upward from the bottom in all but the finest sizes (Fig. 3.14).

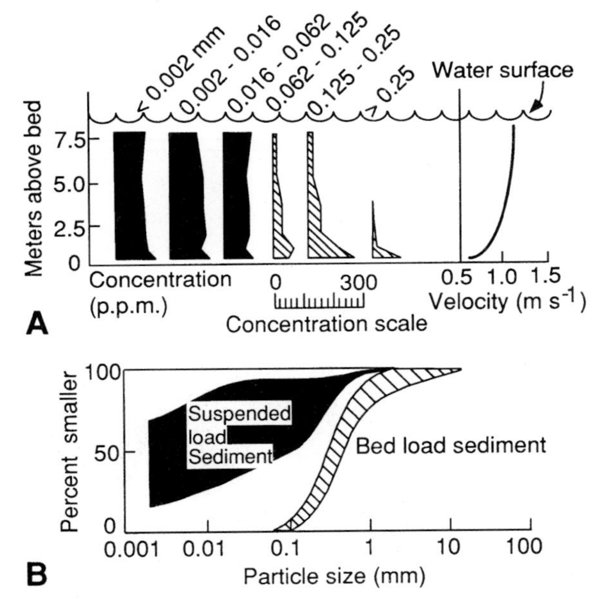

Fig. 3.14A,B. Size distributions in streams: (A) concentration profiles of different size fractions in a stream (Colby 1964, Fig. 7) and (B) size distributions of suspended versus bed load sediment in a stream (from Reid and Frostick 1994, Fig. 4.6). Published by permission of the authors and the Geological Society of London

3.4.2. Waves and Coastal Currents

The circular motion of waves in a sea or lake bottom creates turbulence that entrains particles and keeps them in suspension while currents of tidal, rip, long shore or impinging oceanic currents move them laterally (Fig. 3.16). From Stokes' Law we saw that the smaller the size and density of a particle, the slower will it settle to the bottom, or conversely, the easier will it be entrained or kept in suspension. It is this combination of fall velocity and lateral transport that first segregates pebbles and sand from silt and mud and finally silt from mud to produce a graded shelf with a lacustrine or marine clay mudstone down dip.

The greater the amplitude of a wave, the deeper the agitation of the wave penetrates and the deeper and farther offshore the wave affects a seaward, sloping bottom. Knowing the wave climate (the yearly distribution of wave heights and their average period) of a modern shelf, it is possible to develop a family of entrainment curves for particles of different sizes as a function of water depth (Fig. 3.15). While such curves need to be calculated for each modern shelf (each has, after all, its own wave climate), they all il-

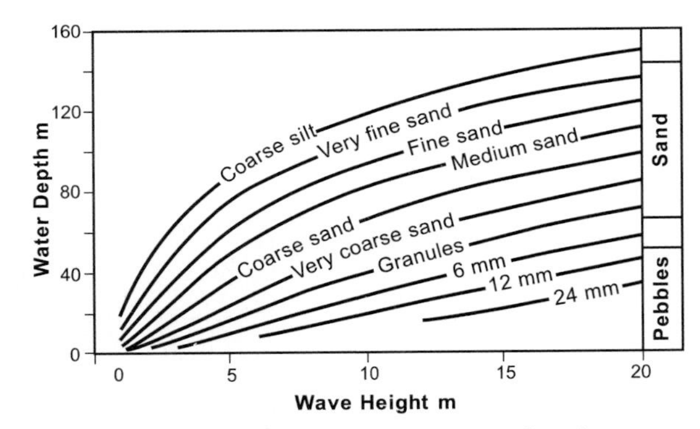

Fig. 3.15. Calculated entrainment curves for the wave-dominated Atlantic shelf of northeastern Brazil based on its modern wave climate and an average wave period of 13 seconds (After Viana et al. 1998, Fig. 5a). Notice how higher waves (stronger storms) entrain larger particles at all depths on a shelf or in a lake. It is this process that produces the graded, equilibrium shelf of inshore sand to offshore distal mud seen so often in ancient deposits. Published by permission of the authors and Elsevier Science

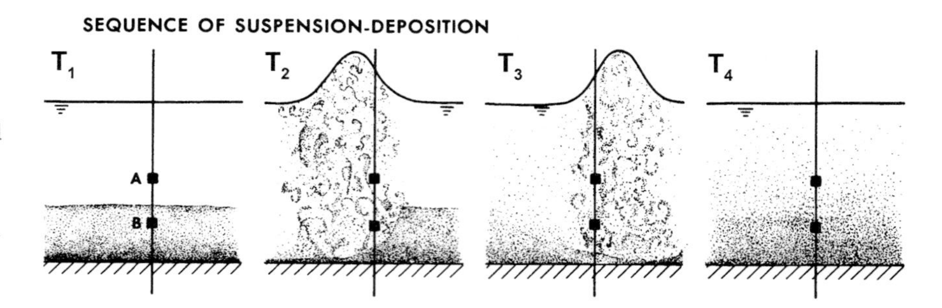

Fig. 3.16. Turbulence under waves resuspends sediment (Wells and Coleman 1981, Fig. 10) permitting coastal currents to move suspended sediment along-, on-, and offshore. Published by permission of the authors and the Society for Sedimentary Geology

lustrate the same tendency – *the stronger the storm, the greater the wave height, the deeper wave base, and the farther seaward sediment can be entrained and transported*. In sum, episodic storm events, like floods on land, seem to be the principal agents of transport and deposition in shallow water in seas and lakes. And, as discussed above, the same storms are likely to remobilize and resuspend sediment on upper slopes creating storm-induced slumps and turbidity currents that carry debris far basinward. Given a seaward sloping ramp or shelf, the effects of wave impingement on the bottom are greatest on the beach

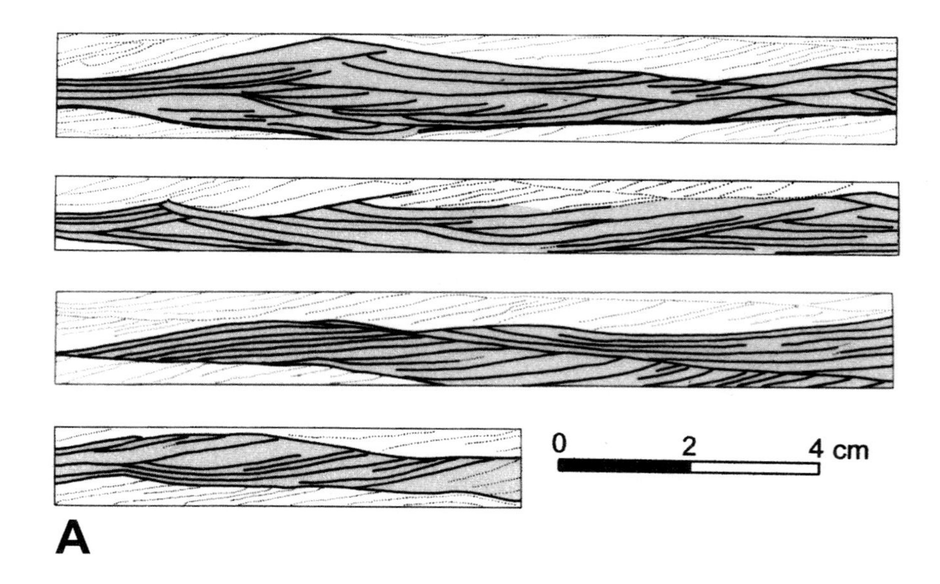

Fig. 3.17A,B. Hummocky bedding ranges widely in thickness: (**A**) thin hummocky beds in a Lower Jurassic muddy shelf from Denmark (Pedersen 1985, Fig. 8), and (**B**) a thick, coarse-grained hummocky bed from the Cretaceous Ferron Sandstone of Emery Co., Utah (Photo by Bob Dott). As a first approximation, the thicker the hummocky bed, the stronger the wave action (stronger storm or shallower water). Published by permission of the author and the Geological Society of London

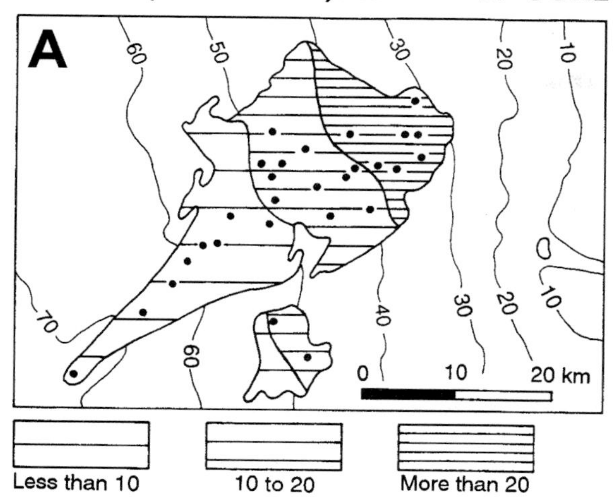

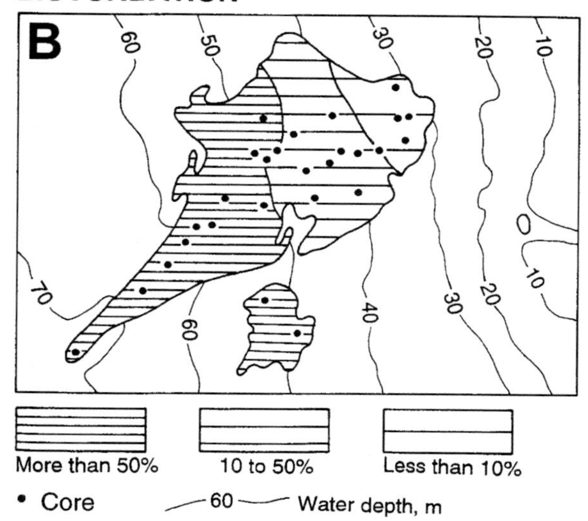

Fig. 3.18A,B. Storm events and bioturbation: (A) storm events are more frequent in shallower water, whereas (B) bioturbation is more intense in deeper water in sloping shelf of the Bay of Biscay, France (after Lesueur and Tastet 1994, Figs. 12 and 13). Published by permission of the author and Elsevier Science

and decrease seaward, as confirmed by many ancient shelf/ramp sequences. In the ancient, it is common to find near-shore interbeds of sandstone and mudstone passing down-dip to distal marine mudstones, perhaps interbedded with limestones. Modern shelves generally lack this pattern because of glacial sea level fluctuations that leave relict off-shore sand deposits as sea level rises.

Storm-wave induced sediment transport forms a bottom density current that flows directly or obliquely down slope and leaves a distinctive graded layer called hummocky cross bedding or hummocky lamination (Fig. 3.17). The resulting bedforms range in size from centimeters to a few decimeters in many marine and lacustrine mudstones. The presence of such structures is the key physical evidence for deposition by storm events in standing water. As a first approximation, the thicker the hummocky unit, the stronger the wave action on the bottom. Between such events are the "fair weather beds" deposited slowly with little transport and deposition, and, bottom oxygen permitting, ample bioturbation by benthonic fauna. As a ramp or shelf deepens seaward, the storm wave base extends to the bottom less and less frequently and consequently bioturbation replaces event beds (Fig. 3.18).

3.4.3. Gravity and Turbidity Currents

Gravity currents are flows that result from a stable fluid being intruded by a denser one (Fig. 3.19). An example is cool air draining down valleys in hilly regions at the end of a hot, windless summer day as the land surface cools. When density contrast is caused by suspended sediment such as clay, silt, sand or even pebbles, the term *turbidity current* is used. Both gravity and turbidity currents play significant roles in understanding mud and mudstone deposition in lakes and seas. Here we focus chiefly on turbidity currents, although in the deep ocean turbidity currents may coexist with contour currents.

The intruding current may be denser than the body of water (Fig. 3.20) and thus follow bottom topography (*underflow or hyperpycnal flow*) as do most turbidity currents, or it may separate from the

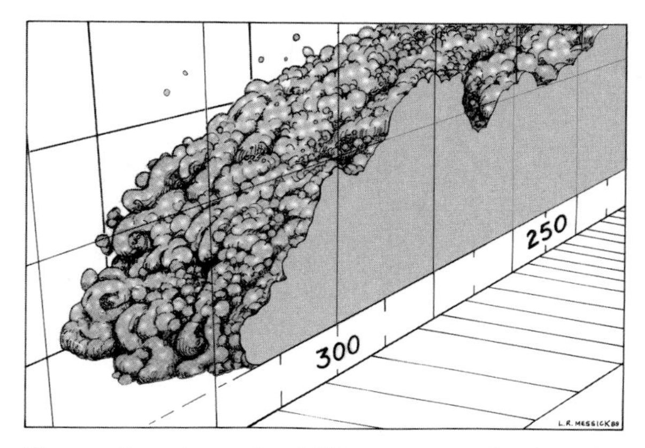

Fig. 3.19. Experimental turbidity current under-flows less dense standing water (redrawn from Garcia and Parker 1989, Fig. 2). Head of flow is highest at front followed by body and tail with separation and diffusion at top interface

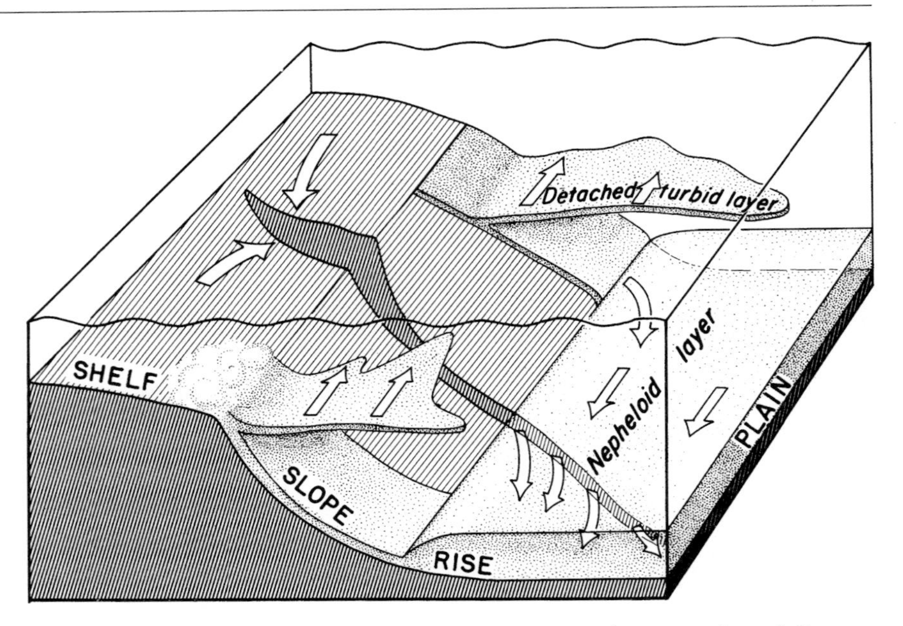

Fig. 3.20. Schematic of major density currents in deep marine and lacustrine basins (Stanley 1983, Fig. 9). Notice how the submarine canyon collects shelf sediment and funnels it directly to deep water, whereas the detached turbid layers derived from the shelf create a hemipelagic rain

bottom as an *interflow* at some intermediate depth. When more dense water is encountered, the flow will stay above it as when fresh muddy water enters saline water or warm water enters a cold lake (*overflow or hypopycnal flow*).

Turbidity currents start with a catastrophic event of some kind, one that suddenly injects sediment into the water or air – sediment laden flood water entering a lake or sea, an underwater slope failure or a large storm that stirs a muddy shelf poised to fail. Once in suspension, this denser, sediment-charged

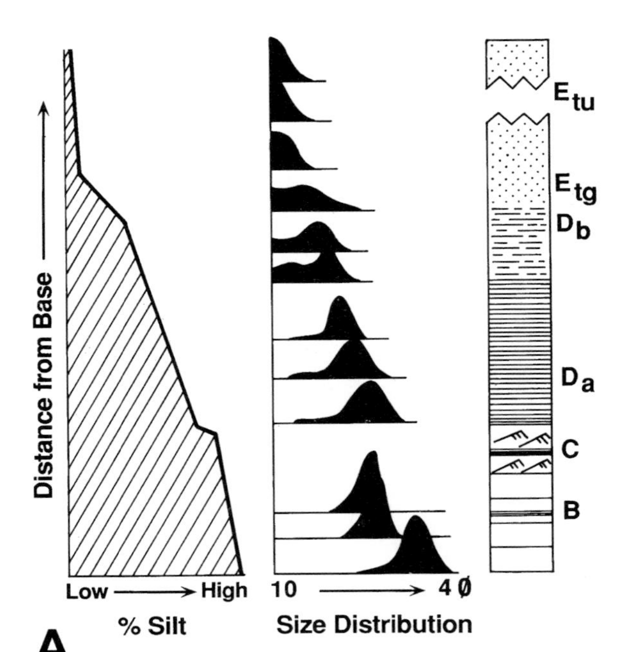

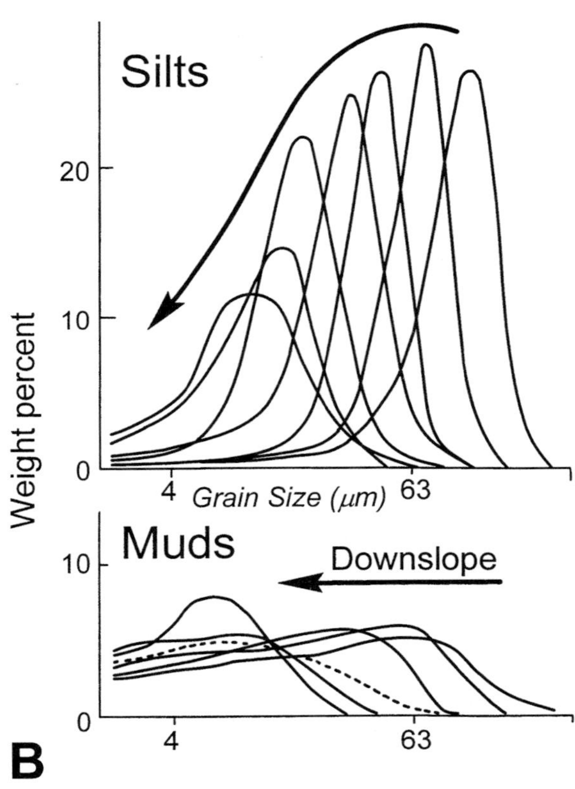

Fig. 3.21A,B. Sequential size distributions produced by turbidity currents:(**A**) vertical size profile of a graded muddy turbidite (Jones et al. 1992, Fig. 23) and (**B**) marked size decline of silt in a turbidite bed in contrast to little decline in its associated mud pair (Stow and Bowen 1980, Fig. 5). Published by permission of the authors, Elsevier Science and International Association of Sedimentologists

water moves down slope propelled by gravity. Such a current has a head, a body and a tail and may last several hours or several days or more. And like a river, it follows topographic lows along the bottom. Early in its course, the current may erode the bottom and acquire more sediment, but as slope and turbulence decrease, deposition occurs with coarse particles first deposited followed by the silts and clays. Up dip, where flow is fast and highly turbulent, a thick graded bed with basal coarse sand is deposited, because the flow is fast and fully turbulent. Down dip, where the flow is weaker and less turbulent, only smaller particles are transported in suspension and thin, fine-grained laminations are deposited. This waning current deposits a bed that is continuous and graded both vertically and horizontally (Fig. 3.21). Such a bed may extend 10's to 100's of kilometers (because of the low fall velocities of fine silts and clay) even on slopes as low as 0.02 to 0.05. In its zone of deposition, size decreases down current (Fig. 3.22). The length of such a deposit, called *run out distance*, depends on initial energy, slope, size of the debris (the finer its size, the longer the distance) and how fast the current is diluted as it moves along the bottom. In a small, steep-sided basin, a turbidity current may have sufficient energy to be reflected from one side back to the other before maximum run out distance is reached.

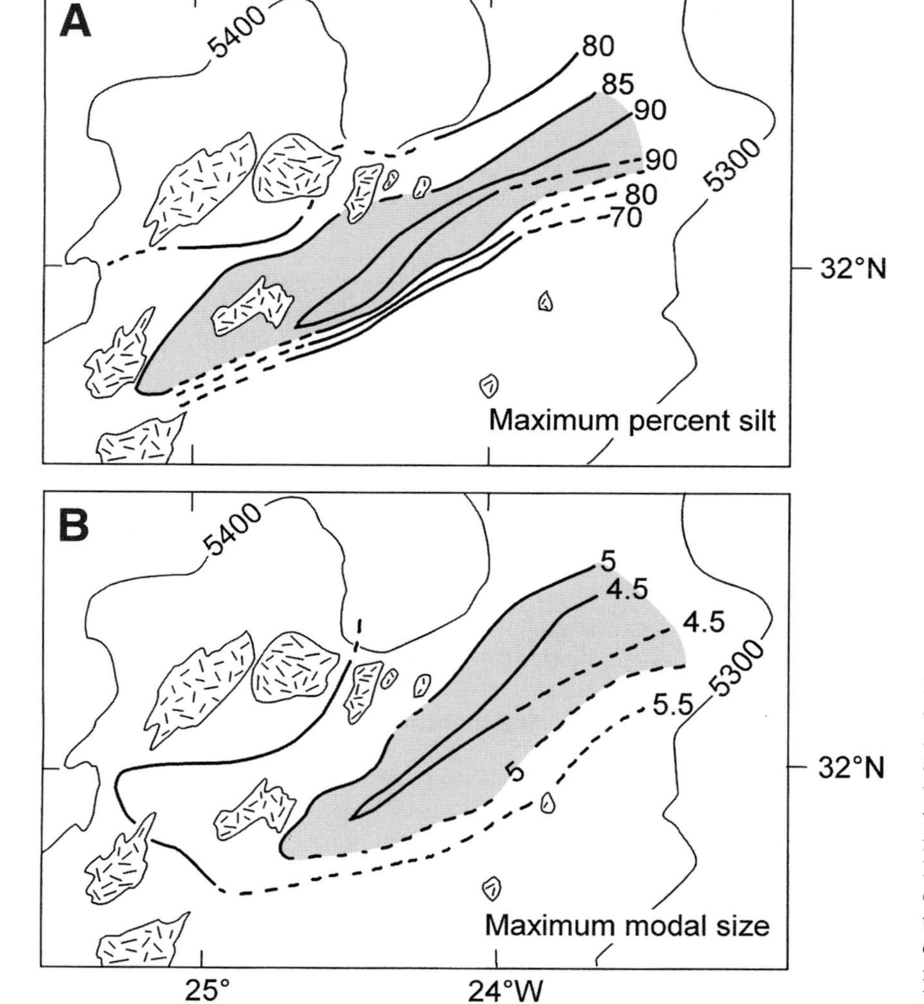

Fig. 3.22A,B. Maximum percent silt and the maximum modal size (phi units) in a modern turbidite bed from the North Atlantic Ocean west of Morocco. Depths in feet (after Jones et al. 1992, Fig. 15). Note elongate geometry and how persistent the thickness is in the transport direction. Published by permission of the authors and Elsevier Scientific Publishing

Although much has been written about the hydraulics of turbidity currents, a simple two-dimensional governing equation provides initial insight to this process

$$\underset{\text{Retarding}}{\tau_o + \tau_i} = \underset{\text{Driving}}{\Delta\varrho\, g\, t\, \alpha}$$

where τ_o is the shear stress at the water-bottom interface, τ_i is the shear stress at the upper interface of the current, $\Delta\varrho$ is the density difference between the turbidity current and the overlying water mass, g is the acceleration due to gravity, t is the thickness of the current and α is the slope angle of the bottom. All the terms on the left side are frictional and retard the current, whereas all those on the right drive it down dip; e.g., the greater $\Delta\varrho$ and α, the faster and more turbulent the flow. In addition, there are formulas that help us roughly estimate the velocity and run out distances of turbidity currents, but which require some assumptions about the thickness of the flow, its slope, the density contrast, and retarding friction (Allen 1997, pp 227–235).

Several observations need our attention. Careful study of sedimentary structures is the first step toward recognizing the deposits of turbidity currents. Hence for muds and silts, we look for millimeter or decimeter sequences of graded laminations and beds and try always to separate event pulses from background sedimentation. Normally, we think of a turbidity deposit as below some topographic high, but smaller and thinner ones are also believed to occur on shallow shelves after large storms and are called *tempestites* (although both have well defined grading, hummocky bedding distinguishes tempestites from turbidites). In addition, it is thought that *interflows*, density flows within the water mass, may be responsible for some of the thin, continuous laminations of silt seen in many mudstones.

3.4.4. Contour Currents

These are currents well below wave base that flow parallel to a slope and are best known in modern deep oceans. In the deep ocean, contour currents are widely recognized along continental rises bordering all the oceans and are especially well documented in the western Atlantic, where many modern deposits called *contour drifts or contourites* have been recognized. These consist of mud, silty mud, silt and fine sand that are mostly terrigeneous, but may have much pelagic material and, exceptionally, may consist mostly of pelagic skeletals. Such deposits are laminated and micro-crosslaminated and have scours as well as bioturbation. Modern contourites occur as thin sheets and mounds and as isolated dunes. See Pickering et al. (1989, pp 219–249) for more details.

As thick mappable deposits, ancient contourites on the continents are mostly controversial. An exception is the Talme Yafe Formation of Cretaceous age in Israel, a fine-grained Cretaceous pelagic carbonate derived from a carbonate bank bordering the Tethyan Ocean (Bein and Weiler 1976). Recalling the broad definition of a contour current as, "Any current well below wave base that flows parallel to its slope", it may be that the ancient contourites are most likely to be found as thin interbeds between turbidites or slump deposits in deep water submarine fans. In other words, after an episodic turbidity current flowed down slope depositing mud, silt, and fine sand over bordering levees and interlopes, these materials may be sorted and reworked by gentle slope-parallel contour currents, if not rapidly buried (Shanmugan et al. 1995). Improved sorting, bed-by-bed contrasts in paleocurrent direction and possible weak scours could all help confirm deposition by persistent contour currents between distal episodic events.

3.4.5. Wind

As discussed in Chap. 2, wind is a significant agent of transport of fines to the oceans, but little has been done to recognize any eolian component in ancient mudstones. Many distal mudstones contain even, single-grain laminations of silt that could easily be of eolian origin. Perhaps some color banding also has this origin. We know that each year many thousands of tons of clay and silt are injected into the atmosphere from deserts and from volcanic eruptions. These sources are major contributors to soil formation on carbonate islands, so it seems reasonable to assume that they are also important in slowly-deposited muds elsewhere.

3.5. Sedimentary Structures

The sedimentary structures of mud and mudstones are most apparent when silts and fine sands are deposited along with mud (Fig. 3.23). When macroscopic structures are seemingly absent, however, X-ray radiography of thinly slabbed core usually reveals structures hidden to the eye. As noted above, the sedimentary structures of fine sands and silts, except for being scaled down, are identical to those

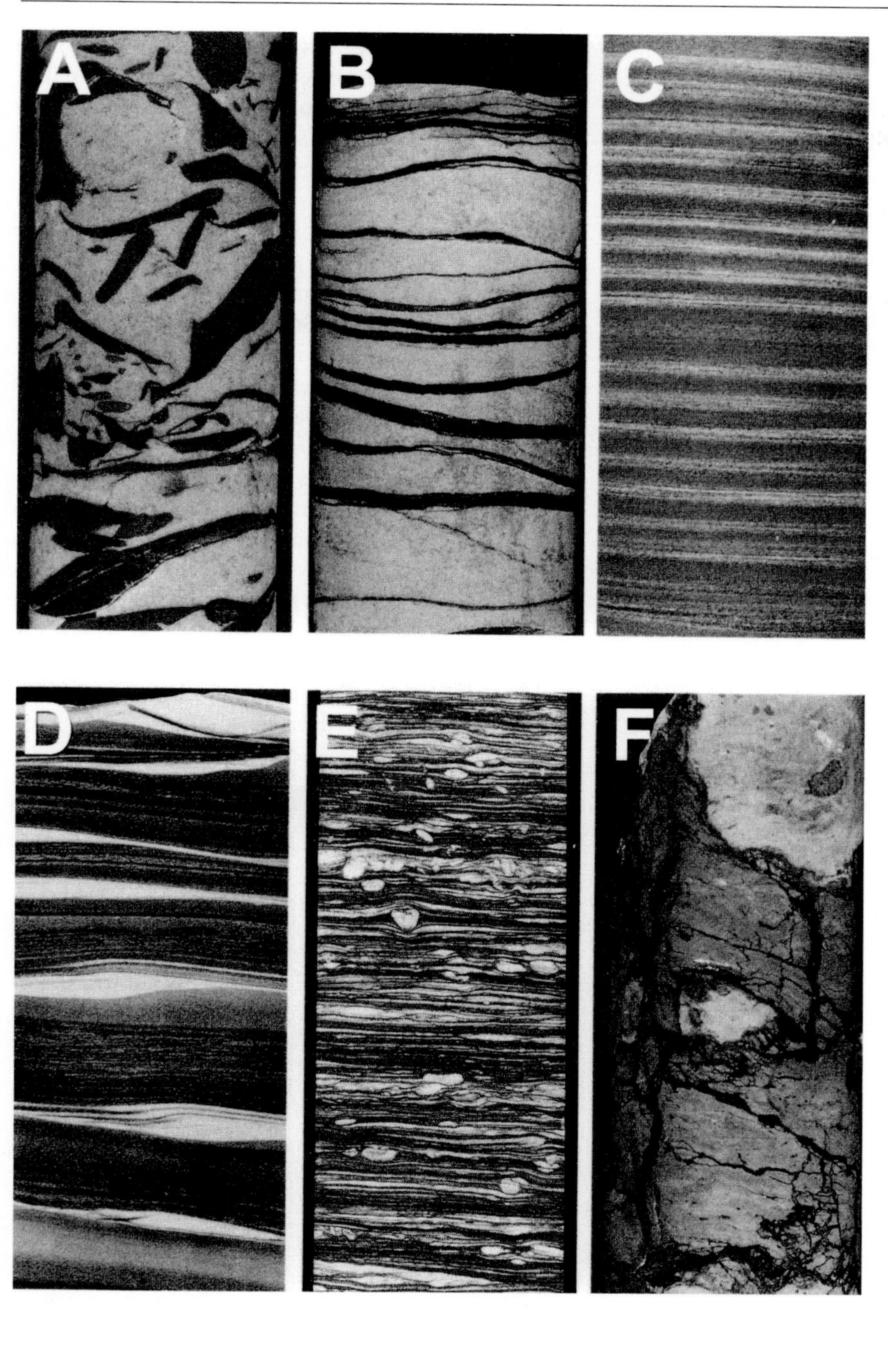

Fig. 3.23A–F. Pennsylvanian coal measures of Illinois Basin in western Indiana illustrate many of the sedimentary structures of mudstones, and their importance in distinguishing shallow water terrigenous environments (Barnhill and Zhou 1996): (**A**) irregularly deformed, shaped and oriented mudstone clasts in sandy debris flow; (**B**) thin, irregularly inclined, mudstone drapes (tidal?); (**C**) even, thinly and rhythmically inter-stratified and graded mudstone and siltstone (tidal); (**D**) mudstone with isolated, cross laminated ripples (distal splay of bay fill); (**E**) thinly laminated, silt-streaked mudstone with some bioturbation (distal delta front?); and (**F**) massive, silty, gray, rooted underclay with large calcareous, white, algal nodules

of the sandstones. An exception is lamination, which is the dominant structure of fine-grained rocks even though much is destroyed by bioturbation and sometimes by mass movement (Table 3.3). Below we first consider the hydraulic structures found in mudstones and then examine those formed shortly after deposition, the so-called *soft-sediment structures*, that include bioturbation and flowage of soft mud. We defer a discussion of the interesting chemical structures found in mudstones to Chap. 6.

3.5.1. Hydraulic Structures

Lamination, stratification thinner than 10 mm, is the dominant sedimentary structure of mudstones and also of most siltstones (Fig. 3.24). The three principal hydraulic structures of mudstones are parallel lamination, cross lamination, and graded lamination. Both clay and silt-sized particles can be parallel laminated and graded, but cross lamination typically occurs only in silt or fine sand. Lamination predom-

Table 3.3. Common sedimentary structures of mudstones and their significance

Processes	Products
Dilute bottom currents	
Distal turbidity currents	Cross lamination/lamination; weak scours; hydraulically contorted lamination; sole marks; grading
Storms	Hummocky bedding & scours
Contour currents	Cross lamination/lamination & some scours
Weak tidal currents	Lamination and small scours
Suspension settling	Lamination thin, even and graded & non-graded
Microbial mats	Wavy, crenulated lamination
Mass movements	Slumps, slides, slurries, and micro-faults (soft sediment deformation)
Bottom feeders	Wide range of bioturbation and trace fossils (types and abundance depend on bottom oxygen and water depths)

inates in mudstones, because muds are deposited principally by weak suspension currents. In the majority of mudstones, lamination is much thinner than the 10 mm upper limit. Schieber (1990) developed a logical sequence for many of the most common hydraulic structures of the fine-grained Precambrian Belt Series of Montana that appears to have wide application (Fig. 3.25).

Only clay minerals are more characteristic of mudstones than lamination. Because of this and, because lamination also occurs in all mudstone environments, it has received the most attention of all the sedimentary structures of mudstones. To appreciate this, see the collected articles in Kemp (see Digging Deeper) and two articles focused on environmental end members – glacimarine laminates (Cofaígh and Dowdeswell 2001) and diatomaceous laminates (Chang et al. 1998).

Lamination occurs in many forms – as clay-on-clay laminations, as thin clay laminations separating well laminated silts to form silt-clay pairs (doublets) and as triplets of silt and mud followed by a thin, chemically precipitated or biochemically deposited

Fig. 3.24. Regular, parallel lamination of fine silt and clay overlies slumped and deformed lamination in splay deposit bordering a Pennsylvanian channel, Franklin Co., Illinois

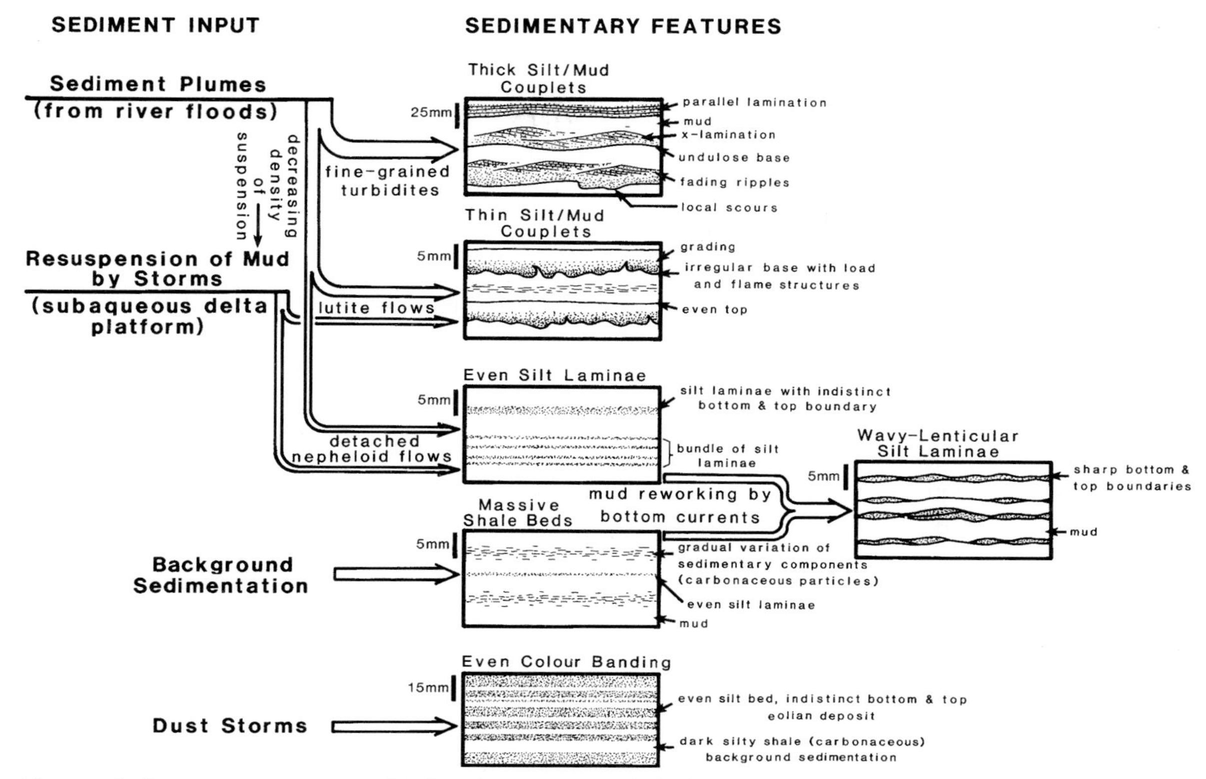

Fig. 3.25. Sedimentary structures and inferred processes in fine-grained Precambrian Belt Series of Montana have wide application to mudstones with minor bioturbation (Schieber 1990, Fig. 4). Published by permission of the author and Elsevier Science

lamination (diatoms, coccoliths, or algal mats). Essential formative requirements for lamination are repetitive variations of terrigeneous, chemical or biological supply all in the absence of bioturbation or liquefaction. Typically, individual laminations are separated one from another by sharp contacts caused by abrupt changes in grain size, possibly by concentrations of coarse micas, heavy minerals, or fine organic matter, but contacts may also be gradational.

Bioturbation is minimal where there is a lack of oxygen (poor ventilation or excessive organic matter) or insufficient food (cold glacial lakes) or very rapid sedimentation. Individual laminations may be even and uniform, wavy, systematically inclined or graded and even or be as thin as just a few silt grains. Laminations only a few grains thick are believed to have been deposited by dilute hemipelagic suspensions or by wind (O'Brien 1989). Lamination may also be continuous, discontinuous and patchy. Such discontinuous laminations can result from primary sedimentation of thin patches of rippled or planar silts moving along a cohesive clay bottom to form flaser bedding, which is very common in mudstones. Graded siltstone-clay pairs, on the other hand, are more likely to be laterally continuous and represent deposition from weak distal turbidity currents. Such graded pairs may have even, flat bases and represent rapid settling from suspension or may have slightly scoured bases and represent bottom density flows. The latter commonly have small flutes whose orientation provides paleocurrent information. Lamination is also present in pure mud, where it is defined by slight grain-size variations and represents a pulsed hemipelagic rain that may be terrigeneous, biogenic or less commonly a chemical precipitate. Where silt is largely absent, polished slabs, X-ray radiography or back-scattered electron images are commonly necessary in the laboratory to demonstrate the existence of lamination and other structures (see Appendix A.1). In purely terrigenous mud, lamination may result from the tails of turbidity currents that deposited sand and gravel up dip, from the pulsed settling of fines from the muddy plume of a delta in flood, or from a possible nepheloid layer. Biological and chemical laminations mostly represent either seasonal or exceptional conditions in the upper part of a water column and are favored by high productivity resulting from warm water temperatures and abundant nutrients. Mudstones commonly show weak to moderately strong, evenly laminated color banding on a millimeter scale

– a poorly understood structure that occurs widely and seems to have several origins. It can form from small-scale pulsed variations of finely disseminated organic matter, perhaps related to climate cycles; it can form from weak, distal eolian contributions of desert dust; it can form from repeated oxygenation events in the bottom water of the basin. Where mudstones are non-laminated, they may represent intense bioturbation or dense mudflows along the bottom (Shanmugam 1997). Detailed thin section study of the argillites of the thick Precambrian Belt Series in the western United States illustrates details (Fig. 3.25) of how weak bottom and suspension currents probably deposited most of this thick sequence. More thin section studies of this kind of sedimentary structure in mudstones should provide many new insights to the origin of mudstones.

3.5.2. Soft-sediment Structures

Convolute lamination, flame structures and load casting are the chief physical soft sediment structures of mudstones. Convolute lamination and flame structure may be produced just below the water-sediment interface by the drag of a passing current on still fluid mud or by down dip gravitational creep or sliding prior to consolidation. Load casts, however, represent the sinking of fine sand and silt into less dense, water-saturated soft mud. Other liquefaction features include mud volcanoes and their feeding pipes. Slump folds, on the other hand, represent failure of a coherent mass – where an over-pressured or over-steepened bed fails, deforms, and moves down slope. Because currents may flow obliquely on a broad shelf, slump folds are likely to be more reliable indicators of down slope direction in a muddy basin than hydraulic structures.

3.5.3. Biogenic Structures

Organisms that live on or in mud and ingest it are the principal enemy of lamination and other hydraulic structures. Bottom dwellers do this by boring clean well-defined holes that may be deep or shallow, by eating sediment, or by stirring it. The search for food dominates this process, but other reasons include concealment from predators, reproduction, etc. Typically, only the traces of these animals are preserved and not the animals themselves and hence the term, *trace fossil*.

The study of these traces is called *ichnology* and the disruption of stratification is called *bioturbation* (Fig. 3.26). The chief control on bioturbation is the availability of oxygen on the bottom. By noting the extent and kinds of bioturbation, its relation to stratification and color (mostly determined by its organic content), one can go far to semi-quantitatively reconstruct oxygen levels in ancient muddy basins (see Chap. 4). But there are other uses of trace fossils: they provide insights to how animals use space, information on the firmness of the bottom, help identify intervals of slow or interrupted sedimentation, and have also been used to infer water depths based on changes in their association as mapped across a basin or vertically through a section. Of all these interesting uses, linkage to paleo-oxygen levels, to color and to organic content have proved the most useful and are emphasized here. Below is a short summary on how to do this.

First, make an overall assessment of the ratio of bioturbation to stratification, and secondly, look carefully at the shapes, sizes and kinds of biogenic structures. As oxygen levels decrease, both size and depth of burrows as well as their abundance decrease so that black muds deposited in oxygen deficient bottom waters have only a few small diameter, shallow burrows and traces, or even none at all, and consequently, lamination is perfectly preserved. Conversely, well-oxygenated muddy bottoms with slow to moderate sedimentation will have maximum bioturbation and virtually no stratification. The rate of sedimentation is also a key factor – a sudden thick pulse of mud or silt with thickness beyond the reach of burrowers will be free of bioturbation, except possibly for escape burrows. Conversely, during a local or regional pause in sedimentation, the bottom will be completely reworked, given the presence of oxygen in the bottom water.

Bioturbation also affects erodibility by affecting the water content and thus density of the mud. Production of fecal pellets during bioturbation can also be significant. Feeding and excretion and consequent bacterial activity during bioturbation is likely to alter water chemistry within the mud and affect the path of diagenesis, especially that of pyrite. For example, a burrow rich in organic matter may be lined with fine pyrite or conversely be free of pyrite, if well oxygenated. Thus the study of bioturbation, along with that of the available body fossils, is always central to the study of mudstones.

Benthonic *microbial mats and films*, although less well understood than trace fossils, are believed to be another type of biogenetic structure, possibly algae related, that are preserved as millimeter-scale wavy-crinkled laminations (Fig. 3.27). These seem to be best developed in the Precambrian before there were

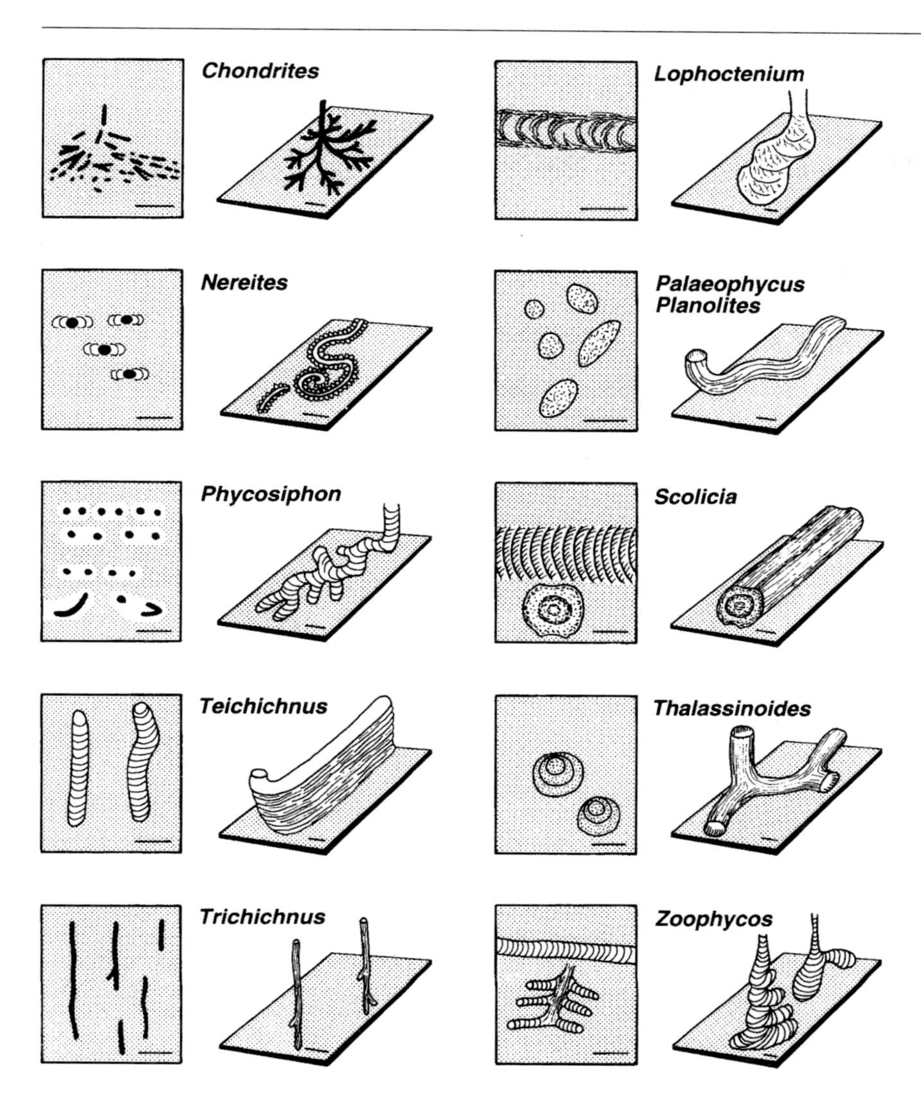

Fig. 3.26. Common trace fossils of mudstones (Wetzel and Uchman 1998, Fig. 3). Published by permission of the authors and E. Schweizerbart'sche Verlagsbuchhandlung

grazing invertebrates and bioturbators (Seilacher and Plüger 1994). They are typically interlaminated with black, silty mudstones and are considered to be subtidal. They can occur in some muddy Phanerozoic basins, especially in shallow hypersaline basins. See Noffke et al. (2001) for criteria distinguishing varieties of these structures.

In all of the above, there is a contest between the fall velocity of the particle and the upward component of turbulence of the fluid, be it a current in water or air, as the particle is transported in suspension. But everywhere and always as the current decays, first the coarsest and finally the finest particles settle on the bottom. In an ancient mudstone this process is preserved in the texture of the mudstone as first the larger (denser) and finally the finer (lighter) particles are deposited. Total silt decreases, silt size decreases, silt laminations decrease in both

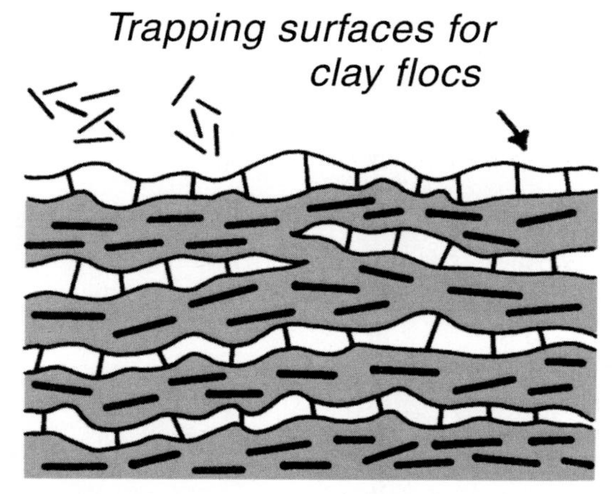

Fig. 3.27. Schematic of microbial mats (redrawn from O'Brien 1996, Fig. 10). These mats produce a crinkly laminated rock, with interbeds of mudstone and limestone. Published by permission of the author and the Geological Society of London

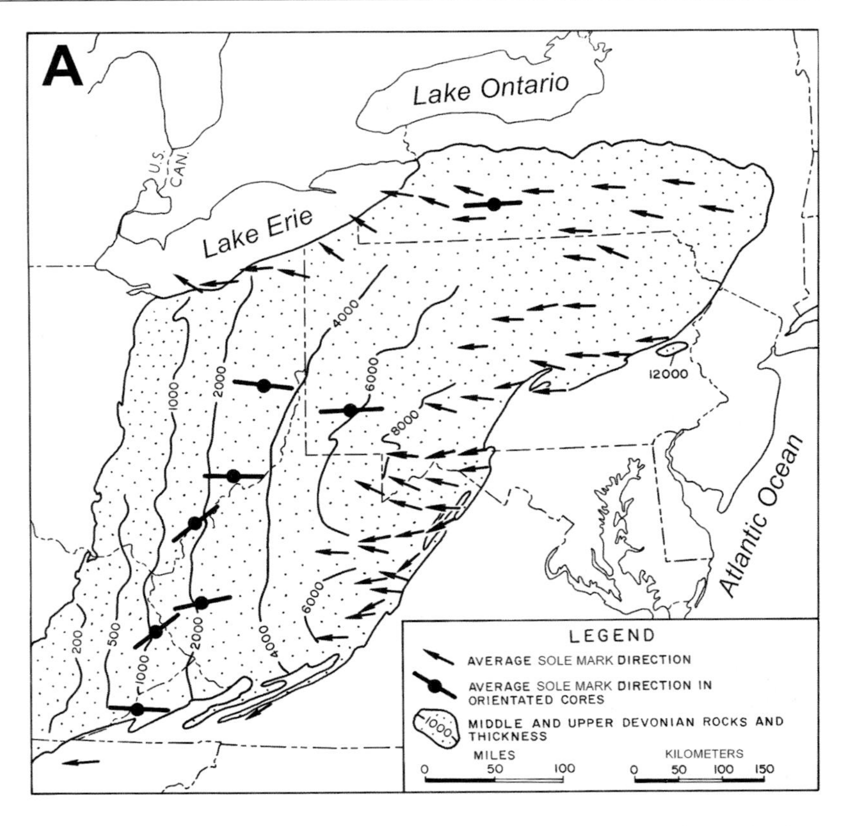

Fig. 3.28A,B. Paleocurrents in a Devonian black shale basin sourced from the east. (**A**) Note uniformity of small flute marks from cores with outcrop paleocurrents; (**B**) varieties of micro-flute marks on thin, millimetric, interlaminated siltstones in black shales. Published by permission of Penn Well

thickness and abundance and they finally totally disappear so that the end result is a clay with only a few fine scattered grains of silt or perhaps even none at all – a tooth paste-like clay. Organic matter transported in suspension has a parallel history in a waning current given an anoxic bottom (Isaksen and Bohacs 1995, pp 32–33). Clearly identifiable woody matter becomes less abundant (and more worn?) down current and pellet and pelloids also decrease in abundance and size, whereas organic matter (lightest of all) increases. Thus the final deposit, given a lack of oxygen, is a laminated, organic-rich mud or clay shale with a few rare, fine scattered silt grains – the ideal source rock for petroleum.

3.6. Paleocurrent Systems

The paleocurrent systems of muddy basins have mostly been inferred from the measurement of sedimentary structures in their associated sandstones (Fig. 3.28) or carbonates and from lateral changes in proportions of sandstone or carbonate to mudstone across a basin or from orientations of invertebrate fossils (Fig. 3.29) or woody particles. So it is fair to ask – do the interbedded mudstones themselves have the same paleoflow direction? Few systematic studies exist to answer this question, but there are some pa-

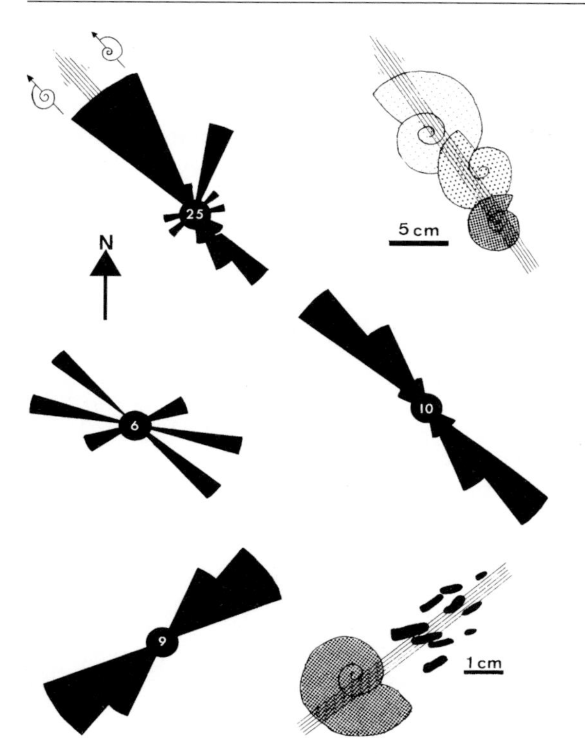

Fig. 3.29. Fossils as paleocurrent indicators in mudstones (after Aigner 1980, Fig. 7). Published by permission of the author and E. Schweizerbart'sche Verlagsbuchhandlung

leocurrent indicators in mudstones (Table 3.3) that help to resolve it, although studies are not abundant. Virtually all the published studies have been made on marine mudstones, which show strong lateral and vertical orientation of paleocurrents. This consistency of orientation occurs in both the distal black shales of turbidite basins as well as the marine mudstones of coal measures. But there are some noteable differences between paleocurrent directions in sandstones and their mudstone interbeds as shown by Crimes and Crossley (1980) for Welsh turbidites. Another conclusion that seems to emerge, but one that needs more study, is that stronger currents not only orient fossils better than weaker ones, but tend to produce parallel orientations of any elongate forms rather than transverse. The final conclusion from these studies is that even the bottoms of anoxic basins had currents strong enough to produce small sole marks and orient pelagic fauna or spores that settled on its black mud.

Paleocurrent studies of muddy basins are most productive when physical structures in the mudstone are systematically mapped and their orientation measured and when orientation of body and trace fossils is systematically recorded. In addition, the resulting directional data always needs to be carefully integrated with regional facies distribution, the sandstone mudstone ratio, and the paleocurrent system of the interbedded sandstones and carbonates. For best results at least some interbedded silt and sand is needed. Examples of paleocurrent studies in mudstones are not numerous, but three of note include Ruedemann's (1897) measurements of graptolite orientation in the Ordovician Utica black mudstone of New York; Schieber and Ellwood's (1988) study of fossil orientation and grain alignment of magnetic fabric; and Cressman's (1985, Fig. 3.9) map of slump fold orientation in the argilites of the Precambrian belt series of Montana.

3.7. Small Scale Sequences and Tidal Rhythmites

Sedimentary structures of mudstones and siltstones provide most information when studied in vertical sequence along with the kinds and abundance of body and trace fossils and organic matter. Add to this trinity scour surfaces in mudstones, also called mudstone-on-mudstone unconformities, and we have all the tools needed to identify cycles of flooding, sea retreat, and erosion, – the key elements of sequence stratigraphy in basin centers (Chap. 8). Scour surfaces are commonly marked by phosphate pebbles, glauconite or fossil lags that separate underlying more organic and clay-rich shales from overlying siltier mudstones or conversely, the abrupt upward termination of a silty sequence with flaser bedding by a pure claystone. Such sequences may range in thickness from a few to many meters.

Here we concentrate on the thinner sequences, show the importance of sedimentary structures to their interpretation and start with lamination – either as doublets or triplets – called *rhythmites* (Figs. 3.23 and 3.24). Rhythmites are, as a rule, best studied in cores, where longer sequences can be investigated. The mudstone component, although commonly subordinate, is nevertheless crucial because it defines the stratification.

Are such cyclic sequences periodic with respect to time, however? There is much debate about this, but by measuring successive thicknesses and recording lithologies, a sequential data set is obtained and becomes the basis for an interpretation. The ideal fundamental unit of time is the *varve* – a couplet of laminations deposited in a year. These were first recognized in glacial lakes in Sweden and consist of silts or fine sands (summer, rapid ice melting and rapid sedimentation) followed by a thinner lamination of mud (winter, frozen lake and slow sedimentation). Scattered, striated *dropstones* in such sequences help establish their glacial origin. Careful field study and calibration by ^{14}C dating provides a 13,500 year record of deglaciation in Sweden.

Varves are also recognized in deep water marine basins and in non-glacial lakes, where the couplet or triplet is defined by annual variations in productivity (summer, warm water and abundant pelagic life or perhaps even a chemical precipitate) whereas with cooler winter temperatures productivity is less and mostly fine mud is deposited. The biologic components in such annual cycles may either be pelagic skeletals (diatoms and coccoliths) or algal mats on the bottom. Secondary pyrite, or possibly manganese, may also develop in the muddy part of such couplets and is thus a common example of an early diagenetic lamination. In the past glacial and nonglacial varves have mostly been counted by hand but increasingly imaging is used (Algeo et al. 1994).

Other frequencies of cyclic deposition are also common in mudstones. Tidal cycles have been identified in siltstone-mudstone pairs (Figs. 3.23A, 3.24, and 3.30). Here the thicknesses of successive pairs are matched against daily and lunar periodicities. Tidal rhythmites in terrigenous deposits consist of stacked laminations as well as sand- or silt-sized ripples with clay drapes. To be considered tidal, a sequence of rhythmites should "wax and wane" in thickness. That is, the rhythmites should have smoothly increasing and decreasing thickness (Fig. 3.30), a property thought to result from a smoothly changing tidal range as the alignment of the Earth and Moon changes monthly and yearly. This characteristic helps distinguish tidal rhythmites from varves or deep-sea rhythmites. In both the rippled and laminated facies, the sand represents deposition during the ebb or the flood, whereas the mud represents deposition during slack water at high tide. The importance of the study of these tidal rhythmites is that they can be made quantitatively as well as qualitatively.

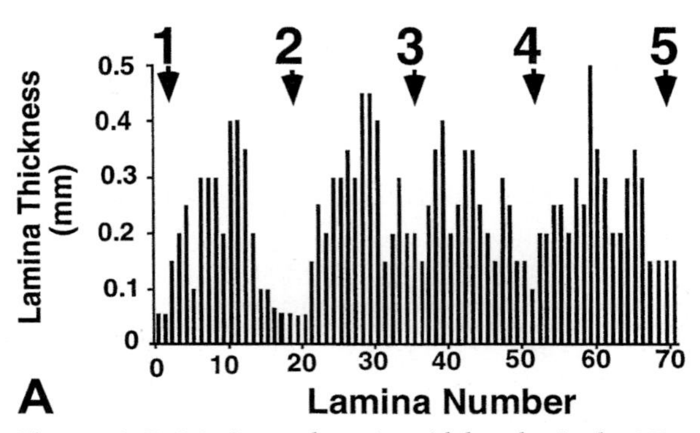

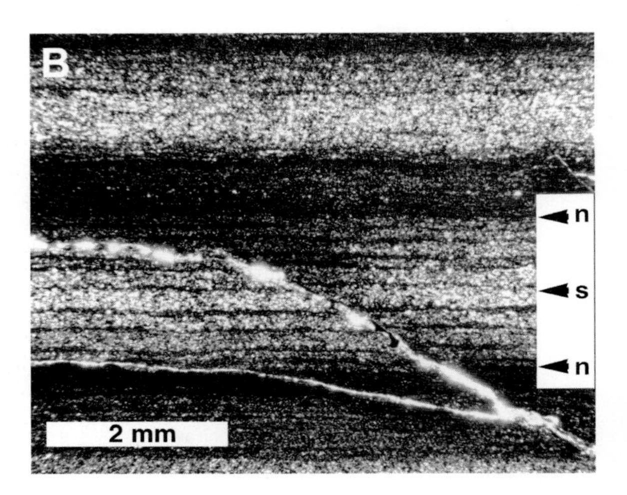

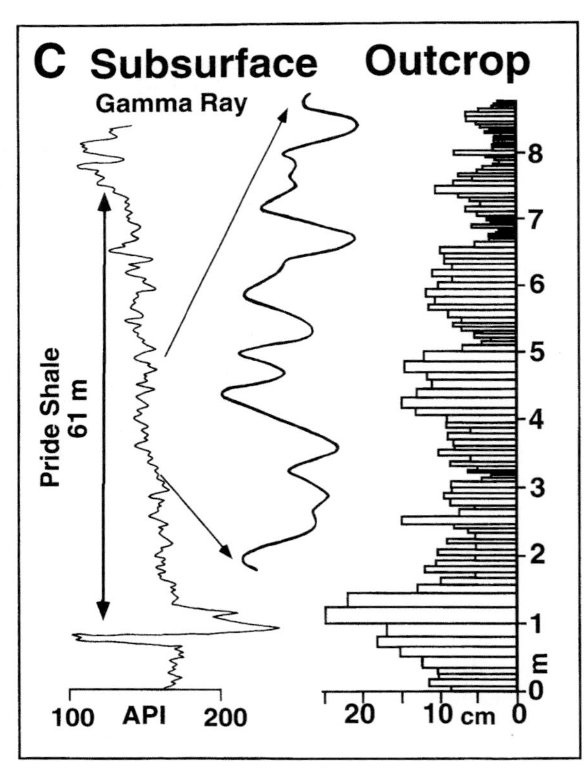

Fig. 3.30A–D. Waxing and waning tidal cycles in the Mississippian Pride Formation of Mercer Co. WV (Miller and Eriksson 1997, Figs. 3, 6 and 7): (**A**) apogee (distant sun and moon) and perigee (proximal sun and moon) cycles; (**B**) thicker bundle of sandy laminites interpreted as annual cycles; (**C**) nearby subsurface gamma-ray log correlated with interval of detailed measurements. Note here how the units fine upward and have some mudstone intervals (*black*); (**D**) shadows highlight four meter-length cycles plus the centimeter cycles that contain the millimeter cycles (see **B**). Published by permission of the authors and the Society for Sedimentary Geology

Quantification can be through graphical displays or time-series analysis using thicknesses of paired siltstone-mudstone couplets (see Box 8.2). With either approach, the problem is to identify the many tidal subcycles ranging from diurnal to multi-year cycles. Care is needed, however, *because making the wrong assumption about the first level of cyclicity invalidates the interpretations of all the others.* Another issue is the possibility of other cyclic processes, such as freshwater runoff from the hinterland, and subtle erosion events. A good example of the quantification of tidal cycles is the study of the spectacular outcrops of tidal rhythmites in the Mississippian Pride Formation that occur in Mercer County in south-western West Virginia (Fig. 3.30), studied by Miller and Eriksson (1997). They recognized five dif-

ferent levels of cyclicity: semi-daily, semi-monthly, monthly, annual, and an 18.2 year cyclicity.

The tidal rhythmite facies seems to be best developed on mudflats and in estuaries with tidal ranges of 2 to 3 m or more. Tidal rhythmites occur throughout the geologic record in paleochannels with estuarine fill as well as in delta front deposits and on tidal flats (carbonate tidal flats are far more common, however). Terrigenous tidal rhythmites are a good example of an environment where mud is commonly subordinate, but has an important role, because it defines in large part the stratification. Sources of much of the above and for more information are Kvale (2003) and Kvale et al. (1998) in Digging Deeper.

Longer cycles of about 19, 98, and 404 thousand years, called *Milankovitch cycles* (Box 8.2), caused by variations in the Earth's orbit, are increasingly recognized in muddy rocks. Such cycles are typically decimeters to meters thick compared to the millimetric scale of daily tidal cycles. These orbital variations produce climate changes that in turn change global temperature, productivity in the water column, rainfall, and erosion. These effects can then be detected as rhythmic variations in mudstone properties such as color, organic content, and grain size (Fig. 3.31), as described in Chap. 8, Muddy Basins.

3.8. Summary

Muds and silty muds accumulate where a supply of fines exists and bottom currents are too weak to erode them, although exceptionally they may accumulate where mud concentration in shallow water is so great that it reduces wave energy. Most mud seems to be transported as aggregates, as floccules and fecal pellets, rather than single clay particles.

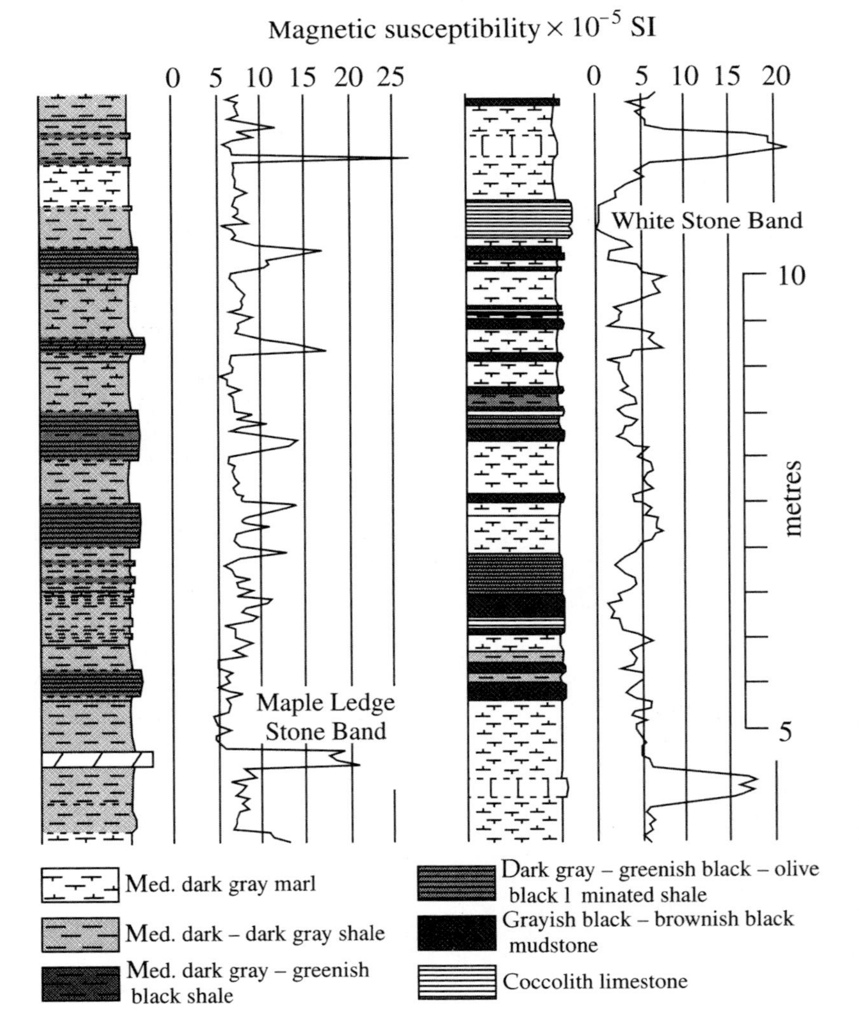

Fig. 3.31. Six different types of mudstones recognized through an interval of about 15 m from bed-by-bed descriptions of two outcrops of the Upper Jurassic Kimmeridge Clay exposed along coast of east Dorset, England (after Weedon et al. 1999, Fig. 8). Magnetic susceptibility curve as well as lithologic sequences suggest cyclic deposition. Published by permission of the Royal Philosophical Society of London

But even granting this, the slow fall velocities of these aggregates permit them to be transported much farther in suspension in comparison to medium and coarse silt and sand, which mostly travel along the bottom as bed load. These great contrasts in fall velocities explain the much greater lateral continuity of mudstones than sandstones and siltstones.

Weak currents abound on the bottoms of most muddy basins as shown by the presence of cross lamination, micro-flutes and scours, graded beds, flame structures, oriented fossils and low-relief discordances that separate coarsening from fining upward sequences defined by sedimentary structures, grain size and organic content. Most probably, weak distal turbidity currents are foremost among all of these currents. Although studies are few, many of these currents seem to be unidirectional and exist in both well and poorly ventilated muddy basins. But, in addition to turbidity and contour currents, vertical hemipelagic, "fall out" sedimentation from nepheloid layers within the water mass as well as deposition from the surface plumes of deltas contributes to many muddy basins and forms muddy halos around landmasses. Much fine silt and clay transported by wind must be present in most mudstones as scattered small silt grains or thin even, one-grain-thick laminations. Microbial mats in muddy rocks are increasingly recognized as a distinct and important sedimentary structure, but clearly need more attention.

Bioturbation can totally destroy primary structures where dissolved oxygen is readily available in the water column and sedimentation is not too rapid. Hence the joint study of primary structures, trace fossils, color and organic content is always necessary for the complete interpretation of the depositional environment of muddy sediments everywhere.

For most fine-grained transport and deposition on the continents, episodic events such as floods, storms along a coast or over a shelf, and turbidity currents produced by slope failures are the dominant processes. In contrast, contour currents have a greater role on the continental rise in the deep ocean. Distinguishing muddy, distal turbidites from contourites remains a problem in ancient fine-grained deposits.

3.9. Challenges

In estuaries, shallow seas, and perhaps in some lakes, much clay deposition occurs via fecal pellets.

In ancient carbonates, such fecal pellets are easy to recognize, but they are rarely distinguished in ancient mudstones. If mudstone pellets could be readily recognized, would it not radically change the interpretation of paleo-environments? Is there a technological advance that would make this feasible, or are concretions the only way to find these structures?

Widespread muddy shorelines are a distinctive feature of the northeast coast of South America today, yet such shorelines are rarely described in the ancient.

Is it not likely that there are important examples in the ancient that have been overlooked? Perhaps the reason they are hard to identify is that key sedimentary structures have not been identified or that their seismic image has not been defined.

It is simplest to assume that the sedimentary structures of mudstones are only scaled-down versions of those of sandstones, but is this really the case?

Would a well-designed flume study of mud deposition confirm or weaken this viewpoint? In any case, there are many questions such a program might explore. For example how do the structures of muds vary with salinity? Or with changes in clay mineralogy? Or with time since the last event?

Interbeds of sandstone and siltstone in a mudstone – do they have the same or different sources as the fine beds?

It is clear that such interbeds have the same origin in the subaerial parts of a delta and the turbidites of a subsea fan, but what about shelfal mudstones? The nearshore of the modern Atlantic continental shelf in many places consists of relict sands with an overlay of new muds. How general is such a pattern?

Long ago Charles Darwin who recognized Saharan dust at sea, commented, "Finally I may remark, that the circumstance of such quantities of dust being periodically blown, year after year, over so immense an area in the Atlantic Ocean, is interesting as showing by how apparently inefficient a cause a widely extended deposit may be in process of formation ... " (Darwin 1845, p 29).

It seems to us that Darwin was saying, more than 150 years ago, that there is a much larger eolian contribution to sediments than one would think. We believe that geologists and soil scientists still under appreciate the magnitude of this contribution. Let's hope that the next 150 years sees faster progress.

References

Adams EW, Schlager W, Wattel E (1998) Submarine slopes and exponential curves. Sedimentary Geology 117:135–141

Aigner T (1980) Biofacies and stratiform of the Lower Kimmeridge Clay (U. Jurassic, Dorset, England). Neues Jahrbuch Geologie Palaontologie Verhandlung 159:324–338

Algeo TJ, Phillips, Jaminski J, Fenwick M (1994) High resolution X-radiography of laminated sediment cores. J Sediment Res A64:665–703

Allen JRL (1984) Sedimentary Structures: their Character and Physical Bases (Developments in Sedimentology). Elsevier, Amsterdam, Vol. 1/2, 593 p and 663 p

Allen JRL (1985) Principles of Physical Sedimentology. Allen and Unwin, London, 272 p

Allen PA (1997) Earth Surface Processes. Blackwell, Cambridge, 404 p

Barnhill ML, Zhou H (1996) Corebook of the Pennsylvanian Rocks in the Illinois Basin. Published by the Indiana Geological Survey for the Illinois Basin Consortium, Indiana Geological Survey (Bloomington). Illinois Basin Studies 3, 19 p

Bein A, Weiler Y (1976) The Cretaceous Tlame Yafe Formation: a contour current shaped sedimentary prism of calcareous detritus at the continental margin of the Arabian Craton. Sedimentology 23:511–532

Chang AS, Grimm KA, White L (1998) Diatomaceous sediments from the Miocene Monterey Formation, California: a lamina-scale investigation of biological, ecological and sedimentary processes. Palaios 13:439–458

Cofaígh CÓ, Dowdeswell JA (2001) Laminated sediments in glacimarine environments: diagnostic criteria for their interpretation. Quaternary Sci Rev 20:1411–1436

Colby BR (1964) Discharge of sands and near velocity relationships in sand-bed streams. US Geological Survey Professional Paper 462-A:47

Cressman ER (1985) The Prichard Formation of the lower part of the Belt Supergroup (Middle Proterozoic), near Plains, Sanders County, Montana. US Geol Survey Bull 1553:64

Crimes TP, Crossley JD (1980) Interturbidite bottom current orientation from trace fossils with an example from the Silurian Flysch of Wales. Jour Sed Petrol 50:821–830

Darwin C (1846) An account of fine dust which often falls on vessels in the Atlantic Ocean. Quarterly J Geol Soc London2:26–30

Einsele G, Overbeck R, Schwarz HU, Unsöld G (1974) Mass physical properties, sliding and erodibility of experimentally deposited and differently consolidated clay muds. Sedimentology 21:339–372

Graf WH (1971) Hydraulics of Sediment Transport. McGraw-Hill, New York, 513 p

Isaksen GH, Bohacs KM (1995) Geological controls of source rock chemistry through relative sea level; Triassic Barents Sea. In: Katz BJ (ed) Petroleum Source Rocks. Springer, Berlin Heidelberg New York, pp 25–50

Garcia M, Parker G (1989) Experiments on hydraulic jumps in turbidity currents near a canyon. Science 245:393–396

Jones KPN, McCave IN, Weaver PPE (1992) Textural and dispersal patterns of thick mud turbidites from the Madeira Abyssal Plain. Marine Geol 107:149–173

Kenter JAM, Schlager W (1989) A comparison of shear strength in calcareous and siliciclastic marine sediments. Marine Geol 88:145–152

Kvale EP (2003) Tides and tidal rhythmites. In: Middleton GV (ed) Encyclopedia of Sediments and Sedimentary Rocks. Kluwer Academic, Dordrecht, pp 741–743

Lesueur P, Tastet JP (1994) Facies, internal structures, and sequences of Modern Gironde-derived muds on the Aquitaine inner shelf, France. Marine Geol 120:267–290

Mantz PA (1978) Bedforms produced by fine, cohesionless granular and flaky sediments under subcritical water flows. Sedimentology 25:83–103

Middleton GV (1976) Hydraulic interpretation of sand-size distributions. J Geol 84:405–426

Miller DJ, Eriksson KA (1997) Late Mississippian prodeltaic rhythmites in the Appalachian Basin: a hierarchical record of tidal and climatic periodicities. J Sediment Res 67:653–660

Minoura K, Osaka Y (1992) Sediments and sedimentary processes in Mutsu Bay, Japan; pelletization as the most important mode in depositing argillaceous sediments. Marine Geol 103:487–502

Noffke N, Gerdes G, Klenke T, Krumbein WE (2001) Microbially induced sedimentary structures – a new category within the classification of primary sedimentary structures. J Sediment Res 71:649–656

O'Brien NR (1989) Origin of lamination in Middle and Upper Devonian black shales, New York State. Northeastern Geol 11:159–165

O'Brien NR, Slatt RM (1990) Argillaceous Rock Atlas. Springer, Berlin Heidelberg New York, 141 p

O'Brien NR (1996) Shale Lamination and sedimentary processes. In: Kemp AES (ed) Palaeoclimatology and Palaeoceanography from Laminated Sediments. Geol Soc Special Publication 116:23–36

Owen MW (1970) A detailed study of the settling velocities of estuary mud. Hydraulic Research Station, Wallingford, p 24

Pedersen GK (1985) Thin, fine-grained storm layers in a muddy shelf sequence: an example from the Lower Jurassic in the Stenlille 1 well, Denmark. J Geol Soc London 142:357–374

Pickering KT, Hiscott RN, Hein FJ (1989) Deep-Marine Environments. Unwin Hyman, London, 416 p

Potter PE, Maynard JB, Pryor WA (1982) Appalachian gas bearing Devonian shales: Statements and discussions. Oil and Gas Journal 80:290–318

Prior DB, Bornhold DB, Johns MW (1984) Depositional characteristics of a submarine debris flow. J Geol 92:707–721

Pryor WA (1975) Biogenic sedimentation and alteration of argillaceous sediments in shallow marine environments. Geol Soc Am Bull 86:1244–1254

Pusch R (1962) Clay particles, their size, shape and arrangement in relation to important physical properties of clay. Statensrad for Byggnades forskning, Handlingar No. 40, 150 p

Pusch R (1970) Microstructural changes in soft quick clay at failure. Canad Geotech J 7:1–7

Reid I, Frostrick LE (1994) Sediment transport and deposition. In: Rye K (ed) Sediment Transport and Depositional Processes. Blackwell Scientific, Oxford, pp 89–155

Ruedemann R (1897) Evidence of current action in the Ordovician of New York. American Geologist 10:367–391

Rust BR, Nanson GC (1989) Bedload transport of mud as pedogenic aggregates in modern and ancient rivers. Sedimentology 36:291–306

Schieber J (1990) Significance of styles of epicontinental shale sedimentation in the Belt Basin, Mid-Proterozoic of Montana, USA. Sediment Geol 69:297–312

Schieber J, Ellwood BB (1988) The coincidence of macroscopic paleocurrent indicators and magnetic lineation in shales from the Precambrian Belt Basin. J Sediment Petrol 58:830–835

Seilacher A, Plüger F (1994) From biomats to benthic agriculture: a biohistoric revolution. In: Krumbein W, Paterson DM, Stal LJ (eds.) Biostabilization of Sediments. Bibliothec and Information Systems der Universität Oldenburg, pp 97–105

Shanmugan G (1997) The Bouma sequence and the turbidity mind set. Earth-Sci Rev 22:201-229

Shanmugan G, Spalding TD, Rofheart DH (1995) Deep-marine bottom-current reworked sand (Pliocene-and Pleistocene), Ewing Bank 826 Field, Gulf of Mexico. In: Winn RD Jr, Armentrout JM (eds) Turbidite and Associated Deep-water Facies. Society Economic Paleontologists Mineralogists, SEPM Core Workshop No. 20, Houston, TX, 5 March, 1995, pp 25–54

Stanley DJ (1983) Parallel laminated deep-sea muds and coupled gravity flow hemi-pelagic settling in the Mediterranean. Smithsonian Contributions to the Marine Sci 19:19

Stow DAV, Bowen AJ (1980) A physical model for the transport and sorting of fine-grained sediment by turbidity currents. Sedimentology 27:31–46

Viana AR, Faugères JC, Kowsman RO, Lima JAM, Caddah LFG, Rizzo JG (1998) Hydrology, morphology and sedimentology of the Campos continental margin, offshore Brazil. Sediment Geol 115:133–157

Weedon GP, Jenkyns HC, Coe AL, Hesselbo SP (1999) Astronomical calibration of the Jurassic time-scale from cyclostratigraphy in British mudrock formations. Philosophical Transact Royal Soc London Series A 357:1787–1813

Wells JT, Coleman JM (1981) Physical processes and fine-grained sediment dynamics, coast of Surinam, South America. J Sediment Petrol 51:1053–1068

Wetzel A, Uchmann A (1998) Biogenic sedimentary structures in mudstones – an overview. In: Schieber J, Zimmerle W, Sethi PS (eds) Shales and Mudstones. E Schweizerbart'sche, Stuttgart, Vol. 1, pp 351–369

Young MG (1982) The Late Proterozoic Tindir Group, east-central Alaska: Evolution of a continental margin. Geol Soc Am Bull 93:759–783

Young RA, Southard JB (1978) Erosion of fine-grained marine sediments: Sea floor and laboratory experiments. Geol Soc Am Bull 89:663–672

Digging Deeper

Burt N, Parker R, Watts J (eds) (1997) Cohesive Sediments. Wiley, New York, 458 p.

Thirty-three technical papers ranging from Settling Velocity, Consolidation, Instrumentation, Deposition and Erosion, Deformation and Waves, to Modeling with much emphasis on instrumentation throughout. Advanced reading for specialists.

Collinson JD, Thompson DB (1985) Sedimentary Structures, 2nd edn. Unwin Hyman, London, 207 p.

A good place to start or review both primary and secondary sedimentary structures.

Ferm JC, Smith GC (1980) A Guide to the Cored Rocks in the Pittsburgh Basin. Geology Department, University Kentucky, Lexington, and Geology Department, University South Carolina, Columbia, p 109.

Beautiful colored pictures of mudstone cores with hints on the use of sedimentary structures and fractures to distinguish different types. Don't log mudstone cores without this handbook!

Hasiotis ST (2002) Continental Trace Fossils; SEPM Short Course Notes no 5. Society Sedimentary Geology, Tulsa OK, 130 p.

Outstanding colored photographs combined with clear text and drawings make this guide a must when you work with continental muds and mudstones. How many of us have observed beetle traces, dinoturbation, ancient termite hotels, and nests, adhesive meniscate burrows, insect larval burrows, and rat burrows? Five stars and a big hole filled.

Johansson M, Stow DAV (1995) A classification scheme for shale clasts in deep water sandstones. In: Harley AJ, Prosser DJ (eds) Characterization of Deep Marine Clastic Systems. Geol Soc London, Special Publication 94, pp 221–241.

Unusual effort at trying to determine the value of shale clasts in deciphering the different depositional processes of deep-water turbidites. Why not expand the scope of this interesting article and include all environments?

Kemp AES (ed) (1996) Paleoclimatology and Paleoceanography from Laminated Sediments. Geological Society Special Publication 116:258.

The best book to learn about lamination contains 17 articles and an informative introduction (see O'Brien below for the best short article).

Kvale EP, Sowder KH, Hill BT (1998) Modern and ancient tides (poster and explanatory notes). Society Sedimentary Geology, Tulsa OK and Bloomington, IN.

This helpful chart illustrates six types of present-day tidal cycles plus slabbed cores of their Carboniferous equivalents. We found this presentation especially helpful for understanding the different kinds of cycles because they are presented together and include graphs and rock images.

Locat J, Mienert J (eds) (2003) Submarine Mass Movements and their Consequences. Kluwer Academic, Dordrecht, p 540.

Some 60 short articles for the well prepared. These are divided into two broad parts – processes (with much on both marine geology and geotechnical engineering) and on regions.

McCaffrey W, Kneller B, Peskoll J (eds) (2001) Particulate Gravity Currents. International Association of Sedimentologists Special Publication 31:302.

Twenty articles arranged in four sections with most about theory and experiment. Advanced reading.

O'Brien Neal R (1996) Shale lamination and sedimentary processes. In: Kemp AES (ed) Paleoclimatology and Paleoceanography from Laminated Sediments. Geological Society Special Publication 16:23–36.

One of the best compact overviews of the sedimentary structures of mudstones available. Has excellent illustrations based on polished sections, thin sections; SEM study. Read it and learn much.

Simpson JE (1997) Gravity Currents, 2nd edn. Cambridge University Press, Cambridge, 244 p.

A well, clearly written, and illustrated text, emphasizing air, but nevertheless a good place to start to begin your study of gravity currents.

Traverse A (ed) (1999) Sedimentation of Organic Particles. Cambridge University Press, Cambridge, 544 p.

Four parts plus a glossary containing 19 papers about modern and Cenozoic examples of transport of organic particles in an unusually wide range of environments – soils, rivers, lakes, bays, carbonate and terrigenous environments. Some applications to pre-Cenozoic deposits.

Winterwerp JC, Kranenburg C (eds) (2002) Fine Sediment Dynamics in the Marine Environment. Elsevier, Amsterdam, 713 p.

Forty-three papers of advanced reading divided into five sections: (1) highly concentrated mud suspensions; (2) flocculation and settling velocity; (3) bed processes of consolidation and erosion; (4) field observation; and (5) modeling. Rich source of information on the physical properties of modern shallow-marine muds.

CHAPTER 4

Role of Oxygen

Oxygen is to varieties of mudstones as it is to varieties of life

4.1. Introduction

The next step in our study is to recognize and understand the key role of oxygen in the depositional environment on the properties of muds and soils. Color is the most evident of these (Fig. 4.1), and organic carbon content is perhaps the most significant, but there are many more. In this chapter we focus on how oxygen controls the chemical, physical, and biological character of the depositional environment. There is an equally important role in early diagenesis, which includes processes within a few meters of the sediment-water interface, which we defer to Burial (Chap. 6). However it is important to recognize that this interface is the locus of significant exchange between pore water and bottom water, so some concepts are discussed in both chapters.

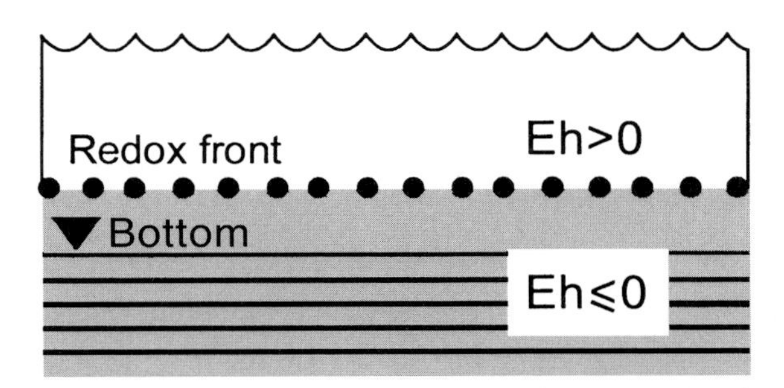

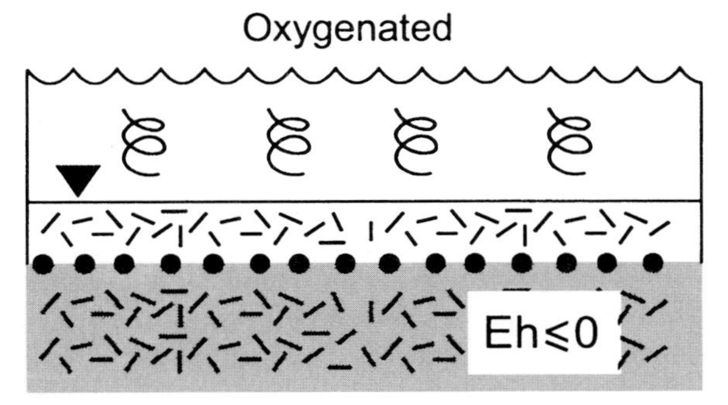

Fig. 4.1. The position of the redox front, the contact between oxygenated and oxygen-deficient waters, is all important for understanding the faunal content, diagenetic mineralogy, sedimentary structures and organic content of mudstones. The same is true of weathering profiles – is the soil oxidized and reddish brown, greenish gray, or dark gray or almost black and reduced?

Before we proceed, we should also be aware of the great significance of dissolved oxygen in bottom and pore waters for the many practical uses of mudstones in the modern world, mostly through the strong effect of oxygen on biological processes in muds. For example, where muds are organic rich, they can become source rocks for petroleum (when sufficiently buried and heated). Organic-rich mudstones are also hosts for major lead, zinc and copper deposits, and contain high concentrations of heavy trace metals such as uranium. On the other hand, for most industrial uses – and there are hundreds, if not several thousand – the less organic matter the better. So either way, much or none, we need to understand how dissolved oxygen in bottom and pore waters affects muds and mudstones.

The balance between organic production in the water column and destruction by both bacteria and oxidation at the sediment-water interface is the key variable determining the oxygen content of bottom water, which controls many key features of mudstones (Fig. 4.2). Oxidation in turn is largely controlled by mixing. Is the water column well mixed, fully turbulent, and liberally exchanging oxygen with the atmosphere or is it stratified by temperature, salinity or geometry and thus isolated from the atmosphere? The other important factor is organic productivity (mostly from pelagic organisms, but also transported woody materials), which supplies food stocks to ever-present oxygen-consuming bacteria.

Organic productivity in lakes and seas is greatest where water is well oxygenated, warm, rich in nutrients and non-turbid (permits sunlight to extend

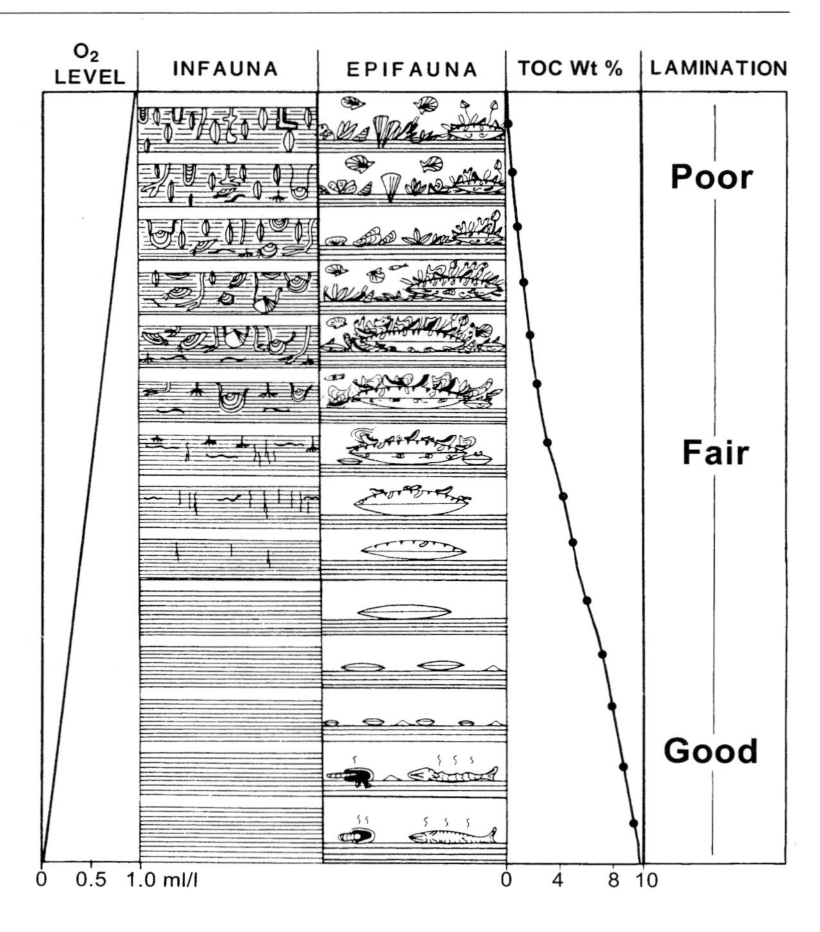

Fig. 4.2. The reciprocal nature of bioturbation and lamination and their dependence on bottom oxygen as seen by Kauffman (1986, Fig. 8). Note also how organic carbon increases as bottom oxygen decreases

deeper into the water column and thus enhance the food chain). The five Great Lakes of the United States and Canada show this well – the most productive for fishing is Lake Erie, the most southern, because it has the shallowest and warmest waters and thus favorable for a rich food chain.

The final concentration of organic matter in the sediment below this water column depends on the relative importance of these two factors plus the amount of dilution by accompanying terrigenous and carbonate debris (Bohacs et al. 2000, p 17):

$$\text{Preserved Organic Carbon} = (\text{Production} - \text{Destruction})/\text{Dilution}\,.$$

There is general agreement that these are the important variables, but much controversy remains over which one is the most important. Demaison and Moore (1980) believed basin anoxia, limited destruction, to be the most critical; Pederson and Calvert (1990) contended that production was the dominant variable; and Sageman et al. (2003) argued for a dominant role of sedimentation rate – dilution. It is likely that each basin is different and that within a basin, the balance of these parameters changes with time. The challenge is quantifying these effects. Another factor in this continuing debate is that some authors, particularly those from the petroleum industry, stress *quality* as much as *quantity* of the preserved organic matter. Anoxic basins tend to produce organic matter with higher hydrogen content, which is more favorable for oil generation and therefore more valuable (Peters and Moldowan 1993, p 8).

Dissolved oxygen in the water column is readily measured in modern streams and seas (Box 4.1) and in this chapter we show how simple field observations and a few measurements help us semi-quantitatively estimate paleo-oxygen levels in ancient mudstones. Finally we will consider some more sophisticated laboratory measurements that can provide more quantitative information.

4.2. Redox Fronts, Organisms, and Mudstone Properties

The amount of dissolved oxygen in the water column and in the pore waters of the mud just below the sediment-water interface is the fundamental con-

Box 4.1. Dissolved Oxygen

In describing mudstones, we often talk about estimating past O_2 levels, whereas in modern sediments we can measure O_2 directly. Because quantitative subdivisions of oxygenation are so widely used, we need to know more about how it is actually measured. The solubility of oxygen in water is dependent upon the partial pressure of oxygen in the air (P_{O_2}), the temperature of the water, and its mineral content. The atmosphere today has a P_{O_2} of 0.2 atmospheres, which was reached sometime during the Paleozoic, and which has stayed at around that level since. Henry's law allows the calculation of the oxygen content of water in contact with the atmosphere: 8.26 mg/l (0.26 mmol/l) at 25 °C. Because oxygen solubility increases with decreasing temperature, at 5 °C its concentration in water increases to 12.0 mg/l. This is important to remember when we consider that glacial ice induces the downwelling of cold, oxygen-rich water to deeper levels in a basin or that, in lakes, winter cooling can produce an overturn bringing oxygen-rich waters to the bottom.

Redox conditions in natural waters are controlled by the processes of photosynthesis and by the bacterial reworking of organic matter. Photosynthetic reactions are a source of free oxygen. As long as O_2 remains available, respiration and organic matter decay are the reverse of photosynthesis and provide a source of CO_2:

$$CO_2 + H_2O \underset{\leftarrow respiration}{\overset{photosynthesis \rightarrow}{\Leftrightarrow}} \text{organic matter} + O_2 \,.$$

Organic matter decay continues by a series of alternative reactions, even when molecular oxygen has been consumed. Denitrification, manganese and iron reduction, sulfate reduction, and methanogenesis all generate electrons, thus lowering the oxidation state of the system, and are the dominant processes in shallow burial in fresh water and marine sediments and also in many groundwaters.

Dissolved oxygen is now directly and accurately measured in the field down to a concentration of about 0.01 mg/l. Below this level, DO can be measured indirectly by measuring Eh with a platinum electrode. At 25 °C and at equilibrium between O_2 and liquid H_2O,

$$Eh = 1.23 + 0.015 \log P_{O_2} - 0.059 \text{ pH} \,.$$

trol on benthic fossil content, degree and kinds of bioturbation, color and pyrite content (Box 4.2). Dissolved oxygen in turn is linked to bottom turbulence and the amount of organic debris in the mud. Key here is the concept of a *redox front* – the contact between oxygen-rich and oxygen poor water, which may be either above the sediment-water interface, below it or at it (Fig. 4.1). When the redox front is in the water column above the sediment-water interface, macro-invertebrates are absent, lamination is well developed, muds are dark colored and organic rich and, in the marine realm, pyrite is common. In contrast, when the redox front is below the sediment-water interface, benthic organisms are present, bioturbation is well developed and muds are lighter colored with only minor organics. Pyrite can be present, even abundant, but forms deeper in the sediment. An intermediate condition prevails when the redox front is at the sediment-water interface or oscillates closely around it, but remember, where there is a local concentration of organic matter in an oxidized mud, there will be a local and isolated redox front around the organic matter.

There are many ways oxygen deficiency may occur (Table 4.1). We can divide these causes of anoxia into three broad categories: restricted water circulation, high organic productivity in the water, and low-oxygen atmospheres/oceans during earlier phases of Earth history. A stagnant or stratified water column, one too deep to be stirred by wind waves, free of strong bottom currents or shielded from surface currents by thermal or salinity-generated density contrasts quickly becomes deficient in oxygen. Such density contrasts normally have sharp, planar contacts referred to as *pycnoclines*.

Or oxygen deficiency may result from overloading of the bottom by organic debris – by excess organic productivity in the photic zone so that dead organisms accumulate on the bottom in such abundance that oxygen *demand* simply overwhelms oxygen *supply;* in short, more producers of organics than consumers. A good analogy is what happens when

Box 4.2. Controls on Pyrite Formation

The formation of pyrite in muds requires three ingredients: organic carbon, iron, and sulfate. That is:

$$\%\ \text{pyrite S} = f(C_{org}, SO_4^{2-}, Fe_{detrital})\ .$$

Any one of these three can be in short supply and therefore control the amount of pyrite formed. In marine sediments, sulfate is always abundant, so the possibilities are that either carbon or iron is limiting. Carbon limitation is the norm, and marine sediments usually show a good correlation of C to S passing through the origin (Berner 1985). In this case, our equation is:

$$\%\ \text{pyrite S} = 0.35 \times (\%\ C_{org})\ .$$

Euxinic sediments (those with toxic, H_2S-laden bottom waters), however, have a shortage of iron relative to C and S (Arthur and Sageman 1994), so pyrite formation is iron limited and C-S plots show S essentially independent of C with a roughly zero slope at about 2% total S. Lime-rich muds can also be Fe-limited and show similar behavior. The controlling equation in this case would become

$$\%\ \text{pyrite S} = a \times \%\ Fe_{detrital}$$

where the coefficient a varies with the mineralogy of the sediment but is in the range 0.3 to 0.5 (see plots in Raiswell and Canfield 1998)

Non-marine sediments, which would include lakes and swamps, are usually low in S and plot along the C axis on C-S plots. They are S limited, but because sulfate concentrations are so low, most reduced S is in the form of organic-bound S (Gerritse 1999) rather than pyrite and there is therefore a weak correlation to C content with a very low slope.

Ideal curves for these relationships are shown beside. Similar plots have been widely used in both modern and ancient sediments as a way of discriminating among the three controlling parameters and thus identifying the depositional environment (e. g. Leventhal 1998).

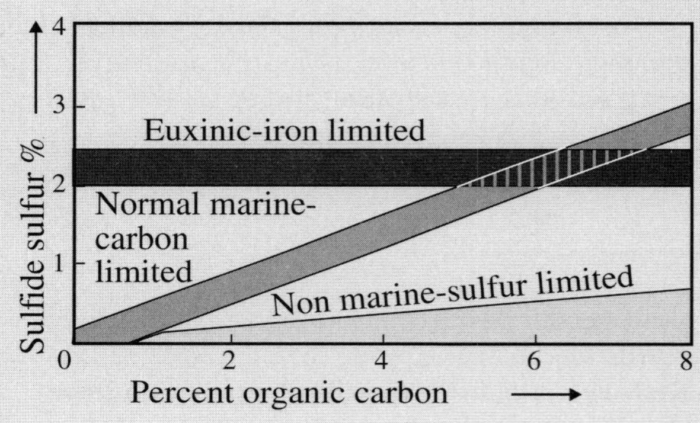

Table 4.1. Conditions favouring anoxia in basins and organic richness in their sediments (after Green, 1985)

Restricted circulation

- Stagnant bottom water in a silled marine basin or abandoned oxbow in non-marine basin
- Stagnant cold bottom water beneath warmer surface water in lakes and inland seas
- Stagnant higher salinity bottom water beneath fresher surface water in lakes and narrow gulfs

Overloading

- Oxygen minimum zone developed between shallow and deep marine water in response to upwelling of nutrients and enhanced surface productivity
- Enhanced surface productivity produced by runoff of nutrient rich waters from rivers to lakes or the ocean (anthropogenic only?)

Changed ocean-atmosphere composition in the geologic past

- In the Archean, a low-oxygen atmosphere produced general anoxia in the oceans below a shallow surface layer (e.g. Holland 2002)
- In the Neoproterozoic, brief episodes of extensive ice cover of the oceans, perhaps even world-wide, may have separated the oceans from oxygen in the atmosphere and led to widespread anoxia (e. g. Kirschvink 1992)

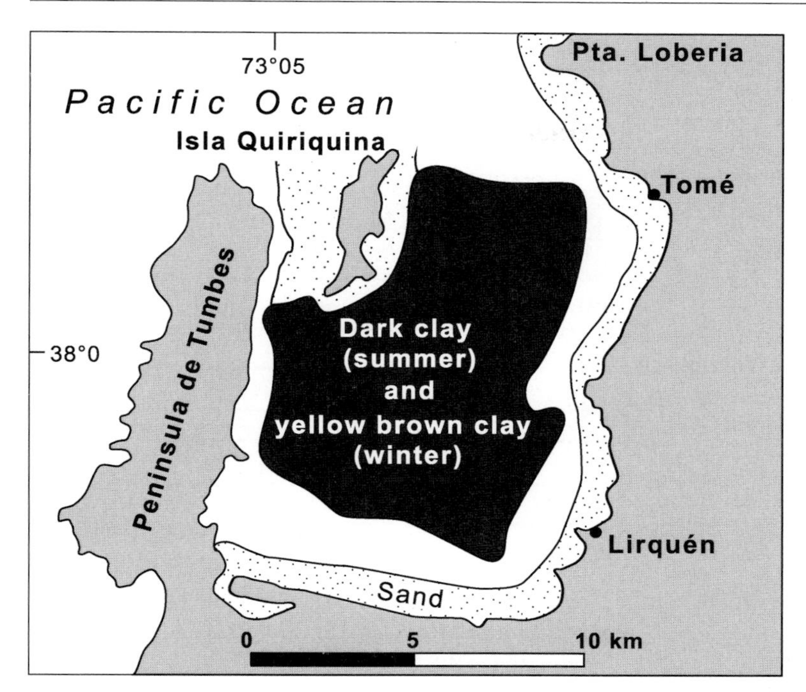

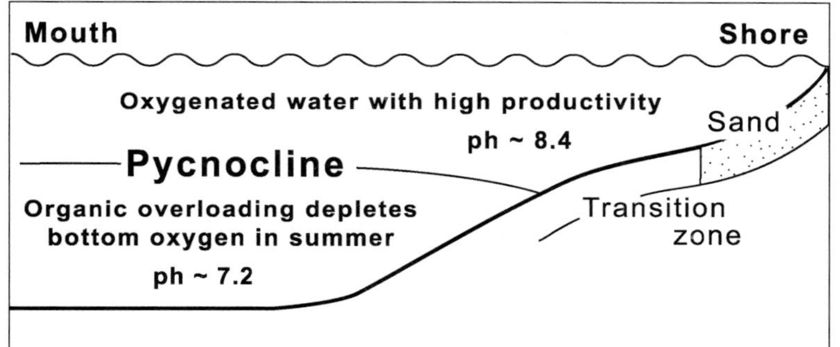

Fig. 4.3. Contrasting summer (*dark gray*) and winter (*yellow brown*) muds in Concepción Bay, Chile, reflect increased productivity in the water column in summer and less in winter (Gallardo et al. 1972, Fig. 4 and Ahumada et al. 1983, Fig. 5). Published by permission of the authors and Academic Press

excess raw sewage is dumped into a small stream at low flow – oxygen demand far exceeds supply and a massive fish kill results. Or consider how a man-made increase in nitrogen content (more fertilizer, sewage, etc) of the waters of a lake increases its productivity (more algae, etc) and thus causes organic overloading and oxygen depletion of its waters. This also appears to be true down current from the Mississippi delta on the shelf of the western Gulf of Mexico, where a summer dead zone is growing areally and also lengthening in duration and believed to be the result of increasing use of agricultural fertilizers (Wiseman et al. 1997). But summer temperatures alone can trigger increased productivity as shown by the muds of Concepción Bay in Chile (Fig. 4.3); here muds are dark colored in summer and yellowish brown in winter (organic overloading depletes oxygen even in these shallow waters).

Finally, if viewed over the long term, the Earth has had past episodes of worldwide anoxia. The Archean generally seems to have been a time when the world ocean was anoxic below the top hundred meters or so (Holland 2002), while the Neoproterozoic is believed to have experienced two relatively brief episodes in which the whole ocean froze over and thus became anoxic (Kirschvink 1992). Both of these times would have seen anoxia extend into environments that would normally be well oxygenated today. This style of anoxia is driven by the predominance of volcanic outgassing of CO_2 over photosynthetic production of oxygen. Local examples still occur in the modern, as in the Lake Nyos disaster, where overturn of CO_2-laden bottom waters displaced O_2 in the local environment and killed more than 1,800 people. The Paleozoic seems to have been a time of intermediate oxygenation of the deep ocean, resulting in an

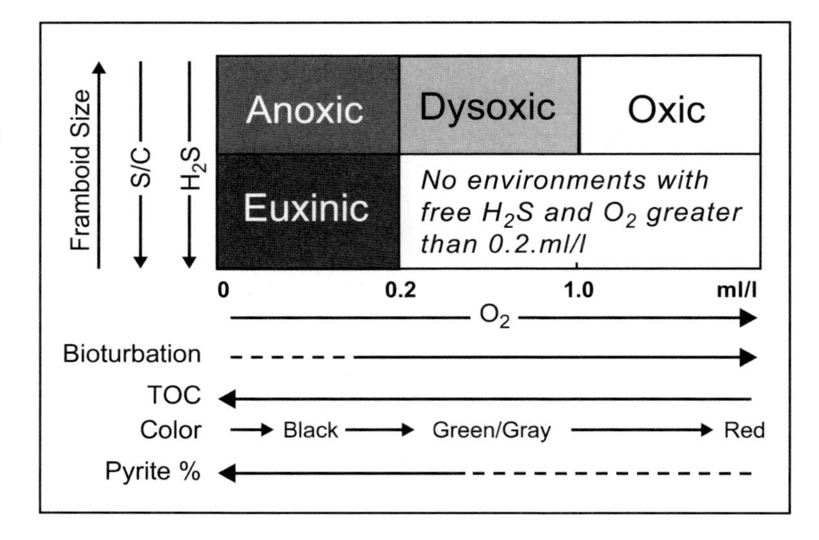

Fig. 4.4. Mudstone environments defined by oxygen and sulfur levels in bottom and pore waters of sedimentary basins. *Arrows* show how rock properties can be used to approximate original oxygen and sulfur levels

enhanced tendency for global anoxic events (Kump 2003).

Although the availability of oxygen is overriding, other factors such as the amount of fines in suspension (turbidity), rate of sedimentation and temperature and light all affect organism abundance and therefore the preservation of organic matter. Of these, sedimentation rate appears to be particularly important for controlling the organic content of mudstones and exerts its influence in two contrasting ways. Under normal oxidizing conditions, an increase in sedimentation rate enhances the preservation of organic matter by burying it more quickly and taking it beyond the zone of bioturbation. By contrast, in restricted, low-oxygen environments, where decay is entirely by anaerobic bacteria, terrigenous sediment serves to dilute organic content.

The interrelationships among dissolved oxygen, organisms, sedimentary structures, color, and organic and pyrite content are fundamental to the interpretation of mudstones. Linking these properties to paleo-oxygen levels was one of the great advances of the last 30 years. Pioneers were Rhodes and Morse (1971) and Byers (1977), who summarized studies in modern environments of dissolved oxygen, benthos, bioturbation and organic content, and most importantly, related them to ancient mudstones. They established terms for three levels of oxygenation in the water column – *oxic*, *dysoxic* and *anoxic* (Fig. 4.4). There is a parallel but distinct set of terms used to refer to biological processes under these different levels – *aerobic*, *dysaerobic*, and *anaerobic*. Thus it is appropriate to speak of a dysaerobic community that lived in a dysoxic environment. Subsequently, these limits have been slightly modified and added to (Fig. 4.5), including the addition of two additional useful terms for biofacies – *poikiloaerobic*, referring to fluctuating conditions, and *exaerobic*, for communities that live at the extreme limit of oxygen at the boundary between dysoxic and anoxic waters. This is a fertile realm for opportunistic species that can tolerate low oxygen and who proliferate wildly on single bedding planes when conditions become ripe for them (see for instance Gaines and Droser 2003). Microfossils also contribute valuable information on oxygen levels. Filter-feeding ostracods are particularly useful, because they proliferate under dysoxic conditions. A high ratio of filter feeders to deposit feeders is commonly used as an indicator of low-oxygen conditions in Mesozoic and Cenozoic rocks, and has been applied to the Paleozoic by Lethiers and Whatley (1994, see especially their Fig. 1).

The numerical values of the oxygen concentration to be used for the boundaries between these categories have been debated (see for example the two sets of values shown in Figs. 4.4 and 4.5). However, the important principle remains – it is possible to examine an ancient mudstone in the field and semi-quantitatively estimate the paleo-oxygen content of its bottom water. Sequence stratigraphy (Chap. 8), because of its emphasis on changes in water depth, goes far to explain variations of paleo-oxygen levels in ancient mudstones.

As dissolved oxygen decreases in bottom waters, fossil diversity decreases, burrowing and bioturbation are less, burrow diameters become smaller and shallower, and benthic organisms become smaller and may be pyritized – the so called dysaerobic fauna. Burrows become both shallower and smaller in diameter because host muds are richer in H_2S and thus

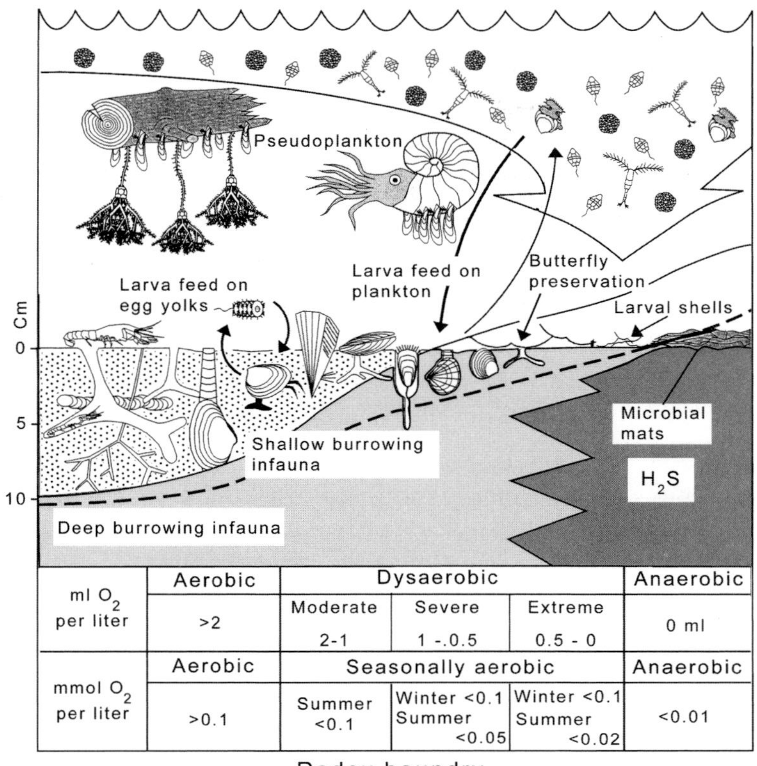

Fig. 4.5. Seasonal variations of oxygen-dependent biofacies in the shallow nearshore (after Oschmann 2001, Fig. 4.3.2.1). Note three features: (1) the complex web of life that lives near the coast, including many diverse burrowers; (2) that anoxic muds can occur at shallow depths; and (3) that redox fronts may oscillate from summer to winter as productivity and temperature change. Published by permission of the author and Blackwell Scientific

more toxic. Consequently, lamination replaces bioturbation (Table 4.2). Similarly, where there is anoxia, multi-segmented fossils such as arthropods or vertebrates like fishes and small amphibians are much better preserved after death, because there are few bottom scavengers – even though a delicious feast is to be had (Fig. 4.5). Combining these criteria, it is possible to map systematically paleo-oxygen gradients either vertically (Fig. 4.6) or laterally in a shaly section. The resulting variations help us better recognize minor cycles (and thus better correlate between wells and outcrops) and better interpret the fauna and depositional environment of a mudstone. Insight to paleo-oxygen levels of a mudstone is always rewarding.

Table 4.2. Detection of declining paleo-oxygen levels in mudstones (after Wignall 1994, p 15–16)

Biological
- Diversity of bottom dwellers declines toward the anoxic/dysoxic boundary
- Shelly faunas and size of fossils decline (deposit feeders replace filter feeders) as burrow depths and diameters decrease with less oxygen

Chemical
- Total organic carbon, TOC, increases as oxygen decreases
- Colors become darker and finally black
- Pyrite increases and phosphates appear

Physical
- Lamination replaces burrowed fabrics as oxygen levels decline
- Stratification becomes thinner and more continuous–cross lamination and scour are less abundant and hummocky bedding is rare or absent

4.3. Water Column Oxygenation and Basin Geometry

Sills and topographic lows of various types favor oxygen restriction, although lower oxygen also occurs on open shelves where upwelling develops (Fig. 4.7). Thus basin geometry, the dynamics of the water column, and the productivity of the surface waters must all be considered. Oxygen restriction may either be local or widespread on shelves, and can be seasonal

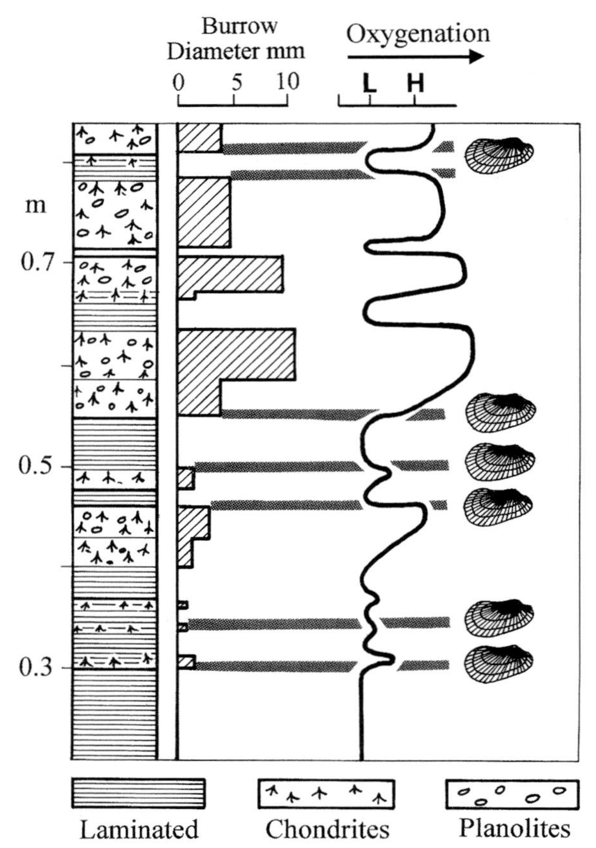

Fig. 4.6. Sequential variations of paleo-oxygen levels in the Miocene Monterey Shale of California inferred from trace fossils, burrow diameters and dense accumulations of the bivalve *Andara montereyana*, which occurs at transitions between anoxic and dysoxic mudstones (Bottjer and Savrda 1993, Fig. 6.1). Sequential variations of paleo-oxygenation in mudstones provide good stratigraphic markers, because they tend to be widespread. Published by permission of the authors and Blackwell Scientific

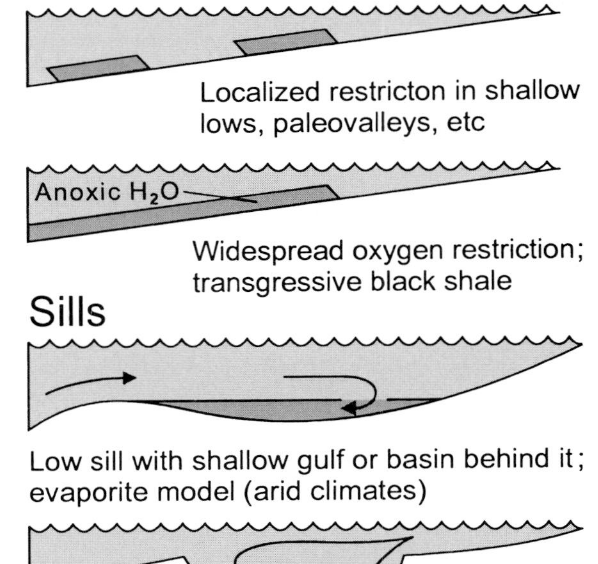

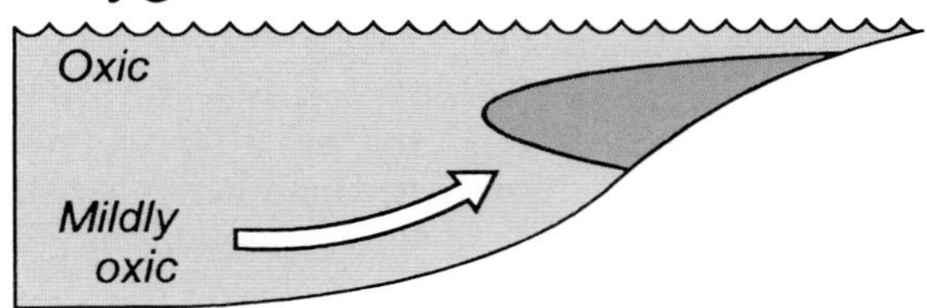

Fig. 4.7. The geometry of a basin, be it large or small, has a major role on bottom anoxia, because it either restricts or facilitates both currents and oxygen exchange with the atmosphere

or permanent. Some modern shelves experience seasonal restriction of circulation related to fresh water inflows, and seasonal stagnation is common in the summer in many lakes. A narrow deep fjord or lake, such as Lake Tanganyika is likely to be permanently anoxic at depth and accumulating thick sections of black mud (Fig. 4.8).

A special case of widespread oxygen restriction is associated with transgressions. During the early phase of a transgression across a wide gently dipping shelf, water depths are shallow and thus wave action is decreased by enhanced bottom friction so that bottom oxygen is reduced. And because water depths on shelves are generally much less than 200 m, a large proportion of the water column is within the photic zone, where organic productivity is maximal; consequently, there is more demand for dissolved oxygen from both living and dead organisms. Collectively, these factors explain why so many ancient organic-rich shales occur on open shelves as well as in deeper basins. Transgression also traps most coarse sediment close to shore, enhancing mud deposition on the shelf. Similarly, shallow topographic lows and canyons provide protection from wave action and thus trap fines along with the low-density organic matter that settles with them. Consequently, transgressive muds and those deposited in protected areas tend to be darker, finer, and have both more clay and

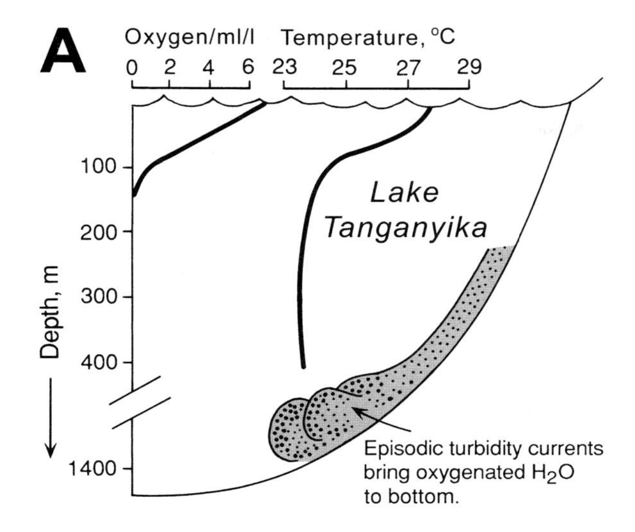

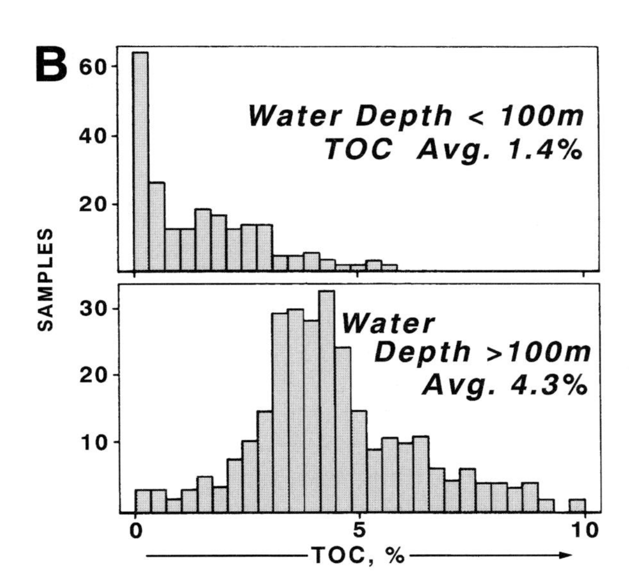

Fig. 4.8. (**A**) Oxygen profile of long, narrow, and deep Lake Tanganyika of the African rift system (Degens et al. 1971, Fig. 2) and (**B**) greater organic preservation in bottom muds below 100 m than those above 100 m (Huc et al. 1990, Figs. 3 and 4). Published by permission of the authors and the American Association of Petroleum Geologists

organics than surrounding muds (remember the inverse relationship of organic matter and grain size discussed in Chap. 3).

Local topographic highs at the entrance to a basin, called *sills*, also restrict water circulation and these may rise only a little above the bottom or far above it. Shallow sills occur in lagoons, in many of the salt basins of ancient epeiric seas (ancient gulfs or embayments) or may result from sand transport parallel to the coast, reef growth or gentle tectonic uplift. Most sills that rise far above the bottom are fault related with the exception of glacially scoured, now flooded paleo-valleys (fjords). The best-known modern example of a silled basin is the Black Sea (with depths of more than 2,200 m) separated from the Mediterranean by the shallow waters of the Dardanelles. Large rift-formed lakes such as Lake Baikal in Siberia or Lake Tanganyika of Africa are likely to have sub-basins separated by sills also.

Silled marine basins are prone to developing very toxic bottom waters rich in dissolved H_2S, and many authors assign to this type of basin a separate status as a "euxinic" environment, a special case of the general "anoxic" environment. The style of pyrite occurrence may help to distinguish this setting. The amount of iron in a mudstone that is tied up in pyrite can be measured by the *degree of pyritization*, DOP, proposed by Raiswell and Berner (1985). This divides pyrite iron by the sum of pyrite iron plus iron that would be available to make pyrite, such as iron oxides. Euxinic basins, because of their high dissolved sulfur, will tend to produce shales with higher DOP values than will other anoxic basins. They suggested, based on modern examples, that a DOP value of about 0.75 separates euxinic from non-euxinic environments, but there is considerable overlap in values. Pyrite size may also be important. Wignall and Newton (1998) report that euxinic environments produce only tiny pyrite framboids (small raspberry-shaped clusters of crystals) with a narrow size range, whereas other environments have, in addition, some larger framboids.

4.4. The Value of Biomarkers

Biomarkers are essentially molecular fossils. They are biochemicals produced by living creatures that, unlike most organics, remain stable through diagenesis and oil migration. Accordingly they have proved highly useful for correlating oils and matching oils to source rocks, in reconstructing the salinity of the depositional environment of the source rock, and in reconstructing the oxygen content of the environment. Biomarker study is a highly specialized branch of geochemistry and some background in organic chemistry is useful to read this literature.

The most widely used biomarkers are *terpenoids*, branched-chain hydrocarbons derived from the phytol side chain of chlorophyll. Under dysoxic conditions of early diagenesis, phytol degrades to the C_{19} hydrocarbon *pristane*, while under anoxic conditions, it goes to the C_{20} hydrocarbon *phytane* (Peters

Box 4.3. Biomarkers

Most organic matter in sedimentary rocks has been transformed beyond all recognition of its original biological source. An exception is a minor but extremely important component – the biomarkers. These are molecular fossils, retaining inherited structures from their parent biomolecules, and therefore useful in much the same was as skeletal fossils. The most useful compounds are hydrocarbons.

Some examples:	Application:
Pristane/phytane	Paleoredox, although influenced by maturation
Homohopane index	Paleoredox, although influenced by maturation
Botryococcane	Unambiguous indicator of fresh water
Oleanane index	Higher plant input (Cretaceous and later)
Gammacerane index	Highly specific for hypersaline conditions
Carotenoids	Indicate lacustrine conditions
Steranes (C_{27-29})	Oil-to-source correlation, also stage of maturation

Source: Peters and Moldowan 1993, p. 147–207.

Another important class of compounds is the porphyrins, multi-ring compounds derived from the tetrapyrrole portion of the chlorophyll molecule. During early diagenesis they loose the phytol side chain, which becomes pristane or phytane, and incorporate a metal into the tetrapyrrole framework, either vanadium or nickel. The nickel-vanadium ratio is controlled by Eh and pH conditions and the availability of H_2S (Lewan and Maynard 1982 and Lewan 1984). Therefore the ratio V/(V + Ni) in extracted bitumen can be used as a paleoredox indicator. Many oils are rich in V and Ni, so this ratio is also useful for oil-to-source correlation.

and Moldowan 1993, Fig. 3.13). Therefore the pristane to phytane ratio of oils or of bitumen extracted from mudstones can be used as one measure of degree of oxidation during early diagenesis. This ratio has been widely applied because it is easy to measure from simple gas chromatography, but it suffers from susceptibility to alteration during thermal maturation of the source rock. The ratio is also a function of the source of the organic matter, whether marine or terrestrial, so rocks of similar thermal history and organic provenance have to be compared. For mudstones within the oil generating window, we suggest high ratios > 3.0 indicate dysoxic deposition, whereas low values < 0.6 indicate anoxic conditions. Values between these ranges are ambiguous and need corroborating evidence from other compounds. Two other useful paleoredox indicators are the homohopane index and the ratio of V-bearing to Ni-bearing porphyrins (Box 4.3).

4.5. Trace Element Indicators

Many metallic elements are enriched in black shales. Accordingly there have been many attempts over the years to relate metal contents of mudstones to oxygenation levels. Cobalt, molybdenum, nickel, uranium, vanadium, cobalt, and to some extent chromium show particular affinities for organic-rich shales. The strong correlation of uranium with organics has been exploited for many years in wireline logging of oil wells where an exceptionally high gamma-ray signal is used to identify black shales. In some basins, this relationship can be quantified to the point of using the gamma-ray log to estimate TOC content of the shales and mudstones (see for example Schmocker 1981).

As paleoredox indicators, trace elements have the significant limitation that most are present in appreciable quantities in all clastic material, so one is looking at a subtle enrichment rather than at a geochemical signal unique to oxygen-deficient conditions. A common strategy has been to use element ratios such as U/Th, or V/Cr in which one element is strongly influenced by oxygenation while another similar element is not. We compared a number of these ratios and concluded that degree of pyritization (DOP), U/Th, V/Cr, and Ni/Co formed a self-consistent set, with increasing values of all four indicating lower oxygen. A problem with this approach is that absolute values of the trace element ratios cannot be defined that separate levels of oxygenation because the detrital contribution for each element varies with the formation being evaluated.

One element that does not suffer from the complication of detrital supply is molybdenum. The Mo content of all sediments except black shales and manganese nodules is uniformly low (Box 4.4). This implies that most Mo comes to the ocean in solution rather than on detritus and that the two Mo-rich sediment types mentioned must acquire their Mo by chemical precipitation. Mo has a strong affinity for dissolved sulfide, so under euxinic conditions, Mo sulfide rapidly precipitates and is greatly enriched in bottom sediments (Crusius et al., 1996). Mo enrichment in black shales has been used by several authors as an indicator of sulfidic bottom water (see Murphy et al., 2000; Werne et al. 2002; Algeo and Maynard,

Box 4.4. Molybdenum isotopes

Recently, the technology to measure Mo isotopes has been developed. Mo offers particular promise for understanding conditions of deposition in sediments because it comes almost entirely from the water column. It is concentrated from basinal waters at the sediment-water interface into two contrasting mineral systems: under oxidizing conditions with manganese nodules of the deep sea floor and under reducing conditions with sulfide minerals (pyrite) in euxinic basins. Examples of Mo abundances in USGS standard mudstones and manganese nodules show these two sinks:

Standard	Source	Rock type	Mo, ppm
SCO-1	Cody Shale	Gray mudstone – marine	1.37
SGR-1	Green River Shale	Black shale – lacustrine	35.1
SDO-1	Ohio Shale	Black shale – marine	134
NOD-A-1	Atlantic Ocean	Mn nodule	448
NOD-P-1	Pacific Ocean	Mn nodule	762

In euxinic basins, the precipitation of Mo as the sulfide is rapid and complete, accounting for the high Mo values in some black shales. For the manganese nodules, the capture is slow and relatively inefficient. Only the very long residence time of the nodules on the seafloor allows high concentrations of Mo to develop.

Mo has several stable isotopes that are fractionated in the sedimentary system. Using the ratio of $^{97}Mo/^{95}Mo$, Barling et al. (2001) found that the deep-sea nodules fractionate Mo, whereas sulfidic systems do not:

Source	Sample type	$\delta^{97}Mo$
Modern examples		
Seawater	Water	1.48
Black Sea mud	Euxinic sediment	1.52
NOD-P-1	Mn nodule	−0.42
NOD-A-1	Mn nodule	−0.63
Devonian example		
SDO-1	Sulfidic black shale	1.02

The modern euxinic muds reflect the composition of contemporary seawater, whereas the nodules preferentially incorporate the lighter isotope ^{95}Mo. Arnold et al. (2004) have proposed that the value of $\delta^{97}Mo$ for seawater is controlled by the relative fluxes of Mo to its oxidizing and reducing sinks:

Mo isotopes in black shale of a particular age = Mo isotopes of seawater of that age = Mo isotopes of river water entering ocean − Flux to deep sea manganese nodules

This suggests that at times like today when the deep ocean is oxidizing, Mo in black shales will be enriched in the heavy ^{97}Mo isotope; conversely, when the deep ocean is too anoxic for manganese nodules to form, ^{97}Mo will be depleted in contemporaneous black shales.

2004 for discussions). Mo can also be enriched under oxidizing conditions by sorbing onto manganese oxides. Manganese nodules in the deep sea grow very slowly, but their long residence time on the seafloor enables them to incorporate large concentrations of trace elements such as Co, Cu, and Mo. Recent work on Mo isotopes (Box 4.4) exploits the isotopic difference between the reducing and oxidizing sinks to estimate paleoredox conditions in the world ocean. This technique is attractive because, unlike the trace element ratios, it is independent of clastic dilution.

4.6. Color and Oxygen

The first observation we usually make from afar about a mudstone is its color, which can range from black to gray to brown, maroon, purple and red and even green or white. Because many of these colors are gradational, many shades of color are possible, but it is well to remember that the most common mudstone color is gray, be it greenish, olive, yellowish, brownish, bluish, light or dark gray. We speculate that the reason for this dominance by gray colors is that dysoxia dominated sediment conditions throughout the Phanerozoic and perhaps the Proterozoic.

Black and red mudstones are the principal end members, and are followed far behind in abundance by the pure green and very rare white mudstones. As with most rock properties, color depends on both primary controls at deposition and later burial history. This is particularly true for argillaceous sediments and rocks, because they trap organic matter, which, when heated, darkens and becomes a key factor in mudstone color. Other factors are the oxidation state of the iron and type of iron mineral host. The overview by Myrow (1990) shows well the complexity of mudstone color (Table 4.3), but note the prominence given to organic matter in determining color, which is confirmed by an examination of the TOC content of mudstones of different colors (Table 4.4).

In the modern ocean, oxygen shows considerable variation with depth (Fig. 4.9). Note especially the prominent oxygen minimum between about 200 to 1,200 m. Below the oxygen minimum layer, typical sea floor mud-sized debris consists of a mixture of land-derived clays and silts (hemipelagics), plus pelagic debris that is both calcareous and siliceous with greater or lesser amounts of organic matter, and usually some minor precipitated minerals. Below a depth of about 4.5 to 5.0 km, increased pressure produces dissolution of all calcium carbonate, defining the Calcite Compensation Depth

Table 4.3. Factors affecting mudstone color (adapted from Myrow, 1990, Table 1)

Principal
- Flux of dissolved oxygen in bottom and pore waters
- Flux of organic matter and whether marine or terrestrial
- Balance of these two determines color through effect on oxidation state of Fe, Mn, and C

Auxiliary
- Grain size – more intense colors in finer fractions; finer grain size reduces permeability and thus favors reduction
- Sedimentation rate – high rates permit quick burial of red and gray muds in dysoxic and anoxic environments
- Availability of SO_4 – controls oxidation of organic matter by bacteria once oxygen has been consumed
- Provenance – fines introduced into basin may be greenish (chlorite, glauconite, possibly illite, etc) or carry red or brown pigments
- Later diagenesis – may darken, discolor or even bleach early-formed colors

Table 4.4. Total organic carbon content of different mudstones (after Hunt, 1996, Table 10-1). Published by permission of the author and W.H. Freeman

Lithology and stratigraphic unit	Weight % TOC
Sandstones	0.03
Red mudstones	
• Chugwater, Colorado	0.04
• Big Snowy, Montana	0.04
Green mudstones	
• Ireton, Alberta	0.11
• Cherokee, Kansas	0.30
• Tertiary, Colombia	0.54
Gray mudstones	
• Frontier, Wyoming	1.2
• Cherokee, Kansas	1.6
• Mowry, Wyoming	3.0
Black shales	
• Woodford, Oklahoma	7.0
• Cherokee, Kansas	8.0
• Bakken, North Dakota	11.0
Calcareous mudstones	
• Alcanar, Spain	4.0
• Antrim, Michigan	6.7
• LaLuna, Venezuela	7.7
• Duvernay, Alberta	7.9
• Nordegg, Alberta	13.0
• Green River, Wyoming	18.0

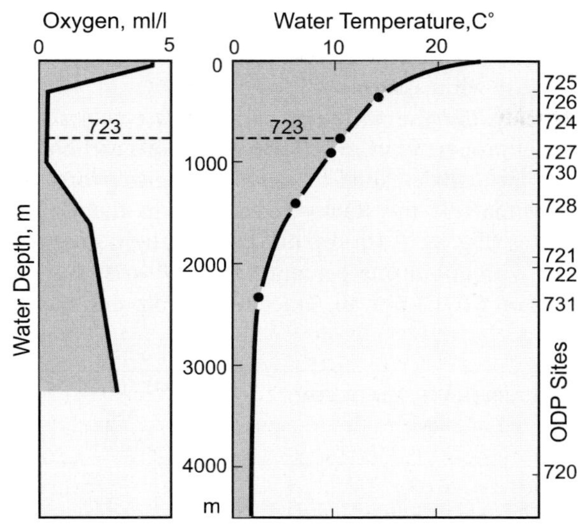

Fig. 4.9. Temperature and dissolved oxygen profiles of the northern Indian Ocean based on eleven sites from the Ocean Drilling Program (after Niitsuma et al. 1991, Fig. 2). Note the well-defined oxygen minimum between about 200 to 1,200 m (OMZ)

(CCD), and bottom sediments are either clay- or silica-rich. Because of the long transit time through the water column, almost all of the organic matter is consumed on the way down at these depths. The resulting slow sedimentation, deficiency of organic material, and mildly oxic deep oceanic waters produced by downwelling of oxygen-rich waters at glacial margins, means that much of the deep ocean floor is covered with brownish-red clay, "red clay" for short. With increasing organic matter, however, dark-colored muds will replace background red clay. In places, there is an intimate interbedding of green clay with this red clay, which implies both an early diagenetic origin of the green color and past episodes of a greater flux of organic material. When this flux is high enough, it produces very fine, black, pyritic muds (sapropels) on the sea bottom (Fig. 4.10).

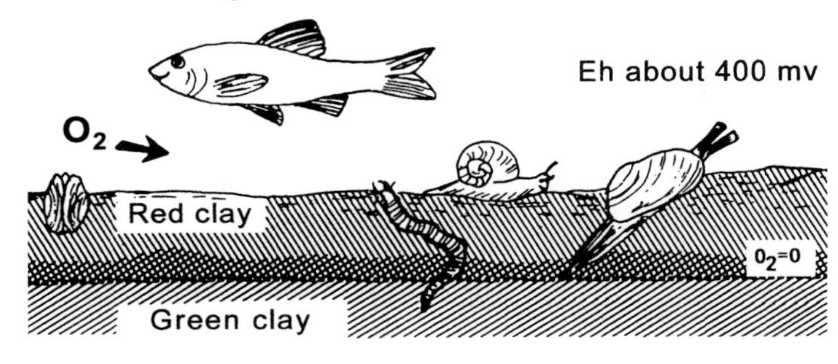

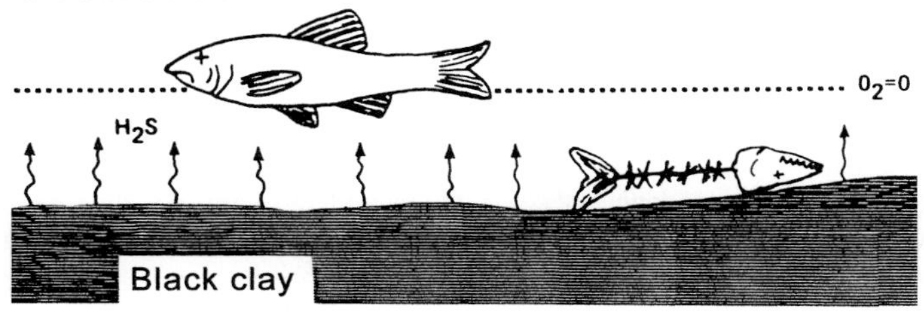

Fig. 4.10. Color variations in deep sea muds have been related to changing rates of dead pelagic debris settling to the bottom (Gardner et al. 1975, Fig. 8). With normal slow rates, muds are reddish brown (slow settling and long residence time at the bottom in a mildly oxidizing water column), but as organic sedimentation increases, first dysoxic conditions develop (green mud) and finally anoxia with black mud (overloading)

We have seen that key initial conditions affecting the color of a mudstone are the thickness of the well-mixed surface layer of the water column (where organic debris can be consumed before reaching the bottom), the amount of terrigenous dilution, and the oxygen level of the bottom. But colors change with burial too as kerogen darkens to a final "burnt black" and as more and more Fe^{III} converts to Fe^{II}. In addition, "color accidents" can happen in burial such as hot oxidizing brines, rising hydrothermal solutions from a nearby intrusion or simple baking by a large nearby sill all of which tend to produce maroon or grayish white colors.

Consider the variation of mudstone colors and other properties in an actual example of a carbonate-to-black shale, shelf-to-basin sequence from the Devonian of the Rocky Mountains in the United States (Fig. 4.11). Up-dip on the shelf, light gray colors with about one percent or less of total organic carbon (TOC) prevail. Traced down dip and basin-

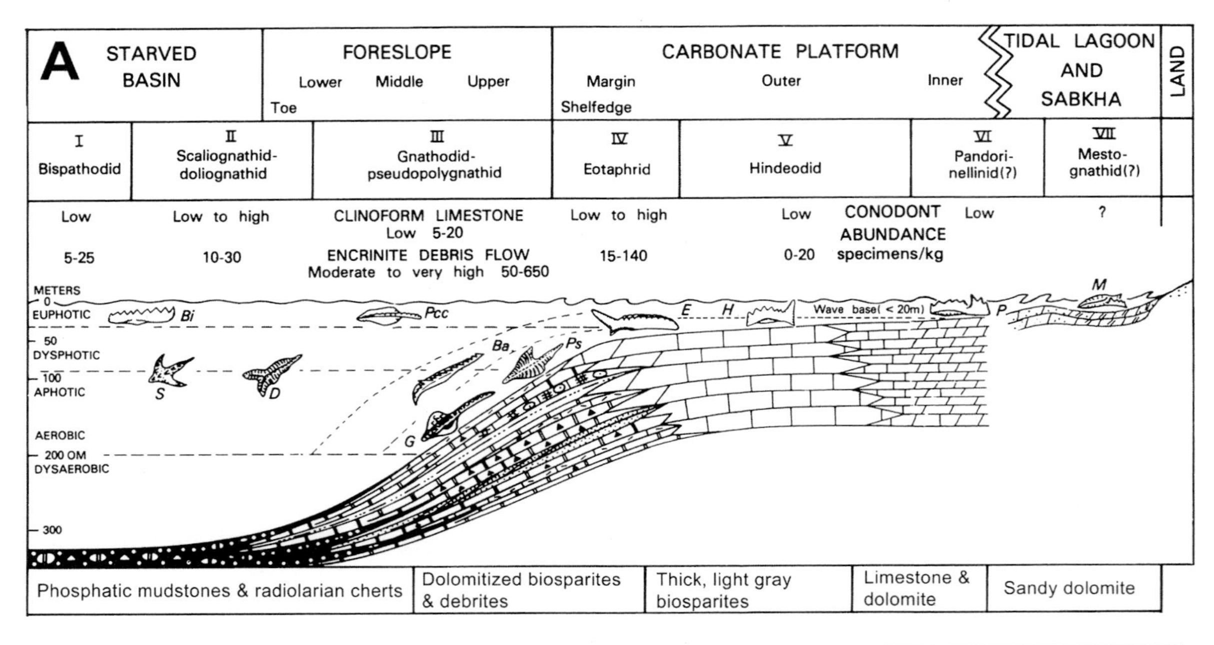

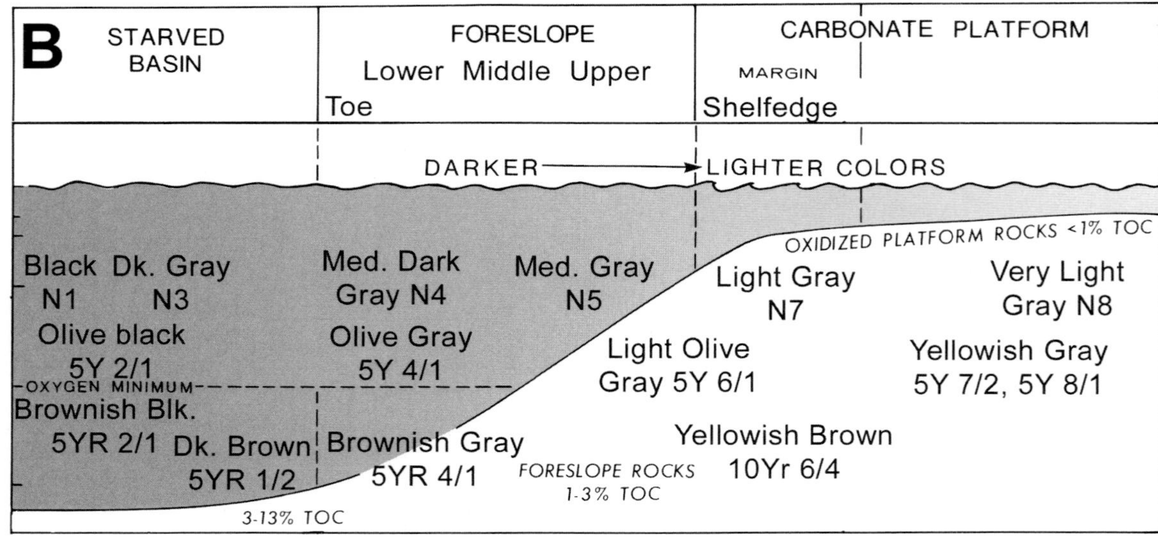

Fig. 4.11. Color changes from a Mississippian carbonate platform to a starved basin in the Devonian Woodman Formation of Utah and nearby states in western United States (Sandberg and Gutschick, 1984, Figs. 14 and 15). In this unusually well documented example, the strong interdependence of color, fauna and lithology clearly stands forth. Published by permission of the Rocky Mountain Association of Geologists

Fig. 4.12. Interlamination produced by weak distal turbidity currents (greenish gray mudstone) that introduced oxic or dysaerobic water into an anoxic basin bottom (*arrows*). Devonian Canadaway Formation (Upper Devonian along Lake Erie near North East, Erie Co., Pennsylvania)

ward, colors become progressively darker and finally brownish black as TOC increases to 15 percent. Field descriptions indicate companion changes in lamination from discontinuous and burrowed on the shelf to thin lamination in the basin. Shelly bottom fauna predominate high on the slope followed by thin, soft-bodied, burrowing fauna down-dip as oxygen levels became lower. Lack of oxygen down-dip also favored pyrite and phosphates in the black shales deposited in the deepest water.

In this and other cases, the shelf edge is the key boundary that separates well-mixed oxic bottom waters from those with less oxygen in the basin down-dip. Waves and coastal currents supply abundant oxygen to shallow shelfal waters, but only some episodic storm events generating weak, distal turbidity currents introduce oxygenated waters at depth. The result is a fine scale interlamination of greenish gray mudstone (deposited by oxygenated turbidity currents) with black shale (hemipelagic mud deposited slowly on anoxic bottom) as seen in a section from the Devonian of northeastern Ohio (Fig. 4.12). Generalizing, the deeper the water depths in a sedimentary basin and the steeper its borders, the more likely are its waters to be stratified and the darker and more organic-rich its shales. Comparison of the colors of modern muds bordering and within canyons cut into continental shelves illustrate well this generalization: muds are darker in the canyons than nearby on the shelf (Stanley 1969, p 12), because of more restricted circulation. A warm climate also enhances surface productivity, which leads to more organic debris settling to the bottom.

At the other oxidation extreme are the red-colored mudstones, which have several origins. Most are continental and their red color is diagenetic; a much smaller proportion are clearly detrital in both continental and marine environments; and a very few are altered hydrothermally or baked. The mudstones of continental redbeds are mostly reddish brown to reddish purple and maroon and commonly are mottled with minor green and grayish white colors. There are also some minor green mudstones in most red bed sequences, most of which are interbedded with red to gray sandstone, anhydrite and gypsum, and rare, thin limestones. Organic matter is virtually lacking, nearly always less than 0.1%, as are most invertebrate fossils. Bioturbation is also rare. The red color of the mudstones is the result of a very fine pigmentation by Fe^{III} oxides, which are believed to be mostly early diagenetic in origin and formed by fluctuating water tables in hot, semiarid to arid climates; i. e., acquired by alternating wetting and drying in the vadose zone and perhaps deeper after deposition by oxidizing pore and ground water (Fig. 4.13). Thus good permeability favors color changes. Red hematitic pigmentation comes from alteration of minerals such as amphiboles, pyroxenes, and biotite as oxygenated ground water seeps through arid alluvial fan, valley fill, bolson, wadi, and sabkha deposits, all low in organic matter. Thus the first cycle debris of igneous rocks eroded and deposited under an arid hot climate is likely to become red after burial.

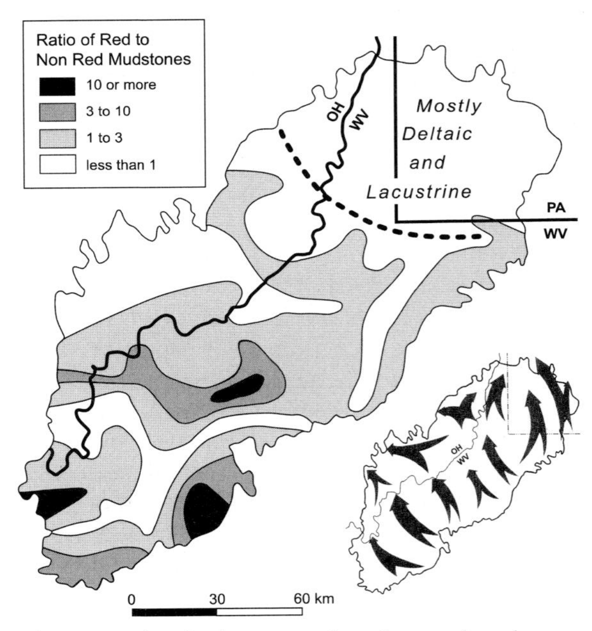

Fig. 4.13. Ratio of red to non-red mudstones, based on 129 sections, in the Permian Dunkard Series (much of it a classic red bed) in the Appalachian Basin. Note increases in fluvial facies to the southeast whereas in the northeastern part of the basin (*dotted line*) deltaic and lacustrine facies predominate and the ratio is less than one (Martin 1998, Figs. 3 and 18). More frequent fluctuations and lower water tables in the fluvial deposits versus persistently higher water tables in the waterlogged deltaic deposits probably explain most of this contrast. *Black arrows* are transport paths

Because transformations between ferric and ferrous iron occur so readily, small variations of dissolved oxygen in pore and ground water produce color changes at all scales. The decimeter to meter mottles of green and white to gray common in red beds represent reduction of Fe^{III} to Fe^{II} oxides around rootlets or other isolated patches of organic matter. Mottles and greenish to gray discolorations may also be well developed along fractures and faults in response to infiltrated reducing waters. All of these cut across stratification and thus are clearly of post-depositional origin. The above interpretation was developed mostly for the siltstones and sandstones of red beds, but seems also likely to apply to their interbedded mudstones.

There are also a few, thin maroon and red mudstones of both shallow and deepwater marine origin in ancient basins. Typically, the shallow, shelfal mudstones are found with interbedded carbonates or sandstones and can be traced up-dip into the thick, classical non-marine red bed sequences described above, or they come from deeply weathered, long-exposed, low-lying arches or old massifs. Marine red mudstones in deep onshore basins are even less common, but have also been described. Seemingly all of these represent detritus rapidly washed into seas with oxidizing water and low organic productivity. Deep-sea *red clay*, which dominates much of the ocean floor remote from land, ranges in color from bright red to maroon. It accumulates at very slow rates at depths generally greater than 3,500 m far from a terrigenous source, has ferromanganese nodules, and consists mostly of fine eolian, volcanic and some cosmic debris. Its red color results from slow sedimentation through a thick, oxidizing water column. Regardless of their origin, red muds and mudstones appear to require at least 2% total Fe and 1.5% hematite, which occur as a fine pigmentation.

In the above discussion of marine red beds, provenance emerged as a factor and provenance certainly does play a role, although its importance is far less than that of either oxygen or organic matter in the depositional environment. Provenance is also a control on the color of some reddish glacial tills (Johnson 1983). The wholesale erosion by continental ice sheets of a source area rich in either red soil or rocks produces a maroon to reddish till. Red desert loess is another example of provenance-related color: winds blowing off the Sahara carry large quantities of red dust to islands in the Atlantic (see Darwin 1846 for a vivid description). How much of the red color of soils in the Bahamas is inherited from this source and how much is generated in situ? Terra rosa soils on limestones are strikingly red because of higher-than-normal hematite resulting from strong drainage and very low organic matter. In arid and semiarid regions, this soil can be eroded into some alluvial muds that retain this red color, organic material being insufficiently abundant to reduce ferric to ferrous iron. In wet regions, reworked red regolith and terra rosa soils rapidly turn brown in alluvium because of admixed organic debris. Another example of mudstone color controlled by provenance is the incomplete weathering of a low-rank metamorphic rock that yields finely divided chlorite and biotite, imparting dark, greenish-gray colors. Or provenance controls may be coupled with oxygen level in a depositional basin as appears to be the case in varves in Sweden, where winter accumulations of mud are brown (perhaps from derivation from a pre- or periglacial regolith) whereas thicker summer varves are gray (more organic matter and more silt).

Most green colors in mudstones are actually gray-green, but there are some intensely green mudstones and some that are bluish-green. Some green marine clays are rich in ferrous iron and magnesium (berthierine and chamosite). They have modern analogs on shallow shelves bordering tropical rivers (Odin 1988, pp 415–418), and in the ancient are found most prominently in oolitic ironstones. They may also occur, however, dispersed in the enclosing mudstones imparting a greenish color. Similarly, finely divided glauconite also contributes to the greenish color of some marine mudstones. For an excellent review of the mineralogy of green clays in mudstones see Velde (2003). Carbonate-rich mudstones tend to have a more bluish-green or bluish-gray tone than do non-calcareous mudstones. Such limestones typically develop a tan weathering rind, suggesting that finely-divided pyrite is the source of the bluish color.

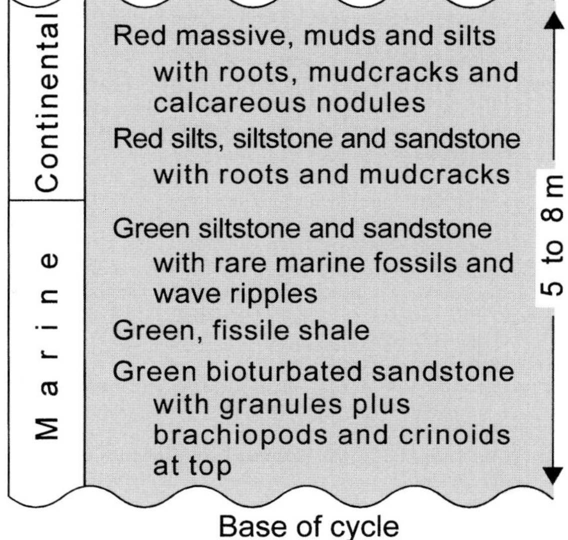

Fig. 4.14. Color as the first guide to depositional environment (after Walker 1971)

Finally, we comment on the age-old question, "How well does the color of a mudstone accurately reflect its original depositional environment?" For a given facies, color can be very decisive as shown by the marine-continental cycles of the muddy shoreline deposits of the Mississippian Catskill delta in Pennsylvania (Fig. 4.14). On the other hand, in the Upper Devonian Canadaway Formation (Fig. 4.12), a slope-deep basin deposit, contrasting mudstone colors simply represent two different processes in the same environment. Moral: *one needs to understand the depositional system fully rather than simply relying on color alone.*

4.7. Summary

The first things most geologists observe about a mudstone are its color and stratification and rightly so, because these two attributes tell us much about both the initial environment of deposition and possible economic significance of the deposit. Color and stratification are also usually the first rock properties we use for correlation.

Paleo-oxygen levels play key roles for both properties. Oxygen levels at or near the bottom control benthic macrofossil communities, bioturbation and lamination, color and pyrite content. Thus oxygen is the principal determinant of megascopic appearance of marine and lacustrine mudstones. Low levels of bottom oxygen can form in two ways: (1) bottom waters become isolated from oxygenated surface waters by differences in temperature or salinity, or (2) low levels may result from organic overloading – great productivity at the surface simply rains so much organic debris to the bottom that oxygen demand far exceeds oxygen supply even though bottom waters are far from stagnant. Under both of these conditions dark colors, lamination and a sparse bottom fauna are the rule and organic content and pyrite are abundant. Conversely, where oxygenated bottom waters prevail, benthic life abounds, bioturbation is extensive, gray colors rule and pyrite is not abundant. Thin maroon to red mudstone beds in marine and lacustrine mudstones are likely to represent detrital clays derived from more or less distant red regolith.

In the Precambrian, lower oxygen levels of the atmosphere and ocean contributed to a general expansion of low-oxygen environments.

In continental mudstones paleo-oxygen levels are also the key control on color and organic content. The maroon to purple colors of the mudstones of red bed sequences appear to be mostly early diagenetic in origin and form by alternate wetting and drying on flood plains in semi-arid to arid climates where little organic matter is buried. But should the continental sequence have had a wet climate and abundant vegetation, gray colors are likely as is true in most ancient coal measures.

Exceptionally, the type of terrigenous detritus can control primary mudstone color and later color can be modified by shallow and deep diagenesis, oxidation in outcrop, proximity to an intrusion, hydrothermal solutions, or hot rising brines from deeply buried sediments.

In sum, oxygen is to mudstone as it is to life – vital and, consequently, always deserving our full attention.

4.8. Challenges

How can the various causes of anoxia listed in Table 4.1 be distinguished in ancient rocks?

There appear to be few practical guides to distinguishing causes of anoxia, in particular the relative effects of overloading and stagnation. The chemical effect at the site of deposition is the same, whatever the cause, so the answer to this question probably involves study of the whole basin and the development of new discriminatory methods.

How can the various measures of paleo-oxygen level be calibrated to one another?

Better calibration of these indices is most critical for intermediate conditions of oxygen depletion, dysoxia. For example, how does one distinguish in ancient mudstones between stable, long-term dysoxia and seasonal fluctuations between oxic and anoxic?

What is the significance of green and blue color in mudstones?

We seem to have a fairly good understanding of the red, black and gray color varieties of mudstones, but is there more information to be gained by a better understanding of the diversity of greenish and bluish mudstones? Are these colors related to slow sedimentation rates? To climate? To non-clay minerals? How much is detrital?

The best indicators of paleo-oxygen levels in the Phanerozoic are bioturbation and body fossils. What is best for Precambrian rocks?

Are there as yet unidentified methods that can help with pre-metazoan mudstones? Rare-earth element anomalies? New isotope systems such as Mo?

References

Ahumada B, Ramón, Rudolph G, Martinez MV (1983) Circulation and fertility in waters of Concepción Bay. Estuarine, Coastal and Shelf Science 16:95–105

Algeo TJ, Maynard JB (2004) Trace-element behavior and redox facies in core shales of Upper Pennsylvanian Kansas-type cyclothems. Chem Geol 206:289–318

Arnold GL, Anbar AD, Barling J, Lyons TW (2004) Molybdenum isotope evidence for widespread anoxia in mid-Proterozoic oceans. Science 304:87–90

Arthur MA, Sageman BB (1994) Marine black shales: depositional mechanisms and environments of ancient deposits. In: Wetherill GW, Albee AL, Burke KC (eds) Annual Review of Earth and Planetary Sciences (Annual Reviews Inc., Palo Alto CA) 22:449–551

Barling J, Anbar AD (2004) Molybdenum isotope fractionation during adsorption by manganese oxides. Earth Planetary Sci Lett 217:315–329

Barling J, Arnold GL, Anbar AD (2001) Natural mass-dependent variations in the isotopic composition of molybdenum. Earth Planetary Sci Lett 193:447–457

Berner RA (1985) Sulphate reduction, organic matter decomposition and pyrite formation. Philosophical Transactions Royal Soc A315:25–38

Bohacs KM, Carroll AR, Neal JE, Mankiewicz PJ (2000) Lake-basin type, source potential and hydrocarbon character: An integrated sequence-stratigraphic-geochemical framework. In: Gierlowski-Kordesch EH, Kelts KR (eds) Lake Basins Through Space and Time. AAPG Studies in Geology 46:3–34

Bottjer DJ, Savrda CE (1993) Oxygen-related mudrock biofacies. In: Wright VP (ed) Sedimentology Review, Vol. 1. Blackwell Scientific Publications, Oxford, pp 92–102

Byers CW (1977) Biofacies patterns in euxinic basins: a general model. In: Cook HE, Enos P (eds) Deep-water Carbonate Environments. Society Economic Paleontologists Mineralogists, Special Publication 25:5–17

Crusius J, Calvert S, Pedersen T, Sage D (1996) Rhenium and molybdenum enrichments in sediments as indicators of oxic, suboxic, and sulfidic conditions of deposition. Earth Planetary Sci Lett 145:65–78

Darwin C (1846) An account of fine dust which often falls on vessels in the Atlantic Ocean. Quarterly J Geol Soc London 2:26–30

Degens ET, VonHerzen RP, Wong H-K (1971) Lake Tanganyika: Water chemistry, sediments, geological structure. Naturwissenschaften 58:229–241

Demaison GJ, Moore GT (1980) Anoxic environments and oil source bed genesis. Am Assoc Petrol Geologists Bull 64:1179–1209

Gaines RR, Droser ML (2003) Paleoecology of the familiar trilobite *Elrathia kingii*: An early exaerobic zone inhabitant. Geology 31:941–944

Gallardo VA, Castillo JG, Vañez LA (1972) Algunas consideraciones preliminaries sobre la ecología bentónica de los fondos sublitorales blandos en la Bahía de Concepción. Bolletin Sociedad Biología Concepción 44:199–190

Gardner JV, Dean WE, Jansa L (1978) Sediments recovered from the northwest African continental margin, Leg 4 l, Deep Sea Drilling Project. In: Gardner JV, Herring J (eds) Leg 41 Abidjan, Ivory Coast to Malaga, Spain, February-April, 1975. Initial Reports Deep Sea Drilling Project 41:1121–1134
Gerritse RG (1999) Sulphur, organic carbon and iron relationships in estuarine and freshwater sediments: effects of sedimentation rate. Appl Geochem 14:41–52
Green AR (1985) Integrated sedimentary basin analysis for petroleum exploration and production. Proceedings of the 17th Annual Offshore Technology Conference Vol. 1, pp 9–20
Holland HD (2002) Volcanic gases, black smokers, and the Great Oxidation Event. Geochimica et Cosmochimica Acta 66:3811–3826
Holloway M (2000) The killing lake. Scientific Am 283:92–99
Huc AY, LeFournier J, Vandenbrouche M, Bessereau G (1990) Northern Lake Tanganyika – An example of organic sedimentation in an anoxic rift lake. In: Katz BJ (eds) Lacustrine Basin Exploration. Am Association Petrol Geol Memoir 50:169–185
Hunt JM (1996) Petroleum Geochemistry and Geology, 2nd edn. Freeman, New York, 743 p
Jones B, Manning DAC (1994) Comparison of geochemical indicators used for the interpretation of palaeoredox conditions in ancient mudstones. Chem Geol 111:111–129
Kauffman EG (1986) High resolution event stratigraphy: regional and global bioevents. In: Walliser OH (ed) Global Bioevents. Springer, Berlin Heidelberg New York, Lecture Notes in Earth History 8, pp 279–235
Kirschvink JL (1992) Late Proterozoic low-latitude global glaciation: The snowball Earth. In: Schopf JW, Klein C (eds) The Proterozoic Biosphere. Cambridge University Press, Cambridge, pp 51–58
Kump LR (2003) The geochemistry of mass extinctions. In: Mackenzie FT (ed) Treatise on Geochemistry Vol. 7. Elsevier, Amsterdam, pp 351–367
Lethiers F, Whatley R (1994) The use of Ostracoda to reconstruct the oxygen levels of Late Palaeozoic oceans. Marine Micropaleontol 24:57–69
Leventhal JS (1998) Metal-rich black shales: Formation, economic geology and environmental considerations. In: Schieber J, Zimmerle W, Sethi PS (eds) Shales and Mudstones, Vol. II. E Schweizerbart'sche, Stuttgart, pp 255–282
Lewan MD (1984) Factors controlling the proportionality of vanadium to nickel in crude oils. Geochimica et Cosmochimica Acta 48:2231–2238
Lewan MD, Maynard JB (1982) Factors controlling the enrichment of vanadium and nickel in the bitumen of organic sedimentary rocks. Geochim Cosmochim Acta 46:2547–2560
Martin WD (1998) Geology of the Dunkard Group (Upper Pennsylvanian-Lower Permian) in Ohio, West Virginia and Pennsylvanian. Ohio Geol Survey Bull 73:49
Murphy AE, Sageman BB, Hollander DJ, Lyons TW, Brett CE (2000) Black shale deposition and faunal overturn in the Devonian Appalachian basin: clastic starvation, seasonal water-column mixing, and efficient biolimiting nutrient cycling. Paleoceanography 15:280–291
Myrow PM (1990) A new graph for understanding color of mudrocks and shales. J Geoll Educat 38:16–20
Niitsuma N, Oba T, Okada M (1991) Oxygen and carbon isotope stratigraphy at Site 723, Oman Margin. In: Prell WL, Niitsuma N et al. (eds) Scientific Results Oman Margin/Neogene Package. Proceedings of the Ocean Drilling Program, Vol. 117. College Station, Texas (Ocean Drilling Program), pp 321–341
Odin GS (1988) Green Marine Clay: Oolitic Ironstone Facies, Verdine Facies, Glaucony Facies and Celadonite-Bearing Facies: A comparative Study. Elsevier, Amsterdam, 445 p
Oschmann W (2001) Oxygen in the ocean. In: Briggs DEG, Crowther PR (eds) Palaeobiology. Blackwell Scientific, Oxford, pp 470–472
Pederson TF, Calvert SE (1990) Anoxia vs productivity: what controls the formation of organic-carbon-rich sediments and sedimentary rocks? American Association Petroleum Geologists Bulletin 74:454–466
Peters KE, Moldowan JM (1993) The Biomarker Guide. Prentice-Hall, Englewood Cliffs, New Jersey, 363 p
Raiswell R, Berner RA (1985) Pyrite formation in euxinic and non euxinic basins. Am J Sci 285:710–724
Raiswell R, Canfield DE (1998) Source of iron for pyrite formation in marine sediments. Am J Sci 298:219–245
Rhodes DC, Morse JW (1971) Evolutionary and ecologic significance of oxygen-deficient basins. Lethia 4:413–428
Sageman BB, Murphy AE, Werne JP, VerStraaten CA, Hollander DJ, Lyons TW (2003) A tale of shales: the relative roles of production, decomposition, and dilution in the accumulation of organic-rich strata, Middle-Upper Devonian, Appalachian Basin. Chem Geol 195:229–273
Sandberg CA, Gutschick RC (1984) Distribution, microfauna and source-rock potential of Mississippian Delle Phosphate Member of the Woodman Formation and equivalents, Utah and adjacent states. In: Woodward J, Meissner FF, Clayton JL (eds) Hydrocarbon Source Rocks of the Greater Rocky Mountains Region. Rocky Mountain Association of Geologists, Denver, Colorado, pp 135–178
Schmoker JW (1981) Determination of organic-matter content of Appalachian Devonian shales from gamma-ray logs. Am Assoc Petrol Geol Bull 65:1285–1298
Stanley DJ (1969) Atlantic continental shelf and slope of the United States – Color of marine sediments. US Geological Survey Professional Paper 529-D:15
Velde B (2003) Green clay minerals. In: Mackenzie FT (ed) Treatise on Geochemistry, Vol. 7. Elsevier, Amsterdam, pp 309–324
Walker RG (1971) Nondeltaic depositional environments of the Catskill Wedge (Upper Devonian) of Central Pennsylvania. Geol Soc Am 82:1305–1326
Werne JP, Hollander DJ, Lyons TW, Sageman DB (2002) An integrated assessment of a "type euxinic" deposit: evidence for multiple controls on black shale deposition in the Middle Devonian Oatka Creek Formation. Am J Sci 302:110–143
Wignall PB (1994) Black Shales. Oxford Science, Oxford, p 127
Wignall PB, Newton R (1998) Pyrite framboid diameter as a measure of oxygen deficiency in ancient mudrocks. Am J Sci 298:537–552
Wiseman WJ, Rabelais NN, Turner RE, Dinnel SP, MacNaughton A (1997) Seasonal and interannual variability within the Louisiana coastal current: stratification and hypoxia: Journal Marine Systems 12:237–248

Digging Deeper

Allison PA, Wignall PB, Brett CE (1995) Palaeo-oxygenation: effects and recognition. In: Bosence DWV, Allison PA (eds) Marine Palaeo-Environmental Analysis from Fossils. Geol Soc Special Publication 83:97–112.

Excellent overview shows how to use chemistry, mineralogy, stratification, fossils and organic matter to reconstruct paleo-oxygen levels. Some background required, but most useful and comprehensive.

Bigham JM, Ciolkosz EJ (1993) (eds) Soil Color: Soil Science Society America, Special Publication 31:127.

Little known to most geologists, this small volume has eight short comprehensive chapters that range from field terms to laboratory measurement (includes stratigraphic and hydraulic influences on soil) and concludes with a discussion of red beds making it most valuable. Twelve colored plates and many informative graphs.

Hunt JM (1996) Petroleum Geochemistry and Geology, 2nd edn. Freeman, New York, 743 p.

Chapter 10, "The Source Rock" has an excellent discussion of organic-rich shales with especial attention to thermal maturation of the organic matter during burial. There is also an interesting section relating the amount of organic matter in muds and mudstones to grain size and the oxygenation of the environment.

Katz BJ (1994) Petroleum Source Rocks. Springer, Berlin Heidelberg New York, 327 p.

Contains 15 case studies of organic-rich shales, both marine and non-marine. Each follows a common format allowing easy comparisons. Each chapter has good coverage of biomarkers and rock-eval parameters as related to depositional setting. The chapter by Isaksen and Bohacs on the Triassic of the Barents Sea has an interesting example of the use of spectral gamma-ray logging in sequence stratigraphy.

Sageman BB, Lyons TW (2003) Geochemistry of fine-grained sediments and sedimentary rocks. In: Mackenzie FT (ed) Treatise on Geochemistry, Vol. 7. Elsevier, Amsterdam, pp 115–158.

Thorough review of early diagenesis in mudstones, with an emphasis on black shales and their C and S chemistry, including stable isotopes. Has three valuable case studies – the modern Cariaco Trench, the Cretaceous Western Interior Basin, and the Devonian Appalachian Basin. Nearly 400 references, making this a valuable resource for finding more information on mudstone geochemistry.

Tyson RV (1995) Sedimentology Organic Matter. Chapman and Hall, London, 615 p.

Comprehensive and well referenced with many useful tables, this book carefully relates all aspects of organic matter to hydraulics and sedimentology facies. For the non specialist, it is essential to read Chapter 1, "The Importance of Organic Matter" and Chapter 24, "Palynofacies in a Sequence Stratigraphy Context". Fundamental reference for the study of muds and mudstones.

CHAPTER 5

Muddy Depositional Systems

Mud is everywhere

5.1. Introduction

Here we examine mud and silt in modern environments and explore how they link together to form the several major depositional systems of fine-grained sediments. Understanding these depositional systems is essential for coping effectively with the different characteristics of mudstones and their stratigraphy both in large basins and for many smaller-scale geotechnical and environmental studies. Why? Because such understanding helps us visualize the probable geometry, lateral extent and properties of a deposit known from only a few wells, outcrops or even a single seismic line.

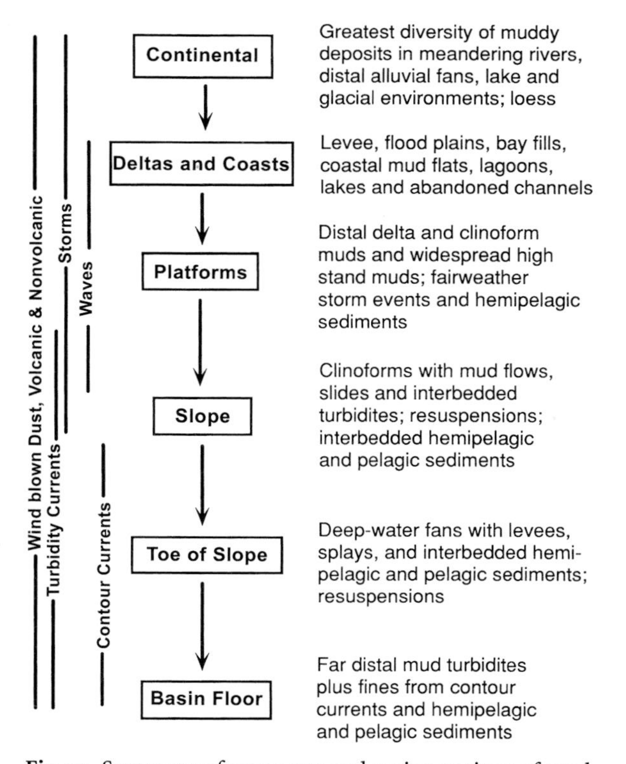

Fig. 5.1. Summary of processes and major settings of mud deposition

Our strategy is straightforward – start at the headwaters of streams and terminate in the deep basin (Fig. 5.1), always linking one environment to the other. Thus here we utilize and combine the information and ideas presented in the earlier chapters on the production of mud and silt, their transport and deposition, and how bottom oxygen affects their properties.

In this scenario we frequently use the concept of *accommodation*, the space in an environment available for deposition. This is particularly appropriate for the deposition of muddy sediments, because they are most likely to be deposited in topographic lows where bottom currents are weak or absent. Clay-sized sediment settles in quiet-water basins large or small, deep or shallow. Clay from suspension drapes over river and tidal bars, is deposited on the far reaches of a floodplain during high water, and in the oxbow lakes of abandoned river or tidal channels. It settles in coastal swamps and on tidal flats in dead water at high tide. Farther from shore, below wave base, clay accumulates on platforms, slopes, toes of slopes and eventually comes to rest in deep water basins (Table 5.1). Thus one can imagine a clay or silt particle starting in the high mountains of a convergent margin, crossing a continent in a large river to a distant passive margin, and then moving along the coast for a 1,000 km or so. One day when a strong winter storm impinges on the coast, the particle is than carried over the shelf edge into deep water and deposited onto oceanic crust. Of course, such a trip happens in many short pulses, most separated by many thousands of years (this is the residence time of a particle in a transport system mentioned in Chap. 2). Here our charge is to gain insight into these resting places and the characteristics of their deposits.

Whether subaerial or subaqueous, all the above resting sites are "low energy environments", protected places where bottom turbulence is minimal, but there are also other ways that mud can accumulate. Suspended sediment concentrations may be so high in the water column that the energy of inshore

Table 5.1. Terrigenous mud – controls, current systems, and selected characteristics of major environments (After Bohacs 1998, Table 1) [Published by permission of the author and E. Schweizerbart'sche Verlagsbuchhandlung]

	Alluvial	Deltaic	Coastal (Beaches, lagoons and mires)	Shelf	Slope/Basin	Overfilled Lakes	Balanced Lakes	Underfilled Lakes
Major controls	Ratio of load to discharge; gradient	Ratio of inshore wave/tidal power to supply	Ratio of inshore wave/tidal power to supply	Water depth, bottom energy and supply	Supply, stability of slope and bottom currents	Supply + $H_2O \gg$ accommodation; minor role for climate	Supply + H_2O = accommodation Very climatic sensitive	Supply + H_2O $\ll$accommodation
Current systems	Overbank flows/shifting channels/floods; suspension in strongly turbulent, uni-directional flows	Diverse depending on type but includes overbank flows, shifting channels, long shore currents, tides and storms	Longshore currents, tides and storms; mud mostly behind barriers except for high-density fluid mud on some coasts	Distal riverine plumes, oceanic currents, storms and tides; surface, midwater flows, and bottom currents	Resedimentation and contour currents on slope and proximal basin; some midwater suspensions	Oscillating, but prograding shore-line Mostly wind-driven, but also flood plumes, storms, turbidity currents (deep lakes), stratification and overturning	Stable shoreline	Stable shoreline
Associated lithologies	Silts, sands and possible peats/coals	Silts, sands, peats, and possible coals	Silts, sands, peats, and possible coals	Silts, sands, carbonates and evaporites also common	Silts, sands and pelagic limestones	Silts, sands and possible peats and coals	Silts and sands; possible carbonates	Silts and sands with carbonates and evaporites
Lateral continuity	Limited by channel cutouts and valley width	Similar to alluvial except for widespread delta front deposits and some bay fills	Limited except for some large lagoons	Widespread lobes and sheets at highstands	Restricted on slope, but widespread in basin	Sheets with channel cutouts on prograding shorelines	Carbonates and organic-rich shales most widespread	Carbonates, evaporites and organic-rich shales most widespread
Organic matter, C_{org}%	Land plants/algae; 2.0 (1.8–500)	Land plants/algae; 6.0 (3.0–34)	Land plants/algae; 9.2 (4.0–77.0)	Marine algae/ some plants: 4.6 (1.1–20.0)	Marine algae; algae mats (?); 7.0 (1.0–27.0)	Land plants algae; 7.0 (0.5–45.0)	Algae/land plants; 15.0 (2.0–30.0)	Algae/bacteria; 1.5 (0.2–15.0)
Type of organic matter	III; HI=150 (230–445)	III, III-I; HI = 280 (170–520)	III, III-II; HI = 188 (35–600)	I/II; HI = 530 (165–825)	II+sulfur; HI = 500 (150–800)	I/II; HI = 600 (50–700)	I; HI = 900 (600–1100)	I; HI = 400 (10–600)
Fossils	Plant debris, spoors, pollen and rare vertebrates and invertebrates	As in alluvial, but also some brackish marine invertebrates	Broadly similar to delta	Open marine benthic and pelagic; some fine plant debris; spore abundance provides distance to shoreline	Open marine pelagics and limited benthics	Modest diversity; plant debris and rare vertebrates	Great diversity of pelagic and bottom fauna; water depth and climate permitting	Sparse, restricted fauna (high salinity or deep water)

Fig. 5.2. Tidal mud accumulates in a small, protected bay along open, high energy, wave-dominated, Pacific coast of Baha California, Mexico. The Pacific lies just to the right of the distant low mountain and has a powerful surf. Irregular surface pattern of mudflat is the result of bioturbation

waves is dampened and fines accumulate, even on open, high-energy coasts. Another exceptional environment is the sediment laden, tidewater glacier with its associated debris flows and turbidites that produce thick, fine-grained glacial marine deposits with dropstones carried to the basin by floating ice. Colluvium and mud-rich debris flows and glacial tills represent still other high-energy muddy environments. And, because of the small fall velocity of mud and fine silt, the mud delivery system is everywhere sensitive to small variations of current velocity. On land, vegetation, early cementation by gypsum or carbonate, accumulation behind an obstacle, or high water tables all trap fines carried by wind and water. Consequently, given a supply of suspended mud, deposition will occur anywhere there is a protected small bay, inlet or basin, including the small ephemeral lake of a braided river system and even in small, protected bays on high-energy, open coasts. Hence small pockets of mud and mudstone can almost always be found even in the most unlikely of environments (Fig. 5.2).

In virtually all of the above environments silt occurs within muds as thin "single grain" laminations, as small, cross laminated ripples and graded laminations, and as scattered grains (Chap. 3). Silt also occurs in soils, perhaps its principal site of origin, and is the dominant component of loess. Silt is also common as individual laminations and thin beds in the upper half of fining upward sandy cycles and in the lower half of coarsening upward sandy cycles. Such sequences can be less than a meter thick in modern alluvium or in a turbidite bed or can be 10s to 100s of meters thick at the front of a delta de-

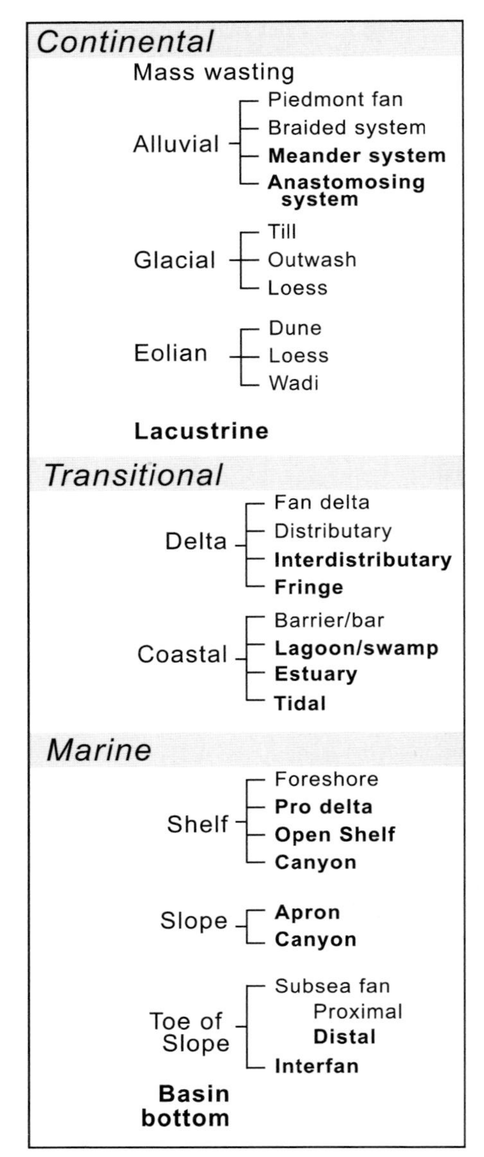

Fig. 5.3. Subdivisions of continental, transitional and marine environments. Environments in *bold face* provide the best traps for mud

bouching into deep water. It is here, in front of deltas, where some of the thickest sections of siltstones are found.

What follows is an outline of the places, processes, and products of the mud-silt delivery system diagramed in Fig. 5.3. Simply think of what follows as a tour or a field trip of the places, processes, and products of mud and silt deposition. For each segment of the delivery system, modern deposits are emphasized and illustrated, although sometimes Tertiary and even Mesozoic deposits serve us better. See Reading (1996) for a complete overview of depositional systems that emphasizes the sandstones and carbonates of these environments, but also provides some information about mudstones.

5.2. Continental Environments

Mud and silt-rich continental deposits include those formed by mass wasting on slopes, alluvium, loess, lake deposits and the diverse facies created by ice sheets. All of these are important for those who work with surficial deposits such as geotechnical and environmental engineers, geographers, city planners, archaeologists, and students of late Pleistocene and Holocene climatic change and history. Although their total volume is only a tiny fraction of that of marine mudstones, probably "insignificant" is a better word, these contemporary continental environments are a critical interface for life.

5.2.1. Mass Wasting on Slopes

Mass wasting (Fig. 5.4), powered by gravity on slopes rather than by running water, is ever present, as is its principal product – *colluvium*. Colluvium is a poorly sorted, unstratified deposit of angular, hillside-derived clasts in a fine-grained matrix. This fine-grained matrix is especially well developed by the humid-climate weathering of granitic, volcanic, carbonate and shaly rocks. *Creep*, the slow transport of colluvium down slope by gravity, moves colluvium to a valley bottom, as do more episodic events such as landslides, debris flows and slope wash. Mud-rich colluvium is of interest to us for two reasons – first of all, its movements are the initial step in the long process of transport of fines to a depositional basin. The higher the mud content of colluvium on a hillside, the easier it is for failure to occur and the more frequent the occurrence of downslope mass movement. Secondly, mud-rich colluvium causes special problems for construction on hillsides. Creep also occurs on the low dips of tidal mud flats, on delta fronts, on many muddy shelves and ramps, on the sides of submarine channels and on slopes beyond the shelf edge. Where slopes are steep and underwater sedimentation rapid, creep is likely to be an active, important process in the downdip transport of mud.

Mud-rich colluvium is one example of a *diamicton* (unconsolidated) or a *diamictite* (consolidated) – a purely descriptive term for a poorly sorted deposit of boulders and pebbles in a fine-grained matrix (Fig. 3.5). The terms diamicton and diamictite could, for example, be used as substitutes for either *boulder till* or *pebbly mudstone*.

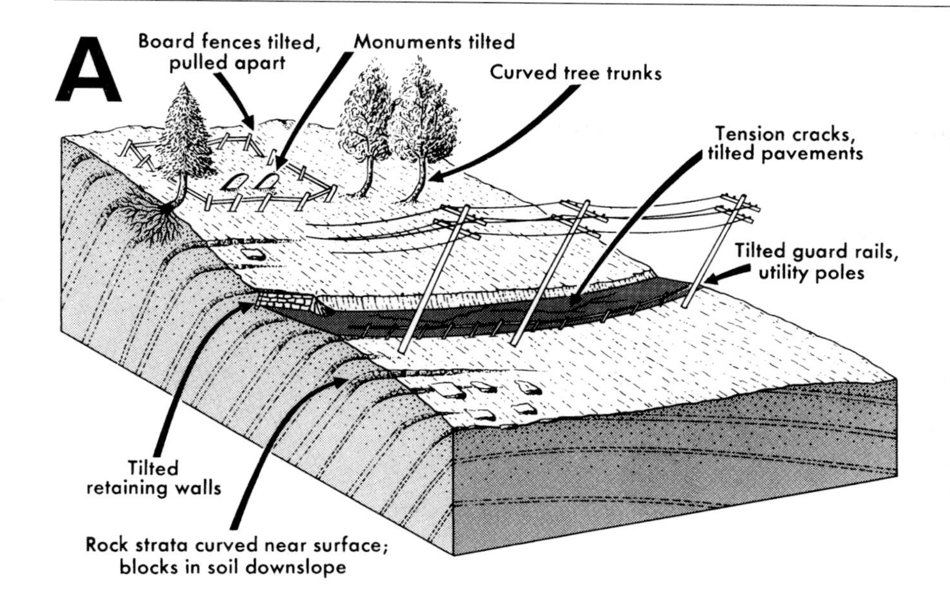

Fig. 5.4A,B. Two extremes of mass wasting on slopes: (**A**) the many faces of hillside creep, a slow but persistent transport of colluvium to headwater streams (Bloom 1998, Fig. 9-5). Published by permission of the author and Prentice Hall; (**B**) heavy rainfall saturates colluvium, creates landslides and debris flows, which fill streams with unsorted rubble and much mud below the high Planalto Escarpment, Rio Grande do Sul, Brazil (Drawn from a photograph by Joel Pellerin). Slope wash and gullying also transport colluvium to valley bottoms as do isolated landslides

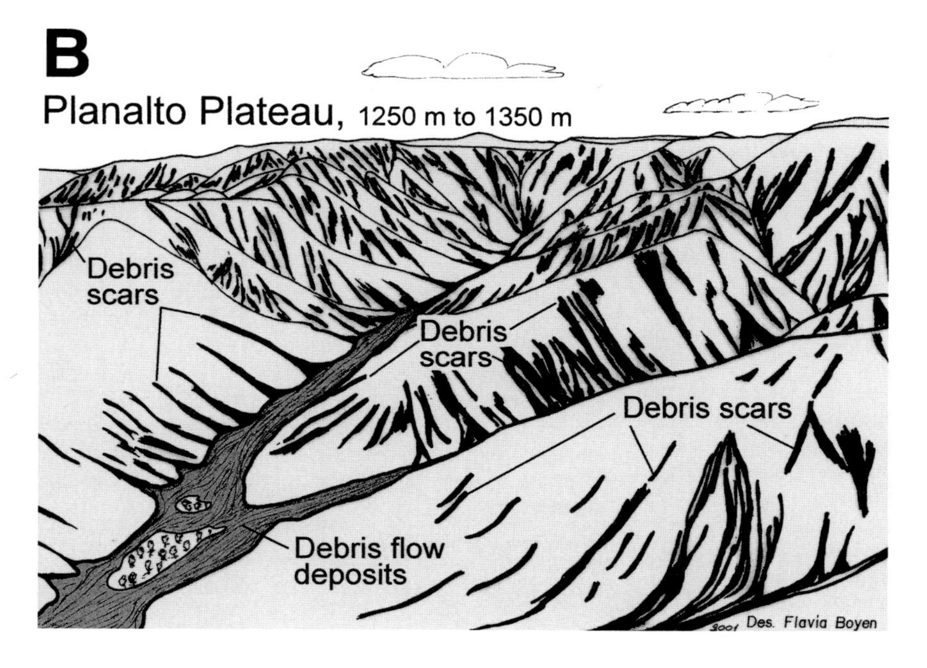

5.2.2. Alluvial

The alluvial valley environment varies greatly in size from a small shallow valley, perhaps with an ephemeral stream, to a valley 10 s to 100 s of kilometers wide such as that of the Ganges, Huang He (Yellow), and Lower Mississippi. The rivers of these valleys can be fairly straight, meandering, anastomosing or braided (Table 5.2). In all but braided streams, there is one strong common denominator independent of stream size (Fig. 5.5) – most of these deposits fine upward from basal gravel (if available) through coarse sand to silt and mud. The depth of the active stream channel controls the thickness of such alluvial fining-upward cycles – the deeper the stream channel, the thicker is a complete alluvial cycle – whereas the thickness of the silt-mud cap itself seems to be more a function of stream gradient and availability of silt and mud in the source region. In or close to the stream channel, sandy bar deposits may be separated by *clay curtains* or *drapes*, thin sheets of mud deposited in the initial, slack water falling stage of a flood that covered a point bar. These thin drapes help separate one point bar from another and

Table 5.2. Fine-grained deposits and stream types

Decreasing Deposition/ Storage of Fines (↓)

Anastomosing
- Mostly clay to fine sand, low ratio of sandy bed load to discharge; both bypassing by suspension and deposition
- Channels of low to moderate stability
- Overbank deposition extensive; flood plain overlies a fine-grained fill with few channel sands. Swamps, lakes and peats common

Meandering
- Intermediate ratio of load (chiefly sand) to discharge with both suspension bypassing and deposition
- Shifting deep to shallow channels of high sinuosity
- Overbank deposition extensive; flood plain overlies a mosaic of mixed sand and fine-grained depositis with point bars, clay sheets and plugs, swamps, lakes and peats

Braided
- High ratio of load (gravel or sand) to discharge with much suspension bypassing
- Shifting shallow channels with low sinuosity
- Restricted flood plain with only a few thin, discontinuous silty clay caps. Much export of dust possible from wide, shallow, multi-channel stream bed

act as barriers to flow in reservoirs of ground water and petroleum and to the mineralizing fluids of some sedimentary ore deposits.

In an alluvial sequence, the lower the stream gradient, the more fines are deposited and the greater the ratio of over-bank mud and silt to channel sands. Finally, even a swamp or lake may develop. Common causes of reduced gradients include actively subsiding tectonic lows, or conversely, a rising fault or anticline; regional back tilting of the river; and ris-

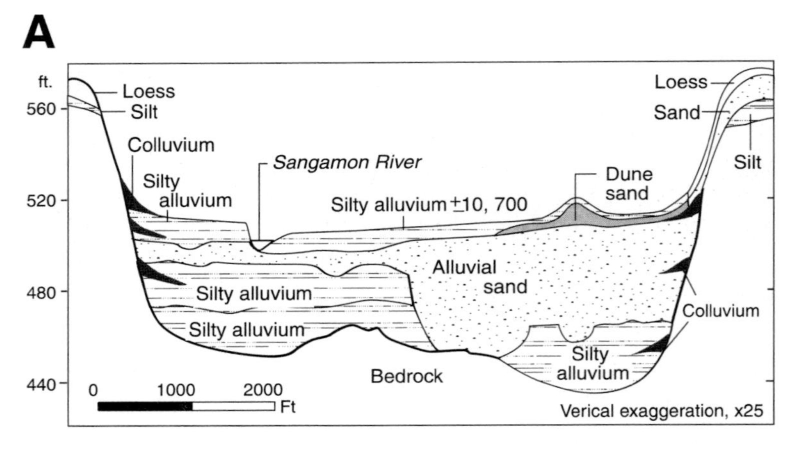

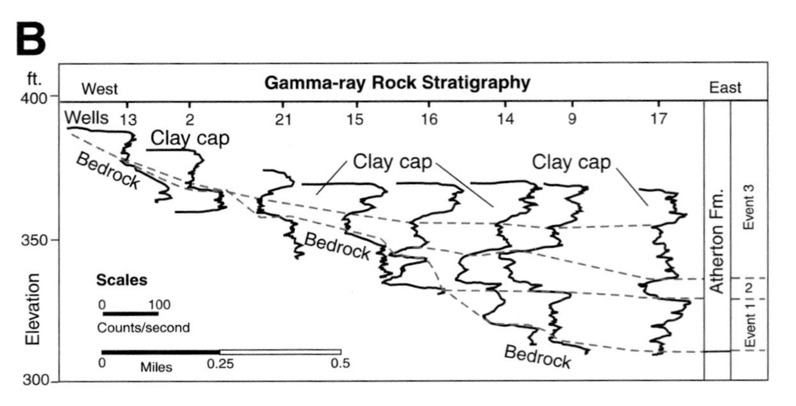

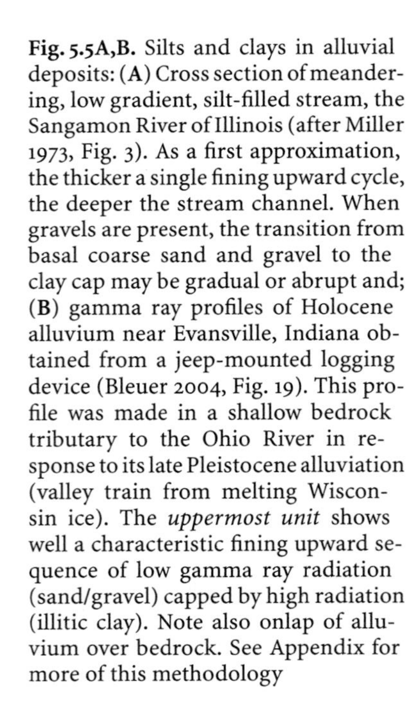

Fig. 5.5A,B. Silts and clays in alluvial deposits: (**A**) Cross section of meandering, low gradient, silt-filled stream, the Sangamon River of Illinois (after Miller 1973, Fig. 3). As a first approximation, the thicker a single fining upward cycle, the deeper the stream channel. When gravels are present, the transition from basal coarse sand and gravel to the clay cap may be gradual or abrupt and; (**B**) gamma ray profiles of Holocene alluvium near Evansville, Indiana obtained from a jeep-mounted logging device (Bleuer 2004, Fig. 19). This profile was made in a shallow bedrock tributary to the Ohio River in response to its late Pleistocene alluviation (valley train from melting Wisconsin ice). The *uppermost unit* shows well a characteristic fining upward sequence of low gamma ray radiation (sand/gravel) capped by high radiation (illitic clay). Note also onlap of alluvium over bedrock. See Appendix for more of this methodology

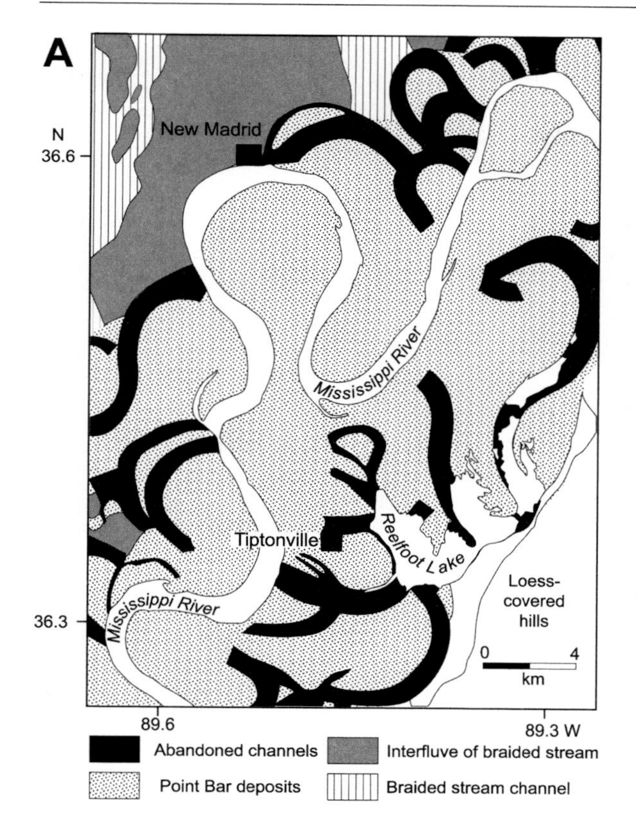

Fig. 5.6A–E. The Mississippi River in far western Kentucky and Tennessee and adjacent Missouri provides a splendid laboratory to study alluvial mud deposition by a major river. (**A**) Distribution of surface environments in study area with abandoned, clay and silt-filled meander loops, called clay plugs, in black (after Fisk 1947 Sheet 9); (**B**) rapid historic changes in a large alluvial river are well documented by the cutoff and filling of Lucas Bend, Carlisle County, Kentucky just below the mouth of the Ohio River. Lucas Bend on the west side of Three Mile Island was for many years the main channel prior to 1930, but was abandoned in the 1940s. The present channel occupies a chute charted as early as 1876. Lucas Bend is filling with silt and mud to form a clay plug. (Information kindly supplied by Robert Saucier, US. Waterways Experiment Station, Vicksburg, Mississippi (photograph by Billy Davis, Louisville Courier Journal)). (**C**) Abandoned course of the Mississippi River southwest of Hickman, Fulton Co., Kentucky. On alluviating floodplains long term residents commonly see and talk about the rapidity with which such abandoned channels fill during floods; (**D**) Reelfoot Lake is shallow, has a muddy bottom, was created during the 1811–1812 activity of the New Madrid Earthquake, and partially occupies abandoned channels; and (**E**) thick, gently inclined clay drape at lower end of a large sandwave of the Mississippi in New Madrid, New Madrid Co., Missouri. Clay drapes range from millimeters in tidalites to several meters or more in thickness and, for alluvial deposits, form significant barriers to flow in the permeable point bar and channel sands (*continued on next page*)

ing base levels. Low-gradient streams tend to be the most mud-rich, and their properties are summarized in Table 5.2 and Figs. 5.6 and 5.7.

Most mud in alluvial sequences accumulates via *vertical accretion* in contrast to the *lateral accretion* of sandy point bars in the active stream channel. Two main modes of mud deposition are (1) *overbank sedimentation*, where mud and silt are deposited during floods as suspension deposits in temporary floodplain lakes and ponds marginal to the main

Fig. 5.6. (*continued*)

channel and (2) *oxbow sedimentation* in perennial lakes formed by abandoned meander loops of the stream channel. It is the deposits of temporary flood plain lakes that produce the sheet-like muds that occur between sandy channel fills. The eventual product of oxbow sedimentation is a *clay plug*, an abandoned stream channel filled by mud, silt and minor sand deposited from overbank sed-

Fig. 5.7A,B. Flood plains: (**A**) Flood stage on the meandering Green River of Western Kentucky highlights its well developed system of low scrolls and flooded, intervening swales, which trap suspension deposits (photograph by Billy Davis, Louisville Courier Journal); and (**B**) cut bank of East Fork of White River in Jackson Co., IN, shows origin of clay and silty clay blocks and clasts that occur at the base of many fluvial sandstones – impinging current easily erodes underlying sand and undermines overlying clay cap

imentation. Seemingly homogeneous to the eye, X-ray radiography reveals many sub-environments (Fig. 5.8), which typically pass upward from riverine to lacustrine to final paludal environments with many oscillations. The thickness of the clay plug directly correlates with stream channel depth, which typically is about 3 to 5 m for smaller rivers, but can be as much as 20 m or more for very large ones.

Where a flood tops or breaks a bounding levee, a *splay* is formed, whose sands extend fan-like into the flood plain, with its distal parts depositing mostly fine silt and mud from either temporarily ponded waters or distal sheet floods. As the channels of a meandering river system migrate laterally, a series of curving, sandy swales (low ridges of abandoned point bar crests) separated by intervening muddy sag forms.

Proximal to the channel on its levee, thin, fine-grained, cross-laminated, mottled brown and gray sands and silts with climbing ripples predominate. Farther away, thin silts are interlaminated with silty muds that finally pass distally into finer gray to brown muds largely lacking silt. Thus the permeability of the silt-clay cap of temporary flood basin deposits generally decreases away from a channel as the mud content increases. Where poorly drained, the term *backswamp* is sometimes used for such distal flood plain deposits (Fig. 5.9). In Devonian and younger deposits, organic matter is abundant (rootlets, leaves and woody debris and perhaps even some peat), and dark gray to black colors are dominant. Where flood plain drainage is better, thin "dark bands" of finely disseminated organic matter and immature soils mark pauses in vertical, overbank sedimentation on flood plains. Scattered twigs, leaves and rootlets oc-

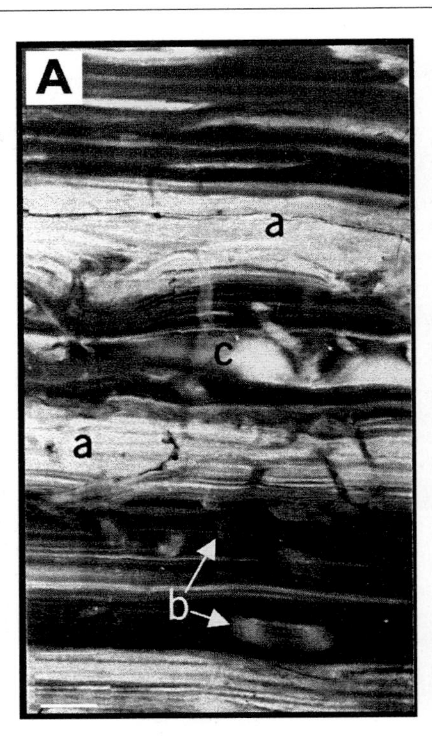

Fig. 5.8A,B. X-Radiographs from well-laminated shallow lake deposits and from poorly-laminated swamp deposits of the lower alluvial valley of the Mississippi River in Louisiana (Coleman 1966, Figs. 7A and 5A). Sediments of the lower part of lacustrine delta (**A**) have thin, well-developed laminae with only a few small burrows. Laminations result chiefly from alternate deposition of flocculated and non-flocculated mud. Also present are *a*) thin laminae of silt; *b*) burrows with abundant fecal pellets; and *c*) incipient $CaCO_3$ nodule. Well-drained swamp deposits (**B**) have carbonate present as *a*) incipient nodules; *b*) carbonate laminae; and *c*) replacements of rootlets

Fig. 5.9. Back swamp along the Great Miami River near its junction with the Ohio River in Southwestern Ohio. Note sloppy mud, standing water, and mud line on trees

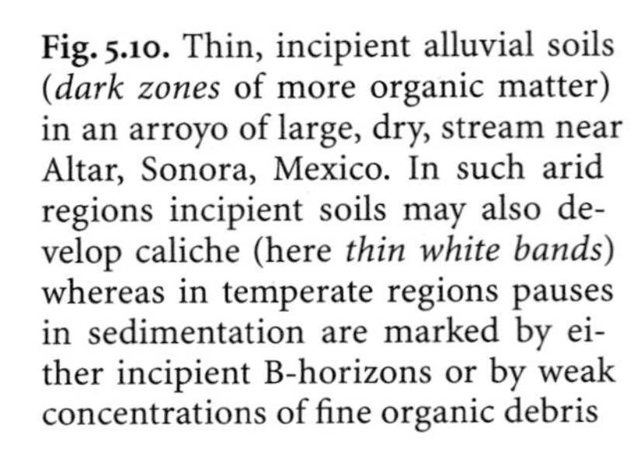

Fig. 5.10. Thin, incipient alluvial soils (*dark zones* of more organic matter) in an arroyo of large, dry, stream near Altar, Sonora, Mexico. In such arid regions incipient soils may also develop caliche (here *thin white bands*) whereas in temperate regions pauses in sedimentation are marked by either incipient B-horizons or by weak concentrations of fine organic debris

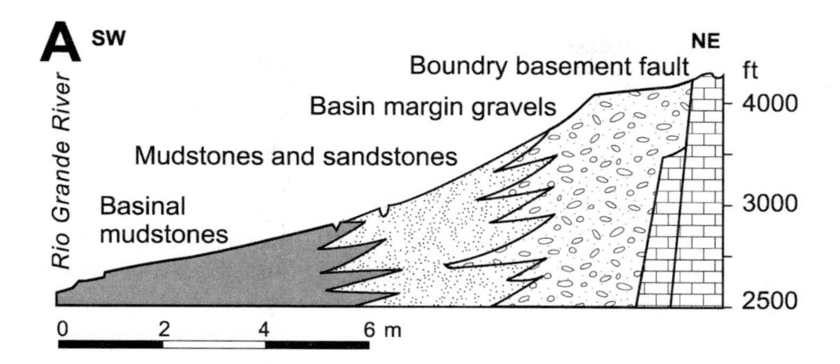

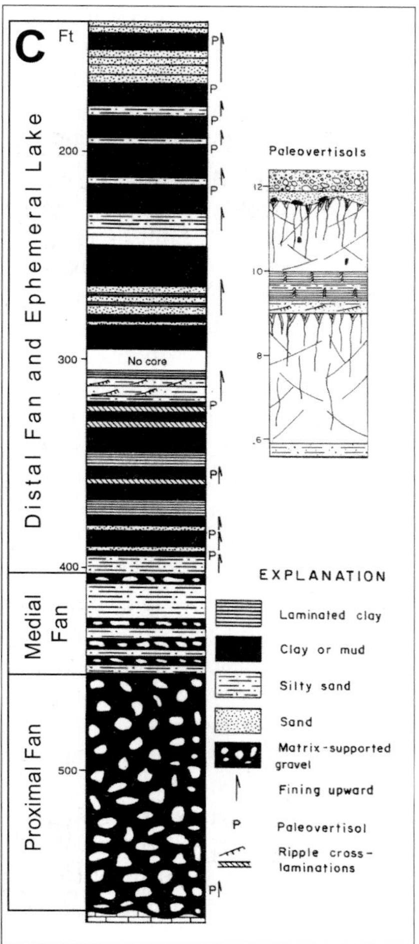

Fig. 5.11A–D. Pliocene fill of Presidio Bolson in Presidio County along the Texas-Mexican border: (**A**) Diagrammatic downdip section (Groat 1972, Fig. 2); (**B**) outcrop of basin-center muddy fill with *inset* of details (*gray beds* are sands); (**C**) similar facies in Plio-Pleistocene fill of nearby Hueco Basin to the north (Gustavason 1991, Figs. 3 and 8); and (**D**) deep desiccation cracks in laminated slightly silty clays of basin floor muds along Hwy 1, Ancash Province, Peru

cur in humid climate alluvium along with mudcracks formed as floodwaters receded. In dry climates, on the other hand, gypsum in the form of crystals, laminations, and veins, or calcite and dolomite nodules are more typical, mud cracks are far more abundant and larger, and brownish to reddish colors prevail. Incipient caliches may also develop. And in arid climates the flat, even surface of the flood plain may be broken by sand dunes so that thin, millimetric sheets of sand and silt as well as dust are likely to be present in the alluvium. See Fisk (1947) in Digging Deeper for his pioneering study of the fine-grained deposits of the lower alluvial valley of the Mississippi River from which much can still be learned.

In both humid and arid climates, paleosols may be present in alluvial deposits and always deserve attention (Fig. 5.10). These are identified by some or all of the following features: mottles, soil peds, rootlets (perhaps now infilled), widening-upward irregular fractures produced by desiccation, concentrations of siderite or calcite nodules (caliche), enhanced clay content, absence of primary stratification, and color changes (Retallack 1988). Waterlogged, glei soils commonly also have oblique fractures (with slickensides). Seasonal changes in soil saturation at the top of the water table related to variable flow in the proximal river are all important in the development of such alluvial paleosols (Aslan et al. 1995) and have been suggested to extend over long distances in some redbeds (see Chap. 9). Shifting water tables alter colors and promote weathering. Where an alluvial sequence is close to sea or lake level, changes in base level are reflected upstream by changes in the water table of the alluvial fill – high sea levels raise water tables and produce water logged, dark colored soils, and swamps, whereas low sea levels produce well-drained, oxidized brownish red, and probably more numerous paleosols (Bohacs and Suter 1997; Howell and Flint (2003) p 171). Thus paleosols can be a useful signal of sea level change, and thus of sequence boundaries. Moreover, the low-stand systems tract may be poorly represented in the preserved sediment record; its best manifestation may be in thick, well-developed oxic paleosols.

Another important site of nonmarine mud deposition is the distal end of alluvial fans that debouch into playa basins or *bolsons*, usually of tectonic origin. Such fans pass down dip from gravel to gravelly sand followed by sand and finally mud at the center of the basin (Fig. 5.11). The greater the suspended load of the entering streams, the more mud-rich its fill, especially where a basin is mostly closed.

The characteristics of these mudflats differ with climate. Where arid or semi-arid, the mudflat (playa or inland sabkha) has both a dry and saline facies. The dry mudflat, possibly with a thin saline crust, is mud cracked, yields flat mud chips and may also have thin random patches of sand or silt, whereas the saline mud flat is saturated with brine and consequently choked with gypsum or halite that almost completely destroys initial lamination. Gypsum occurs as single crystals, thin lenses, and veins, and perhaps even as laminations and beds. Its abundance and that of other saline minerals is related to the local evaporation/precipitation ratio and closure of the lake. Some inter-bedded, thin siltstones are nearly always present and represent distal flood events. A few scattered rootlets may be present and are typically calcified. Sheet wash and some deposition from temporarily ponded waters during floods are the dominant processes, followed by deflation. Clay dunes consisting of silt- and sand-sized aggregates or peds locally occur on some alluvial mud flats, sabkhas, and lagoonal margins in arid and semi-arid climates. The dunes consist of soil peds as well as detritus derived from the breakup of mud curls (There is also a Cretaceous "saw dust" sand composed of such peds). In either case, frequent wetting and drying seems essential (Bowler 1973). In wet climates, on the other hand, an alluvial fan with a through-flowing stream will have frequent floods and be bordered by swamps and peat. Dark colored muds rich in plant debris, rootlets, and abundant bioturbation prevail, and long-lasting lakes may develop, as are found in many rift systems.

Pedogenic carbonate beds, caliches, in mud-rich alluvial or coastal deposits occur in semi-arid or arid climates. These paleosols, develop at low stands with low water tables, are prized stratigraphic markers and, ideally, should correlate with the "shale-on-shale" unconformities in basin centers.

5.2.3. Glacial

Vast quantities of mud and silt are generated by glaciation and deposited on land as tills, in ice-margin lakes, in lakes formed by an alluviating main

Table 5.3. Mud in glacial deposits

On shore			Off shore
On or under glaciers	At or beyond ice front		Deposition from floating ice
	Proximal	Distal	
Supra glacial debris Abalation till, debris flows (flow tills) and rare, poorly-sorted, water-transported gravels Englacial till Lodgement and melt-out till with striated and faceted clasts	Debris flows (flow tills), rafted till and poorly sorted muddy gravels; ice-dammed lake deposits, Gilbert deltas, and varves	Well washed gravels and sands fine down valley train which forms lake deposits in tributaries; loess; fines exported to sea or lake	Rainout till from floating ice, "till deltas" and tongues, debris flows, slumps and turbidites all interbedded with normal lacustrine/marine deposits with characteristic drop-stones and scattered clots of till

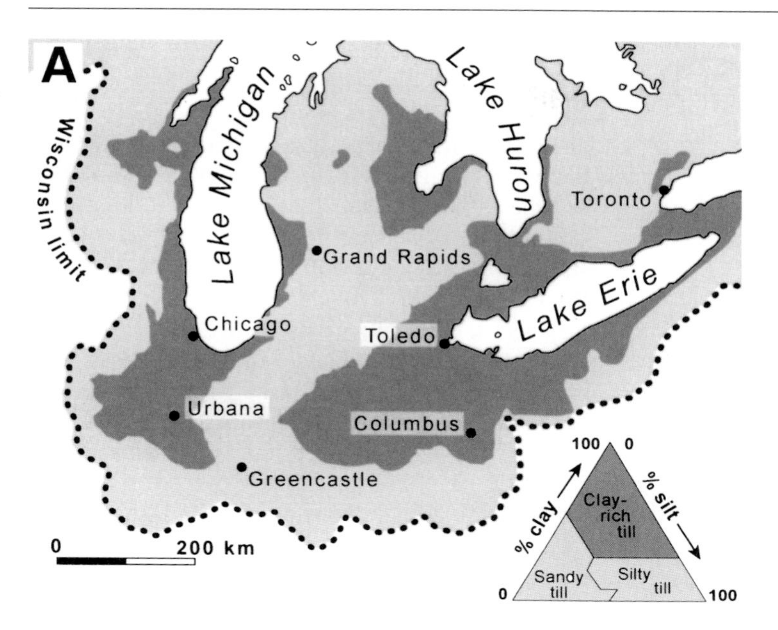

Fig. 5.12A,B. Clay-rich tills: (**A**) Wisconsin tills in the Great Lakes region of southern Canada and eastern Midwest are clay-rich close to the lakes whose scouring provided much of the clay for the tills (after Soller 1982, Fig. 5) and (**B**) details of clay-rich Wisconsin till capped by lacustrine silts on south side of Lake Erie east of Sandusky, Ohio

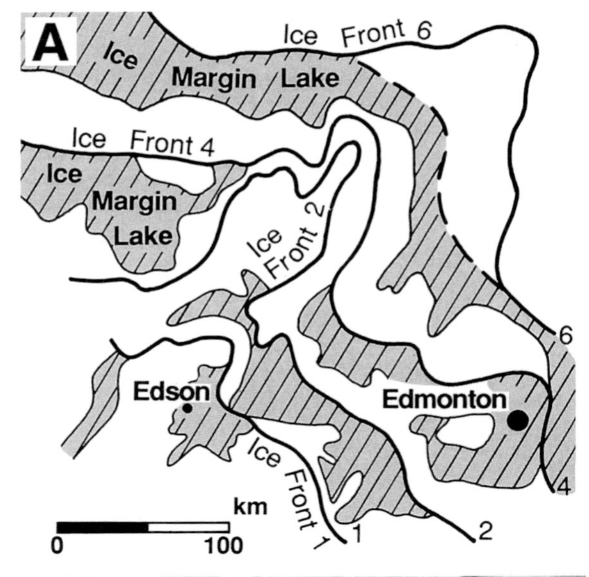

Fig. 5.13A–C. Glacial lake deposits exhibit a wide range of stratification styles: (**A**)Clay and silt-rich lake deposits formed in front of Wisconsin ice sheet as it retreated down regional slope in western Canada (Quigley 1983, Fig. 6b); (**B**)even, varve-like laminations with dropstones in lake below prodelta deposits of Wisconsinan age at Scarborough, Ontario; and (**C**)thicker laminations of silty clays higher in the prodelta sequence (Eyles and Clark 1988, Fig. 4A). Published by permission of the authors, Pergamon Press, and Geological Society of America

outlet channel (valley train) that dams tributaries, and as wind-blown silt or loess (Table 5.3). Much larger volumes are transported to the ocean where an active continental margin is glaciated (Pacific border of southern Chile, British Columbia, and Alaska) or where the glaciated hinterland of a large river drains to a passive margin or to a small sea (Mississippi River Basin and Gulf of Mexico; the Volga and the Caspian Sea). The glacial erosion of sedimentary rocks, especially easily eroded mudstones and carbonates, greatly enhances the glacial production of clay and silt, but much of the glacial fine-grained sediment may also be derived from the initial erosion of deep, long-weathered, residual preglacial soils on cratons.

Mud-rich glacial deposits onshore include lodgment and ablation (meltout) tills, mud-rich debris flows at the front or on the surface of ice, and the fine-grained deposits of temporary glacial lakes. *Lodgment tills* (Fig. 5.12) are dense, clay-rich, tough, well-compacted, unstratified deposits formed at the base of a moving mass of ice and consist of unsorted, angular boulders, cobbles, and pebbles in a gritty muddy, unstratified matrix. Abundant woody material may also be present, as well as some minor lenses of silt and sand. With more water, as occurs at the top of or at the front of a glacier, fines are removed and the deposit becomes a somewhat better sorted *ablation till*, which may pass down dip into a muddy debris flow or muddy water-transported gravel or sand proximal to the ice front. All these deposits can be difficult to distinguish from weathered colluvium, but helpful criteria include the much greater lithologic variety of clasts in tills, the presence of striated clasts in tills, and the characteristic glacial shape or faceting of its clasts. In addition, the association of glacial outwash and laminated clays with dropstones is compelling.

The northern hemisphere contains almost 90% of the world's lakes larger than 500 km^2, most of which are of glacial origin. Lake deposits formed in response to glaciation are abundant and of several types: those formed by deep glacial scouring, by morainic dams where the ice sheet retreats down regional slope (Fig. 5.13), and in tributaries blocked by melt- water-induced alluviation in a major glacial sluiceway (valley train) during deglaciation (Fig. 5.14). All are excellent traps for mud and silt and, if the blocked valley is later sufficiently well drained, excellent for agriculture. Ice contact lakes are identified by the presence of coarse-grained deltas, debris flows and deformed sediments at the ice front plus dropstones and small clots of till deposited from

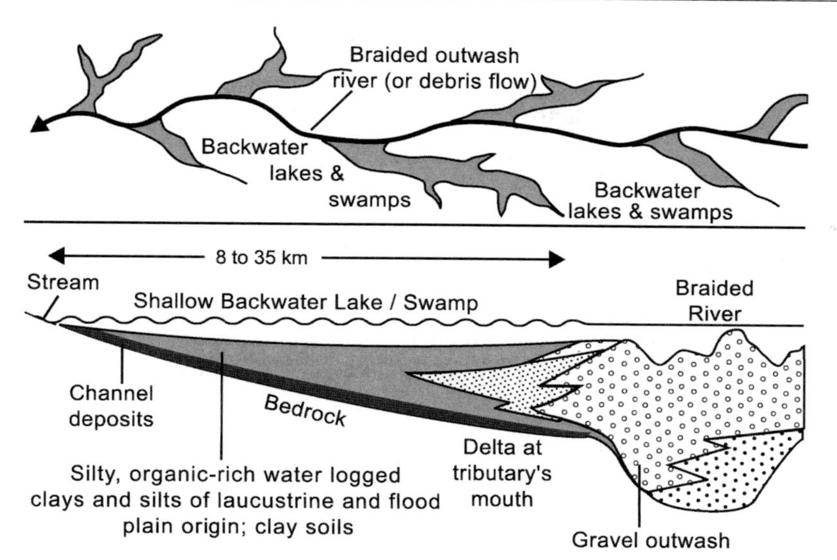

Fig. 5.14. Backwater, fine-grained lake deposits of tributary valleys occur in different geologic settings. They are common along outwash valley trains beyond glacial limits (Fraser 1994), but also may form in mountainous regions after a catastrophic debris flow or volcanic mudflow. They are also typical of many of the tributaries of the Amazon-Solimões River in Brazil. See Fraser for examles along the Wabash River between Indiana and Illinois

floating ice far from the glacier. Distal deposits are mostly composed of laminated mud and silt interbedded with thin turbidites. Where this lamination can be shown to be annual, the term *varve* is used. In contrast, lake deposits formed in tributaries dammed by glacial outwash lack dropstones and tend to have a low-energy delta at their mouth with sand and silt content decreasing up-tributary. In such deposits, plant debris, gley soils and rootlets are abundant – the result of high water tables.

Glacially influenced sequences vary depending on their proximity to the sea and to local relief. A late Wisconsinan age glacial sequence bordering the west side of Hudson Bay in Canada illustrates well the products of an ice sheet retreating into a marine basin (Fig. 5.15). From base to top we have basal till followed by ice retreat and invasion of the sea (flooding of isostatically depressed crust) and subsequently a gradual rebound, all capped by final, seaward-prograding alluvial deposits. Along mountainous,

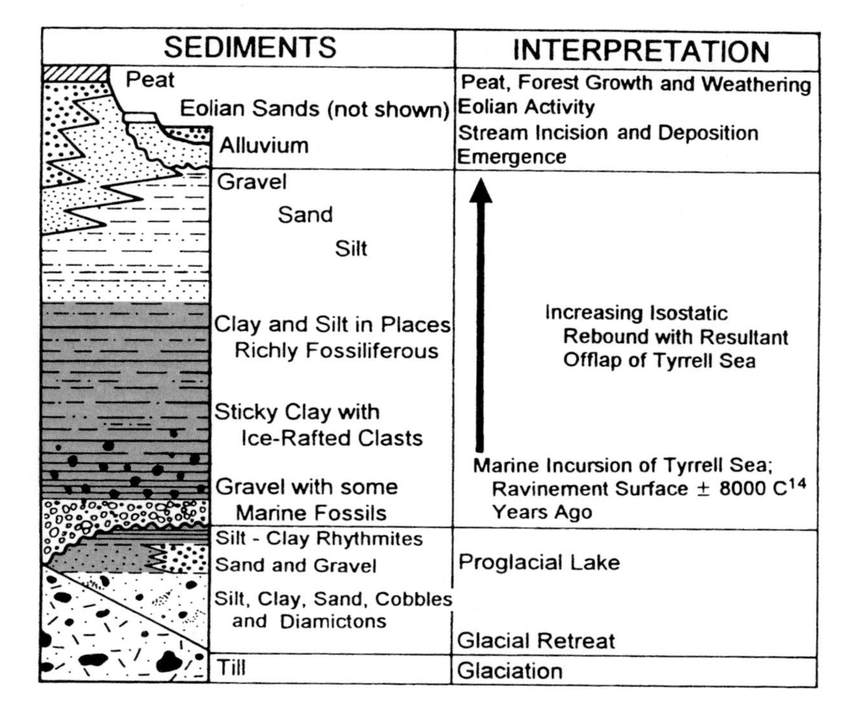

Fig. 5.15. Idealized glacial rebound sequence along the western margin of Hudson Bay in Manitoba, Canada (Skinner 1973, Fig. 20). Note coarsening-upward sequence as basal marine deposits pass upward into fluvial sands and gravels and eolian sands. Published by permission of the Geological Survey of Canada

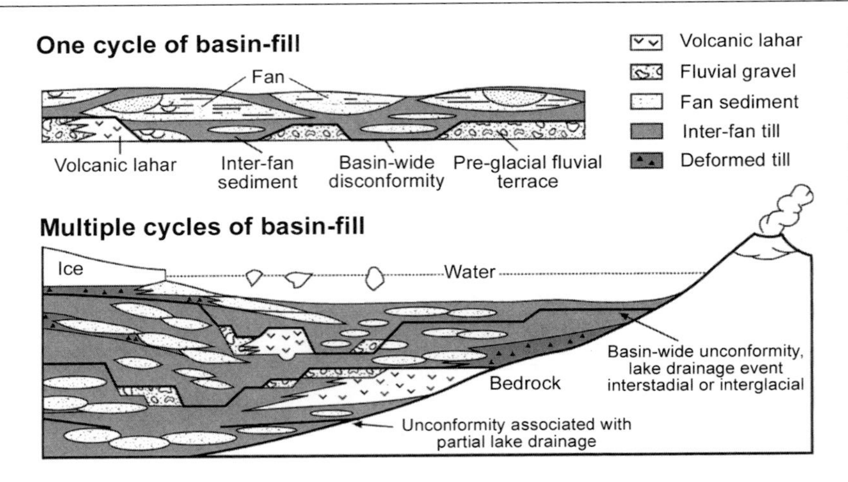

Fig. 5.16. Cycles of glacial basin filling in a deep Alaskan valley along the Copper River, one whose section includes a wide range of depositional environments. (After Bennett et al. 2002, Fig. 26). Published by permission of the authors and Pergamon Press

glaciated, active margins or the high escarpments of glaciated passive margins, thick glacial marine turbidites and mudstones with dropstones and masses of slurried till dropped from icebergs accumulate offshore in deep water (see Chap. 8). Quite different is the fill of an intermontane, glacially-filled basin of an accretionary margin, where a deep river valley was blocked by ice (Fig. 5.16). The resultant deposit is much thicker and, while it has some clays, silt and sand are more common as are slump deposits and turbidites.

5.2.4. Loess

Loess is a dominantly silt-sized (20–50 micron), weakly coherent, poorly stratified to massive homogeneous, blanket deposit of eolian origin. It may also contain scattered shells, especially gastropods, plus pollen and perhaps even some macroflora. Fine sand and clay are also common. Loess typically has a light brown or buff color (although Saharan loess is reddish), has some rootlets, is porous, and where fresh, much is calcareous. Loess may contain small calcareous concretions. Where it is thick, loess forms steep bluffs, because of good permeability. Mineralogically, the composition of loess varies with provenance so that its silt fraction may range from virtually pure quartz to a quartz-feldspar mixture to one rich in volcanic debris along active, arid continental margins, such as the Andean coast of Chile and Peru.

Two broad types of loess are recognized: *periglacial loess* (Fig. 5.17) derived from the deflation of glacial outwash valley trains and *lee dessert loess* (Fig. 2.10), deposited downwind from a desert or semiarid region. Periglacial loess thins exponentially and becomes younger downwind from its well-defined, linear riverine source. Down wind, clay content increases and, because sedimentation rates

Fig. 5.17. Glacial loess: bluff at Chester, Illinois exposes loess derived from nearby Pleistocene outwash deposits of the Mississippi River

also decline downwind, soil-forming processes leave a greater imprint on a thinner loess. Paleosols are recognized in loess by their reddish and brownish colors, leaching and greater clay content and are prized stratigraphic markers over thousands of square kilometers. Dark, organic-rich bands in loess, like those of alluvium, represent reduced sedimentation rates.

Loess is best known in Pleistocene deposits and not well reported in older deposits, where it deserves more attention, although Edwards (1979) has described a Proterozoic example. Pleistocene loess is widespread in areas of continental glaciation and downwind from deserts and occurs on all the continents except Antarctica. Maximum known thickness is 200 m in central Asia, but typically loess is less than 30 m thick and distally may be only 1 or 2 m or less. Far from its source, loess makes subtle contributions to many soils and lacustrine and marine deposits as eolian dust. See Ruhe (1984) and Lazarenko (1984) for summaries for North America and Central Asia. See Chap. 2 for more discussion on loess derived from deserts.

5.2.5. Non-glacial Lakes

These predominate over glacial lakes in the ancient, and there are many reasons to study their fine-grained deposits. After marine mudstones, lacustrine mudstones and siltstones are second in abundance. Perhaps about 20% of today's petroleum comes from lacustrine source beds. In addition, lacustrine deposits provide some of our best insights to paleoclimates (they trap a changing record of sedimentation closely related to climate). Below we briefly summarize their modern occurrences and processes, but also turn to the ancient – another example of understanding the modern through the lens of the past. We do this because such an approach provides a unity that the great variability of modern lakes obscures (Bohacs et al. 2000; Gierlowski-Kordesch and Kelts 2000 in Digging Deeper). Cohen (2003, in Digging Deeper) is also highly recommended.

Non-glacial lakes have a wide range of processes (Fig. 5.18) and include the largest and deepest lakes in the world. These deep lakes contain a great range of terrigenous environments including deltaic, coastal, shelfal, and deep-water (turbidite) environments, whereas carbonates and evaporites are more likely to occur in shallow lakes with low sediment input. Both, however, can be anoxic, especially in summer. Deep lakes typically are anoxic, and processes such as turbidity currents (Fig. 5.19) mirror well those of the marine realm despite the lack of tides, contour currents, and salinity. Many ancient lakes had long lives and produced thick deposits that make lacustrine mudstones second in volume to marine mudstones. There are also many small and ephemeral lakes in the deflation hollows of semi-arid regions, in sinkholes, and in dune-, landslide-, or volcanic-dammed rivers. Here we concentrate on deposits of the larger non-glacial lakes, because they are most likely to be preserved in the geologic record and

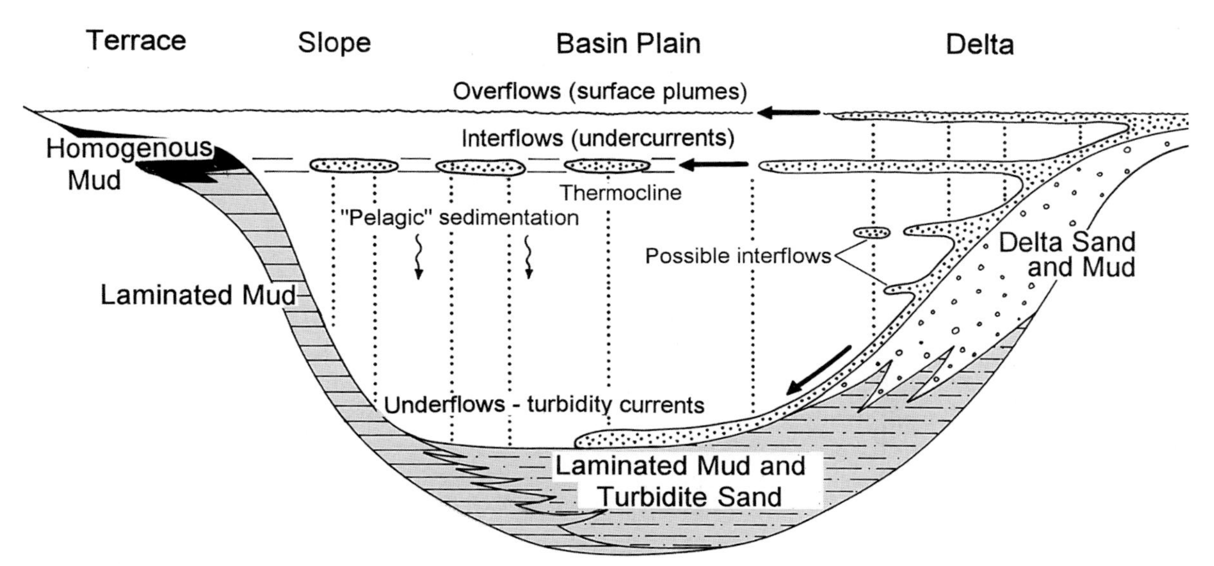

Fig. 5.18. Well-studied, deep, glacially scoured Swiss lake illustrates the key processes that occur in deep lakes, including summer time deposition of pelagic carbonates (after Sturm and Matter 1978, Fig. 10). Published by permission of the authors and the International Association of Sedimentologists

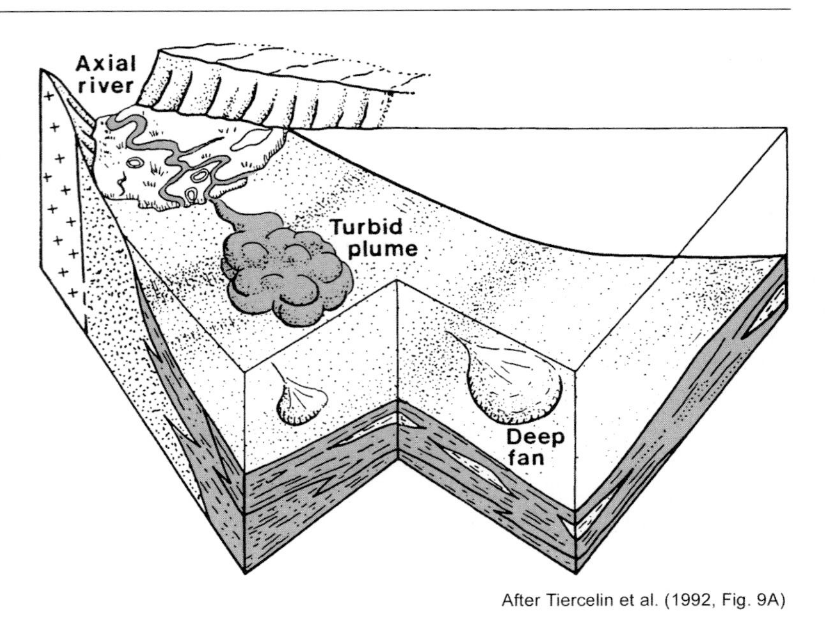

Fig. 5.19. Schematic of turbidite plume of small river that supplies mud, silt and sand to Lake Tanganyika in Africa (after Tiercelin et al. 1992, Fig. 9.A). Published by permission of the authors and Innovative Interfaces, Inc.

are all-important as source beds for petroleum. Because many such lakes are long lived and quickly respond to rainfall changes, their fine-grained sediments contain some of the best records of continental paleoclimates.

Lake deposits occur in a variety of settings, but are most common in rift, intramontane, and foreland basins. Ancient lacustrine deposits of rifts and transtensional basins with rapid subsidence are especially likely to be both thick and well preserved. Examples of major lakes in rift valleys include Lakes Malawi and Tanganyika in Africa, the Dead Sea of Israel and Jordan, and Lake Baikal in Siberia. A striking pull-apart lake is Lake Izabal in Guatemala. Large lakes with thick deposits are also common within orogenic belts; for example Lake Valencia in Venezuela occupies an intramontane low. In this setting, tectonic movements produce a rapid reorganization of drainage that outpaces the erosive power of streams (Dickinson et al. 1988). Lakes and swamps are common in many foreland basins, when subsidence exceeds sedimentation and the sea cannot enter; modern examples include eastern Bolivia, Paraguay and Mar Chiquita and others in Argentina and in the wide foreland basins in front of the great suture along the south side of the "stans" of central Asia. Many of the lakes of foreland and intermontane basins are likely to be shallow and some may have carbonates, whereas those in rifts may be either shallow or deep.

The ratio of basin subsidence to sediment plus water input is the fundamental control on the deposits of lakes both modern and ancient. Why are these factors so important? There are two reasons. First, the space available for deposition (accommodation) determines water depth (a big factor in oxygen deficiency), depositional environments, and facies. Secondly, the flux of water through the lake determines its biota and salinity (remembering that dense saline bottom water seals the bottom from oxygen). Here we note that accommodation is mostly the result of tectonics, whereas it is climate (rainfall and evaporation) that controls the flux of water into the lake and evaporation from it and also affects transport of sediment into the lake.

Three types are recognized (Fig. 5.20 and Table 5.4). One end-member is an *overfilled* lake, which has a persistent outflow because the inflow to it exceeds the amount of evaporation from its surface; the intermediate case is a *balanced filled* lake, which is broadly similar to a shallow marine environment with mixed carbonate siliciclastic facies and possibly some evaporite interbeds; and the opposite end-member is an *underfilled* lake, which contains a variety of evaporites and mudflats interbedded with sheet-like fluvial deposits and even dunes. Lake Baikal is an example of a modern overfilled lake (its outlet is the 5,550 km long Yenisey River), Lakes Tanganyika and Victoria are balanced-filled lakes and the Dead Sea and Lake Chad in the Middle East and Africa are examples of modern underfilled lakes. We note here that when accommodation is small, the lake will be shallow and its deposits are likely to be widespread, fine-grained and consist of many thin sequences, but when accommodation is great, thick

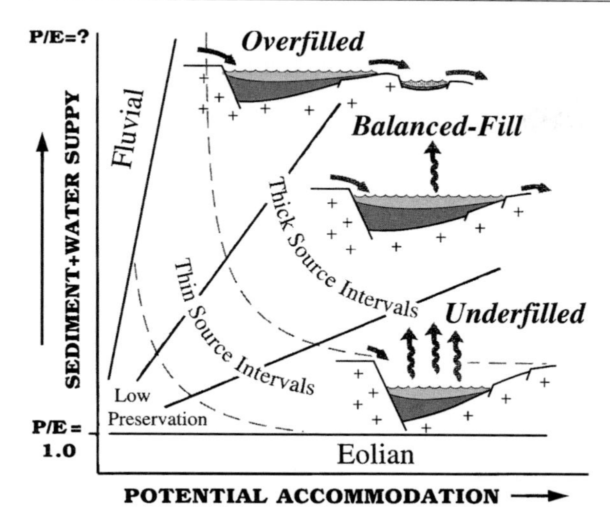

Fig. 5.20. Overfilled, balanced, and underfilled lakes depend on the proportions of sediment supply, water supply, and basin subsidence (after Bohacs et al. 2000, Fig. 7). See Table 5.4 for details. Published by permission of the authors and the American Association of Petroleum Geologists

sections with mass movements and turbidites are the rule, because water is deep. Deep or shallow, however, when lake waters are stratified, anoxia is likely and dark, organic rich muds are deposited. In seasonal climates where surface productivity is greatest in summer, millimeter-scale pelagic carbonates may be interlaminated with dark, organic rich muds. During its life span, a lake may oscillate between the end members of overfilled and underfilled as rainfall, terrigenous supply, and subsidence vary.

The overfilled lake, on the one hand, is mostly filled with terrigenous debris and has relatively few biological or chemical deposits. An underfilled lake, by contrast, is likely to have far more chemical deposits and carbonates, some of which form on sabkha-like mudflats. Oscillations of lake levels and shoreline are also more frequent in underfilled than in overfilled lakes. Consequently, vertical alternation of contrasting facies – dolostones, algal mats, evaporites (or pseudomorphs and crystal molds of evaporite minerals in muds) interbedded with mudstones and siltstones – occur far more commonly in underfilled lakes. Underfilled lakes, therefore, are likely to have more and better defined climatically controlled cyclic deposits and more subaerial breaks in their deposits than either balanced or overfilled lakes. Ideally, transitions to mud should be more abrupt in saline lakes, because of greater salinity (more intense flocculation) than in open lakes. In addition to the evaporitic carbonates found in lakes of arid and semi-arid regions, freshwater lakes in temperate and tropical climates commonly have thin, calcitic laminations formed by upwelling (overturning) of cold, nutrient-rich water into the photic zone producing a skeletal rain of ooze on the bottom at millimetric scales. See Bohacs et al. (2000) for full discussion of overfilled, balanced and underfilled lakes including their contrasting hydrocarbon potentials.

Table 5.4. Characteristics of end member lakes. After Bohacs et al. (2000, Tables 1 and 2). Published by permission of the authors and the American Association of Petroleum Geologists

Types	Stratigraphy and Facies	Lithology and Organics
Overfilled	Mostly progradational poorly defined sequences. High fluvial input.	Mudstone, marl, coquinas coal and coaly shale, etc. Fresh water biota, low to moderate TOC, plus kerogen Types I and II. Oil and gas generator.
Balanced Fill	Mixed progradational and aggradational; distinctly expressed sequences.	Marl, mudstone, kerogenite carbonate, etc.; salinity tolerant biota, low to moderate TOC. Mostly Type I kerogen and oil prone.
Underfilled	Principally aggraditional, high-frequency, thin, wet-to-dry cycles. Low fluvial input.	Mudstone, kerogenite, evaporite, carbonate, etc. Low biotic diversity, dominantly low TOC and Type I kerogen. Moderate to high sulfur oil.

Fig. 5.21. Hyper-arid, ephemeral lake in Death Valley, California (Hunt and Mabey 1966, Fig. 3)

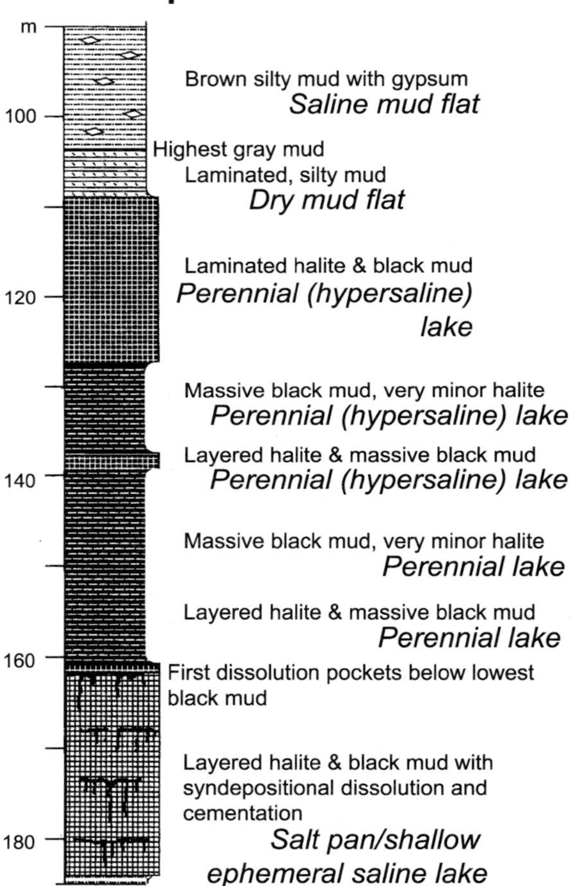

Fig. 5.22. Massive black muds interlayered with halite in a core from Badwater Basin of Death Valley (after Roberts et al. 1994, Fig. 4). Black muds in core quickly turn green upon exposure to air. Change in color is most likely the result of oxidation of iron monosulfide. Published by permission of the authors and the Society for Sedimentary Geology

Below we present three contrasting case histories that illustrate some of the above principles and the great variations of lacustrine deposits.

Lake Tanganyika in the East African Rift System (Fig. 5.19) is the seventh largest lake in the world, and of all the rift lakes, is the second largest and also the deepest (1,471 m). It is well studied (Tiercelin et al. 1992) and, because of its size and great depth, has sedimentary environments ranging from deltas and fan deltas and even some narrow carbonate platforms to slope and deep-water turbidites. Because of its warm climate near the equator, the lake is rich in nutrients and therefore has a high productivity. Combine high surface productivity with steep sides and great depth, and it is not surprising that waters are anoxic below 100 to 200 m and the sediments have up to 12% organic matter (TOC). Lake Tanganyika has northern and southern parts, each a half graben, separated by a mid-lake high, that produces steep slopes and central deep basins. Marginal deposits include deltas at the ends of the lake and fan deltas along its sides as well as some shallow-water deposits and, on its steeper slopes, channel and canyon systems with turbidites fed by cold, high-suspension streams sourced in rift-shoulder highlands. The largest rivers follow the axis of the rift and form the axial deltas at the ends of the lake. On the slopes and in the central parts of the lake, there is a widespread blanket of fine-grained organic-rich mud of three types: (1) interstratified diatom and clay laminations, (2) homogeneous green to black clay containing some minor silt as well as abundant plant debris, sponge spicules, and diatoms (the chief contributor), and (3) homogeneous muds with fecal pellets (flaky muds). Total organic matter in this organic facies range up to 12% and Rock-eval hydrogen index values reach 600 (Appendix A.4). Thus in this deep rift basin, there are potential source rocks close to bordering reservoirs, ideal conditions for large petroleum fields. Hence this rift fill provides a good model for ancient deep rift basins that developed in the warm climates that prevailed in much of the geologic past.

Far different are the late Pleistocene sediments underlying the 8 × 15 km salt pan of Badwater Basin in Death Valley (Figs. 5.21 and 5.22), a fault-bounded intermontane basin in southern California (Roberts et al. 1994). This basin is the lowest point in the western hemisphere and has an annual average rainfall of only 4.1 cm. It has recorded temperatures up to 120 °F (49 °C) – the driest and hottest place in the United States. See Hunt and Mabey (1966) for geomorphic and tectonic background. In the lower half of a 185-m long core taken in this basin, four facies were recognized: (1) perennial lake, comprising massive black mud and layered halite, the latter precipitated from the water column; (2) salt pan/ephemeral lake, marked by halite with vertical dissolution structures; (3) saline mudflat made up of brown silty mud with displacive sulfate salts; and (4) dry mudflat with laminated, desiccated silty mud. All of these muds contain some carbonate, but differ in significant ways from one another in the four facies. The black muds of the perennial lake are dense and massive with rare wavy laminations, but have a fine-scale pelloidal texture. This black mud has very low organic matter (TOC is only 2%) and turns greenish-gray when exposed to air as its iron monosulfide is oxidized – a reminder that it is not always TOC that makes muds and shales black. The salt pan/ephemeral lake muds fill dissolution cavities and vugs, vary in color from brown to red to green to gray and from massive to mottled to well-laminated and are associated with halite that is mostly well bedded. Muds of the saline mudflats are mottled, poorly stratified, mud-cracked, contain scattered gypsum crystals and nodules, and range in color from brown to red to gray and green. Muds of the dry mudflats have fairly well-defined laminations of clay, carbonate, silt and fine sand, are mud cracked and scoured, and are variable in color (light gray to gray-green to white and black). See Eugster and Hardie (1978) for details about the petrology and chemistry of these evaporate- associated muds. In sum, the core at Badwater Flats provides a good model of an underfilled, hyper-arid lake deposit, whose interbedded halites were both precipitated in its water column and later within the mud.

The Newark Supergroup (Smoot and Olsen, 1988) is Early Mesozoic in age and extends over 5,000 km along the eastern side of North America, where it is preserved in 20 isolated onshore grabens and more offshore (Fig. 5.23). Some of these basins persisted as long as 30 Ma. Thickness ranges up to 9,000 m and mudstones, which are bounded by sandy fans and deltas, are a major part of the fill. Among the fossils are whole fish and fish debris, complete and fragmented reptiles and their footprints, plus clams, ostracodes, gastropods, pollen, spores and plants.

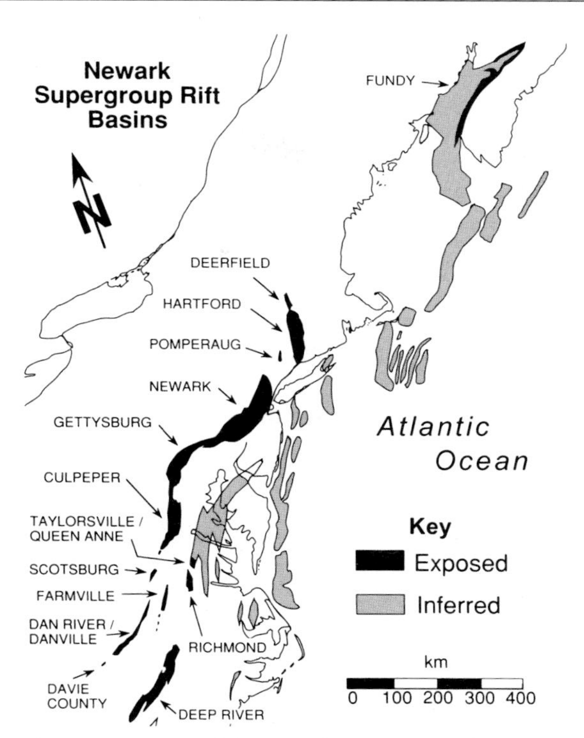

Fig. 5.23. Triassic rift basins of eastern United States and Canada contain thick fills and many lake deposits (Olsen et al. 1990, Fig. 1). Milankovitch cycles are common in these deposits (see Fig. 8.36). Published by permission of the authors and the American Association of Petroleum Geologists

These Triassic basins are of interest because of the regional variations in their mudstones, the Milankovitch cyclicity of their sedimentation patterns, and the great area over which these rift lakes had broadly similar sedimentation, even though most basins were isolated and far apart. Finely-laminated black shales are present, but much more abundant are red and gray mudstones. These are, however, unequally developed – in New England and Nova Scotia only mud-cracked, red massive mudstones with scattered thin lenses of sand occur, whereas burrowed and rooted gray mudstones are dominant in the south, suggesting a wetter climate. Cycles are recognized as follows: shallow water (thin bedded, silty mudstone with desiccation cracks and wave-rippled sandstone), deeper water (finely laminated organic-rich black shale and thin-bedded gray mudstone), and shallow water (massive mudstone plus some with root casts and some sandstones). Salt casts and dolostones are also present in minor amounts and were formed during closed lake phases. These help identify Milankovitch cycles, which are typically one to 15 m or more thick. Less rainfall produced contracting, closed saline lakes whereas more rainfall produced expanding, open freshwater lakes.

5.3. Transitional Environments

The transitional fine-grained environments include low-energy deltas of several types. Fine-grained sediments of these deltas grade into shelfal deposits and into diverse coastline deposits. Many estuaries also have abundant fine-grained deposits. Muddy coastal deposits are best developed in small lakes, seas, and protected gulfs, where waves and coastal currents are weaker than those of an open ocean (with some notable exceptions). And, of course, without a supply of fines to the coast, fine-grained deposits will be absent.

5.3.1. Deltas

Rivers are the source of virtually all marine and lacustrine muds and thus the delta is a key starting point for transitional environments. The delta is here defined as a *distinct accretionary deposit, both subaerial and subaqueous, at the mouth of a river*. Of the two, the subaerial part of a delta is commonly the most variable, because it responds more readily to variations in both sea level and climate; i. e., to changes in the depth of the water table, changes in relative sea level, and to the local ratio of precipitation to evaporation. It is these deltas (and their commonly associated submarine fans) that form the depocenters of many important ancient marine mudstones. There are three well-defined end member types – *river-dominated, tide-dominated, and wave-dominated deltas* (Figs. 5.24). The type of delta depends on the relative magnitude of fluvial input versus reworking by coastal currents, inshore wave power and tidal range. The more energetic the body of water, the less likely is the delta to be river dominated and mud rich. Mud-rich deltas are also called *low-energy* deltas and sand- and gravel-rich deltas *high energy*. Mud-rich deltas are most common in lakes, gulfs, seas, and small oceans both today and in the past. They are especially well developed along the margins of ancient epeiric seas (restricted inshore

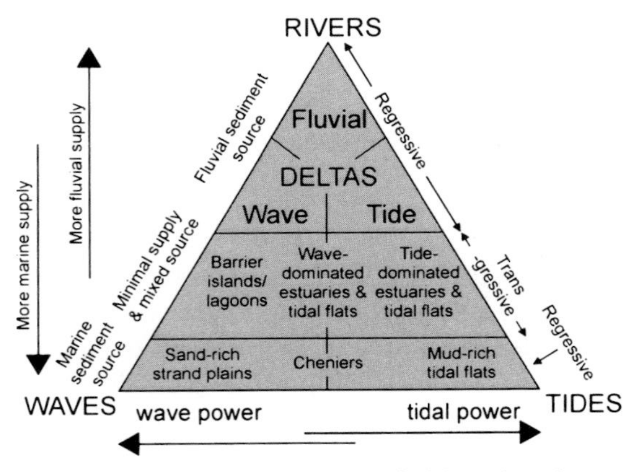

Fig. 5.24. Three end members of deltas (Reading and Collinson 1996, Fig. 6.17). Published by permission of the authors and Blackwell Science

wave power) rather than along large open oceans, ancient or modern. Where tidal range, coastal currents, and inshore wave power are insufficient to winnow silt and mud, a prograding, bifurcating, mud-rich, Mississippi-type delta forms. Where tidal range is high, the delta, rather than protruding seaward, is a funnel-shaped estuary with sandy elongate bars. Most of the fines in a tide-dominated delta are deposited in front of the estuary proper or carried along the coast. And where inshore wave power or coastal currents are high, the wave-dominated delta consists of coalesced sandy beach ridges that extend but little from the coastline. Mud and silt are transported away from the mouth along the shelf or across it beyond the shelf edge into deep water to supply a submarine fan (deltas of the Niger and the São Francisco Rivers).

In addition to the above factors, we need to add three others. The first is the size of the river: the longer the river, the lower is its gradient near its mouth and the finer are its deposits (Orton and Reading 1993, Fig. 3). The second is the synchroneity of its flood stage with stormy periods. Should these coincide, mud can be transported far from the river's mouth by wave-generated, high suspension bottom currents, which reduces the importance of the visible surface plume (Bentley 2002). The contrast in density between the river plume and the water of the receiving basin is also a factor. Cold, dense, muddy river water sinks to the bottom readily in a warm lake, whereas it is more likely to mix effectively where the receiving water is of similar density. On the other hand dense saline water in the basin will keep the delta plume at the surface.

Thus four factors need consideration for major muddy and silty rivers – *the energy regime of the coastline, the size of the delivery system to it, the*

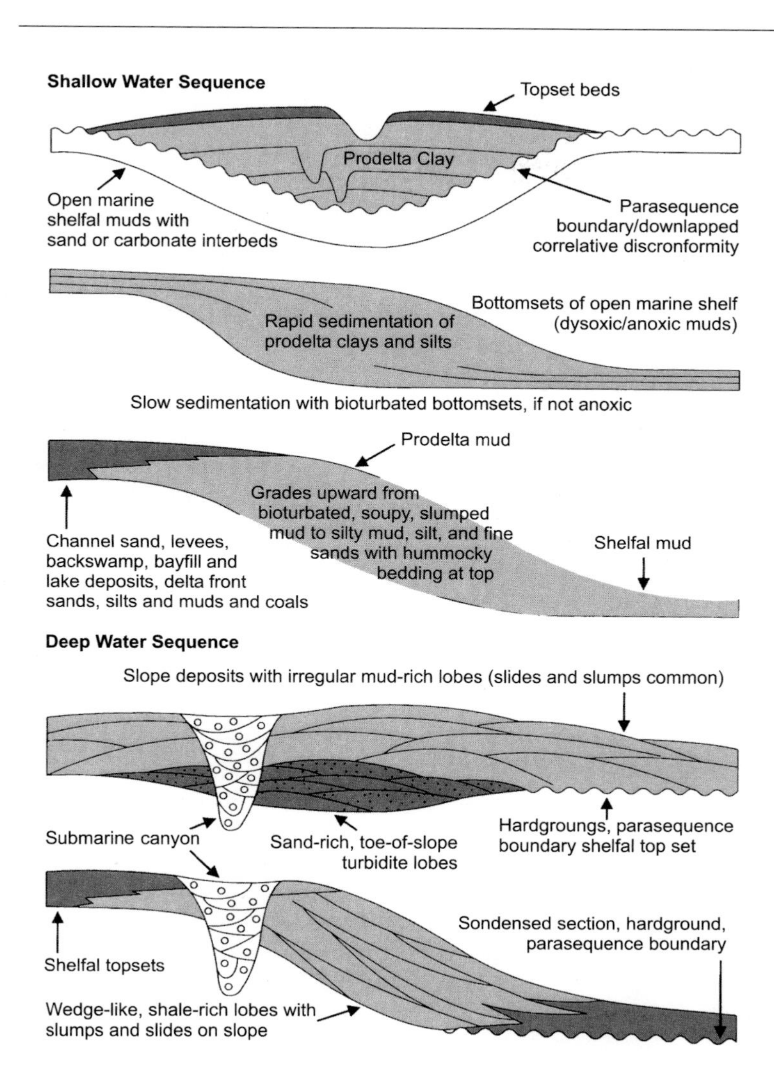

Fig. 5.25. Broad similarities between shallow water deltas and deep sea fans. Parasequence boundary marks the erosional-hiatal contact of one sedimentary cycle with another

water stratification of the basin, and the timing of flood stage and stormy periods. As we will see later in Chap. 8, there is yet another, the geometry of the receiving basin. Terminology has long been standard for deltaic and coastal deposits (Fig. 5.25).

Good examples of large mud- and silt-rich modern deltas are the Lena, Chao Phraya, Orinoco, Volga and Mississippi; good examples of silt-rich deltas are the Yangtze (Changjiang), Yellow (Huang He) and Brahmaputra-Ganges. Where rivers are large, such as the Mississippi and the Congo, there is likely to be a coupled submarine canyon so that there is, in reality, an *integrated fluvial-marine mud/silt delivery system* that extends from the alluvial valley far into deepest water (Fig. 5.26). At low stands of sea level, the river will extend to the shelf edge where its deposits can become over-steepened and fail and thus contribute sediment directly to an associated canyon and subsea fan. Or silts and muds may be directly delivered to the slope and beyond. Thus at lowstands the river, canyon, and submarine fans are all closely coupled, but at highstands, the delta is well inboard, sediment is trapped on the inner shelf (or far up dip on the craton) and both canyon and submarine fan are largely inactive and disconnected from the river.

Consider now a river-dominated delta with a river heavily charged with mud and silt in suspension (Fig. 5.27). Some mud and silt will be trapped in the subaerial parts of the river-dominated delta in lakes, swamps, abandoned distributaries, and as interdistributary bay fills. The rest moves to the delta-front environment near the mouth or farther off shore to its medial and distal clinothem as prodelta deposits. Turbid plumes deposit much of the prodelta deposits (Fig. 5.28). Tidal deltas also trap some of their rivers' load of mud, which is deposited at slack water away

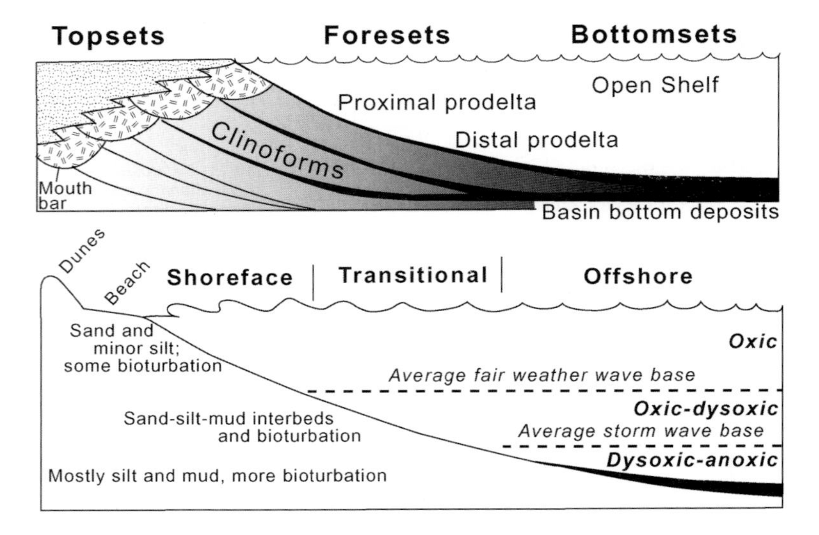

Fig. 5.26. Facies distribution and summary of processes on an idealized clinoform

Fig. 5.27. Bifurcation of Mississippi River (flow from upper right to lower left) at Pass á Loutre, Plaquemines Parish, Louisiana (Hrabar and Potter 1967, Fig. 10). Note straight channels and abundance of lakes separated by swamps. At the bifurcation the river is only about one meter above mean sea level. Published by permission of the authors and the American Association of Petroleum Geologists

Fig. 5.28. Turbid plume of fresh water of Rio Parnaiba, where it meets the south Atlantic Ocean, Piauí, Brasil. Plume curves to left, because of powerful, northwest-flowing South Equatorial Current

from the main tidal channels – stored on bordering tidal mud flats or at special dead water nodes – while the rest is exported seaward, down current. In contrast, in the wave-dominated delta all the mud and silts are exported far along the coast or across the shelf and directly into deep water, while the sand is concentrated along the shore in a series of closely spaced beach ridges.

The river-dominated, mud-rich delta has many diverse environments of mud deposition, although prodelta foresets and bay fills are major traps. Such a delta forms a classic clinoform with *top-*, *fore-* and *bottom sets* (Fig. 5.25), which, when thick enough, are easily seen on seismic images. In a river-dominated delta, mud accumulates in much the same way as it does in a muddy submarine fan between inter-distributary channels and on the distal part of the fan, the fan fringe. From this point of view, the overall architectures of a mud-rich submarine fan and a delta have much in common. For both, the greater the accommodation, the thicker the individual cycles of the delta and its total thickness.

Let's first think of the topset part of a mud-rich delta. Here, beyond its levee-bounded distributaries, are inter-distributary bays that trap silt and mud. These fines pass upward from shallow marine inshore shelfal muds with shells, bioturbation, and perhaps thin hummocky cross stratification, into siltier brackish and finally freshwater muds rich in terrigenous organic matter and plant rootlets as swamps replace open bays in wet climates. In this transition more and more rippled and laminated silts and fine sands are deposited.

In the distal subaerial part of the delta, crevasse splays play the key role. As a crevasse splay extends farther into a bay, its channel progressively bifurcates, currents weaken, fines fill an ever-wider area of the bay, and deposit alluvial mud and silt over brackish and marine deposits. In broad terms, the processes of wave and storm action and marine bioturbation on a shallow inshore shelf are gradually replaced first by flood-related suspension deposits as the splay first develops followed by increasing proportions of traction-transported fine silts and sands. In so doing, a marine fauna is replaced upwards by first a brackish and then a freshwater fauna. In a delta with a high subsidence rate, this infill stops when a distributary is abandoned and the sea re-occupies the bay. Thus bay fill muds range from marine to brackish to fresh water and, along with peats and some thin splay-deposited sands, form a strongly *cyclic sequence relatable to relative sea level* (Table 5.5). Within such cycles there may be

Table 5.5. Relative sea level and deltas (after Coleman and Roberts 1988, Fig. 32

Relative Sea Level Curve	Lower Alluvial Valley	Delta Plain	Continental Shelf	Continental Slope	Deep Basin
HIGHSTAND	Overbanking of back-swamp muds; meander belts	Delta switching along shore; maximum inshore exposition of brackish muds	Muds mostly on inner shelf; condensed section on middle and out shelf	Minimal mud deposition with only hemipelagics; condensed section	Thin pelagics or hemipelagic muds only – "starved basin" and condensed section
		← TSE	← TSE		
Rising Sea Level – Transgression	Meandering streams replace braided streams	Deltas migrate updip and inshore; pebble lags on TSE	Rapid infill of channels followed by peats and transgressive, anoxic muds	Mud deposition reduced as deltas migrate landward	Hemipelagic muds dominent
	LSE				
LOWSTAND	Weathered surface with roots, kaolinite, silica and siderite; coarse lags in channels; fines to deep sea	Weathered surface with paleosois incipient to mature; channels export fines seaward	Muds carried directly by streams and deltas to middle/outer shelf; erosion on inner shelf?	Slope aprons and fans have declining activity; more hemipelagic muds	Fans and deep basins have declining activity; more hemipelagic muds
Regression →	Rapid and deep entrenchment	Regressive deltas move seaward; channel deepening upstream	Channels/canyons cut across shelf; fines carried beyond shelf edge	Canyons erode inward and are active conduits; deposition on slope and beyond	Maximum activity of fan systems on slope and aprons
HIGHSTAND	Same as at top				

special event beds formed by either a large flood or a major storm. Such beds tend to be both distinctive and widespread and thus good local markers. A storm event typically has a scoured base, lags, and current bedding. The lateral continuity of bay fill deposits is interrupted by the random presence of sand-filled channels formed as distributaries switched.

Such topset sequences tend to thicken downdip and prograde over earlier clinoform deposits. Where little sand is delivered to the coast, only silty muds and silts mark the shoreline, although the same clinoform structure prevails. At any time, such a progradational sequence may be interrupted either by delta switching (abandonment) or by a relative rise in sea level. With a relative rise of sea level, transgressive marine shelf mudstones blanket the underlying deltaic system and thus provide effective seals for the sandy reservoirs of distributary channels, splays and bar mouth sands.

The principal site of mud-silt deposition in low-energy deltas is, however, in the basin in front of the distributaries, where prodelta muds accumulate. Floods produce the plumes that deposit these muds. These plumes of suspended sediment are buoyant and occur at the surface (*hypopycnal*), or at intermediate depths (*interflows*), or else are denser than seawater and sink to the bottom (*hyperpycnal*) to flow along it as density-driven underflows (Fig. 5.28). The submerged plumes can be either channelized or blanket-like on the delta front. Muds deposited in this environment are prone to resuspension by storms waves and tidal currents. Furthermore, rapid deposition (producing high pore water pressures in the muds), and reduced mud density due to methane gas all favor slope failures and resedimentation.

The most distal prodelta mud accumulates slowly near the outer limits of the delta front. Here the plume of the river mouth is thin, intermittent and dilute and produces only a thin deposit that downlaps onto normal marine shelfal muds or a condensed, possibly shell-rich, interval. The resulting deposit is thin and bioturbated so that a mottled, homogeneous, somewhat silty claystone possibly containing scattered marine fossils results, and there may even be thin carbonates near the limits of such plumes. Minor storm beds may be present with hummocky cross lamination, but in the absence of much silt or sand, can be hard to recognize.

Viewed as a vertical sequence, the overall structure of the prodelta begins with a thin, bioturbated, shelly mud at the base. Closer to the source and higher on the prodelta slope, sedimentation rates are higher, thickness of the prodelta mud increases, and silt is more common both as scattered grains in the mud and as individual laminations. Some of these are cross-laminated and have hummocky interbeds produced by storm events. Higher still on the delta front, rates of sedimentation increase, muds are siltier, silt is coarser, fine sand is more abundant and cross-lamination, convolute lamination, hummocky cross lamination, slumps and small-scale cut and fill structures prevail as bioturbation decreases. In both marine basins and saline lakes, rapid sedimentation of mud on the upper delta foreset is enhanced by flocculation. And at the very top of the clinoform a distributary-mouth sand (commonly silty to fine grained) develops. Although the entire sequence fines seaward and coarsens upward, the prodelta sequence may contain thin, fining upward parasequences, probably related to decaying flood pulses, that interfinger with sea-level induced interbeds of marine shelfal mudstones.

The instability of the top of the delta front of a large muddy delta debouching into deep water is a key factor in the subsequent transport of mud into deeper water. Slides and slumps, caused by oversteepening of the upper slope and excess pore water pressure (rapid sedimentation or methane gas generated from the decay of organic matter) transport mud into deeper water as either slope failures or far-traveling turbidity currents (Fig. 3.8). Failures occur on slopes as low as 0.2 to 0.5 degrees. Cyclic loading by storms commonly triggers such events as does a rapid fall in lake or sea level or an earthquake. Still another consequence of the low strength of delta front muds are *mudlumps* in the vicinity of distributary mouth bars, where sand sinks into soft mud and forces it upward as a diapir (Figs. 6.8 and 6.9). Where a low energy delta progrades to an abrupt shelf edge or steep ramp, perhaps even a continental margin, resedimentation is dominant (mass movements and turbidites) resulting from rapid sedimentation on a steep, unstable slope and forward growth of the delta is limited; conversely, in the interior of a craton far from a continental margin forward growth of deltas is great and mass movements and turbidites are much less common, and the delta migrates far up and down regional slope.

5.3.2. Estuarine Deltas and Fjords

Drowned river mouths – estuaries – contain funnel-shaped deltas that develop in flooded valleys. The limits of their fill are placed at the landward end of

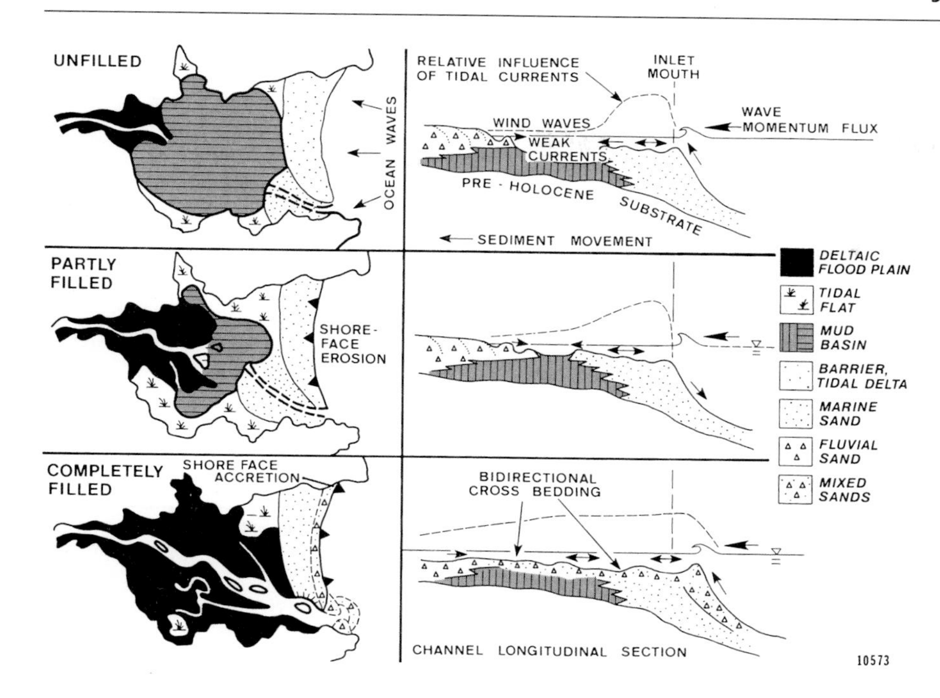

Fig. 5.29. Holocene wave-dominated estuaries in Australia, each with a coastal barrier at its mouth (Roy et al. 1980, Fig. 5). Such barriers create ideal mud traps even along high-energy coasts. Published by permission of the authors and Elsevier Science

the tidal facies and the seaward end of the coastal facies (Dalrymple et al. 1991, p 1131). These features range in size from the estuary of the Amazon (160 wide ×300 km or more long) to small ones only a few kilometers in width. There are two end members: tide dominated and the wave dominated. The tide-dominated, estuarine delta consists chiefly of elongate sand bars, some of which are islands capped by mud, and virtually all of the fines in the estuary are exported seaward. Some fines may accumulate as a submarine delta in front of the estuary mouth analogous to the distributary mouth bar of the river-dominated delta, some may be carried far down current along the coast, and some may be carried across the shelf.

The other end member, the wave-dominated estuarine delta, has a coastal barrier at its mouth that traps significant fines behind it (Fig. 5.29). And in both, muds are trapped in coastal swamps or mud flats along the sides of the estuary (Fig. 5.30). In sum, it is the ratio of supply of fines to the intensity and kind of currents that determines estuarine

Fig. 5.30. Estuarine mud flat on a tide-dominated coast (tidal range 16 m) at Rio Gallegos, southern Patagonia, Argentina, another mode of mud deposition under high-energy conditions. Published by permission of the authors and Elsevier Science

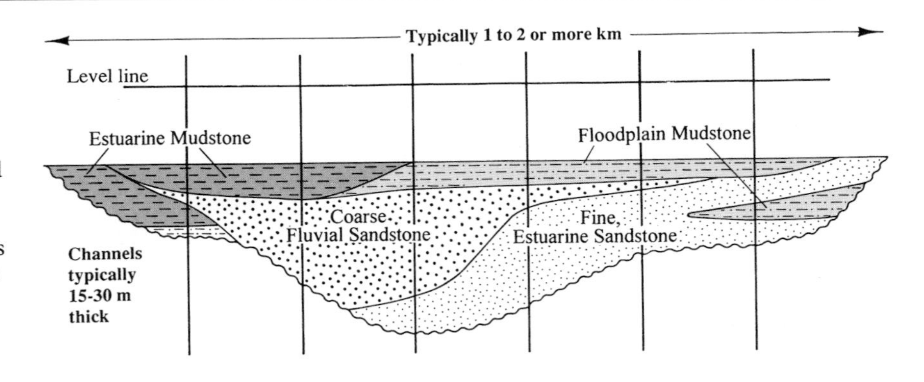

Fig. 5.31. Cross-section of a Pennsylvanian estuary in the Midcontinent of the United States, whose fill contains much estuarine fine sandstone, siltstone and mudstone (after Krystinik and Blakeney 1990, Fig. 5). See Table 5.6 for the varieties of mudstones recognized in this estuary. Published by permission of the authors and the Rocky Mountain Association of Geologists

Table 5.6. Fine-grained facies in a Pennsylvanian estuarine fill (Adopted from Krystinik and Blakeney 1990)

Characteristics	Flood Plain	Estuary	Marine
Color	Green to olive drab and rarely reddish; dark greenish gray, where gradational to estuarine deposits.	Black to dark gray	Black
Petrology/ structures	Silty to sandy, rooted, carbonized plant remains common; waxy texture. Poor to fair lamination; coarse to fine terrigenous organic matter	Silty to sandy, micaeous, platy to fissle with finely disseminated organic debris; glauconite and burrows common. Poor lamination	Platy to fissle and strongly pyritic. Sideritic laminations, some fossils. Rare glauconite. Organic matter, 0.5 to 3.0%, is mostly terrigenous.
Associated lithologies/ environments	Fine-grained interbeds of cross laminated siltstones and sandstones deposited as overbank splays. Thin coals in ponds	Siltstones with sea-ward dipping ripples, some with clay drapes. Minor tidal creeks and muddy estuary	Partly filled valley forms linear, poorly ventilated, low area on bottom of shallow, epeiric, strongly stratified sea

morphology and where its mud is deposited. Weak tidal periodicities (typically thin interlaminations of clay and silt) have also been found in some outer estuarine deposits. A Pennsylvanian paleovalley fill in the midcontinent of the United States illustrates well the contrasts between floodplain, estuarine and capping marine mudstones as a fluvial channel was submerged and filled during transgression by the sea (Fig. 5.31 and Table 5.6).

Although of glacial origin and generally much deeper, fjords also have wave, tidal, and fluvial subtypes (Syvitski 1986). They accumulate turbidites and mass flow deposits in dysoxic to anoxic environments, because they are deep, narrow, and generally silled. See Chap. 8 for one probable ancient analogue to the fill of modern fjords.

5.3.3. Coastlines

Mud accumulates in bays and in lagoons behind barriers on coastlines where both tidal currents and inshore wave power are minimal. It also accumulates on open coastlines where the supply of suspended sediment is sufficient to dampen inshore wave power. Such "open coast" mud deposits are mostly likely

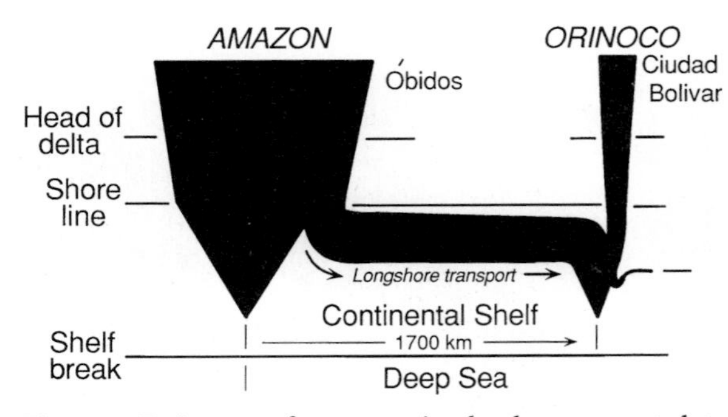

Fig. 5.32. Estimates of comparative loads transported to the Atlantic Ocean by the Amazon and Orinoco Rivers of northern South America and along the coast between them (Meade 1994, Fig. 8). Published by permission of the author and Pergamon Press

Table 5.7. Muddy tidal flat, Sonora (after Meckel 1975)

Subtidal	Intertidal	Supratidal	Evaporite Pan
Lithology and Thickness			
Dominantly silty clay, brownish gray; some thin partings of silt and very fine sand; 20 ft. (6.1 m)	Chiefly silty clay with thin, irregular to lenticular laminae of silt and very fine sand; well sorted, micaceous; 30 ft. (9.1 m)	Mostly silty clay; thin crusts and pockets of halite and gypsum common; virtually no admixed terrigenous sand; variable thickness	Relatively pure evaporates; chiefly halite with some gypsum at edge of pan; 3 ft. (0.91 m)
Structures			
Thin laminations, typically lenticular and irregular	Lamination common, some small-scale cross bedding in sand; burrows common	No distinct lamination or bedding	Poorly defined
Biota			
Not sampled	Fine shell hash in sandy laminae; complete mollusk shells in mud	Rare shell fragments	None

to be found downcurrent from a large muddy delta (Fig. 5.32) or where many small rivers drain a dissected landscape with deep soils and thick saprolite. Along such coasts, suspended sediment concentration may effectively dampen inshore wave power and tidal currents. Supply becomes less important only where there is a barrier – usually a barrier bar, spit, or long narrow gulf such as the head of the Gulf of California (Table 5.7) or a reef or tectonic high that shelters the coastline from waves, tides or inshore currents.

Coastal currents, waves, and mud supply are key variables for coastlines. Assuming an adequate supply, coastal currents and mud transport can be thought of in four ways (Bently 2003); (1) low-salinity riverine plumes with low concentrations of mud that occur off river mouths during floods; (2) coastal currents of normal salinity with dilute muddy suspensions; (3) bottom currents of normal salinity supercharged with mud that act much like deepwater turbidity currents, and (4) currents of normal salinity supercharged with mud (Bently 2003). These four processes are useful to keep in mind when reading the examples discussed below and also when working with ancient mudstones (Chap. 8).

The coastal mudbelt of northeastern South America extends some 1,600 km northwest of the mouth of the Amazon. The river brings some 500 million tons per year of sediment, mostly clay, to the Atlantic of which about 250 million tons moves along the coast. About 150 million tons moves in suspension and the rest as large migrating mud banks. As well expressed by Meade (1994, p 37), it is fantastic to imagine a clay flake eroded from an Andean slope,

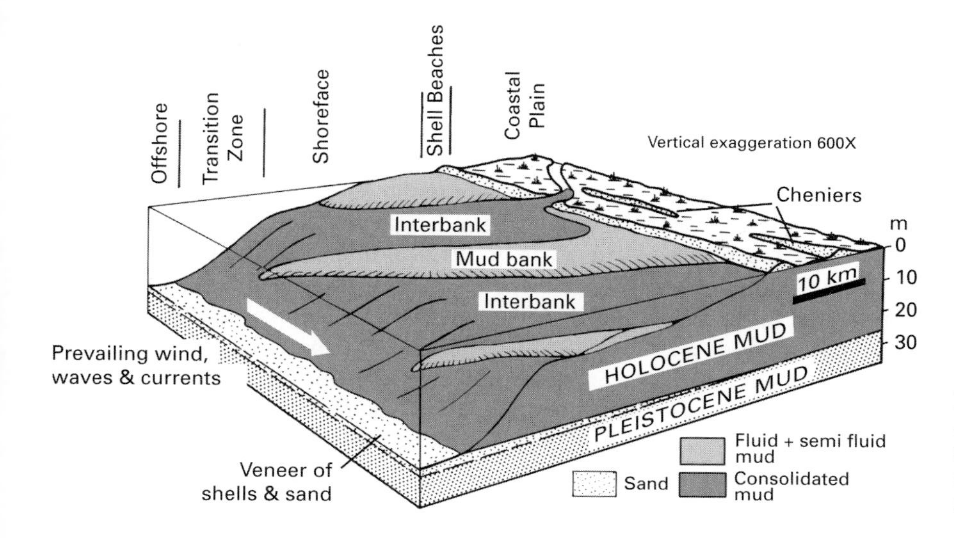

Fig. 5.33. Block diagram of migrating mud banks in Holocene mud belt of Suriname (after Rine and Ginsburg 1985, Fig. 3)

Fig. 5.34A–E. Muddy Suriname coastline near Suriname River in northeastern South America: (**A**) Low altitude view of soupy, muddy tidal flat with logs and stumps; (**B**) eroding coastline seen from a shell beach covered with fallen trees; (**C**) coastline with migrating bedforms (of shells)? above soupy mud; (**D**) looking seaward from a shell beach across a wide, intensely burrowed mud flat to the North Atlantic Ocean, Jim Rine holding machete for scale; and (**E**) Vibrocoring in the shallow, muddy foreshore. Note the many fallen trees, typical of eroding muddy segments of the coast of Surinam

spending perhaps several thousand years in transit in the alluvium and swamps bordering the Amazon, reaching the Atlantic, and finally being carried along the coast by the Guyana Current, a start to finish distance of some 6,000 km (Figs. 5.33 and 5.34). Here the coastline is exposed to strong, onshore northeast trade winds, mean tidal range is 1.8 to 2.8 m, and there is the powerful, offshore, coast-parallel Guyana current flowing to the northwest. Storms are rare and, on this open, high-energy coast occurs the world's longest inshore Holocene mud belt. Much of it is more than 20 m thick, and extends seaward for 30 km. This mud belts has been well described in Surinam by Augustinus (1978), Wells and Coleman (1978) and by Rine and Ginsburg (1985). Seaward of this muddy coast is a relic, current-swept shelf composed of sand and shells.

The muddy coastline of Surinam consists of alternating large migrating mud banks 50 to 60 km long ×10 to 20 km wide ×5 m thick separated by inter-bank areas of erosion. These banks migrate about 1.5 km per year, are obliquely tied to the shore, prograde parallel to the shore, have weak internal northwest dipping seismic reflectors and consist of semi-consolidated, "soupy" fluid mud (Table 5.8). Over such banks, waves rarely reach the shoreline. The fluid muds on the banks are very fine, vary in thickness from 0.1 to 1.0 m, have low strengths, and

C

D
Burrowed Mud Flat
Shell Beach

E

Table 5.8. Characteristics of some fluid muds (after Wells 1983, Table 2)

Characteristics	Louisiana	Surinam	Korea
Dimension of mushoals, km			
Alongshore	1–5	10–20	1–30
Offshore	0.5–3	10–20	5–50
Width of intertidal zone, km	0.1–0.3	2–5	5–30
Fluid mud thickness, m	0.2–1.5	0.5–2	0.1–3
Bulk density	1.15–1.30	1.03–1.30	1.20–1.30
Surface features	Generally absent	Small drainage channels	Extensive drainage channels; scour depression
Mean particle size, μm	3–5	0.5–1	6–11

have about 1 to 2% TOC. It is this slurry, with concentrations nearshore of 100 to more than 10,000 mg/l, which effectively dampens and reduces the waves produced by the strong northeast trade winds that impinge the coast. Tidal mud flats are well developed along the coast. On the banks large flow slides move seaward on dips as low as 0.03 to 0.08 °C and thus return proximal muds seaward (Wells et al. 1980). Between these moving mud banks are interbank areas where mud is better consolidated, fluid muds are largely lacking, and inshore wave energy is not damped so that waves reach the shoreline, erode it, and shell beaches are common. Sedimentary structures in the mud banks include parallel to subparallel wavy laminations, micro-crosslamination, minor scour and fill, some distorted layering, and featureless, massive muds. Bioturbation seems better developed in interbank areas, which also have small discontinuities.

Smaller in scale, but broadly similar is the muddy coastline of western Louisiana, formed by the westward moving mud stream of the Mississippi and, more recently, the Atchafalaya deltas. Here tidal range is low, only about 0.5 m. In the Holocene a transgressive, shore parallel, mud wedge 6 to 8 m thick and 24 km wide was deposited over a dissected Pleistocene surface formed at a Wisconsin lowstand of sea level. Pulsations of supply of fines caused shifts in the shoreline – the shoreline migrates seaward when supply is abundant when a western distributary of the Mississippi is active, and landward when supply is reduced when an eastern distributary is dominant. Such erosional intervals are marked by abandoned low, muddy beach ridges called *cheniers* (Fig. 5.35) several meters in height. Mud flats front this plain while behind it lie brackish to saline muddy swamps. Thus the name, *chenier plain*. Pools of fluid mud accumulate during fair weather offshore in the lower shore face or inner shelf, but are carried shoreward by storms – deeper wave base plus water piled against the shore (coastal set up) – and deposited as a mud drape over shell lags on the beach or flood the swampland behind it. Here, unlike Surinam, major storms and hurricanes play an important role in open coastal mud accumulation.

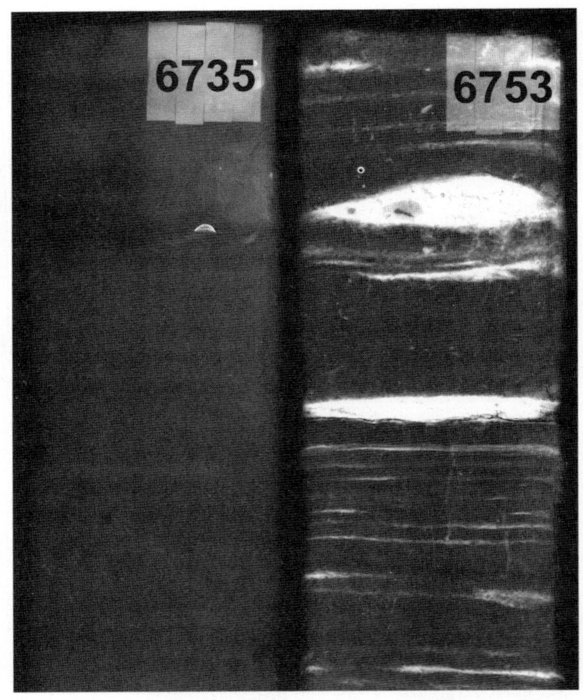

Fig. 5.35. X-radiographs of vibrocores of clays from the chenier plain of southwestern Louisiana down current from the Mississippi delta. Massive clay at left is almost silt free (clay shale equivalent), whereas radiograph at right has silty streaks and several silt ripples (*white*) indicative of some wave action – possibly the passage of cold fronts with strong winds and coastal flooding (Coastal Studies Institute, Baton Rouge, LA)

On the southwest coast of Korea there are extensive tidal mud and fine sand flats even though tidal range is 5 to as much as 9 m and there are intense winter storms (Wells et al. 1990). Compositionally, these Korean tidal flats are very variable ranging from about half mud and silt to some sand and even gravel; in addition, some mud flats abut against rock outcrops. This example also differs from others in that many short rivers draining a deeply weathered Precambrian landmass provide the fines. At Namyang Bay in Korea, the intertidal deposits are made up of 40 to more than 90% mud. The surface varies from smooth to slightly undulating with shallow scours and mounds. Crab burrows and rare concentrations of shells are also present. Mud accumulates on such tidal flats for two reasons: as flood tides move landward, velocities finally decrease and sand and silt are deposited, but mud in suspension is carried farther in a thin sheet of water from which it finally settles. Secondly, ebbing tidal currents erode non-cohesive silt and sand easier than cohesive mud. Thus given a source of mud, tidal action tends to concentrate fines inshore as vertical accretion deposits.

Ancient equivalents of similar deposits are not well documented in the literature. Walker (1971) provides one good example from the upper Devonian of Central Pennsylvania, where green marine mudstones and banded siltstones pass upward into red mudstones and siltstones with rootlets and mud cracks without a sand body at the paleoshore (Fig. 4.14).

Lagoons are shallow bodies of variable salinity water along coasts that are connected to the ocean, but separated from it by an extensive, shore-parallel barrier of sand or coral reef. Along terrigenous, low-tidal-range passive margins, sandy barrier islands can be 10's to 100's of kilometers long with only a few inlets where tidal range is small. Such protected shallow basins provide excellent traps for mud, which accumulates either along lagoonal shorelines or in its deeper waters. Rivers are the source of most of the mud entering the lagoon, whereas storms bring fine sand and silt to its ocean side via washover fans and flood tide deltas. Mudflats, swamps and saline marshes commonly fringe much of a lagoon. Climate modifies lagoonal muds and their associated sediments: in humid climates, there are bordering swamps, muds will be dark colored and rich in terrestrial organic matter, and a fauna that ranges from marine to freshwater but is dominated by brackish species. Bioturbation is likely to be extensive, unless the bottom lacks oxygen, and silts will have wave and tidal structures. In contrast, muds of an arid climate lagoon will be lighter in color, may have dispersed single crystals or masses of gypsum and halite (or grade into beds of it) so that bedding is disrupted, carbonate content probably will be higher, and perhaps parts of the lagoon will be bordered by carbonate sabkhas. A subaerial mudflat is distinguished from nearshore muds by many more mudcracks, clay chip conglomerates, possible thin, wind transported sand patches and by rootlets and plant debris plus contrasting styles of bioturbation. In arid climates, subaqueous lagoonal muds will have more abundant evaporite minerals than their subaerial equivalents.

5.4. Marine Environments

The marine environment is the principal trap for fine-grained deposits, which accumulate on *shelves, ramps, slopes* (also called *slope aprons*) and in *deep basins* on the continents and the *deep sea* with great lithologic diversity – from coastal silts and silty muds of the distal shoreface to the pelagic, almost silt-free deposits of the deep sea. This great diversity results mostly from three factors: (1) great variation in water depths and under-water relief from shallow foreshore to shelf break to slope to a deep basin perhaps with steep fault-bounded margins, (2) great variations in oxygen levels, and (3) more faunal impact on marine fine-grained deposits than in most continental fines – a few lakes and lagoons excepted. In addition, in the deepest parts of the ocean, calcium carbonate is dissolved so that the normal carbonate pelagic contribution to marine mud is missing and non-calcareous deep-sea red clays or siliceous oozes are deposited. Another distinctive factor is the much greater lateral continuity of marine mudstones than those deposited in the coastal zone or in continental environments. Such mud blankets accumulate during transgressions and early highstands as the shelf widens and shorelines retreat landward.

Here we trace this dispersal system sequentially seaward into deeper water. The reader should recognize, however, that where shelves are narrow or where slopes and canyons come close to river mouths, mud may be transported directly into deep water and that during a rapid sea level rise virtually all mud may be trapped inshore with none transported across the shelf or ramp to a slope or deep basin beyond, so that a thin *omission surface* or *hardground* marked by bioturbation, pyrite, glauconite, and phosphatic nodules may develop seaward of the trapped mud.

5.4.1. Shelves

Shelves dip seaward from the shoreface at low angles of 0.1 to 1.0° to a shelf break, which typically occurs at about 200 m or less of water depth. Shelves are built on continental crust, rim most of the continents, and vary greatly in width from a few kilometers along active and faulted continental margins to more than 200 km along passive margins (in ancient basins on continents the term *platform* is often used). The head of a submarine canyon may extend onto a shelf and become an important sluiceway for transport of mud and sand into deeper water. A distinct shelf break does not always exist, however, and where there is a uniform, regular transition into deeper water the term *ramp* is used instead. Shelf width is largely determined by tectonics: on stable cratons and passive margins shelves can be very wide and thus progressively reduce inshore wave power, whereas on active margins shelves are narrow (or even non existent); consequently, oceanic currents and waves can impinge directly on the shoreline and prevent mud and silt accumulation except in protected gulfs and bays (Fig. 5.2). Shallow, wide epicontinental seas on the continents – common in the geologic past, although not today – also belong here, although it should be noted that, because of Pleistocene and Holocene sea level variations, many of the deposits of modern shelves are not in equilibrium with present current regimes and hence do not serve as good examples for most ancient muddy shelves or epicontinental seas. This contrasts with slope and deep sea deposits where relative water depths changed much less in the Pleistocene and, consequently the major processes and deposits seen today in these environments provide better models for their ancient equivalents.

The availability of fines and the current regime of the shelf determine whether or not it will be muddy. The current regime of shelves includes rip and tidal currents, currents produced by both fair and storm waves that impinge the bottom, geostrophic and intruding oceanic currents and internal waves (Fig. 5.36). To these we need to add seaward-extending riverine plumes (Fig. 5.28) because either at the surface, at intermediate depths within the water column, or less commonly along the bottom, they bring virtually all mud to a shelf. With a rise of sea level, many fines are trapped in estuaries and the lower part of river valleys so that sedimentation of fines on the shelf is reduced until such traps are filled. During such intervals only the finest material will be transported in suspension to the shelf or beyond – a process called *dynamic bypassing* – and, because sedimentation rates are slow, organic-rich muds are likely to accumulate (an explanation for many black shales being most organic rich at their base). Like deltas, shelves may be tide, wave, or river (supply) dominated.

Storms affect a muddy shelf in two ways. First, the suspended discharge of its bordering rivers is greatly enhanced (Fig. 5.37 and Chap. 3). Second, storms lower wave base on the shelf and resuspend much of the bottom sediment to form a density current that flows along the bottom into deeper water, perhaps even beyond the shelf edge. Where the Coriolis effect deflects such currents, the term *geostrophic current* is used. Storm-generated beds, *event beds*, with their distinct hummocky stratification of fine sand and silt

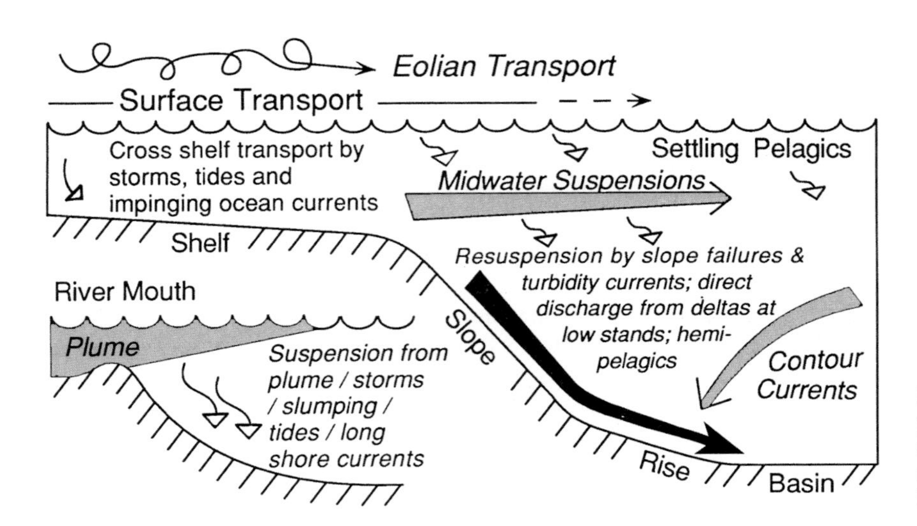

Fig. 5.36. Schematic of principal processes that transport sediment along and cross a shelf and slope into a deep basin (after Stow 1981, Fig. 1)

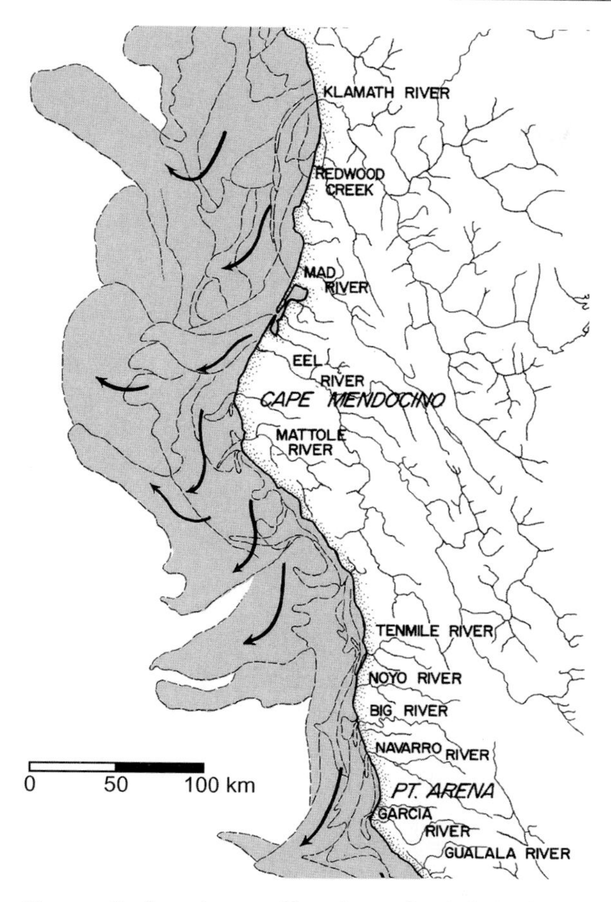

Fig. 5.37. Surface plumes of less dense, flood-derived, surface plumes of muddy floodwaters along the coast of California (after Griggs and Hein 1980, Fig. 7). Published by permission of the authors and the University of Chicago Press

are characteristic of all shelves, be they muddy, sandy, carbonate or mixed (Fig. 3.17). Some shell lags and small scours may also result from such storm events. The distal parts of such storm-induced currents finally deposit a silty mud, perhaps with a slightly erosional base, in otherwise homogeneously bioturbated hemipelagic muds derived from suspension during fair weather. Key here is the idea that most of the time sedimentation on the shelf is slow, so there is ample opportunity for bioturbation by bottom fauna (oxygen permitting) and only exceptionally – rapid deposition from rare storm events – will bottom sediment escape bioturbation and primary structures be preserved. And, of course, the deeper the shelf, the less frequent are the effects of storms. The Atlantic shelf off France shows this well (Fig. 3.18 and Table 5.9).

Mud deposits on shelves have many geometries. They may be mostly down current from a delta (where currents largely parallel or subparallel the shoreline), may totally blanket the shelf (where riverine supply is great and shelf currents weak or because of a recent sea level rise), may be restricted to linear belts (separated from the shore by silts or sands or perhaps concentrated inshore as along the northeast coast of South America), or have an irregular distribution pattern (perhaps related to topographic lows or relic preservation from a different current regime). And locally much of the mud transported by a river to the ocean may directly enter a canyon and this bypass the shelf as is true for the Congo River (Eisma 1988). Or strong oceanic currents may exceptionally

Table 5.9. Depth-related mud facies, Aquitaine inner shelf, France (after Lesueur and Tastet 1994)

Characteristics	Inner	Central	Outer
Water depth, m	30 to 40	40 to 55	More than 50
Basal contacts	Erosional; gutters and incisions	Erosional or sharp with incisions	Sharp or erosional
Contacts per meter of core	More than 20	10 to 20	Fewer than 10
Thickness of mud sequences, cm	1 to 10	2 to more than 10	More than 10
Character of basal part of sequence	Fine sand and coarse silt	Medium to coarse silt	Fine silt
Stratification/thickness (cm) of basal part of sequence	Lenticular, hummocky cross stratification (HCS) 5–15 cm thick and planar stratification 0.1–3.0 cm thick	Planar (0.2–3.0) with some HCS	Planar (0.2–0.5) faintly laminated
Character of upper part of sequence	Clay hummocks with gradual transition to silty clay	Gradual transition to clayey silts	Gradual transition to silty clay
Biological mixing, %	More than 20	20 to 80	70 to 90

impinge on a shelf and sweep it free of mud even where supplied by rivers. Such ocean currents may extend completely across a shelf or only partially, so that a shelf may have a distinct inshore mud belt (as along the northeast shelf of South America) beyond which are relict sands and possibly some carbonate deposits. In the absence of such a current, the outer limit of a mud belt may simply represent the depositional limit of prograding delta-derived muds prior to lobe abandonment, climatic change or an abrupt sea level rise, which traps most mud farther inshore. Typically on graded shelves unaffected by intruding oceanic or tidal currents, the mud belt will be separated from the shore by sands and silts of the shoreface.

In ancient deposits, the distinction between *inner* and *outer* shelf or ramp is a useful one. An inner shelf is recognized as sandier with more and thicker hummocky cross beds and scour surfaces – more proximal storm events – than an outer shelf with fewer and thinner storm bed events in its deeper waters. Thus more mud accumulates farther from shore in deeper water. Bioturbation will also generally increase seaward as long as bottom oxygen is available. Here weak distal bottom currents generated by such storm events deposit thinly interlaminated siltstones and muds. Expressed differently, the inner shelf is more influenced by coastal process whereas the outer shelf responds more to oceanic processes.

On a shelf or ramp, cycles of shallowing upward (thicker, coarser beds and less clay) or deepening upward (thinner, finer beds and more clay) sediments, with companion faunal and trace fossil changes, define the parasequences of sequence stratigraphy (Chap. 8). The boundaries of these parasequences are flooding surfaces (Fig. 5.38), which form many of the widespread marker beds of the marine realm. Oxygen content of the water column is closely related to such cycles.

The organic content of modern shelfal muds and their ancient equivalent deserves our special

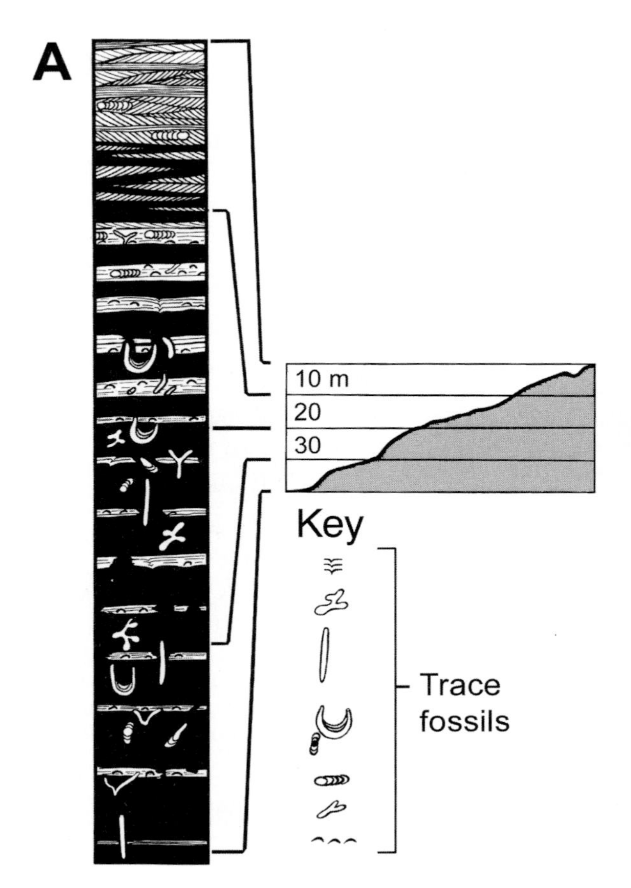

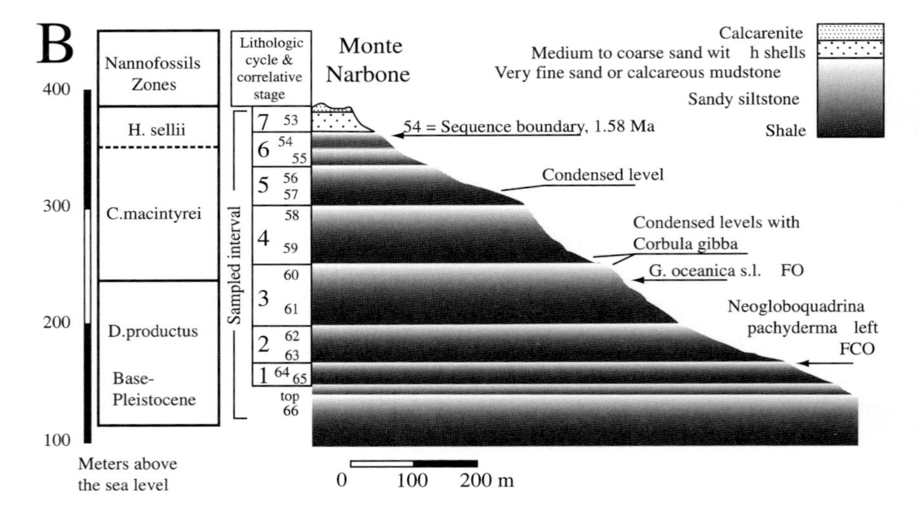

Fig. 5.38A,B. Two shallowing upward (shoaling), muddy shelf sequences: (**A**) Regressive inshore sequence along the muddy German coast of the North Sea (Reineck et al. 1968, Fig. 15) and (**B**) shoaling upward cycles in Plio-Pleistocene deposits at Monte Narbone, Sicily (Catalano et al. 1998, Fig. 18). The coarsening-upward motif is common in mud deposition (see also Fig. 5.40. Published by permission of the authors and the Society for Sedimentary Geology)

attention for several reasons. First, understanding the processes that control organic production and preservation on the shelf has wide application to all mudstones, because so many formed on shelves. There are two other important reasons as well – modern shelfal muds have about three times more TOC than modern, deep ocean muds and, secondly, black, organic-rich mudstones occur in many ancient epicontinental deposits, especially before the Tertiary. Thus the importance of the question, "What are the conditions that control oxygen distribution in shallow shelfal waters and those of former epicontinental seas?"

The organic content on the shelf is greater than in the deep sea for several reasons. Much of the water column of the shelf receives nutrients from riverine runoff and thus has a high organic productivity. Furthermore, water depths are relatively shallow so that transit times to the bottom are short and organic particles are less likely to be either oxidized or eaten enroute to the bottom. Finally, sedimentation rates are high in comparison to those of the deep ocean so that organic material is quickly buried. Remember that it is long transit times through mildly oxidizing water and slow sedimentation that form the widespread red clays of the deep sea. During nonglacial times in Earth history, however, it seems likely that bottom waters were everywhere low in oxygen, including the deep sea.

There is still another aspect of the organic richness of shelfal mudstones to consider. Most of the richest ancient shelfal black shales were deposited at times of rising relative sea level. Rising relative sea levels trap fines inshore and thus minimize the dilution of organic matter settling to the bottom. Thus sedimentation rate has a complex and contradictory effect on organic richness of muds: rapid sedimentation protects the organic matter but also dilutes it. Lowered oxygen in the bottom water, combined with lowered sedimentation rates, seems to be necessary to produce significant organic enrichments.

Isolation of oxygen-deficient bottom waters from well mixed and oxygenated surface waters on shelves occurs most readily in its topographic lows – in a canyon crossing a continental shelf or perhaps in a topographic low above a subsiding syncline or graben. In such lows oxygen supply will be restricted and either dysoxia or anoxia will develop so that organic matter will be easily incorporated into the mud. An additional factor to remember is that topographic lows are also traps for low-density organic matter, because it is here that bottom currents are weakest (Chap. 3). Excluding coastal upwelling (Chap. 4 and below), anoxia usually starts in a topographic low and spreads outward from it along the bottom as bottom oxygen decreases.

On shelves with temperate and polar climates, winter storms lower wave base and mix the water column better than in the summer. Therefore anoxia was probably very seasonal in many ancient shelves and epicontinental seas. Seasonal anoxia is also perhaps the most logical explanation for the millimetric lamination of dark and gray mudstone or mudstone and pelagic carbonate seen in many ancient mudstones.

Coastal upwelling provides yet another mechanism for both high organic productivity and preservation. It occurs along the east side of large oceans where deep, cold, well-oxygenated nutrient-rich bottom waters impinge on the upper slope and shelf. Good examples occur today off Peru and southwest Africa. As such cold, nutrient- and oxygen-rich water enters the photic zone, a superabundance of marine microorganisms results; upon death these accumulate on the bottom, and through excess oxygen demand, create bottom anoxia or dysoxia. Such a zone is typically elongate parallel to the shelf break or shoreline and its margins may have marked concentrations of phosphatic nodules, much chert (diatomites and radiolarites), and a rich microfauna (calcareous dinoflagellates, coccoliths, pelagic foraminifera, etc). Although some dark, organic-rich mudstones may have formed this way, most probably did not, because most seem to have formed in ancient epicontinental seas rather than along the margins of continents.

Three examples illustrate many of the above generalizations.

Hudson Bay (Fig. 5.39) provides an excellent modern model for a high latitude (52 to 64° N), large (longest dimension is over 1,000 km) and shallow (most in less than 200 m) epeiric sea that flooded an old craton. Isostatic depression of the crust by a continental ice sheet, which followed the topographic lows of the preglacial Great Arctic River that drained into the Labrador Sea, plus geologic structures produced the broad saucer-like shape of Hudson Bay.

The bay has a well-defined, counterclockwise current system that maintains normal salinities throughout most of it. The bay is frozen in winter (thus there are many dropstones in fines especially near shore), and has an almost classic gravel/sand transition offshore to silt and mud in the deepest parts of the bay. Mud and silt colors range from gray to greenish gray in shallow water to reddish brown in deepest water (slow sedimentation in oxidized seawater as in the deep ocean). In parallel, highest TOC

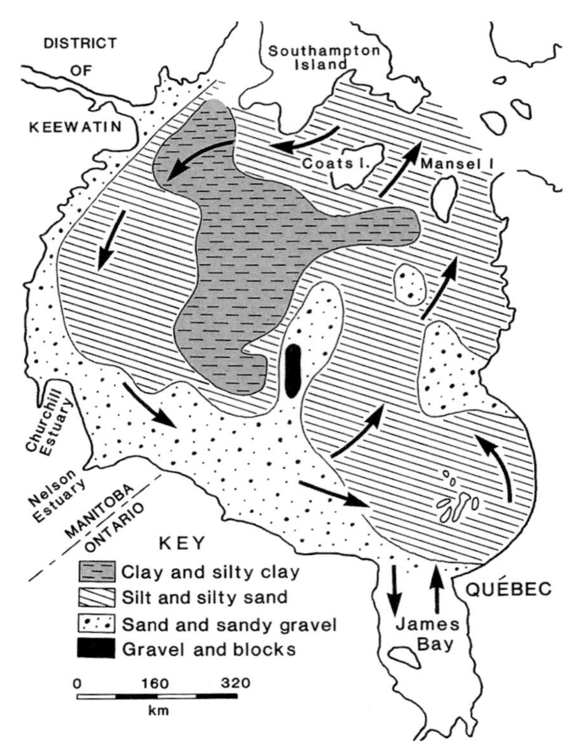

Fig. 5.39. Postglacial Hudson Bay provides good model for many mudstones of cratonic and pericratonic basins. After Pelletier (1969, Figs. 8.1, 8.7, and 8.9). Published by permission of the Geological Survey of Canada

values occur in shallower waters where they reach as much as 3.93%, although the average value is a very low 0.57% (low productivity in cold, but oxygenated water). The clay mineral assemblage of illite-chlorite closely reflects bedrock. Over 200 biologic species live in Hudson Bay. Surely there are many ancient, high-latitude, intra-cratonic basins for which the distribution of fines in Hudson Bay is a good analog (Pelletier 1969).

The Cretaceous Kaskapau Formation (upper Colorado Shale) of Alberta consists chiefly of dark gray marine mudstone with some thin marine sandstones (Fig. 5.40). It was deposited between 73 and 91 Ma ago in a vast epicontinental seaway as it deepened and transgressed an underlying delta of Cenomanian-Turonian age (Wallace-Dudley and Leckie 1995). The Kaskapau Formation, which ranges in thickness from nearly 500 to 700 m in thickness, has a distinct clinoform structure (Fig. 5.40). The Kaskapau has eight mudstone-dominated facies defined by proportions of sandstone and siltstone, degree of bioturbation, kind and abundance of trace and body fossils, and foraminifera plus TOC and HI values. These facies show that the mudstones of Kaskapau were deposited between fair and storm wave base – lower shoreface to open marine – and that oxygen levels were normal to slightly dysoxic as seen by both TOC values and trace fossil assemblages. Sedimentation of the mudstone was slow as demonstrated by the abundance of trace fossils in most of the siltstones and mudstones. Coarsening upward cycles of 2 to 4 m are dominant as would be expected on a prograding, constructional, muddy shelf. The Kaskapau Formation contains three thin internal sandstones bounded by disconformities. The basal disconformity of a sandstone represents a sudden fall of sea level and its upper one a transgressive surface of erosion.

Overlying the Kaskapau Formation and conformable with it is the First White Speckled Shale,

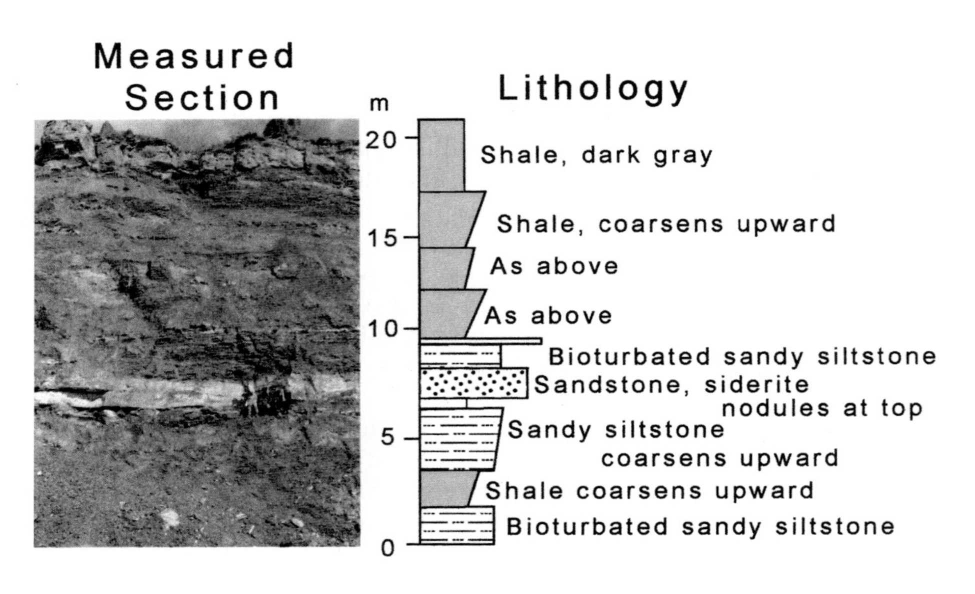

Fig. 5.40. Cretaceous Kaskapau Formation (Upper Colorado Shale) of Alberta, Canada (after Wallace-Dudley and Leckie 1995, Fig. 23). Note coarsening upward (mudstone→siltstone) cycles. Published by permission of the authors and the Geological Survey of Canada

which is up to 90 m thick, has typical TOC values of about 5 to 6%, is finely laminated, dark gray, contains both a shallow water foraminiferal assemblage of arenaceous forams and a deep water pelagic assemblage plus aggregates of coccoliths and rhabdoliths that form its white specks. The First White Speckled Shale represents a rapid, worldwide transgression, which caused a 250 m deepening of the Western Interior Seaway about 91.5 Ma ago. The First White has its highest TOC values near its base, is very widespread, and is a potential oil source bed where deeply buried. There are common features of transgressive systems tract shales. Conversely, the Kaskapau Shale, the high-stand systems tract, was deposited on a prograding constructional shelf, where slow sedimentation and better-oxygenated waters consumed most of its organic matter. Thus, it is not a source rock.

Now consider the mudstones and siltstones of an epeiric, shelf-to-basin transition in a Devonian-Mississippian sea in eastern Kentucky, where abrupt but small, perhaps 35 to 45 m, changes in water depth at the edge of a wide shelf produced significant lateral changes in mudstones and siltstones over a short distance of 10 to 15 km or less (Pashin and Ettensohn 1987). Here the widespread Berea-Bedford Formation abruptly passes laterally from shelfal siltstones and silty greenish-gray mudstone into a thin section at the top of the basinal Ohio Shale, a famous, widespread black shale (Fig. 5.41).

The sheet siltstones of the Berea shelf are mostly medium to thick bedded and have hummocky bedding with tops that are characteristically wave rippled. These prominent wave ripples are very uniformly oriented and their strike reflects the strike of the original shelf, whereas current ripples within siltstone beds and grove casts and flutes on their bases are at right angles and show that the dip of the shelf was to the southwest. Also present are siltstones with convolute bedding and load casts plus some interlaminations of greenish-gray mudstone. Closer to the shelf edge swaley bedding is dominant and load casts are largely absent. Both the shelf and shelf-edge facies overlie and pass into a mudstone-slope facies of distal tempestites, which in turn pass down slope to greenish-gray mudstones and thin turbidite siltstones with a fauna whose shells are commonly preserved only as molds or have been pyritized – a dysoxic environment that is transitional to the anoxic environment of the black shale at the base of the slope. Discordant with the thin turbidites and the green mudstone facies is a narrow siltstone feeder channel trending down dip that contains flow structures and clasts. In front of the delta, shales are well laminated silty, brittle and pyritic and are, overall, grayish black, but contain some thin laminations of greenish gray mudstone that represent distal mud turbidites that briefly introduced oxygenated waters into anoxic deeper waters beyond the shelf break. TOC values of the black shales reach 10 to 14%, whereas TOC values of the greenish-gray mudstones typically range from only 1 to 2 percent. The brachiopods *Lingula* and *Orbiculoidia*, some conodonts and *Tasmanites* are present in the black shale as are greenish-gray mudstone filled burrows, made by temporary "doomed pioneers" that extend downward into black shale beds. At the top of the siltstones is a thin condensed section of bluish-gray shaly siltstone with phosphate nodules, glauconite and much bioturbation – an omission surface. This surface represents rapid deepening and starvation. This surface is overlain by another black shale, the

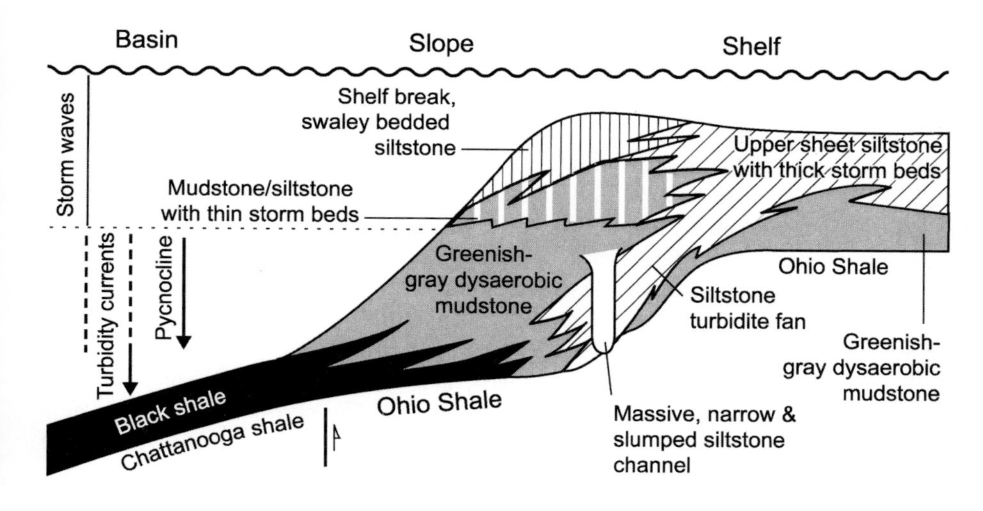

Fig. 5.41. Clinoform structure of sand-rich Devonian Mississippian Berea-Bedford Formations of eastern Kentucky at their southwestern limit beyond which are the black shales of the Ohio Shale (Pashin and Ettensohn 1987, Fig. 9). Published by permission of the authors and the American Journal of Science

Fig. 5.42A–C. Styles of mudstone-carbonate interbeds: (**A**) Nodular discontinuous and irregular bedded wackestones with about 60% light gray mudstone in the Upper Ordovician near Richmond, Madison Co., KY; (**B**) thin, dark, phosphatic mudstones separating finely laminated dolostones of the Permian Irati Formation in Goías State, Brazil; and (**C**) silty mud filling the interstices of oyster coquina of Tertiary Patagonia Formation in Chubut Province, Argentina. Normally, we only study the carbonate component of deposits such as these. Would not the mudstone component also yield valuable information (*continued on next page*)

Sunbury Shale. Here it is thin, typically 8 to 10 m thick, widespread and is organic rich with TOC values as high as 11 to 15 percent. This epeiric shelf-to-basin transition is instructive, because it consists mostly of siltstones and greenish-gray mudstone that in a few kilometers pass from a normally oxygenated bottom environment through dysaerobic to anoxic as the sea bottom deepened below storm wave base. It also passes abruptly vertically into basinal black shales that swept rapidly across shallow-water clastics in the turnaround to the transgressive systems tract.

Carbonate interbeds are common in many ancient mudstone-rich basins, especially in middle and low latitude shelf and platform deposits, some of which (Fig. 5.42) lack good modern analogues. So here part of our understanding comes from ancient deposits. Thus we should be aware that transitions, either vertical or lateral, of mud to carbonate are far from being well understood.

Interbeds vary with the type of carbonate factory and basin geometry. The carbonate interbeds may be pelagic in origin, resedimented, in situ, and some even are thought to be diagenetic. All form within

Fig. 5.42. *(continued)*

or along the sides of shaly basins and nutrient-rich, mud-free water is essential for all. Basin geometry (tectonics) is also a factor, because it determines relief in the basin and has much to do with water circulation within it. Benthic carbonate factories, the reef, bank and platform carbonates, all develop in shallow water whereas pelagic carbonates are independent of water depth (down to the carbonate compensation depth). There are a number of possible origins for carbonate interbeds in muddy shelf, slope, ramp, and basin floor deposits (Table 5.10). In all of these, full attention to the kinds, life conditions of the bottom (paleoecology), and burial conditions (taphonomy) of the fossils in both the mudstones (Brett and Allison 1998) and in the interbedded limestones is essential.

Table 5.10. Origins of carbonate interbeds in mudstones

- Pelagic rainouts, periodic (seasonal or longer cycles) or random
- Mass transport (turbidity currents, slides, slumps and debris flow) from high carbonate rims into adjacent basin floor
- Storm redistribution and separation of shells and carbonate sand and silt from mud on a mixed terrigenous/carbonate shelf (hummocky bedding)
- Thin lags of shells (plus concretions and pyrite) ranging from patchy to continuous on ravinement surfaces (may grade to hard grounds basinward)
- Shell beaches – onshore wash-over fans and shore face sheets both with limited lateral extent
- Algal nodules in underclays and closely spaced calcareous concretions
- Pedogenic carbonates (caliche/crusts) developed in muds and mudstones in arid and semi-arid coastal and lower alluvial valley deposits

Pelagic rainouts of "carbonate snow" and calcareous tests of microfossils are common in mudstone, marlstones and muds deposited in the Mesozoic and younger. They are also present in some earlier Paleozoic mudstones either as distinct beds or diluents to terrigenous mud. Millimetric carbonate laminations in lakes and some marine deposits are often thought to be seasonal – carbonate deposited in summer (when surface productivity is high and bottoms deficient in oxygen) followed by mud in winter (lower surface productivity and greater terrigenous supply). This model for carbonate interbeds, seasonal or not, mixes carbonate ooze and thin shelled pelagic skeletals with terrigeneous mud in all proportions from little to virtually pure biogenic carbonate laminations or beds. In the deep ocean, however, below 3 to 5 km, total dissolution of pelagic carbonate particles occurs. But the possibility of distal turbidity currents importing carbonate mud-, silt- and fine sand into the basin should also be considered, especially where a mudstone-rich basin is bordered by a high carbonate rim. Such a rim is also the source of distinctive exotic carbonate slide blocks, slumps, and debrites found in some deep-water mudstones. Flank beds of reefs along the margin of a nearby deeper muddy basin also shed such deposits, but on a smaller scale.

On mixed carbonate-terrigenous shelves and on upper slopes and ramps, deep storm waves can impinge on the bottom and rework the unconsolidated skeletal debris of carbonate banks and swells. This reworking not only tends to separate carbonate skeletal grains from terrigenous and carbonate mud, but also sorts and grades the carbonate fraction by size to separate its grains into characteristic graded hummocky and laminated interbeds. Some thin shelly beds also occur as lags (along with reworked concretions and nodules of pyrite, phosphate, ironstone, bones, etc.) at the base of transgressions (the transgressive surface of erosion in the terminology of sequence stratigraphy) as semi-continuous or continuous marker beds. Traced basinward, such beds are likely to pass into either "shale-on-shale" unconformities or hard grounds. Shell beaches are common in the tropics along many terrigeneous coasts. These are the source of thin, wash-over fans landward and thin shelly sheets in the proximal near shore, both of limited extent. Algal nodules in underclays and carbonate concretions cemented together to form thin beds are other less common examples.

Depending on basin geometry and type of carbonate factory, several of the above possibilities can occur together. Certainly more attention to the fossils and their state of preservation seems compelling for future studies of mudstones interbedded with carbonates.

In the above mixed terrigeneous carbonate associations, the carbonates often provide the best, most easily recognizable evidence of cycles of basin filling through hard grounds (recognized by burrowing and borings), submarine weathering, and authigenic minerals such as pyritic and phosphatic nodules, encrustations and replacements of fossils, and by the presence of glauconite. Such carbonate hardgrounds or omission surfaces are keys to correlation. Where carbonate interbeds are absent, shale-on-shale breaks are recognized by slight disconformities and higher clay contents in addition to lags of phosphatic nodules, pyrite and reworked fossils. Such shale-on-shale time breaks and carbonate hardgrounds are widespread and can be recognized on seismic images. Thus they are useful for correlation and for unraveling the history of basin filling.

5.4.2. Slopes and Toes of Slopes

Like their updip shelves, mud-rich slopes and toes of slopes are found on passive margins with big rivers. Slopes extend from the shelf break to basin floor and typically have dips of 2 to 4°, although some are as steep as 7° or more. Very broadly, slopes can be thought of as forming an *apron* along the sides of a basin or continent. The processes that transport sediment into deep water beyond the shelf edge are flood-induced distal riverine plumes, resuspension of shelf fines by storms plus mass movements, and turbidity currents on the slope itself (Fig. 5.36). Of these, turbidity currents and mass movements are the principle ones. Mass movements include rock falls, slides, slumps, debris flows and creep. Excluding turbidity currents, the processes of subaerial and subaqueous mass movements are essentially identical. And, as on subaerial slopes, creep is virtually always active, especially where deposition is rapid.

Proximal, high-energy turbidity currents (Fig. 3.19) have high densities and transport both pebbles and coarse sand. Conversely, distal, low-energy turbidity currents are dilute with low density and only carry fine silt and mud. While net transport is mostly down slope, oblique transport and along the slope also occurs. Slope-parallel transport is the result of deep-water, basin-wide or ocean-wide circulation driven by wind, or salinity, or temperature differences (thermohaline contour currents) that impinge on the slope and at times even the shelf itself (Chap. 3).

Failure on slopes occurs where they are steep and where sedimentation is rapid (less time for compaction and escape of pore water). Failure is also favored by a rapid fall in sea or lake levels (silts and muds become over pressured and methane gas hydrates exsolve more easily into pore water, additionally reducing shear strength). Or failure may be triggered by a passing storm (cyclic loading by passing waves), and, of course, by seismic shocks during earthquakes. Any of these processes may cause liquefaction or detachment of a solid mass on the slope and start it down slope into deeper water. In this process part or all the sediment is progressively mixed with water and part or all of the mass is transformed into a turbidity current whose most distal and weakest flow carries mud in suspension far out onto the basin plain. As turbidity currents flow into deeper water, their dilute upper portions may separate along density contrasts to form thin, detached sheets (interflows) of suspended fines called nepheloid layers (Fig. 3.20). Nepheloid layers also form during storms and transport mud beyond the shelf edge depositing it as a thin, fine sheet or film on the bottom.

The slope itself grows basinward with a *clinoform* structure as sediment is washed over the shelf edge. Along continental margins, contour currents parallel to its strike redistribute and smooth a surface made irregular by the topography of mass wasting (Fig. 3.8). Slopes may be only a few kilometers wide in basins located on the continents as in the Berea-Bedford example of eastern Kentucky or 10's of kilometers wide on some passive margins.

5.4.3. Submarine Fans and Aprons

On slopes dominated by terrigenous sediment, where deposition is more rapid than on mixed terrigenous-carbonate shelves, the mass movement-gully-canyon system of the slope is replaced by submarine fans a few kilometers to hundreds in length (and in the deep ocean up to a thousand or more). Unlike aprons, submarine fans have a single point source fed by a channel or canyon. Such fans have channelized flows in their proximal and medial parts that follow and at times overflow channel-levee systems plus distal lobes with widespread sheet flows. Slides and slumps may also be present on the proximal and medial parts of the fan, but turbidity currents dominate its distal part. In contrast, aprons have a line source supplied by randomly scattered slope failures and debris chutes below the shelf edge. These transport material in mass and also generate turbidity currents. Hence the apron lacks a well-organized channel distributary system and instead consists of disorganized slumps, slides and flows interbedded with some turbidity current deposits, some hemipelagic and perhaps even some contour current deposits.

It is useful to distinguish between mud-rich (more than 70% mud), mixed mud-sand (70 to 30% mud) and sand-rich fans and aprons (less than 30% mud). Mud-rich fans are the most abundant of the three types. Fans and aprons are mud rich, when fed by a large, low-gradient river transporting mostly fines (Chap. 3), whereas they are sand-or gravel rich, when fed by short, high gradient rivers draining a tectonic highland. On both distal fans and aprons, turbidity currents are weaker and more dilute and deposit thinner, less silty muds as their energy is dissipated by bottom friction and dilution with the overlying water column. Hemipelagics are everywhere deposited on these fans and aprons between flows; criteria have been developed to distinguish such muds from those deposited by final deposition from dilute turbidity currents (Table 5.8), a distinction not always easy to make, however. Should there be a long pause between random resedimentation events, a *mud blanket* (Fig. 5.31) will accumulate over all or part of the fan or apron. Such a deposit makes a good seal for sandy channel fills. Another source of thick individual mud beds in deep small basins with turbidite sands could be the reflection of a turbidity

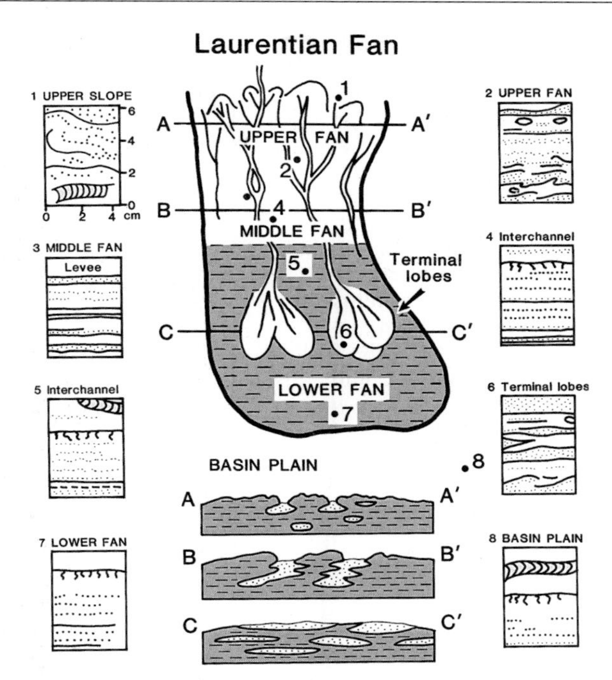

Fig. 5.43. Generalized turbidite model and resulting deposits (after Stow 1981, Fig. 13). Published by permission of the author and the American Association of Petroleum Geologists

current back and forth from one side of the basin to the other (Pickering and Hiscott 1985, Fig. 17). Thick uniform muds in deep basins, if deposited by a single turbidity current tend to be silty, fine upward and be weakly laminated and, regardless of origin, have been called *unifites*. In addition to the above mud-dominated deposits, there may be some far-traveled debris flows derived from a slope failure. These are thin but distinctive and traceable (Fig. 3.5).

A number of facies models outlining the distribution of sand and mud within submarine fan deposits have been proposed none of which, however, seem to be universal (Fig. 5.43). The sensitivity of density currents to slight differences in bottom topography (differential subsidence due to active tectonism, crevassing of channel margins, and compaction) appears to be responsible for this variability, in exact analogy to meandering rivers, which always seek and follow topographic lows (Fig. 5.44). See Stow and Mayall (2000) for a review of models of sub-sea, turbidite fans. Here we recognize only a first order distinction between *proximal*, *medial*, and *distal* (upper, medial, and lower) submarine fan deposits. Proximal submarine fan deposits typically consist mostly of stacked sands, have many debrites, slides, slumps and subordinate mudstone. Mounded and chaotic seismic reflectors also are found here. Me-

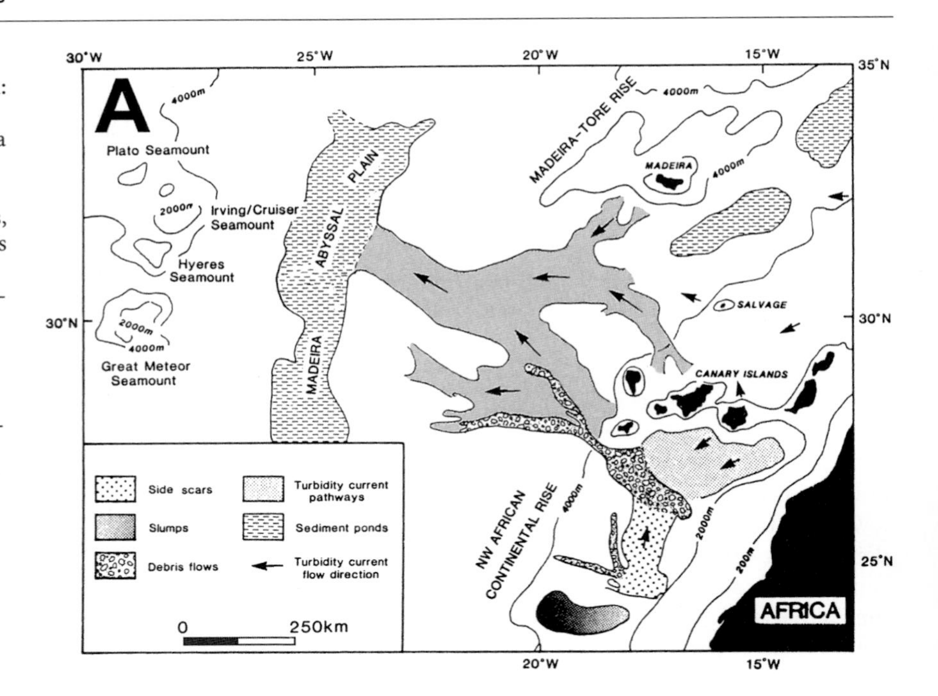

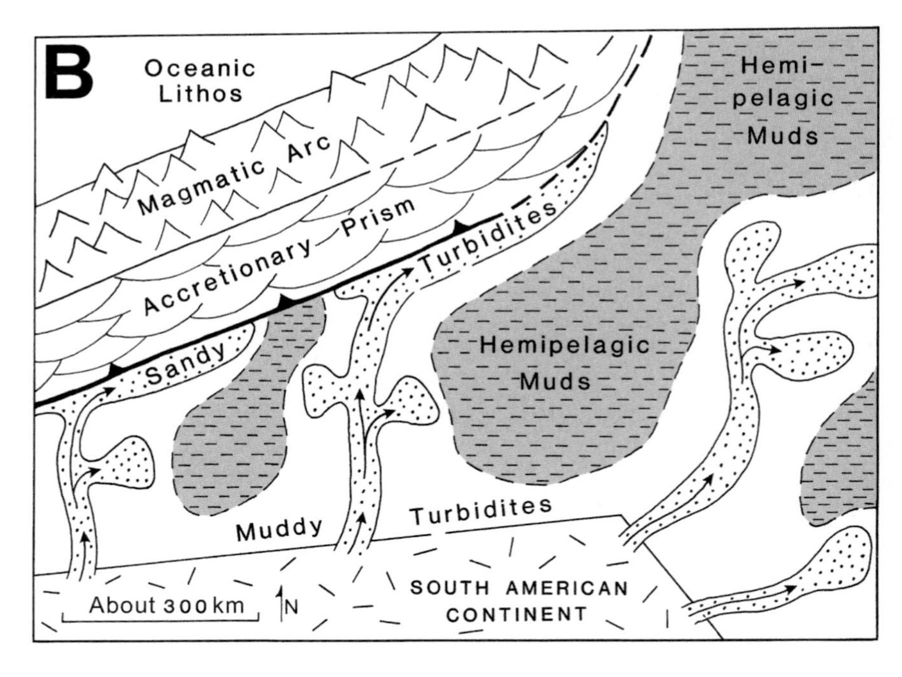

Fig. 5.44A,B. Regional patterns of turbidite deposition: (**A**) Late Quaternary turbidites off the coast of Africa where there are, in addition to preferred pathways of turbidites that follow lows, slumps scars and debris flows (after de Large et al. 1987, Fig. 1) [Published by permission of the authors and the Geological Society of London] and (**B**) generalized interpretation of turbidite fairways in Eocene deep water sediments of the Lesser Antilles forearc basin (after Speed 1986, Fig. 3)

dial deposits have more mud separating well-defined, levee-bounded, sand-filled channels with appreciable interchannel deposits of mixed sand, silt, and mud (but with still a few debrites the result of exceptionally large slope failures with long run-out distances). In distal fan deposits (Figs. 5.45 and 5.46), on the other hand, mud and some silt are dominant, and there are only a few shallow channels bounded by low, silty levees. Hemipelagics are common and there even may be a few thin pelagic limestones. Because fan area increases rapidly down dip from the base of slope, fan thickness rapidly decreases. The Miocene mudstones of coastal California reveal well a mixture of distinctive turbidite and hemipelagic interbeds based on five characteristics some of which can be seen in the field (Table 5.11).

The submarine fan of the Mississippi is but one of several that has been studied in detail (Bouma et

Table 5.11. Miocene slope mudstones, California (adapted from Slatt and Thompson 1985)

Property	Turbidite	Hemipelagic
Associated lithology	Sandstone	None
Color	Olive-gray	Tan
Stratification	Laminated with distinctive thin silt laminations	Massive with only a few wispy silty streaks
Mineralogy	More kaolinite and chlorite than smectite; pyrite; TOC 0.4 to 0.6% with some marine organics	More smectite than kaolinite and chlorite; pyrite absent; TOC 0.6 to 0.8% all terrestrial
Fossils	Low total abundance, but some shallow water calcareous forams	Some forams, mostly arenaceous and benthonic

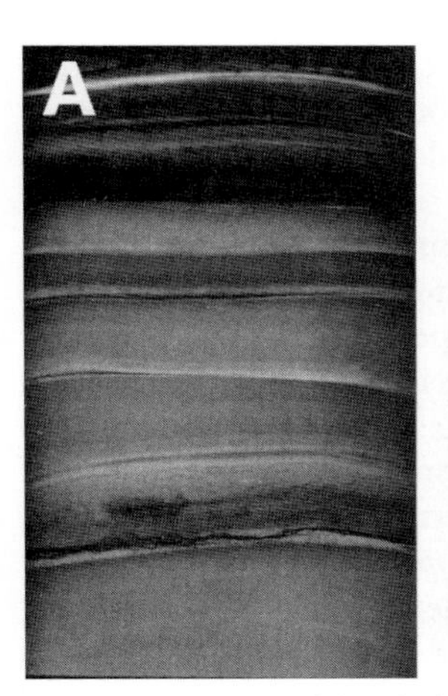

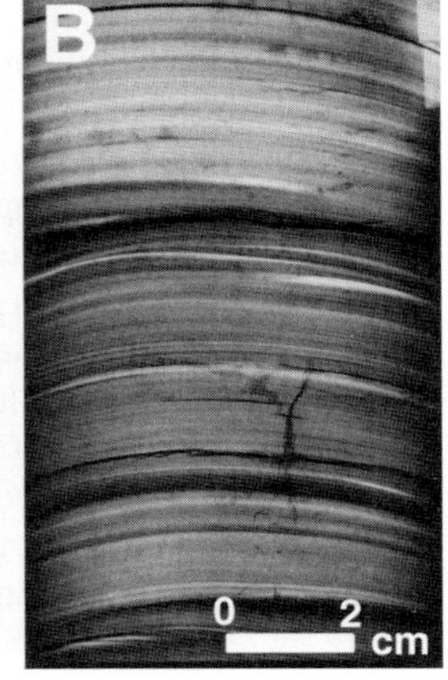

Fig. 5.45. Features of distal, deep-water fan deposits of the Mississippi River as revealed in X-radiographs of cores. Here, even far down the fan there are thin graded silts and regular laminations largely free of bioturbation (Coleman et al. 1986, Fig. 10)

al. 1986). Reading and Richards (1994) and Richards et al. (1998) provide other valuable details and were the source of much of the above. See also Bouma and Stone (2000) in Digging Deeper.

5.4.4. Contour Currents and the Deep Ocean

Geostrophic currents that follow the contours of continental slopes and rises are called *contour currents* and in the modern ocean form large, generally fine-grained ripples, waves, mounds and sheets collectively called *contour drifts*, whose thickness may be 150 m or more (Table 5.12). These drifts are best known from Cenozoic deposits drilled by the deep sea drilling project. Contour currents have velocities of 10 to 30 cm/sec and are produced by global oceanic

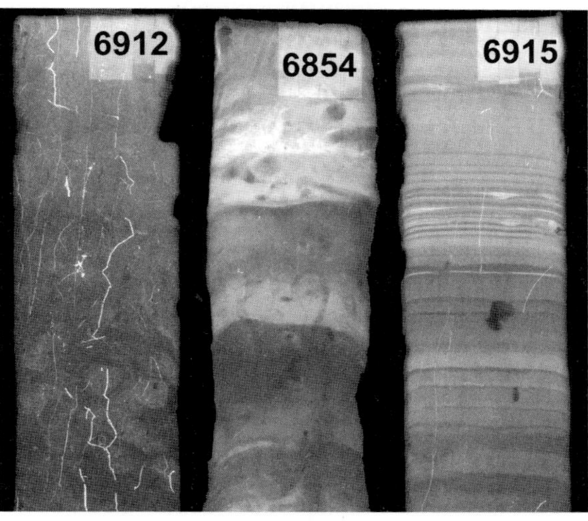

Fig. 5.46. X-radiographs of modern, deep water muds in the Gulf of Mexico from the Garden Banks Lease Area, Blocks 873 and 877 about 150 miles offshore, SSE of Beaumont, Texas (Courtesy of Harry Roberts, Coastal Studies Institute); Picture 6912, coarsely burrowed almost massive hemipelagic clays with near vertical pyrite (*white*) filaments in 4,670 ft of water; Picture 6854, distinctly graded silts with sharp bases separated by hemipelagic clays in 5,200 ft of water, and Picture 6915, thinly laminated silty clay to clay with thin strands of pyrite (*white*) in 4,670 ft of water. Not shown are massive and structureless remolded mudflow depositions

Table 5.12. Characteristics of contourites (adapted from Stow et al. 1998, Table 2)

Occurrence
- Thick, uniform sequences commonly interbedded with distal turbidites in oceanic and deep-water basins along continental margins

Texture
- Silty muds plus silts and fine sands and rare gravel lags in deep-sea channels and straits

Fabric
- Silt and sand grains tend to be oriented parallel to slope where not bioturbated

Composition
- Mixed biogenic (benthonic/pelagic) and terrigenous including volcanic debris

Structures/cycles
- Shallow scours and many gradational contacts. Biogenic mottling and poorly defined stratification impart strong homogeneity with weakly defined decimetric cycles

circulation patterns caused by the temperature and salinity contrasts of polar and tropical water masses. The greater the contrasts of temperatures of these water masses, the higher are velocities. Hence good development of contour drifts seems to be related to cooler than normal polar temperatures. Bottom topography such as straits and channels usually accelerates flow and may even locally produce gravel lags, although mud, silt and fine sand are dominant. Such deposits occur in water depths from more than 2,000 m to as little as 300 m in the modern ocean.

Fine-grained contourites in ancient, deep-water onshore basins are difficult to distinguish from the deposits of distal turbidity currents. Some helpful criteria include the presence of shallow erosional scours in contourites, flow along rather than downslope, and possibly higher TOC values than in turbidite muds. Two additional criteria include a general absence of thick beds and a lesser degree of vertical cyclicity (Table 5.12). Stow et al. (1998) report on twelve ancient examples, mostly silty terrigenous contourites, some of which have abundant silt-sized skeletal grains. Although contour drifts form 5 to 15% of modern deep-sea deposits, only a few have been recognized in ancient deposits. Nonetheless, they deserve consideration as an alternative to distal, fine-grained turbidite deposits.

Another type of fine-grained deposit that may be interbedded with either contourites or distal turbidites or by itself extend over vast areas of the ocean bottom are sediments composed chiefly of pelagic skeletal organisms. These may consist of calcareous coccoliths (clay size), foraminifera (mostly silt size), and pteropods plus siliceous radiolarians, and diatoms. Foraminifera predominate among these tests. All these organisms, which chiefly live in the uppermost 100 m of the water column, at death form a slow rain of debris, which falls to the bottom where, along with some terrigeneous debris (often eolian), to form *oozes*, *marls* and sands. Calcareous deposits are more abundant than siliceous ones on the modern seafloor, but below 3 to 5 km depending upon latitude, calcite tests are dissolved at the Calcite Compensation Depth, CCD with aragonite dissolving at shallower depths. Consequently below this depth, only siliceous tests and terrigenous debris accumulate on the bottom. Here red clays, some with Fe and Mn nodules, are widespread. Because deep sea water is mildly oxidizing, travel time to the bottom is long, and because sedimentation is slow, organic matter is oxidized both as it falls though the water column and at the sediment-water interface. Consequently, deep-sea terrigenous clays tend to be reddish

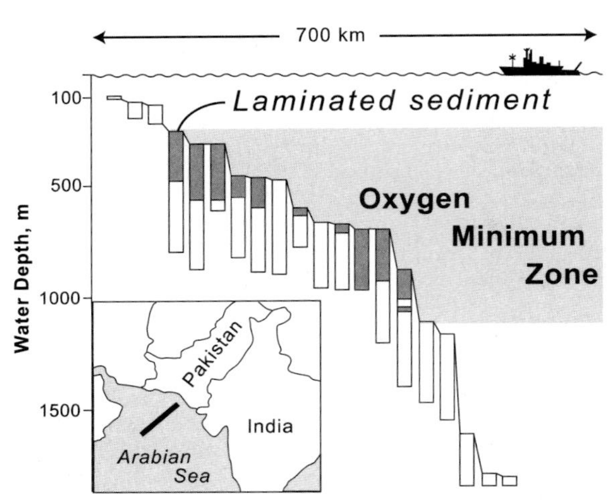

Fig. 5.47. Lamination preserved on sea bottom where oxygen minimum zone intersects deep continental slope off Pakistan and India (after Schulz et al. 1996, Fig. 3C). Published by permission of the authors and the Geological Society of London

brown, unless there was a special productivity event at the surface, which produced organic overloading on the seafloor. On deep slopes, especially on those with upwelling (Fig. 5.47), enhanced productivity causes oxygen minimum zones to develop, which restrict bottom life so lamination is preserved, and organic-rich muds are deposited (Fig. 5.48).

Pelagic deposits grade into or are interstratified with terrigenous deposits on continental margins and rises and slope aprons, on the "far sides" of basin plains having a terrigenous source, and on isolated sea mounts. *Sapropels*, organic-rich mud deposits, may be interbedded in response to anoxia. Where there is relief on the sea bottom, pelagic oozes are resedimented by mass movements and turbidity currents. Pelagic sediments and red clays also occur in some deep-water basins on the continents and may also be found closely associated with some black shales. Cyclicity, defined by variations of clay content (variations of supply linked to climate), bed thickness, $\delta^{18}O$ and microfauna, is fairly common and often ascribed to Milankovitch cycles (Chap. 8). See Reading (1996, Chap. 10) for more about deep marine sedimentation.

McCoy and Sancette (1985) provide a useful overview based on over 1,200 cores of the fine-grained surface sediments of the North Pacific Ocean. They mapped three sediment types: (1) calcareous ooze and marl, (2) terrigenous and pelagic clay, and (3) biosiliceous ooze and mud over the four major basins of the North Pacific Ocean. These units are fully described and their physical properties such

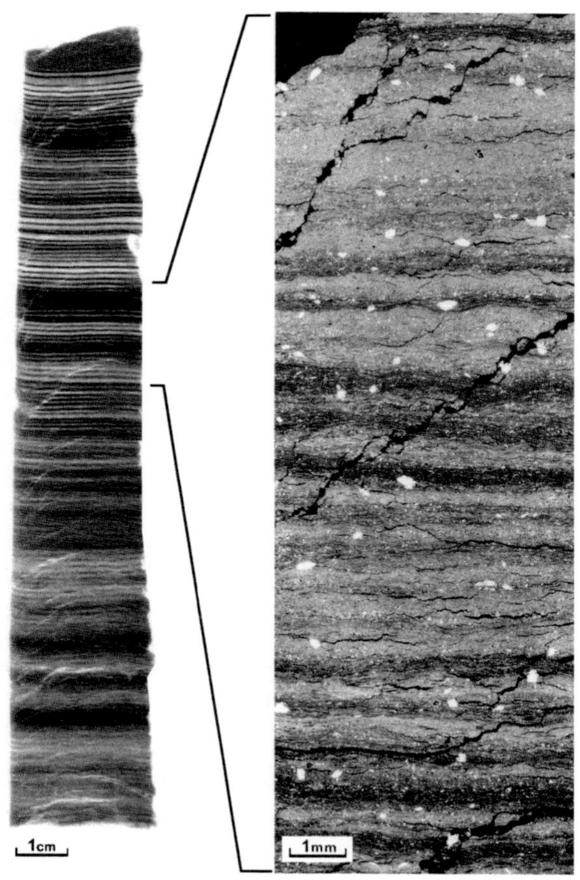

Fig. 5.48. Near perfect lamination-Holocene/late Quaternary deep-water sediments from the Santa Barbara Basin off southern California (Bull and Kemp 1996). In *left* X-radiograph, white laminations are diatom oozes and dark laminations are organic-rich clays, whereas in the *right-hand* Back Scatter Electron Image (BSEI), colors are reversed. Core taken from bottom of deep anoxic basin, where thick diatom laminations represent more productivity and thinner ones less (lesser upwelling of cold water). Note total lack of bioturbation. Although insignificant in this part of the core, fine silt, representing onshore seasonal rains, also form such laminations. Published by permission of the authors and the Geological Society of London

as grain size, water content, porosity, and biogenic contents summarized. Six sedimentary provinces are recognized. The floor of the North Pacific Ocean could be a model for the scattered slivers and pieces of ocean crust that are preserved in orogenic belts.

A notable pelagic deposit is *chalk*. Chalks are hard to friable, light gray, fine-grained micritic limestones consisting of pelagic grains (coccoliths and some foraminifer plus some siliceous microfossils). Trace fossils are common. Chalks were deposited as oozes, chiefly in the Upper Cretaceous and Paleocene, when sea levels were high. Chalks are best known in Western Europe and the Western Interior seaway of North America, but also occur on plateaus in the ocean above the carbonate compensation depth. The chalks of the greater North Sea region were deposited both on shallow shelves and deeper water and range in thickness from 50 to more than 1,000 m, whereas in the Western Interior, deposition was on a shallow shelf. Although many chalks have purities of close to 100% $CaCO_3$ (end members of clay-free sedimentation), clay is admixed with chalk to form marlstones and is also present as distinct, continuous laminations over wide areas (a good example of even a little clay in a sedimentation system having a marked affect). As clay sedimentation increases, chalks→marlstones→calcareous claystones.

5.5. Summary

Most terrigenous mud and silt is dispersed and deposited in oceans, seas and lakes by short term, high energy, episodic events – floods, storms and waves, and by turbidity, contour and tidal currents. In the atmosphere vast quantities of fines are directly injected into the atmosphere by volcanoes and by dust storms. Most of the continental environments, however, are either simply conduits or temporary storage sites for mud and silt in transit to either a lake, sea or ocean, the principal sinks for fine-grained, terrigenous debris and for the accumulation of associated fine-grained carbonates. Everywhere quiet water, either deep or shallow, is needed for the deposition of mud, except where its concentration in the water column is so high, as on some deltas (and coastlines down current from them) that inshore wave energy is dampened. And everywhere, because fall velocities are small for clays and fine silts, changes in even weak currents can markedly alter dispersal patterns within and between environments. Organic matter follows the same depositional patterns, making mudstones the source of most hydrocarbons.

Distinguishing the muds and mudstones of one environment from another starts with the universal tools of stratification and lithology, body and trace fossil associations, facies associations, and position in both the basin and in the local cycle of sea level oscillations. It is always important to ask, "Where does the mudstone fit in a cyclic sequence of falling-, low-, rising-, and high sea levels (low stand, transgressive and highstand deposits)?" But what is special to muds and mudstones and makes them so different from sands and sandstones is *their sensitivity to oxygen*, which affects body and trace fossils, stratification, color, abundance and type of organics (TOC and HI), and the development of a wide range of diagenetic reactions, especially involving iron minerals. The abundance, size, sorting and abrasion of pollen and spores (their sedimentology so to speak) is also best studied in mudstones and is a most helpful discriminator of depositional environments. And, as we will see in Chap. 6, changes in the maturity of kerogen with depth reveal burial history. Thus insight to the dispersal of mud and what happens to it during burial as water is expelled is all-important for understanding its stratigraphy across a basin.

Research opportunities in years to come are numerous. Perhaps foremost among these are more detailed studies of mud and silt in modern environments, because today these are far outnumbered by those of carbonates and sandstones. Underlying such research is the question, "Should we be thinking of separate facies models for large deposits of mud or silt or are they simply grain size variants of existing ones based on sand-sized terrigenous or carbonate material?" Or perhaps the question is better asked in reverse, "What do mudstones tell us about existing facies models of sandstones and carbonates?" "Can mudstones, for example, help us better identify their sub-environments, or even better, help provide us with improved insights to the geometry and location of interbedded sandstone and carbonate bodies of economic interest (the amount and type of their porosity, permeability and mineralization)?" More study of the different types of carbonate-mudstone interbeds is vital to answer this question.

For answers to all of these questions, one fact seems clear. At their best, such studies should combine standard sedimentological examination with detailed study of body and microfossils, trace fossils, and organic petrology and chemistry, relating each to the other and to the sequence stratigraphy of the deposit.

5.6. Challenges

What is the origin of carbonate interbeds in mudstones?

This question immediately invites many more. For example, "What are the most common types? How many different processes are involved? Where are 'good modern equivalents'? Or will most of our understanding come from the study of the ancient?" We rank these questions among our most challenging of all in the study of mudstones.

Mudstone types (choose any classification you wish) and tectonic setting.

Is dependency strong or weak? While we suspect local depositional environment is the control (unlike sandstone composition) very little data exists. But see Chap. 8, Provenance, about dependence of some aspects of mudstone chemistry on tectonics.

How well, if at all, do the sedimentary structures and textures of muds and mudstones of different environments distinguish one environment from another when evidence from interbedded lithologies is unavailable?

This has rarely been attempted. Could it be that most of the physical processes of suspension transport are everywhere the same?

Environmental discrimination of modern muds by biomarkers.

Some well-selected, comprehensive studies designed from inception to integrate physical sedimentology with organic chemistry including biomarkers would go far to help sedimentologists better appreciate biomarkers as environmental discriminators and thus make more use of them.

The silt content of modern muds – does it provide a broad guide to their environmental discrimination?

It is commonly said that fluvial muds are siltier than marine ones. For example, in a given sequence studied by wireline logs, will silt percentage determined from the self potential, sonic or gamma ray logs discriminate between marine and non marine or their subenvironments? And beyond specific basins can we develop general environmental discrimination for mudstones from geophysical logs?

References

Aslan A, Autin WJ, Törnquinst TE (1995) Holocene to Wisconsinan sedimentation, soil formation, and the evolution of the Mississippi flood plain, southern Lower Mississippi Valley (SLM). In: John CJ, Autin WJ (eds) Guidebook of Geological Excursions, Geological Society America, Annual Meeting New Orleans, La. 6–9 November 1995, Guidebook 3 (Basin Research Institute, Louisiana State University, Baton Rouge, La, 70803), pp 61–93

Augustinus PGEF (1978) Actual development of the chenier coast of Surinam (South America). Sedimentary Geology 26:91–114

Bennett MR, Huddart D, Thomas GSP (2002) Facies architecture within a regional glaciolacustrine basin, Copper River Alaska. Quaternary Science Reviews 21:2237–2379

Bentley SJ Jr (2002) Dispersal of fine sediment from river to shelf: process and product. Gulf Coast Association Geological Societies Transactions 52:1055–1067

Bentley SJ Jr (2003) Wave-current dispersal of fine-grained fluvial sediments across continental shelves: the significance of hyperpycnal plumes. In: Scott ED, Bouma AH, Bryant WR (eds) Siltstones, Mudstones and Shales: Depositional Processes and Characteristics. Society Sedimentary Geology/Gulf Coast Association Geological Societies, Joint Publication, pp 35–48

Bleuer Ned (2003) Slow logging, subtle sequences. Indiana Geological Survey, Reports of Progress

Bloom AL (1998) Geomorphology, 3rd edn. Prentice-Hall, Englewood Cliffs, New Jersey, 482 p

Bohacs KM (1998) Contrasting expressions of depositional sequences in mudrocks from marine to non marine environs. In: Schieber J, Zimmerle W, Stehi P (eds) Shales and Mudstones. E. Schweizerbart'sche Verlagsbuchhandlung Stuttgart, pp 31–78

Bohacs KM, Suter J (1997) Sequence stratigraphic distribution of coaly rocks: fundamental controls and paralic examples. American Association Petroleum Geologists Bulletin 81:1612–1639

Bohacs KM, Carroll AR, Neal JE, Mankiewicz PS (2000) Lake-basin type, source potential and hydrocarbon character: an integrated sequence-stratigraphic-geochemical framework. In: Gierlowski-Kordesch EH, Kelts KR (eds) (2000) Lake Basins Through Space and Time. AAPG Studies in Geology 46:3–34

Bouma AH, Coleman JM, Meyer AW et al. (1986) Initial reports of the Deep-Sea Drilling Project, Vol. 96. US Government Printing Office, Washington, DC, 824 p

Bowler JM (1973) Clay dunes: their occurrence, formation and environmental significance. Earth-Science Rev 9:315–338

Brett CE, Allison PA (1998) Paleontological approaches to the environmental interpretation of mudrocks. In: Schieber J, Zimmerle W, Sethi PS (eds) Shales and Mudstones. E Schweizerbart'sche, Stuttgart, pp 301–349

Bull D, Kemp AES (1996) Composition and origins of laminae in late Quaternary and Holocene sediments from the Santa Barbara Basin. In: Kemp AES (ed) Paleoclimatology and Paleoceanography from Laminated Sediments. Geological Society, Special Publication 116, pp 143–156

Catalano R, DiStefano E, Sulli A, Vitale FP, Infuso S, Vail PR (1998) Sequences and systems tracts calibrated by high-resolution bio-chronostratigraphy: the central Mediterranean Plio-Pleistocene record. In: de Graciansky PC, Hardenbol J, Jacquin T, Vail PR (eds) Mesozoic and Cenozoic Sequence Stratigraphy of European Basins. SEPM Special Publication 60:155–177

Coleman JM (1966) Ecological changes in a massive fresh-water clay sequence. Transact Gulf Coast Assoc Geol Soc 16:159–174

Coleman JM, Roberts HH (1988) Sedimentary development of the Louisiana continental shelf related to sea level cycles: Part 2 Sedimentary sequences. Geomarine Lett (Special Issue) 8:63–108

Coleman JC, Bouma AH, Roberts HH, Thayer PA (1986) Stratification in Mississippi Fan cores revealed by X-radiography. In: Bouma AH, Coleman JM, Meyer AW et al. (eds) Initial reports of the Deep-sea Drilling Project, Vol. 96. US Government Printing Office, Washington, DC, pp 505–518

Dalrymple RW, Xaitlin BA, Boyd R (1991) Estuarine facies models: conceptual basis and stratigraphic implications. J Sediment Petrol 62:1130–1146

Dickinson WR, Klute MA, Hayes MJ, Janecke SU, Linden ER, McKittrich MA, Olivares MD (1988) Paleogeographic and paleotectonic setting of Laramide sedimentary basins in the central Rocky Mountain Region. Geol Soc Am Bull100:1023–1039

Edwards MB (1979) Late Precambrian glacial loessites from north Norway and Svalbard. J Sediment Petrol 49:85–92

Eisma D (1988) Transport and deposition of suspended matter in estuaries and near shore seas. In: Lerman A, Maybeck M (eds) Physical and Chemical Weathering in Geochemical Cycles. Kluwer, Amsterdam, pp 127–146

Eugster HP, Hardie LA (1978) Saline lakes. In: Lerman A (ed) Lakes: Chemistry, Geology, and Physics. Springer, Berlin Heidelberg New York, pp 237–293

Eyles N, Clark BM (1988) Storm-Influenced Deltas and Ice Scouring in a Late Pleistocene Glacial Lake. Geol Soc Am Bull 10:793–809

Fisk HN (1947) Fine-Grained Alluvial Deposits and their Effects on Mississippi River Activity. US Army Corps of Engineers Waterways Experiment Station, Vicksburg, Mississippi 1:82; 2:maps

Fraser GS (1994) Sequences and sequence boundaries in glacial sluiceways beyond glacial margins. In: Dalrymple RW, Boyd R, Zaitlin BA (eds) Incised-Valley Systems: Origin and Sedimentary Sequences. Society Sedimentary Geology, Special Publication 57, pp 337–357

Griggs GB, Hein JR (1980) Sources of Dispersal and Clay Mineral Composition of Fine-Grained Sediment off California. J Geol 88:541–566

Groat CG (1972) Presidio Bolson, Trans-Pecos Texas and adjacent Mexico: geology of a desert aquifer system. Texas Bureau Economic Geology Report Investigations 76, 46 pp

Gustavason TC (1991) Arid basin depositional systems and paleosols: Fort Hancock and Camp Rice Formations (Pliocene-Pleistocene), Hueco Basin, West Texas and adjacent Mexico. Texas Bureau Economic Geology Report Investigations 198, 49 p

Howell JA, Flint SS (2003) The parasequences of the Book Cliffs succession. In: Coe AL (ed) The Sedimentary Record of Sea-Level Change. Cambridge, Cambridge University Press, pp 158–178

Hrabar SV, Potter PE (1967) Lower West Baden (Mississippian) sandstone body of Owen and Greene Counties, Indiana. Am Assoc Petrol Geol Bull 53:2150–2160

Hunt CB, Mabey DR (1966) Stratigraphy and structure, Death Valley, California. US Geological Survey Professional Paper 494A:162

Krystinik LF, Blakeney BA (1990) Sedimentology of the upper Morrow Formation in eastern Colorado and western Kansas. In: Sonnenberg SA (ed) Morrow Sandstones of Southeastern Colorado and Adjacent Areas. Rocky Mountain Association Geologists, Denver, Colorado, pp 37–50

de Large CJ, Jarvis I, Kuiypers I (1987) Geochemical characteristics and provenance of Late Quaternary sediments from Madeira Abyssal Plain, North Atlantic. In: Weaver PPE, Thomson J (eds) Geology and Geochemistry of Abyssal Plains. Geological Society. London (Special Publication) 31, pp 147–165 (Blackwell Scientific Publications, Oxford)

Lazarenko AA (1984) The loess of central Asia. In: Velicko AA (ed) Late Quaternary Environments of the Soviet Union. University of Minnesota Press, Minneapolis, pp 125–131

Lesueur P, Tastet JP (1994) Facies, internal structures and sequences of the modern Gironde-derived muds on the Aquitaine inner shelf, France. Marine Geol 120,:267–290

McCoy FW, Sancetta C (1985) North Pacific sediments. In: Nairn AEM, Stehli FG, Uyeda S (eds) The Ocean Basins and Margins, Vol. 7A, Pacific Ocean. Plenum Press, New York London, pp 1–46

Meade RH (1994) Suspended sediments of the modern Amazon and Orinoco Rivers. Quaternary International 21:29–39

Meckel LD (1975) Holocene sand bodies in the California delta area, northern Gulf of Mexico. In: Broussard ML (ed) Deltas, Models for Exploration. Houston Geol Soc, pp 239–266

Miller JA (1973) Quaternary history of the Sangamon River drainage system, central Illinois. Illinois State Museum Reports Investigation 27:36

Olsen PE (1990) Tectonic, Climatic, and Biotic Modulation of Lacustrine Ecosystems; Examples from Newark Supergroup of eastern North America. In: Katz BJ (ed) Controls on Distribution of Lacustrine Source Rocks through Time and

Space. American Association Petroleum Geologists Memoir 50, Tulsa, OK, pp 209–224
Orton GJ, Reading HG (1993) Variability of deltaic processes in terms of sediment supply with particular emphasis on grain size. Sedimentology 40:475–512
Parson KA, Brett CE, Miller KB (1988) Taphonomy and depositional dynamics of Devonian shell-rich mudstones. Palaeogeography Palaeoclimatology Palaeoecology 63:109–139
Pashin JC, Ettensohn FR (1987) An epeiric shelf-to-basin transition: Bedford-Berea sequence, northeastern Kentucky. Am J Sci 287:893–926
Pelletier BR (1969) Submarine physiography, bottom sediments and models of sediment transport in Hudson Bay. Geological Survey Canada Paper 68–53:100–135
Pickering KT, Hiscott RN (1985) Contained (reflected) turbidites from the Middle Ordovician Cloridorme Formation, Quebec and Newfoundland. In: Leggett JK, Zuffa GG (eds) Marine Clastic Sedimentology. Graham and Trotman, London, pp 190–211
Quigley RM (1983) Glacial lacustrine and glacial marine clay deposition: a North American perspective. In: Eyles N (ed) Glacial Geology. Pergamon Press, Oxford, pp 140–167
Reading HG (ed) (1996) Sedimentary Environments, Processes, Facies, and Stratigraphy, 3rd edn. Blackwell Science, Oxford, 688 p
Reading HG, Collinson JD (1996) Clastic coasts. In: Reading HG (ed) Sedimentary Environments, 3rd edn. Blackwell Science, Oxford, pp 154–231
Reineck HE, Dorges J, Gudow S, Hertwech G (1968) Sedimentologie, Faunenzonierung, und Faziesabfolge vor der Ostküste der Innerdeutschen Bucht. Senckenbergiana Lethaea 49:261–309
Retallack GJ (1988) Field recognition of paleosols. In: Reinhardt J, Singleo WR (eds) Paleosols and Weathering through Geologic Time: Principles and Applications. Geol Soc Am (Special Paper) 216:1–21
Rine JM, Ginsburg RN (1985) Depositional facies of a mud shoreface in Surinam, South America – A mud analogue to sandy, shallow marine deposits. J Sedimentary Petrol 55:633–652
Roberts SA, Spencer, RJ, Lowenstein TK (1994) Late Pleistocene saline lacustrine sediments, Badwater Basin, Death Valley California. In: Lomando AJ, Schreiber BC, Harris PM (eds) Lacustrine Reservoirs and Depositional Systems. SEPM Core Workshop 19, Denver 12 June 1994 (Society Sedimentary Geology), pp 61–104
Roy PS, Thom BG, Wright LD (1980) Holocene sequences on an embayed high-energy coast: An evolutionary model. Sedimentary Geology 26:1–19
Ruhe RV (1984) Depositional environment of Late Wisconsin loess. In: Porter SC (ed) The Late Pleistocene. University of Minnesota Press, Minneapolis, pp 130–137
Schulz H, Von Rad U, Von Stackelberg U (1996) Laminated sediments from the oxygen-minimum zone of the northeastern Arabian Sea. In: Kemp AES (ed) Palaeoclimatology and Palaeoceanography from Laminated Sediments. Geol Soc London (Special Publication)116:185–207
Skinner RG (1973) Quaternary stratigraphy of the Moose River Basin, Ontario. Geol Survey Canada Bulletin 225:77
Slatt RM, Thompson PR (1985) Submarine slope mudstone facies, Cozy Dell Formation (Middle Eocene), California. Geomarine Lett 5:39–45
Smoot JP, Olsen PE (1988) Massive mudstones in basin analysis and paleoclimatic interpretation of the Newark Supergroup. In: Manspeizer W (ed) Triassic-Jurassic Rifting. Elsevier, Amsterdam, Part A, pp 249–274 (Developments in Tectonics 10)
Soller DR (1982) Text and references to accompany "Map Showing the Thickness and Character of the Quaternary Sediments in the Glaciated United States East of the Rocky Mountains". US Geol Survey Bulletin 1921, 54 p
Speed R (1986) Geologic history of the Barbados: A preliminary Synthesis. Transactions 11th Caribbean Geological Conference Barbados, 20 to 26 July, 1986, pp 29:1–29:11
Stow DAV (1981) Laurentian fan: morphology, sediments, processes and growth patterns. Am Assoc Petrol Geol Bull 65:375–393
Stow DAV, Mayall M (2000) Deep-water sedimentary systems: new models for the 21st century. Marine Petrol Geol 17:125–135
Stow DAV, Faugères J-C, Viana A, Gonthier E (1998) Fossil contourites: a critical review. Sedimentary Geology 115:3–31
Sturm M, Matter A (1978) Turbidites and varves in Lake Brienz (Switzerland): deposition of clastic detritus by density currents. In: Sturm M, Tucker MA (eds) Modern and Ancient Lake Sediments. International Association Sedimentologists (Special Publication) 2, pp 147–168
Syvitski JPM (1986) Estuaries, deltas, and fjords of eastern Canada. Geoscience Canada 13:91–100
Tiercelin JJ, Soreghan M, Cohen AS, Lezzer K-E, Bourouillec J-L (1982) Sedimentation in large rift lakes: Example from the Middle Pleistocene modern deposits of the Tanganyika Trough, East African Rift System. Bulletin Centre Recherche Exploration-Production Elf Aquitaine 16:83–111
Walker RG (1971) Non deltaic depositional environments in the Catskill clastic wedge (Upper Devonian) of central Pennsylvania. Geol Soc Am Bull 82:1305–1326
Wallace-Dudley KE, Leckie DA (1995) Sedimentology and source-rock potential of the Lower Kaskapau Formation (Cenomanian-Lowermost Turonian), northwestern Alberta. Geological Survey Canada Bulletin 490, 60 p
Weimer RJ (1982) Developments in sequence stratigraphy: foreland and cratonic basins. Am Assoc Petrol Geol Bull 76:965–982
Wells JT (1983) Dynamics of coastal fluid muds in low-, moderate-, and high-tide range environments. Canadian J Fisheries Aquatic Sci 40:(Suppl):130–142
Wells JT, Coleman JM (1978) Longshore transport of mud by waves; northeastern coast of South America. Geologie en Mijnbouw 57:353–359
Wells JT, Coleman JM, Prior DB (1980) Flowslides in muds on extremely low angle tidal flats, northeastern South America. Geology 8:272–275
Wells JT, Adams CE Jr, Park Y-A, Frankenberg EW (1990) Morphology, sedimentology, and channels processes on a high-tide range mudflat, west coast of South Korea. Marine Geol 95:111–130
Whelan T III, Coleman JM, Suhayda JN, Garrison LE (1975) The Geochemistry of Recent Mississippi River delta Sediments: Gas concentration and sediment stability. Offshore Technology Conference Number 7, 3:71–84

Digging Deeper

Bennett MR, Huddart D, Thomas GSP (2002) Facies architecture within a regional glaciolacustrine basin: Copper River Alaska. Quaternary Sci Rev 21:2237–2279.

This is a profusely illustrated and very detailed sedimentological study of an intermontane basin whose exit was blocked by ice. Consequently, its basin floor has a thick fill of diamictons and lacustrine sediments – gravels, laminated silts and sands and volcanic layers. These were mostly deposited as basin floor fans by slumps, debris flows, and density and turbidity currents, and in some smaller basins as marginal fans. An excellent Pleistocene model for the study of ancient glacial deposits of intermontane basins. Fig. 26 notable.

Bentley SJ, Roberts HH, Rotondo K (2003) The sedimentology of muddy coastal systems: the research legacy and new perspectives from the Coastal Studies Institute. Gulf Coast Association

Geological Societies / Gulf Coast Societies Society Economic Paleontologists Mineralogists Transactions 53:52–63.

A most instructive, well-illustrated historical review of studies starting in the early 1950s of Louisiana's muddy coast and shelf. This review shows how our ideas and insights have expanded in 50 years, and the great importance of direct observation of modern muddy sediments over an extended period. Another lesson is the importance of new instrumentation for the field study of fine-grained sediments in three dimensions. Read this for an integrated summary of processes in a large delta

Bouma AH, Stone CG (eds) (2000) Fine-grained Turbidite Systems. American Association Petroleum Geologists Memoir 72, 342 pp.

Twenty-eight chapters (many based on North American examples) provide broad insights to fine-grained turbidites – from modeling to case histories and methodologies. On the inside of the front cover is a helpful graphic key to 33 topics covered in the 28 chapters of the memoir. Informative illustrations, many in color.

Cohen AS (2003) Paleolimnology. Oxford University Press, Oxford, p 500.

Fourteen impressive chapters, a glossary, and over 2,000 references span all aspects of lakes new and old. Essential reference if you study lakes.

Coleman JM (1966) Ecological changes in a massive, fresh-water clay sequence. Transact Gulf Coast Geol Soc 16:159–174.

Pioneering paper uses continuous X-ray radiography to discover all the fascinating details of seemingly massive fresh water clays with only minor intercalations of peat and silt including many diagenetic structures as well as primary ones. Fluctuations in water level of swamp greatly affected lithology and diagenesis. Many informative photographs.

Davis HR, Byers CW (1993) The role of bottom currents and pelagic settling in the deposition of shale in an oxygen-stratified basin: the study of the Mowry Shale (Cretaceous) of Wyoming. In: Caldwell WGE, Kauffman EG (eds) Evolution of the Western Interior Basin. Geol Assoc Canada (Special Paper) 39, pp 177–178.

Short, but telling description and interpretation of fine-grained lithofacies and inferred processes. The Mowry Shale has hummocky bedding and lamination in western Wyoming but graded bedding and lamination and much less bioturbation in eastern Wyoming where water was deeper, shales more organic rich, and there was less bottom oxygen. Orientation of slump folds confirms an easterly dip of this shelf. Excellent model for both processes and lithologies of an oxygen-deficient, muddy epeiric shelf. Compare with Wallace-Dudley and Leckie (1995).

Fisk HN (1947) Fine-grained alluvial deposits and their effects on Mississippi River activity. US. Army Corps of Engineers Waterway Experiment Station, Vicksburg, Mississippi 1:82; 2 maps.

One of the classics of all time and far, far ahead of its time too. Rich in data and cross sections (70 large plates). The fine-grained deposits of the Mississippi are divided into meander belt, back swamp, braided stream and delta plain- most with subdivisions. We think that more studies of modern muds on the scale of those made by Fisk are needed.

Froehlich AJ, Robinson GR Jr (eds) (1988) Studies of early Mesozoic basins of the eastern United States. US. Geol Survey Bull 1776:433.

Twenty-five short articles of which 13 pertain to sedimentary rocks and include facies and environments, studies of kerogen, inorganic geochemistry, clay mineralogy and palynology. Rich source of diverse quantitative data for early Mesozoic lacustrine fill.

Gierlowski-Kordesch EH, Kelts KR (eds) (2000) Lake Basins Through Space and Time. AAPG Studies in Geology 46:648.

Abundantly illustrated collection of articles on modern and ancient lakes and their deposits has 60 articles organized in 9 parts. The first chapter is an overview with colored plates that is followed by specific examples arranged by ages – Carboniferous to Permian, Permian-Triassic to Jurassic, Early to Middle Cretaceous, Late Cretaceous to Paleocene, Miocene-Pliocene, and Quaternary. A landmark source of information about lakes deserving your attention.

Healy T, Wang Y, Healy J-A (2002) (eds) Muddy Coasts of the World: Processes, Deposits, and Function. Elsevier, Amsterdam, 542 p.

Twenty-one, short, moderately technical, well-referenced chapters cover the world geographically with a muddy coast inventory and summarize both the processes and characteristics of muddy coasts. Perhaps of most interest are "Mudshore Dynamics and Controls: Biochemical Factors Influencing Deposition-Erosion of Fine-grained Sediment"; "Research Issues of Muddy Coasts"; and "Morphodynamics of Muddy Environments along the Atlantic coasts of Coasts of North and South America". The chapter "Geographic Distribution of Muddy Coasts" is an extremely valuable compilation with over 600 references.

Hardie LA, Smoot JP, Eugster H (1978) Saline lakes and their deposits: a sedimentological approach. In: Matter A, Tucker MC (eds) Modern and Ancient Lake Sediments. International Association Sedimentologists, Special Publication 2, Blackwell Science, Oxford, pp 7–41.

In our opinion, the most insightful article about the sedimentological processes and deposits of a saline lake (more than 5,000 ppm dissolved solutes) and its bounding environments. This classic study stresses modern examples, describes 10 facies from marginal fan to basin center, includes both chemistry and biology, and presents some ancient examples. Descriptions are detailed and always linked to processes. Important starting point.

Kelts K, Talbot MR (1990) Lacustrine carbonates as geochemical archives of environmental change and biotic/abiotic interactions. In: Tilzer MM, Serruya C (eds) Large Lakes. Springer, Berlin Heidelberg New York, pp 288–315.

Valuable paper reviews stable isotopes and summarizes their significance for the study of lacustrine deposits. Excellent discussion of lacustrine carbonate facies and how their different isotopic signatures contribute to the study of ancient lake deposits, which are commonly either carbonate rich or have mudstones interbedded with carbonates.

Kolb CR, Van Lopik JR (1975) Depositional environments of the Mississippi River delta plain. In: Shirley ML, Ragsdale JA (eds) Deltas in their Geologic Framework. Houston Geol Soc, pp 18–61.

Well-written, easy-to-read description of the muds and silts of the Mississippi Delta with emphasis on their engineering properties (Fig. 16 provides an unusual, very complete graphic summary). The central idea of the paper is that the depositional environment controls engineering properties. Outstanding classic.

Mansfield GR (1938) Flood deposits of the Ohio River, January-February 1937. US. Geological Survey Water Supply Paper 838:639–736.

Remarkable early "just-after-the-event" description of the alluvial deposits of the 1937 flood – the largest of the 20th century in the Ohio Valley. Distribution and characteristics of deposited mud especially informative. Reading this is time well spent.

Pedersen GK (1985) Thin, fine-grained storm layers in a muddy shelf sequence: an example from the Lower Jurassic in the Stenlille 1 well, Denmark. J Geol Soc London 142:357–374.

Remarkable for the recognition of 14 types of lamination, which help distinguish a sandier, more scoured, cross laminated inner shelf association from an outer shelf association where much of the lamination is graded and considered to have been deposited from distal storm deposits.

Pickering KT, Hiscott, RN, Hein FJ (1998) Deep-marine Environments. Unwin Hyman, London, 416 p.

Subtitled, "Clastic Sedimentation and Tectonics", this book has 12 well-illustrated chapters that combine both onshore and marine geology, outcrops and cores, seismic images and field mapping. Particularly valuable is Chapter 3, which contains descriptions of seven major facies of which 14 subfacies are fine grained, terrigenous and biogenic. Excellent.

Retallack GJ (1997) A Color Guide to Paleosols. Wiley, New York, 175 p.

One hundred and thirty six beautiful and informative color photographs each with a complete explanation plus chapters on recognition, burial diagenesis, interpretation and methodology. Don't study continental deposits without this book.

Tyson RV, Pearson TH (1991) Modern and ancient continental shelf anoxia: an overview. In: Tyson RV, Pearson TH (eds) Modern and Ancient Continental Shelf Anoxia. Geol Soc London (Special Publication) 58:1–24.

Clear analysis of a complex problem that has long troubled stratigraphers and sedimentologists. Table 2 clarifies terminology and Tables 4 and 5 summarize dysoxic and anoxic facies of ancient shelves and theories thereof. While controversies still remain, this paper goes far to resolve many of them. Strong emphasis on engineering properties (Fig. 16 provides an unusual, very complete graphic summary).

Wang Y (1983) The mudflat system of China. Canadian J Fisheries Aquatic Sci 40:(Supplement 1)160–171.

Describes the mudflats of China's coast, which are chiefly nourished by the Yangtze and Yellow Rivers, and discusses how deforestation, dike and dam building over 2,000 years have altered the supply of fines to the coast and thus the position of its shoreline.

Wright LD, Wiseman WJ Jr, Yang Z-S, Bornhold BD, Keller GH, Prior DB, Suhayda JN (1990) Processes of marine dispersal and deposition of suspended silts off the modern mouth of the Huang He (Yellow River). Continental Shelf Res 10:1–40.

Hydraulics of concentrated silt dispersal by density-driven underflows is well explained.

CHAPTER 6

Burial

Biology drives shallow burial reactions, temperature drives late

6.1. Introduction

Burial in sediments occurs because subsidence of the basin or sea level rise creates accommodation space that allows new sediment to accumulate on top of that previously deposited. The weight of this overlying sediment plus the higher temperatures encountered with depth in the earth transform the initial muddy sediment into a lithified mudstone. In the process, solid organic matter is converted to mobile oil and gas; water is expelled that drives the migration of petroleum or metal-bearing fluids; mineral transformations occur that produce (or destroy) economic deposits; and the record of earth surface conditions in the past is altered, perhaps to become unrecognizable. For these and other reasons of pure curiosity, the burial of muds has received great attention from both industrial and academic researchers.

We have separated our discussion into physical and chemical aspects. This approach reflects a separation in the way most workers have addressed questions of burial and hence the nature of the literature on this subject. Within these categories, we further subdivide burial processes into those occurring very early in the burial sequence, within the top few meters of the sediment pile, and those occurring much later, after a considerable accumulation of overlying sediment (Fig. 6.1). Chemically, little reaction seems to occur in the transition interval between these two depth ranges. Shallow burial is characterized by rapid changes in porosity and by vigorous biological reactions. Furthermore, because these processes involve exchange between the mud and the overlying water, the environment of deposition exerts strong control. Deeper burial sees a slower, more linear rate of compaction and abiologic reactions such as generation of hydrocarbons and the transformation of less stable minerals like Ca-plagioclase to more stable varieties like albite and microcline. Because these later processes occur far from the surface, the environment of deposition exerts little influence; instead, the rate of burial and the geothermal gradient of the basin control events.

A good way to develop insight into the burial behavior of mudstones is to contrast them with sands. Muds and sands are very different in their responses to burial because of the much greater compressibility of mud. Typical modern muds have porosities of

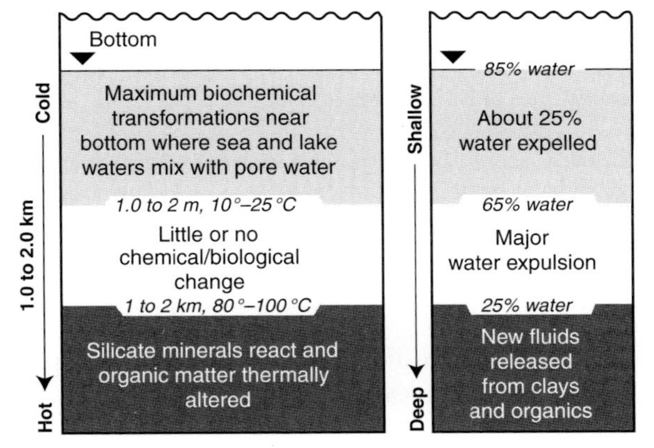

Fig. 6.1. A framework for the study of burial

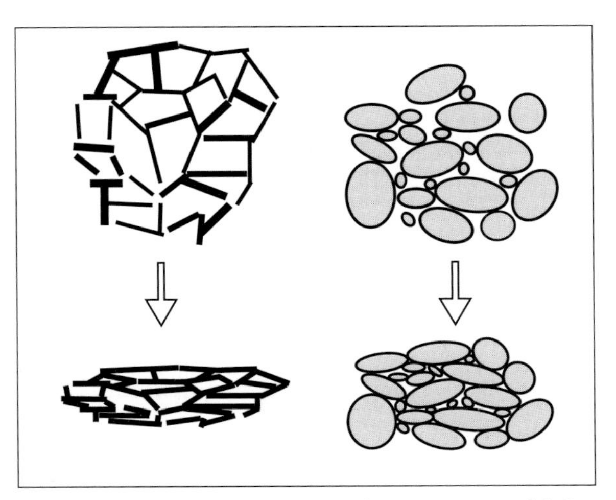

Fig. 6.2. Differential compaction of clay versus sand (after Velde 1996, Fig. 5)

Table 6.1. Mud to mudstone with burial

Physical Processes	Chemical Processes
Shallow	
Driving force is gravity; rapid expulsion of pore water in first few meters (especially in non-flocculated muds), but differences in texture and rate of sedimentation cause wide contrasts in compaction curves, which are generally negative exponentials; fabrics range from open and variable to less common tight and parallel; void ratios typically 1.5 to 3.0 or more; some fractures form early during pore water expulsion to interface	*Open system* – Driving force is bacterial generation of H_2S, CO_2, CH_4 from organic matter degradation: • early Fe sulfides • dissolution of aragonite and precipitation of Ca, Mg and Fe carbonates • Fe(II)↔Fe(III) conversions • early phosphates at interface • evaporite minerals (if hypersaline) • amorphous silica from sponges, diatoms, radiolaria converts to opal CT
Deep	
Driving force is gravity; compaction slow and approximately linear except where rapid burial prevents expulsion of pore water (overpressured mudstone); expelled water carries silica, ions, organic acids that go into sandstones and carbonates to produce quartz overgrowths, albite, carbonate cements or secondary porosity; most mudstone fabrics strongly subparallel and tight; void ratios < 1.2; only effective porosity is in fractures	*Largely a closed system* – Driving force is heat: • smectite and kaolinite converted to illite and chlorite; release of H_2O, H_4SiO_4, Ca^{2+} and Na^+; K^+ added • higher crystallinity and better ordering of clay minerals • unstable silica polymorphs to quartz • plagioclase converted to K-feldspar • conversion of organic matter to darker, less volatile kerogen; expulsion of organic acids, gas and oil

Table 6.2. Contrasts in burial of mud and sand

Mud	Sand
Compaction	
Great: 60 to 80% initial porosity (mostly platelets deposited as open flocs) goes to less than 40% in first few 100 m so much pore water expelled	Modest to little (where early cemented): deposited as single, touching, rigid grains rather than open flocs
Deep burial framework	
Tight, subparallel, felt-like fabric of clay minerals admixed with silt, pellets and organic matter with 10 to 20% porosity or less and negligible permeability (except for microfractures)	Rigid framework with point, concavo-convex and sutured contacts protects original and secondary porosity and permeability unless filled with cement; matrix-rich varieties intermediate with mudstones
Interaction with fluids	
Much fluid expelled in first several hundred meters plus some deeper (smectite → illite/chlorite + H_2O), but mostly a *closed system* below a few hundred meters (*aquicludes/aquitards*)	Moderate fluid expulsion at first, then little (rigid framework); mostly an *open system* at all depths unless cemented (*aquifers/reservoirs*); fluids added from adjacent mudstones or circulating brines
Secondary porosity	
None except for microfractures (no rigid framework; clay mineral transformations do not create pores)	Common – dissolution of framework feldspars, rock fragments and cements by hot acid solutions from mudstones or by fresh water from outcrop
Mineral transformations/timing	
Early: pyrite; carbonate; phosphate; glauconite *Late*: Smectite and kaolinite → illite + chlorite + H_2O; organics → organic acids + CO_2 + hydrocarbons; carbonate and silica exported to sandstones or limestones	*Growth*: Some early carbonate; quartz/feldspar overgrowths at depth, late carbonate, secondary matrix, clay mineral transformations *Dissolution*: Secondary porosity in carbonates and feldspars

70 of 80%, which are reduced with deep burial to only a few percent. Permeabilities also become negligible at depth. In contrast, sandstones have initial porosities of no more than 50%, but may have as much as 25% porosity at deep burial (Fig. 6.2). Sands have higher permeabilities than muds in both the modern and the ancient. Well-sorted sands range from 1 to 100 darcys, whereas clays are only in the

range 10^{-6} to 10^{-3} darcys (Fetter 1994, Table 4.6). After burial, sandstones retain appreciable, but highly variable permeabilities, whereas for mudstones, matrix permeability is effectively zero with values as low as 10^{-11} (Hunt 1996, p 268). Thus fluid flow in mudstones is through their fractures rather than through their pores.

Contrasts in framework explain the difference in porosity behavior. Newly deposited mud consists of an open framework of clay mineral platelets with some silt, sand, pellets and perhaps a few percent of organic matter and bacterially-generated pyrite. Unless this open framework is "frozen in" by the growth of a concretion, it collapses and deforms gradually as burial proceeds. For sands, the framework consists of dense, semi-elliptical rigid grains, well packed together and dominated by quartz. Furthermore, organic matter is almost totally absent (Table 6.1). Such a framework is less subject to collapse under pressure and instead porosity is lost by infilling of pores by mineral cements. These contrasts produce the differences between mudstone *confining layers* and *reservoir seals* on the one hand and sandstone *aquifers* and *petroleum reservoirs* on the other (Table 6.2). With deep burial and time, the diverse clay minerals of a mud are converted to mostly illite and chlorite, whereas most of the initial grains of a sandstone survive, because mineralogically stable quartz forms the greatest part of the framework. The term *diagenesis* is used to describe all these post-depositional chemical and mineralogical changes.

First, we examine the transformation of mud into a consolidated mudstone starting with the first few meters of burial and, then deep burial to 1,000 s of meters. We emphasize that present depth, the *path* taken to that depth (rapid or slow burial), geothermal gradient, rate of sedimentation, and later tectonic accidents (folding, faulting or shearing) are key factors for the physical and mineralogical changes experienced by mudstones.

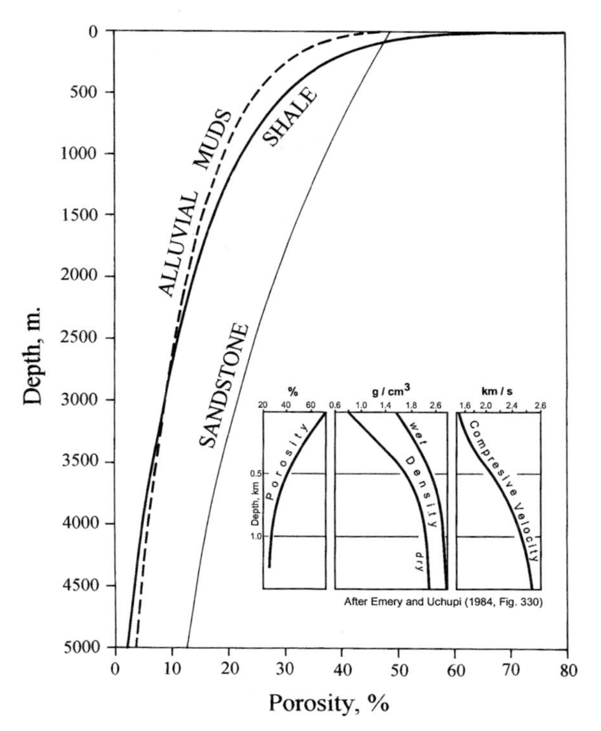

Fig. 6.3. Porosity-depth curves for deep and shallow burial (Mackey and Bridge 1995, Fig. 6; Emery and Uchupi 1984, Fig. 330). Published by permission of the authors and the Society for Sedimentary Geology

6.2. Physical Processes: Compaction and Deformation

The compaction of mud reduces its volume and expels large volumes of water upward and outward. Consequently, the porosity and permeability of the mud decrease with depth, whereas density and seismic velocity both increase (Fig. 6.3). This process is directly tied to the initial, open fabric of mud. Many muds are deposited as open aggregates or floccules (Fig. 3.2), because of ionic attraction between the edges and faces of the clay platelets. Broken bonds along the edge of a clay platelet show a positive charge to the surrounding solution while the large flat surface displays a negative charge. Therefore the platelets tend to clump in a "house-of-cards" structure. This open structure is responsible for the great porosity, water content, and low shear strength of mud. These features in turn cause muds to fail easily by creep, flow and sliding on the most gentle of slopes (Chapter 3). Water continues to be expelled as burial depth increases, with a companion decrease in porosity. The void ratio decreases from perhaps 3.0 to 1.5, and the clay mineral platelets become more parallel. In addition, ovoid fecal pellets become flatter, organic matter is greatly compressed and both the clay minerals and organic matter deform around scattered quartz grains and early nodules. The result is a somewhat open but felt-like fabric with perhaps 30 to 50% porosity, but with much variation. It is this variability that makes it difficult to find a single mathematical curve that adequately represents the compaction process from the sediment-water interface to great depths (Box 6.1). But compaction is more than just physical compression and rotation of

Box 6.1. Porosity, void ratio, and permeability

Compaction of a sediment is commonly measured either by change in its *porosity*, $\emptyset$, or by changes in its *void ratio, e*, both of which decrease with depth of burial whereas density and sonic velocity increase. Porosity is defined as volume of voids divided by *the total volume*, whereas void ratio is defined as volume of voids divided by *volume of solids*. These measures are related to one another by

$$\emptyset = \frac{e}{1+e} .$$

Void ratio is commonly used by geotechnical engineers whereas porosity is preferred by geologists. Compaction with depth is often roughly approximated by a negative exponential

$$\emptyset = \emptyset_o e^{-ad}$$

where $\emptyset_o$ is initial porosity, a is a constant for a particutar data set, and d is depth. But a linear relation also is useful for some older, more compacted mudstones,

$$\emptyset = \emptyset_o - ad$$

where a is the estimated decrease in porosity per 100 m for a particular basin.

Another parameter closely associated with the compaction process is *permeability*, k, the capacity of a sediment to transmit fluids rather than simply contain or store them. Permeability like porosity, also decreases with depth of burial. Permeability, k, is defined by Darcy's law as

$$k = Q/iA$$

where Q is discharge (volume/time), i is hydraulic head (gradient), and A is the cross sectional area of the flow. Permeability is a function of both the percent of pores in a sediment and the ease with which fluids can flow through them. Even well compacted mudstones may have up to 10 to 20% porosity, but negligible permeability and form aquicludes or seals, because, while there are many small pores, they are poorly connected. Thus when mudstones do have permeability, microfractures probably are responsible.

platy materials, because at depth mineral transformations and the growth of new minerals combine to reduce porosity and bind the aggregates together so that porosities of only a few percent survive in old mudstones.

6.2.1. Shallow Burial

The early stages of burial of a mud see rapid expulsion of water and reduction in bulk volume. The amount of this early compaction that an ancient mudstone experienced can be measured from sedimentary structures. Examples are the marked increase in the spacing between laminae when they pass into early-formed concretions, the folding of originally vertical dikes and burrows, and ball-and-pillow structures. Stratification preserved in large concretions is typically 4 to 6 times thicker inside than outside of the concretion (Fig. 6.4). Compaction also causes near vertical burrows and early dikes in mud to deform, typical values ranging from 30 to 45% (present length divided by original length and corrected for an estimate of percent of ductile material). Ball and pillow structures occur where denser sandstones or fine-grained carbonates were deposited over less dense mud (see Ettensohn et al. 2002 for examples). Shaking, by earthquakes or pressure pulsing by wave action, instantaneously dewaters the mud which allows portions of the overlying coarse bed to

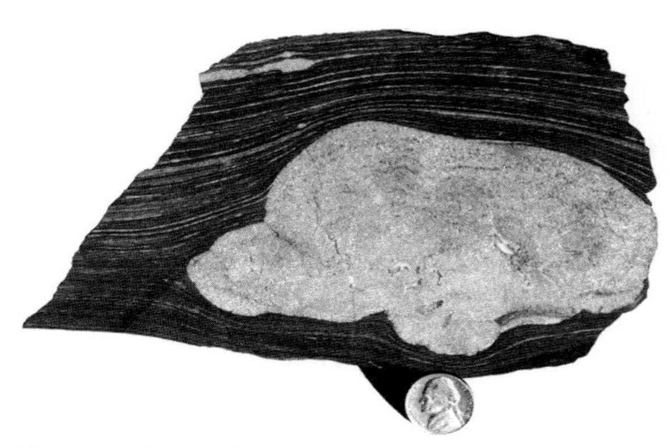

Fig. 6.4. Differential compaction around a concretion

Table 6.3. Skeletal preservation and burial at interface (adapted from Brett and Allison (1998, Table 1))

	Rapid/Episodic Burial	Intermediate Burial	Slow Burial
High energy			
Fragile/branching	Mostly intact	Much fragmentation	Destroyed
Bivalves	Mostly intact, but rarely in situ	Partly intact; some fragmentation	Abraded and disarticulated fragments
Multiple segments	Mostly intact, but rarely in situ enrolled trilobites, etc.	Partly intact; some size sorting	Disarticulated and sorted
Low energy			
Fragile/branching	Intact	Some fragmentation	Much fragmentation and corrosion
Bivalves	Intact; some in situ	Mostly disarticulated valves	Disarticulated; minor fragmentation and corrosion
Multiple segments	Mostly intact; some in situ	Partly articulated, but not sorted	Disarticulated, but not sorted

sink down into it. The texture of the surface of the pillows provides insight to the state of consolidation of the underlying mud at the time of deposition of the coarser bed. If the bottom of the pillow is perfectly smooth, the mud was too soft and soupy to support either body fossils or burrowers, but if the bottom of the bed or pillow is fluted or has casts of burrows from resting and grazing trails, the mud was already semi-consolidated. A special case of shallow dewatering is syneresis, which produces polygonal fractures in mudstone. These are commonly sand filled and resemble desiccation mud cracks but lack the turned-up edges and form in demonstrably deep-water sediments. Like ball-and-pillow, they may be seismite layers, with rapid dewatering and sand injection being produced on the seafloor by seismic shock waves (Pratt 1998).

Where a mudstone contains fossils, their preservation – broken or intact, sorted or corroded – provides rapid and easy qualitative insight into how long the fossil assemblage remained at the sediment-water interface before final burial. The longer at the interface, the slower the rate of sedimentation and burial, and the greater the chance for organic sheaths that bind calcareous segments to decay, for currents and predators to break or attack skeletons, or for dissolution to occur on the sea floor (Table 6.3).

Variations in the rate of shallow compaction are mostly caused by differences in the texture and fabric of the mud, and the amount of intermixed fine calcareous debris. The finer grained a mud, the greater its water content and the more it compacts. The fabric of flocculated muds permits more water to be held than non-flocculated fabrics, so flocculated muds compact more unless the fabric is held open by admixed silt grains or by early cementation (Almon et al. 2002). Bioturbation seems also to reduce compaction of flocculated muds, perhaps by enhancing early carbonate cementation or by mixing in more silt grains. Because of the solubility of calcite and the rapid kinetics of its reactions (low energy thresholds for initiation of dissolution and precipitation), shear strength increases more rapidly downward at shallow depths in calcareous muds than in non-calcareous muds (Kenter and Schlager 1989). More organic matter in mud also results in more compaction.

But drainage is important too and is provided by thin interbedded sandstones. Where these are absent and sedimentation is rapid, pore water cannot escape easily. Consequently, compaction is incomplete and the mudstone is said to be *overpressured*. Overpressuring causes a mud or mudstone to have a low density compared to the compaction equilibrium value for its burial depth, which makes the mud unstable. Thinking of a mudstone section as consisting of incremental units of deposited mud is helpful here. Suppose, for example, that the pore water of the last deposited mud, call it unit 1, freely communicates with the water at its sediment-water interface. Under these conditions the pore water pressure in unit 1 will be entirely hydrostatic. But now suppose a new mud bed, call it unit 2, is deposited before the compaction of the mud of unit 1 is complete so not all of the pore water of unit 1 can escape. Now the pore water pressure of unit 1 is greater than hydro-

Fig. 6.5A,B. Overloading of unconsolidated deltaic mud: (**A**)idealized block diagram of deformed Bedford Shale (Mississippian) underlying and adjacent to linear deltaic distributary in northern Ohio (Pashin and Ettensohn 1995, Fig. 29), and (**B**)penecontemporaneous microfaulting in Berea Sandstone formed by its subsidence into the uncompacted muds of the Bedford, Cleveland and Chagrin Shales

static and the mud of unit 1 is *overpressured*. Stated differently, rapid sedimentation inhibits the escape of water from a mud (protects porosity and thus reduces density), which causes it to be mechanically unstable.

Severe overpressuring occurs in mudstones at shallow depths in response to rapid deposition of thick sand beds on top of the mud. One case is a low energy delta where a new distributary channel deposits a thick sand body on soft, water-rich mud, which then deforms and rises to the surface producing mud lumps. Here the heavy sand body can be considered the trigger that starts deformation in rapidly buried muds already prone to overpressuring and creep (Figs. 6.5A and B). Still another possibility is an esker rapidly deposited over glacial clays (Fig. 6.6).

Muds are particularly prone to soft sediment deformation compared to other types of sediment. Soft,

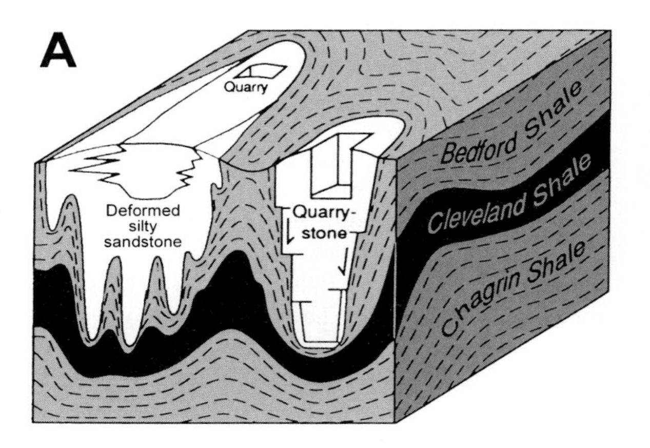

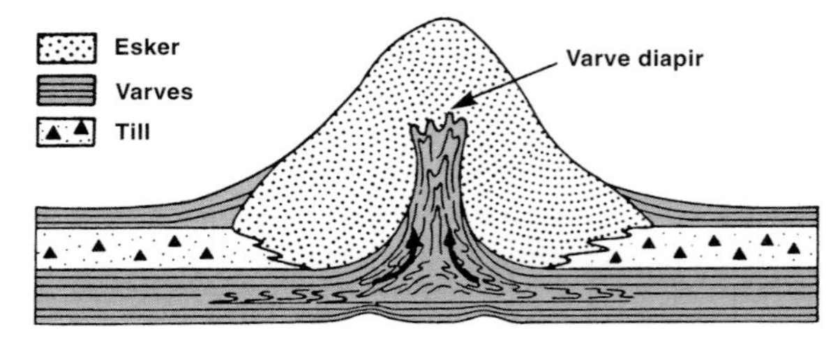

Fig. 6.6. Overloading of unconsolidated proglacial mud: diapiric intrusion of deformed varves into an esker (Banerjee and McDonald 1975, Fig. 1). Published by permission of the Society for Sedimentary Geology

wet, and plastic muds deform readily in response to even small shear stresses. Mud creeps down slopes as low as 0.03 to 0.08° on land where the water table is near the surface, on exposed tidal flats and underwater on delta fronts, on the sides of channels and beyond into deep water. Both mud and silt easily liquefy on horizontal scales of a few decimeters to many hundreds of meters through *dynamic cyclic loading* by storm waves or seismic shocks (or by rapid change of the water table in alluvial deposits). Cyclic pressure pulsing of the pore water of shallowly buried sand, silt, and mud and may also exsolve methane gas from the pore water. Such exsolution reduces the effective pressure and is a cause of liquefaction, failure (slumps and slides) or unwanted shifting of piers and piles. Storm wave effects extend to depths of 120 m or so in the Gulf of Mexico and were responsible for a number of rig failures during hurricane Camille (Sterling and Strohbeck 1973; Prior and Coleman 1980, p 26).

Methane appears to be particularly important in slope instability of marine muds. Shallow muds of the Mississippi delta can have methane concentrations above saturation, producing bubbles, which leads to very low shear strength (Whelan et al. 1975). In deeper water where the pressure is higher, methane actually occurs as a crystalline solid in the form of methane hydrates. These hydrates, being solids, are a form of early cementation of the mud and lend it strength, but they are very sensitive to pressure variations, and the drop in sea level associated with glaciation appears to be sufficient to cause reversion of the solids to methane gas with consequent loss of stability and hence an increased frequency of submarine slumps and slides (Paull et al. 1995).

6.2.2. Deeper Burial

Somewhere between 100 and 1,000 m, the "transition zone", the rate of compaction slows dramatically and the curve of porosity change with depth becomes approximately linear. Below this transition, increasing burial pressure collapses more and more floccules and the individual platelets become more parallel. The mudstone fabric now has a tighter, better-defined felt-like structure and, although original floccule structure can survive, it is now much denser than at shallow depths. Void ratios are normally less than 1.2 and sonic velocities increase as density increases.

A striking feature of deeper burial is the appearance of reversals in the porosity/depth curve. That is, a zone is reached where porosity and accompanying pore pressure suddenly increases. Because such overpressured zones can be extremely dangerous in drilling for petroleum, much research has been devoted to understanding this phenomenon. One process generating extra water pressure at depth is the transformation of smectite to illite, which converts mineral-bound water to free water in pore spaces of the mudstone (Fig. 6.7). It has been suggested that high geothermal gradients promote overpressure, because they accelerate both the smectite-illite transition (Dutta 1986) and organic maturation. Spencer (1987) argued that, in the Rocky Mountains, petroleum generation was a major cause of overpressuring through the transformation of solid kerogen to liquid and gas hydrocarbons. Overpressure has also been suggested as a process producing fracture systems in mudstones, some sufficient to create reservoirs for petroleum (Capuano 1993), or conversely, such fracturing might destroy a petroleum reservoir by introducing permeability into a mudstone seal.

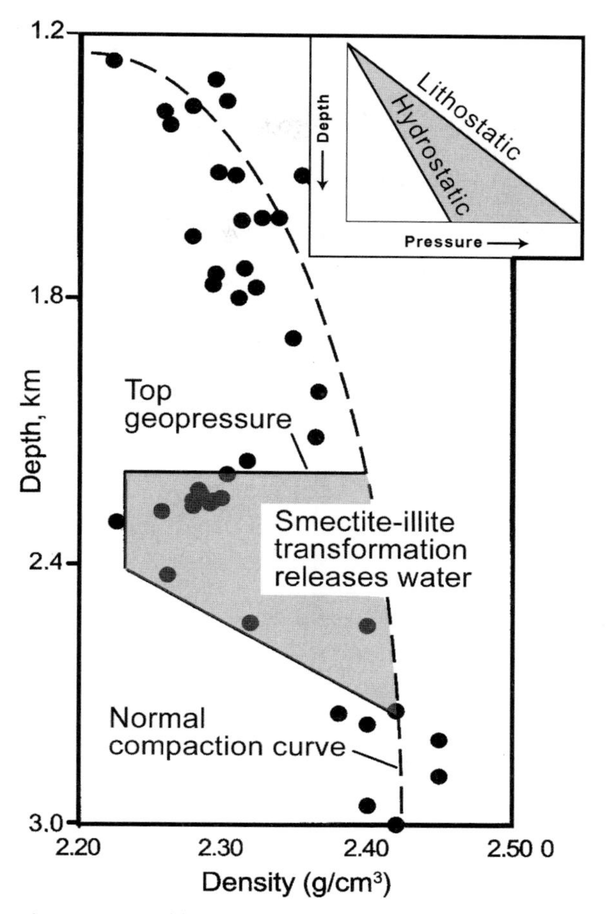

Fig. 6.7. Normal burial pressure and zone of overpressure associated with smectite-illite transition (after Kim et al. 2001, Figs. 1 and 12). Published by permission of the authors and the Association of Gulf Coast Geological Societies

In modern deltaic deposits, overpressured mud can cause smaller mud diapirs (Fig. 6.8) that can rise through the section and reach the surface to form mud volcanoes (Fig. 6.9) or "mud lumps". On passive margins, these structures, along with salt diapiric intrusion, produce complex structures in what otherwise would be layer-cake stratigraphy and thus generate petroleum traps without tectonic deformation. On active margins, regional tectonic squeezing or shearing produces mud-mudstone diapirs on a grand scale some of which rise 100 s to 1,000 s of meters in the section. Mudstone diapirs, like their salt counterparts, also cause highs on the sea or lake bottom that deflect turbidity currents so that there is commonly a close inverse relationship

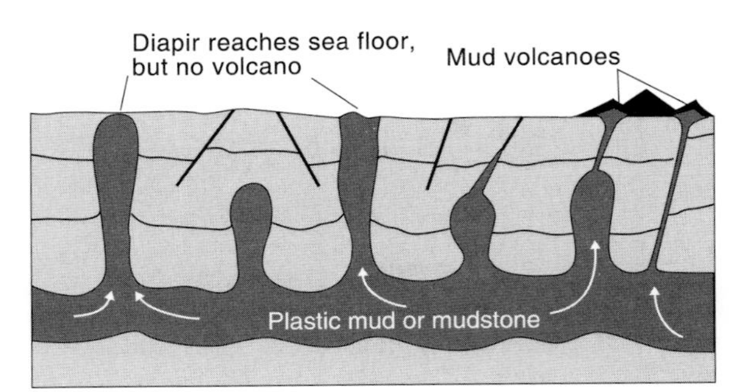

Fig. 6.8. Basic types of mud diapirs (after Milkov 2000, Fig. 2). A variety of responses are possible, from piercement domes to overpressured slurries of mud flowing along faults. Published by permission of the author and Elsevier Science

Fig. 6.10. Slickensides in a soft Cretaceous mudstone core, offshore Brazil (Courtesy of PETROBRAS), and from a mudstone in a fault along Pine Mountain in eastern Kentucky

between position of the diapiric high and adjacent, elongate turbidite sand bodies. Higher in the section, the weak doming caused by the rising mudstone or salt diapir may be sufficient to trap petroleum in a sand body that crosses the high. Mudstone overpressuring occurs at both deep and shallow depths and has many different causes and consequences. It is therefore essential to be aware of and understand it.

Of all the sedimentary rocks, mudstones are the most incompetent, with the exception of bedded salt. Deformation ranges in scale from slickensides (Fig. 6.10) to thickening in the nose of folds (Fig. 6.11) to giant mud-cored anticlines along continental margins (Fig. 6.12). Mudstones provide the lubricants for low angle thrust faults, and on a small scale, may smear along a fault to make it a seal rather than a conduit for fluids. Still another example is a deep road cut with rocks dipping gently toward the cut (bringing water to it) and interbeds of shale at its bottom. This is an invitation to trouble, particularly if the shale should have a large proportion of expandable clays or a fracture network that transmits water. The higher water pressures at the base of the cut will cause the overlying rock mass to slide along the mudstone layer toward the cut. Spectacular slope failures of this kind occur along the southern California coast where volcanic ash beds dip seaward (Watry and Ehlig 1995).

Unlike cemented sandstones and carbonates, an indurated mudstone, when subject to the same lateral stress parallel to bedding, shears and fails easily because its original sub-parallel, platy clay-mineral

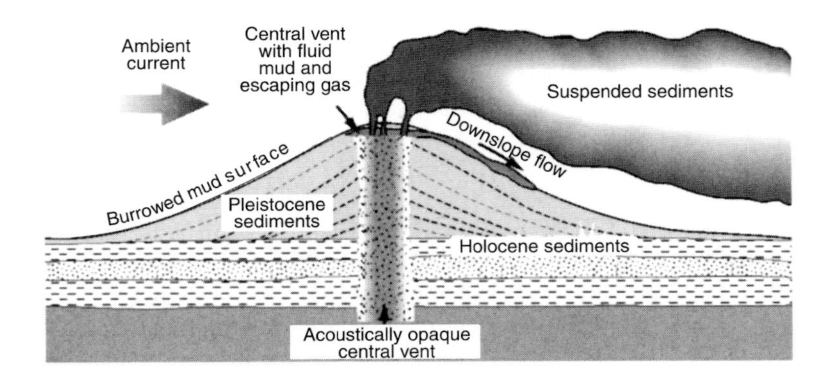

Fig. 6.9. Schematic representation of mud volcanoes on the sea bottom off the front of the Mississippi delta in the Gulf of Mexico (Roberts 2001, Fig. 2). The mounds are 5 to 10 m above the seafloor and vent suspended mud and methane. Published by permission of the author and the American Geophysical Union

Fig. 6.11. Tectonic thickening of the Ordovician Diecke Bentonite at the convergence of three bedding plane thrusts in a fold of the Appalachian Mountains at Hagan, Lee County, VA, USA (After Miller and Brosgé 1954, Fig. 14)

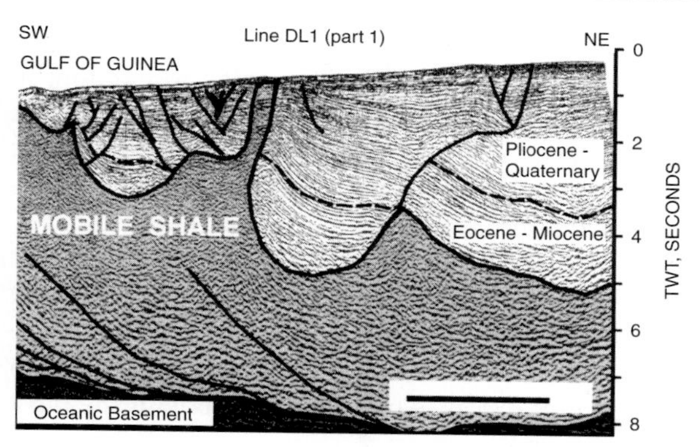

Fig. 6.12. Giant mass of mobile Cretaceous mudstone sliding seaward underlies faulted and deformed prodelta deposits (light) of Niger delta. (After Wu and Bally 2000, Fig. 10). Published by permission of the authors and the American Geophysical Union

structure actually is now more planar through burial and physical compaction. Consequently, one platy mineral, in spite of the intergrowths and outgrowths described above, slides easily over the other (and internally cleaves and slips along more weakly bonded sheets), so even weak shear stresses cause differential movement. Put some expandable clay minerals in the system and increase pore pressure and a hard mudstone becomes a plastic paste, tectonically speaking – and is the reason that the soles of many low angle overthrusts follow mudstones. Small bedding plane thrusts (listric faults) are common in lithologies with interbedded mudstones in faulted and folded areas and illustrate well the often little recognized slip that

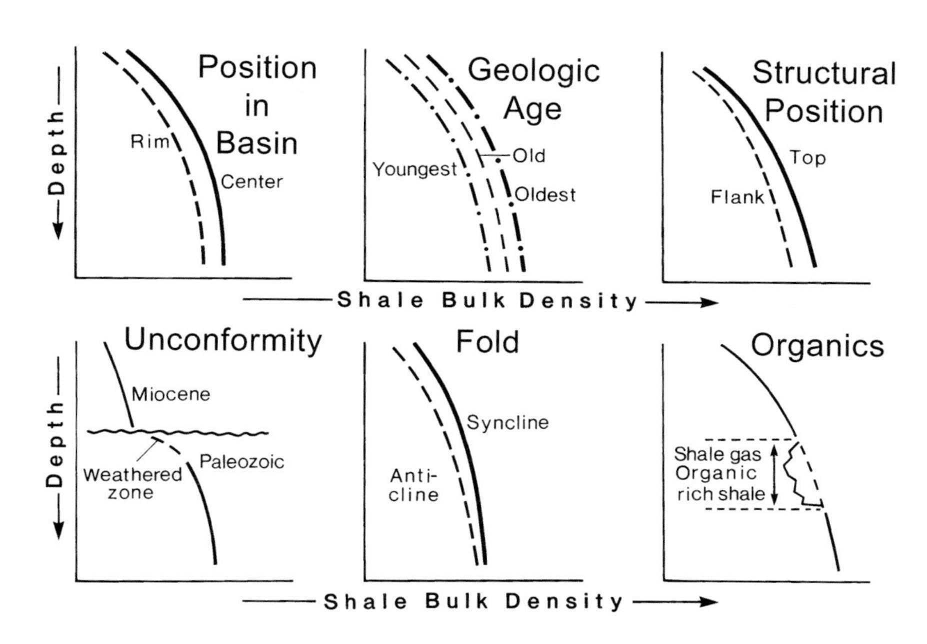

Fig. 6.13. Changes in mudstone density (compaction) with different geologic settings (after Fertl 1977, Figs. 6, 7, 8, 9, 10, and 11)

occurs in mudstones. Where subject to shear stress, a small listric fault dies out parallel to bedding in the up-stress direction. The thinning of mudstones along the axes of folds provides still other examples.

But there are other controls on mudstone density. These broadly relate to its location in a basin, its later structural history, its location with respect to fold axes, whether or not there is an unconformity above it (Fig. 6.13), and finally its age. Mudstones are generally denser in the center of a basin than on its margins, because basin centers contain the most complete and thickest stratigraphic sections (more overburden pressure). Structural position within a basin is another factor. For example, consider buried mudstones that underlie a broad, low dome 10's of kilometers in diameter. Here the same mudstone in the subsurface will be denser on the flank of a dome than near its top, because burial depth is greater on the flank. Unconformities also strongly affect mudstone density, because mudstones below an unconformity once had a now-eroded overlying section and are *over-consolidated* with respect to present overburden. A melted ice sheet has the same effect on glacial clays and tills, because the weight of a vanished ice sheet makes such clays over-consolidated, even though near the surface. They are ideal for construction: an easy-to-dig, but stable clay. Or consider the top of a mudstone as it is traced from an anticline into its nearby syncline – in the syncline compression enhances density at the top of the mudstone, whereas on the anticline tensional fractures at the top of the mudstone reduce it (remember, however, that this relationship is exactly reversed at the bottom of the mudstone as it is traced from an anticline into a syncline). And finally there is some published data to show that mudstone density increases at a given depth simply in response to age (Fig. 6.14).

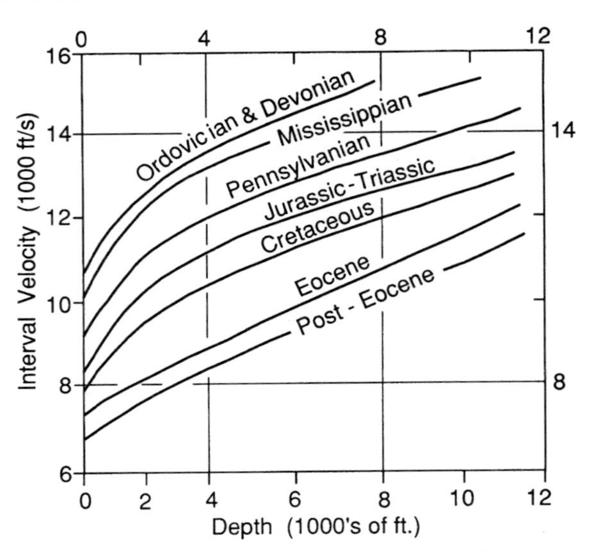

Fig. 6.14. Seismic velocity of mudstones – a good proxy for their density – increases systematically with age and depth (Faust 1951, Fig. 1)

The effects of compaction, thermal subsidence, thrust loading, etc. can be modeled quantitatively with *subsidence history curves* (Box 6.2). The simplest case is using one deep well in a basin to develop a one-dimensional model, which is very useful for categorizing the tectonic setting of the basin. Two-dimensional and three-dimensional models using a series of wells along transects provide a basin-wide picture of subsidence that is valuable in petroleum and mineral exploration. Because so much of the subsidence is caused by compaction of the mudstone layers, an understanding of the compaction behavior of mudstones is critical for correct construction of such curves.

Box 6.2. Subsidence History Curves

A subsidence history curve (Angevine and Heller 1987) is a standard technique for the study of sedimentary basins and mudstones because, when combined with knowledge of heat flow, it permits us to estimate *how long* and at *what temperatures* a mudstone was buried. In other words, it permits us to separate the effects of *time* of heating (when and how long) from its *intensity*. Klein (1991) provides detailed instructions for how to construct such curves, and commercial software packages are also available. To make a depth-subsidence curve two things are required in addition to a detailed lithologic description of the well or the outcrop. First, careful age dating, usually at least to plus or minus two million years of both the mudstone itself and of all the stratigraphic units above it. This is especially true for those units directly above and below an unconformity so we can estimate the duration of this break. The other needed ingredient is a good estimate of the geothermal gradient of the basin, which varies somewhat with its type. A gradient of 21 to 25 °C/km is fairly typical although much higher local values are known in some rift basins and volcanic provinces.

6.3. Chemical and Mineralogical Changes

Burial of mudstones drives two important chemical processes: re-equilibration of the mineral components of the mud and transformation of its organic matter. These processes in turn cause both the migration of water and hydrocarbons. Because temperature increases with depth, the initial, random, detrital and disequilibrium mineralogical mixture (unweathered, high-temperature crystalline debris mixed with weathered debris) interacts and readjusts as it is descends into the hotter subsurface. At the same time, the oxygen and nitrogen rich humic organic matter of freshly-deposited mud is transformed into hydrogen-rich kerogen. Both of these transformations, as well as simple compaction, drive fluids from mudstones. Water expelled at depth from muds as they are buried is rich in dissolved ions that migrate into adjacent sandstones, where they may dissolve or precipitate carbonate cements and attack framework rock fragments and feldspars (or alternatively precipitate new quartz or feldspar as overgrowths on the framework grains). Much of the dolomitization of limestones, particularly along fractures, may be related to introduction of dissolved Mg and Fe in waters derived from mudstone compaction (e.g., Wickstrom and Gray 1989). This water also carries organic acids that are important for generating secondary porosity in sandstones and serves as the main drive for upwards petroleum migration in basins (Pittman and Lewan 1994, Chap. 1). In this section, we explore the chemical changes that occur as a mud is buried. That is, how is a smectitic, porous and soupy mud with immature organic matter converted to a dense, low permeability mudstone dominated by illite-chlorite and kerogen?

What are the driving forces of chemical and mineralogical change with increased burial? There are two end members: shallow, near-surface changes driven by rapid, biologically-mediated reactions and deep, long term, higher temperature reactions (Fig. 6.1). Within one or two meters of the sediment-water interface, the principal driving force is exchange between the sediment pore water and the overlying bottom water of the basin and thus indirectly with the atmosphere (or directly with the atmosphere in the case of alluvial muds). Commonly oxygen demand in the mud exceeds oxygen resupply from the atmosphere, which leads to a succession of chemical and mineralogical changes as other sources of oxidants are pursued by the microbes living in the mud. But in deep burial, the situation is different. Once depths of several hundred meters are achieved, temperatures are high enough for kinetic barriers to be overcome and to suppress most bacterial activity. The principal driving force becomes the equilibration of the diverse set of detrital and authigenic minerals and organic matter that were inherited from deposition and early diagenesis. Here, important processes include the formation of new minerals, solution-reprecipitation reactions, replacement, and recrystallization. And as the composition of the initial mixture varies, so will the final product.

Thus we recognize two contrasting styles of mineralogical change with burial: *biologically-driven early diagenesis* and *chemically-driven late diagenesis*. We start with subaqueous marine burial of mud, because it is by far the most common starting point for mudstones, and continue through subaerial burial of mud in alluvial plains. Finally we consider the special case of burial transformations in volcanic ash layers in muddy sequences.

6.3.1. Shallow Burial

Chemical changes in water-deposited muds during shallow burial (down to depths of a few meters) depend mostly on the character of the bottom water. Critical variables are salinity and the amount of oxygen (Fig. 6.15). Marine environments are saline with little variability spatially in salinity levels. Modern surface seawater ranges only from a low of about 33.8 g/kg (parts per thousand or ppt) near the south polar ice cap to a high of 36.4 ppt in the Mediterranean (Van der Leeden 1990, Table 3-60). Lacustrine muds, on the other hand, experience a wide range

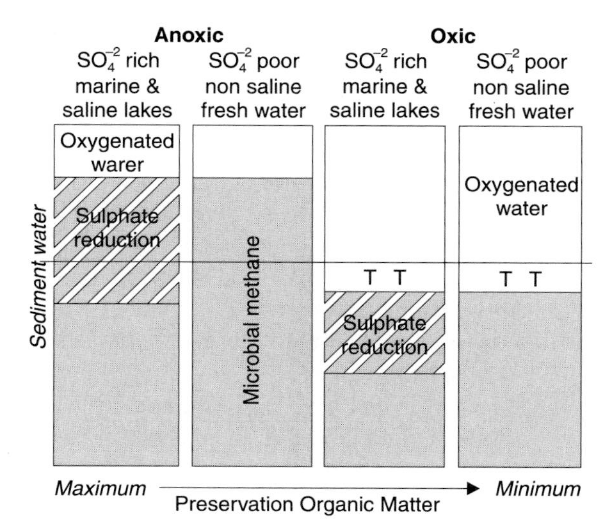

Fig. 6.15. Anoxic and oxic bottoms and preservation of organic matter (after Curtis 1987, Fig. 1). T's represent turbulence

of salinities, from 0.1 ppt for most glacial-associated lakes to 13 ppt for the Caspian Sea to 280 ppt for the Dead Sea (Lerman 1978). Alluvial muds on flood plains can also be thought of in similar terms, although here the frequency of wetting and drying of the mud and the average depth to the top of the water table are dominant controls on the minerals, textures and permeabilities that appear in the final mudstone products.

We distinguish four scenarios, each with subtypes:

1. Marine – Overlying bottom waters are saline (34–37 ppt) and rich in sulfur (2,700 ppm).
   a) Normal marine – The bottom water contains free oxygen and the sulfur is present as sulfate. Mineral products are calcium and iron carbonates and, under upwelling areas, calcium phosphates.
   b) Dysoxic marine – Bottom water low in oxygen. Mineral products are iron and calcium carbonates with some pyrite.
   c) Anoxic marine – Oxygen absent. Where some of the sulfur is present as toxic hydrogen sulfide, the water is referred to as euxinic. Mineral products are calcium carbonates and pyrite. Iron carbonates are absent.
2. Estuarine – Overlying water experiences strong fluctuations in salinity and oxygen. River borne clays exchange cations with seawater. Diagenetic minerals highly variable.
3. Lakes – Overlying bottom water can be low to high salinity, but is very low in sulfate compared to seawater.
   a) Low salinity lakes – Bottom oxygen levels are low, except in high latitudes. Siderite and calcite concretions form, and some vivianite (Fe phosphate) occurs.
   b) High salinity – Bottom oxygen levels are low unless the lake is very shallow and wind waves aerate the bottom. Early diagenesis tends to be dominated by precipitation of exotic, highly alkaline carbonate minerals such as trona or nahcolite or even borates.
4. Alluvial soils – The mud has direct contact with the atmosphere.
   a) Humid climate – Low salinity water. Minerals are iron oxides in well-drained areas, vivianite and some pyrite in poorly-drained areas.
   b) climate – Water varies in salinity. When high, calcium carbonate accumulates in the soil as *caliche*. If very arid, gypcretes and silcretes also form.

Normal to Dysoxic Marine. The great majority of modern muds and most ancient mudstones, at least since the Archean and perhaps even then, seem to have experienced early diagenesis in this setting. The series of reactions and their mineralogical products are remarkably constant. There is little variation with the source of detrital components of the mud, with the rate of deposition, or with the depth of the water except at very low sedimentation rates in the abyssal regions of the oceans. The reaction sequence is controlled by the progressive consumption of oxidants by organisms as they digest the organic matter in the sediment (Fig. 6.16). Close to the sediment-water interface, oxygen can diffuse into the mud a short distance from the overlying water column, but as mud continues to accumulate, replenishment of oxygen by diffusion becomes increasingly difficult and, finally, interstitial waters become anoxic. Bioturbation, however, can extend oxygenation deeper than can diffusion alone: where animals stir, dig, or tunnel into bottom muds and silts, they create pathways for oxygenated waters to enter and facilitate diffusion around such conduits. The zone of intense bioturbation extends about 10 cm below the sediment-water interface in modern oxic sediments. A distinction between oxic and dysoxic bottom waters is that bioturbation, and hence the depth of oxygen penetration into the sediment, is reduced in the dysoxic case; faunal diversity is much lower; and sometimes the size of fossils is reduced (so-called "dwarf faunas").

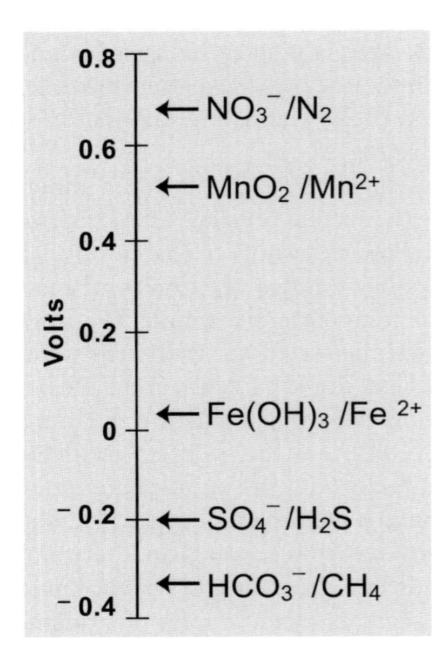

Fig. 6.16. Sequence of reduction reactions encountered with depth during shallow burial (or with lateral distance in many aquifers). Redrawn from Langmuir (1997, Fig. 11.10)

Oxygen depletion is produced by the decay of organic matter in a process analogous to burning wood in air, which likewise consumes oxygen. Organic matter in sediments consists of *labile components* such as carbohydrates and proteins and *refractory components* such as the lignin fraction of wood or the outer cases of spores and pollen grains. Decomposition of the labile components is rapid, and, in warm temperatures, they may entirely disappear in a few days. We can express this reaction using acetic acid to represent the complex universe of easy-to-metabolize organic matter:

$$CH_3COOH + 2O_2 \rightarrow 2H^+ + 2HCO_3^- \,. \qquad (6.1)$$

An important consequence of this reaction is that it generates bicarbonate, which favors the precipitation of carbonate minerals. For marine-derived organic matter, which comes mostly from algae, this reaction is a fair representation of the decay of all of the organic matter reaching the bottom. For land-derived organic matter, which has a high proportion of indigestible components from woody plants, reaction pathways are different. This refractory organic matter may escape biological degradation during early diagenesis and give rise to a mudstone that is richer in total organic matter than would normally be expected. Other exceptions to the normal pathway also result from extremes of water depth and sedimentation rate. As discussed in Chap. 4, organic matter is almost completely destroyed during the long passage through the water column to most deep-sea sediments, so that oxygen-depleting diagenesis never gets started. Also for very slow sedimentation rates in shallow water, bioturbation is so efficient that again all but the most refractory organic matter is destroyed before bacterial processes can take hold.

Because of the high calcium content of seawater, the bicarbonate generated by reaction (6.1) precipitates mostly as calcite during shallow burial. In sediments beneath dysoxic bottom waters, where the general level of oxygen in the pore waters is less, iron and manganese become mobile. Accordingly, some iron carbonate in the form of siderite tends to form in this setting and incorporates some manganese in solid solution.

After dissolved oxygen in the pore waters is exhausted, the biota change to sulfate as a source of oxygen:

$$CH_3COOH + SO_4^{2-} \rightarrow 2HCO_3^- + H_2S \,, \qquad (6.2)$$

which produces H_2S as well as bicarbonate alkalinity. If reactive iron is available – as it always is in terrigenous muds in the form of iron oxide coatings on the detrital grains – much of this H_2S is converted to pyrite. Such pyrite occurs as disseminated single crystals, as concretions, or as minute *framboids* (from "framboise", French for raspberry), which are tiny globular aggregates of crystals up to several tens of microns in diameter.

If any digestible organic matter remains after the sulfate in the pore waters is exhausted, a new group of bacteria take over that generate methane by carbonate reduction:

$$2CO_2 + 4H_2 \rightarrow CH_4 + H^+ + HCO_3^- + H_2O \,. \qquad (6.3)$$

This reaction is one of two pathways for *microbial methanogenesis*, and the dominant one in marine ecosystems (Kvenvolden 1999). Such biogenic methane, when trapped in porous muds causes overpressuring at shallow depths and also obscures high-resolution seismic reflections causing "white-outs". Release of this methane to the surface may lead to spectacular "mud volcanoes" on the sea floor up to 30 m high, cemented by carbonates derived from bacterial degradation of the released hydrocarbons (Roberts 2001). Before the abundance of shallow overpressuring was realized, serious drilling problems were encountered from "blowouts" that could shift or capsize drill platforms or even set them on fire. Shallow seismic "sparker" profiles now enable dangerous areas to be delineated.

Anoxic Marine. This setting is characterized by very low oxygen bottom waters that tend to develop toxic levels of H_2S, in which case they are referred to as *euxinic*. Often this process is seasonal or controlled by other periodic renewals of the bottom water, so a basin alternates between anoxic and euxinic or between dysoxic and anoxic. The Baltic Sea is the classic example of this behavior (Huckriede and Meischner 1996), which has also recently been reported from the Gulf of Mexico (e. g., Wiseman et al. 1997). During euxinic episodes, the normal sequence of oxidants is truncated and sulfur rather than oxygen dominates the system. Even though some H_2S is present in the bottom waters, enough to be toxic to bioturbating organisms, abundant SO_4^{2-} is still available, and, because there is no competition for organic matter from the more efficient oxygen-using organisms, sulfate reducers have a free hand and pyrite becomes the dominant diagenetic mineral. In fact, so much sulfide is formed that it overwhelms the available reactive iron and these environments are said to become iron limited (Chap. 4). Therefore no iron is available to go into carbonate minerals and only calcite, not siderite is found. As discussed in Chap. 4, the pyrite framboids formed under euxinic conditions are smaller and more uniform in size than those from anoxic and dysoxic settings (Wignall and Newton 1998).

Methanogenesis also occurs in this environment, and, because of the absence of an infauna consuming the organic matter, is more likely to be significant here than under normal marine conditions. However, because both sulfate reduction and methanogenesis are inherently inefficient modes of organic matter decomposition, much organic matter tends to survive early diagenesis in this environment to produce the organic-rich shales that become the source rocks of petroleum.

Estuaries. This is the zone of transition from freshwater to marine conditions, and diagenesis partakes of aspects of both regimes. One process that is characteristic of this setting is clay ion exchange. Clay minerals, with the exception of kaolinite, have uncompensated cation substitutions in their lattice structures that produce a net negative charge on the clay. This charge is balanced by sorption of cations like Na^+. K^+, Ca^{2+}, or Mg^{2+} to exchange sites between the layers of the clay (Appendix A.3). River clays contain dominantly Ca^{2+} in exchange positions, but seawater has a very high ratio of Na^+/Ca^{2+}. The consequence is that on entering the estuarine environment, clays rapidly exchange bound Ca^{2+} for Na^+ in solution. This process continues during shallow burial in marine environments and is one of the dominant controls on the composition of pore waters in the top few meters of the sediment column (Drever et al. 1988).

Low-salinity Lakes. Like oceans, lakes tend to produce sandy facies near shore and muddy facies in deeper water. Carroll and Bohacs (2001) have referred to these as the *fluvial-lacustrine facies association* and the *fluctuating profundal facies association*. It is the fluctuating profundal association that contains most of the mud, so this is our focus here.

Shallow burial of mud in lakes differs from the ocean in a lower abundance of shelly organisms and a much lower concentration of sulfate in the water. The low sulfate concentration makes pyrite rare compared to marine muds, unless a marine transgression brings seawater on top of an originally non-marine mud or peat, as happens in some cratonic basins to produce high-sulfur coals. Another distinction between oceans and lakes is that lakes commonly receive a much higher nutrient load and therefore experience a much higher algal productivity in surface waters. The decomposition of the dead algae as they settle produces anoxia in deep water and a dominance of laminated mudstones. Even if not visible by ordinary means, X-radiography commonly reveals a fine lamination, related in some cases to alternating layers of flocculated and dispersed clays (Coleman 1966).

Although anoxic conditions are more prevalent in lakes than oceans, lakes do not experience euxinic conditions with highly toxic H_2S in the bottom water because of their low concentration of sulfate. When anoxia prevails in these freshwater lake bottoms, the dominant geochemical process is the generation of methane by fermentation,

$$CH_3COOH + H_2O \rightarrow CH_4 + H^+ + HCO_3^- . \qquad (6.4)$$

The restriction of other pathways of organic matter degradation means that the methane bacteria have more of the labile fraction of the organic matter available, and methane genesis occurs via fermentative degradation of the organic matter rather than by reduction of CO_2 as in marine sediments. Because of this dominance of methanogenesis over sulfate reduction, iron carbonates and iron phosphates rather than iron sulfides dominate the authigenic minerals. As was the case for euxinic marine basins, organic degradation is very inefficient under these circumstances and major petroleum source beds can be formed. Interbedded carbonates are rather common in this environment, as in the Green River Shale, and evaporative carbonate minerals can also occur at the tops of drying cycles.

High-salinity Lakes. Even large lakes at times in their history may develop high salinity. Carroll and Bohacs (2001) have referred to the sediments from this stage of lake development as the *evaporative facies association*. The mud deposits of this facies are much thinner than in the profundal facies and they commonly show signs of subaerial exposure, such as sand-filled mudcracks, wind-blown clay chips and sand streaks as well as crystals, masses or lenses of salts and sometimes dolomite/evaporite lithologic transitions. The deeper-water portions of these basins can contain thin but laterally extensive organic-rich shales that are important source beds for petroleum (bottom anoxia caused by saline density stratification). Because aridity both suppresses land plants in the area surrounding the lake and also reduces inflow to the lake, little woody organic matter is contributed to the sediment, and oil-prone algal-derived material dominates, making these attractive source beds.

Table 6.4. Contrasting diagenesis of wet and dry Mississippi swamp (after Roberts et al. 2002, p 61–62)

Mostly wet	Mostly dry
Processes – Stagnant water rarely deeper than 0.5 to 1.0 m, floods excepted; decomposition of organic matter blocked by overloading so most initial biota preserved	*Processes* – Submerged only during high floods so organic matter destroyed by early oxidation and leaching
Products – Organic-rich, dark to black clays with minor laminations of woody debris and thin peats, root burrows and some gastropods and ostracodes; charophytes common, but diatoms rare; pyrite and FeS common (0.5 to 3.9%) as small cubes, framboids and root replacements in association with vivianite; siderite, calcite, and dolomite as nodules and rim cements.	*Products* – Moderately organic-rich, medium to dark gray clays; mostly massive, porous; pyrite rarely exceeds 1.0%; present as globular masses and incrustations on rootlets; vivianite scarce whereas nodules of $CaCO_3$ very abundant as are those of iron oxide (common around rootlets); siderite rare.
Significance – Low permeability and thus good seals	*Significance* – Relatively porous with high permeability and thus poor seals

Alluvial Deposits. There is a continuum in the degree of flooding of soils:

$$\text{Lakes} \rightarrow \text{poorly-drained swamps} \rightarrow \text{wet soils} \rightarrow \text{arid zone soils}.$$

Swamps differ from lakes in having root growth that disrupts the good lamination characteristic of lakes. For swamps, there are two end-members based on the frequency of wetting whose resulting mudstones have strongly contrasting permeabilities, as illustrated in Table 6.4 for swamps of the Mississippi delta. Transitional from swamps to better-drained soils are water saturated gley soils (histosols) consisting of compact, blue-gray to gray-green, rooted soils that have amorphous oxy-hydroxides, siderite, vivianite, and pyrite as iron minerals (Retallack 1997, pp 30–31). Such soils form the underclays of many coal beds. Better-drained alluvial soils are largely shades of brown and reddish brown, with goethite the dominant iron mineral. Areas with very strong drainage, such as in the tropics or on karstic limestones in temperate regions, produce bright red hematite-bearing soils (oxisols and terra rosa). In arid climates, where movement of soil water is dominantly upward, thick accumulations of fine-grained calcium carbonate, caliche, and even silica (silcrete) and gypsum (gypcrete) can form.

Many early soil features (zoning, rooting, bioturbation, iron mineralogy, slickensides, expansion fractures) are preserved or only modestly altered during burial. Therefore paleosols in ancient mudstones are a rich source of information about paleoclimate and paleo-water table fluctuations.

Volcanic Ash Beds. Deposition of volcanic ash into a muddy environment creates a special set of circumstances, because the ash is nearly always different in composition from the normal terrigenous debris being contributed to the basin. Also it has a high proportion of reactive constituents such as glass, calcic plagioclase, pyroxenes, etc. Some unknown proportion of the background sedimentation in a basin will be this volcanic material, but occasional large eruptions, even hundreds of kilometers distant, will produce nearly pure ash beds, which are important time markers (Spears 2003). We distinguish two end-member styles of alteration: marine and terrestrial. Ash falling into a marine basin will experience diagenesis in a saline solution high in potassium, among other ions. The clay mineral product is then a smectite-illite and the rock itself is termed a *bentonite*. With time and burial, these beds tend to acquire extra potassium from circulating basinal brines and become *K-bentonites*. Both types are widely used in long-distance correlation (Kolata et al. 1996). Ash falling into a swamp on the other hand, experiences strong leaching in a low salinity, acidic environment. The clay mineral product is usually kaolinite, and the resulting rock is termed a *tonstein* (from the German "clay-stone"). Tonsteins are common features of coal measures and can form distinctive seams within coal beds, such as the one in the Fire-clay Coal of eastern Kentucky (Fig. 6.17), and in several coals of the Illinois Basin. These clay seams extend over wide areas, and are used for long-distance correlation. In sum, ash falls become *diagenetic opposites* depending on the environment into which they settle. They differ from most other fine-grained deposits in that the same initial composition turns into a radically different final product because of differences in the chemistry of the water in the shallow diagenetic environment.

Fig. 6.17. Kaolinite-rich tonstein in a coal from eastern Kentucky (Courtesy Kentucky Geological Survey)

Concretions and Nodules as Recorders of Shallow Burial Conditions. Concretions are secondary, elliptical to spherical, hard, typically calcareous masses with sharp contacts, and are common in many muds and mudstones, where they range in diameter from a few centimeters to up to two or three meters. The term *nodule* is used for smaller concretion-like bodies, often phosphatic or pyritic. There is a wide and diverse literature about concretions and nodules, with much of it, especially in recent years, focused on their stable isotopes. Despite this attention, it remains to be seen how best to use concretions to understand mudstones. A survey of this literature suggests to us that most concretions form early in the biologically active zone that exists at the top of the sediment column, but below the zone of vigorous bioturbation. We think of this zone as the *concretion factory*. We mean by this that the vast majority of the concretions and nodules in marine mudstones form in the restricted, narrow zone of intense microbial activity found within the top meter or so below the sediment-water interface. We recognize of course that some concretions form later and that most early concretions are subject to small amounts of additional growth, either to the outer rim or in pore space remaining within the concretion or in fractures.

It is common to see a zonation of nodule and concretion type with water depth. In cratonic basins, this may be a response to decreasing oxygenation (Fig. 6.18). The boundary between anoxic water with H_2S and oxic water is the site of intense microbial activity and hence is a favored site for mineral precipitation (Maynard 2003). Variations on this theme would be extremely slow deposition on submarine highs under dysaerobic conditions, which favors phosphate nodules. Conversely, slow deposition in the deep-sea under highly oxidizing conditions favors manganese nodules.

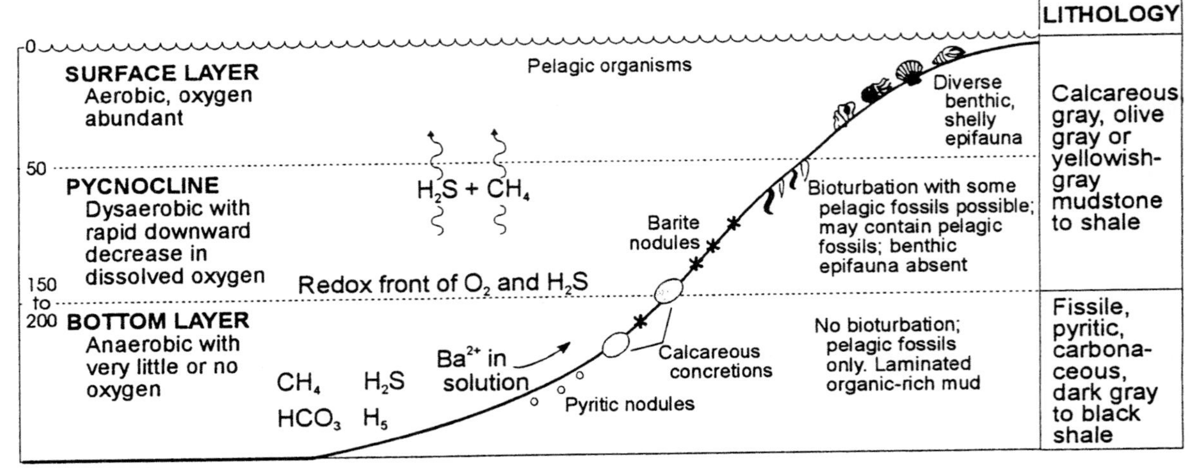

Fig. 6.18. Concretions are good indicators of early diagenetic conditions in muddy sediments, which in turn tend to reflect bottom water oxygenation (After Clark and Mosier 1989, Fig. 5)

Fig. 6.19A–D. Concretions in mudstones have many different shapes and sizes: (**A**) Large calcareous concretions in a marine Cretaceous marlstone from Morocco preserve early horizontal lamination (El Albani et al. 2001, Fig. 4); (**B**) small, closely spaced siderite concretions in Pennsylvanian Breathitt Formation at Barbourville, Knox Co., Kentucky; (**C**) large, isolated calcareous, sideritic concretion in marine Nancy Shale Member of the Mississippian Borden Formation along US 421 south of Big Hill, Rockcastle Co., Kentucky; and (**D**) concretions so large and hard in the Devonian-Mississippian Ohio Shale at Columbus, Ohio that they were pushed aside during highway construction (*continued on next page*)

On the outcrop scale, nodules and concretions tend to follow specific horizons and, if calcareous and abundant, can pass laterally into a nodular limestone. In a thick muddy interval, different horizons have concretions that differ in their size and type; i. e., concretions vary more between than within horizons, again pointing to formation during shallow burial. Where concretions have internal fractures of non-tectonic origin, the term *septarian* is used. Many but not all concretions nucleate around fossil debris, which is typically well preserved and is more evidence for early formation (e. g., Baird et al. 1985). Other evidence for early formation close to the surface of the sea bottom includes:

1. arching of the stratification of the enclosing shale above and below the concretion (Figs. 6.4 and 6.19)
2. greater thickness of the stratification inside rather than outside it (4 to 6 times or more)
3. unstable minerals better preserved inside rather than outside of the concretion (evidence for a sealed environment)
4. preserved borings and sessile bottom dwellers on the surface of the concretion (a hard isolated substrate surrounded by a soft muddy bottom)
5. "card house" clay mineral fabrics indicating as much as 80 to 90% pre-compaction cement.

Fig. 6.19. (*continued*). Figure 6.19A published by permission of the International Association of Sedimentologists

Thus it is clear that most concretions grow by cementation of a host mud during its early, open porosity stage, although some replacement of silicate minerals occurs, making precise estimates of the original porosity difficult (Scasso and Kiessling 2001). Exceptionally, concretions are also concentrated by currents along scour surfaces or long-term omission surfaces and rolled about by waves. The flatness ratio of a concretion has sometimes been used to estimate its proximity to the bottom – spherical concretions are believed to have formed at or close to the bottom whereas flatter ones are considered to have formed farther below it. Growth has commonly been assumed to be concentric, but this has been challenged (e.g., Raiswell and Fisher 2000).

Because the bulk of a concretion forms close to the sediment-water interface, concretions aid in environmental reconstruction and sequence stratigraphic interpretations of basin history. By contrast, late infills of pores and fractures in a concretion are logically unrelated to stratigraphic position, but should carry information about the composition and movement history of late fluids. The stable isotopes carried by the concretion minerals constitute a valuable recorder of environmental conditions at various stages in burial history, and three common isotopic systems – C, O, and S – can readily be employed. The chemistry of these isotopes is reviewed in Appendix A.5.

Carbon isotopes record microbial processes, and $\delta^{13}C$ depends on the type of bacteria that predomi-

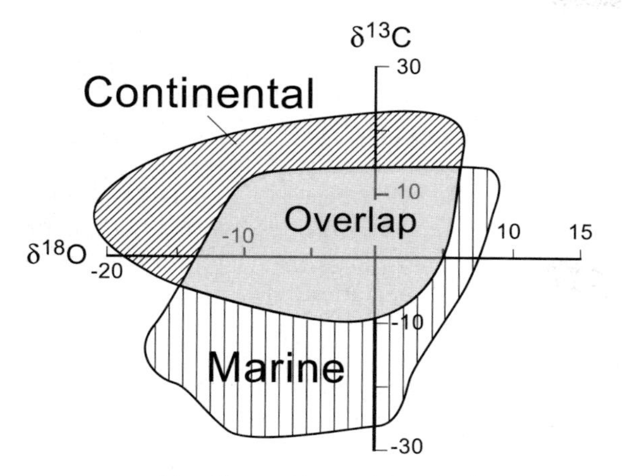

Fig. 6.20. Oxygen and carbon isotopes of concretions in continental and marine mudstones of the Cretaceous of Alberta (after Merewether and Gautier 2000, Fig. 18, from data of Mozley and Wersin 1992)

nate – whether sulfate reducers (who produce light carbonate C depleted in ^{13}C) or methanogens (who produce light methane but heavy carbonate C). Mineralogicaly, calcite concretions seem to be entirely marine, whereas siderite appears in both marine and terrestrial deposits. Marine siderites tend to be isotopically lighter than marine siderites, although with major overlap of the two groups (Fig. 6.20), which reflects the greater importance of sulfate reduction compared with methanogenesis in marine muds vs. freshwater muds.

There appears to be a sequence stratigraphic control on carbonate concretions. In shales of transgressive systems tracts of the Colorado Group of western Canada (Bloch et al. 1999) concretions are entirely calcitic and have an average δ^{13}C of -11.4 permil, indicating formation in the zone of sulfate reduction. Highstand mudstones, by contrast, have almost entirely siderite and C is heavier, averaging -1.4 permil. This difference in isotopic composition indicates a much larger contribution from methane bacteria to the siderite concretions. In transgressive calcite concretions, δ^{13}C shows only modest variation, but high-stand siderites have a wide range, from -20 to $+10$ permil. This wide range suggests that there were large fluctuations in salinity during highstand and hence in the amount of SO_4^{2-} in the water.

Carbonate minerals also carry an oxygen isotopic signal, which responds to temperature and water source rather than to microbial populations. δ^{18}O in carbonates is therefore useful in reconstructing the depositional environment when unaltered shell material is available (e. g., Azmy et al. 1998). For concretions, however, δ^{18}O commonly re-equilibrates with higher-temperature pore fluids during deeper burial. For example, Scasso and Kiessling (2001) found that δ^{18}O in Jurassic concretions from Antarctica correlates well with thermal indicators like vitrinite reflectance and illite content of mixed-layer illite-smectite, despite the fact that the concretions themselves formed during shallow burial. Consequently, the oxygen isotopic composition of concretions, particularly the late fracture fillings, can be valuable in determining later basin history, but should be used with caution in environmental reconstructions.

Pyrite nodules are exceedingly common in mudstones and are produced via bacterial sulfate reduction. δ^{34}S values record the rate of bacterial diagenesis of organic matter by the sulfate-reducers – the faster the bacteria work, the less they discriminate ^{34}S from ^{32}S. The resulting pyrite is isotopically heavier (enriched in ^{34}S). For modern marine sediments, there is a good correlation between sulfate reduction rate and isotopic composition of the resulting pyrite (Goldhaber 2003, Fig. 5). Many factors can affect reduction rate, but rate of sedimentation is the most important in marine systems, based on independent measurements with short-lived isotopes like ^{210}Pb. This observation is a basis for a method for estimating sedimentation rates in marine mudstones (Table A.10, Appendix):

$$\log \omega = 1.33 - 0.042 \Delta\delta^{34}\mathrm{S} \qquad (6.5)$$

where ω is the sedimentation rate in cm/yr and $\Delta\delta^{34}$S is the difference in isotopic composition between seawater SO_4^{2-} and pyrite S. Calculations in Appendix A.7 using this equation suggest that highstand gray mudstones of the Devonian-Mississippian of the Appalachian Basin were deposited at about 8 times the sedimentation rate of transgressive black shales, and that nearshore Pennsylvanian black shales from the Illinois Basin were deposited at 1.5 times the rate of the offshore black shales.

Sedimentation rate is an important variable in sequence stratigraphy. In the Cretaceous of western Canada (Fig. 6.21), rapidly-deposited mudstones of the high-stand systems tract have heavy sulfur compared to the more slowly-deposited shales of the transgressive systems tract. Applying (6.5), the highstand mudstones were deposited nearly five times faster than the transgressive shales. Note also in Fig. 6.21 that the slowly-deposited shales are richer in organic matter. The highstand mudstones have a mode of about 1.5% TOC (much of it woody),

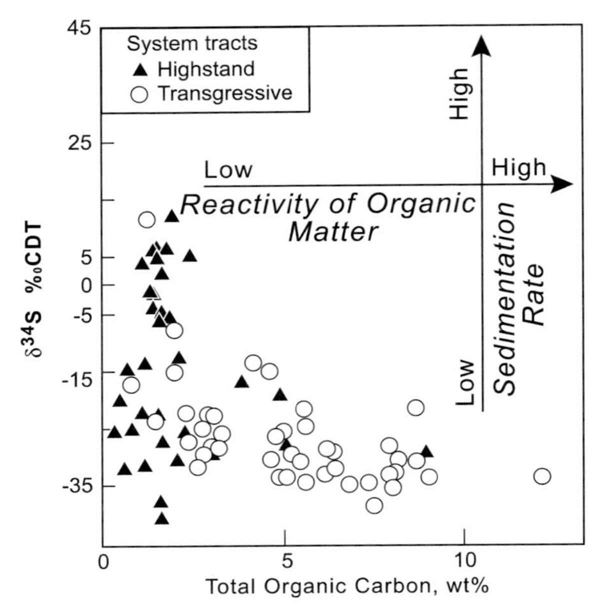

Fig. 6.21. Sulfur isotopes of pyrite in marine shales and mudstones of transgressive and highstand systems tracts (after Bloch et al 1999, Fig. 30). Published by permission of the authors and the Geological Survey of Canada

whereas the transgressive shales have a mode of 5% TOC (much of it marine). Thus the transgressive shales would make much better source rocks for petroleum (see also Fig. 8.15).

There are many other aspects of the mineralogy of concretions that might relate to sequence stratigraphy. Below we suggest some possibilities. Phosphatic nodules and glauconite occur along omission surfaces, where sedimentation rates are extremely low, thus the common occurrence of phosphate and glauconite in offshore, bioturbated hardgrounds. Therefore these diagenetic minerals are good guides to maximum flooding surfaces (shoreline far inland so that little mud in suspension reaches the offshore). Barite nodules commonly occur in phosphatic, organic-rich shales. Thus they dominantly reflect upwelling waters, but their build-up in the sediment also indicates very low sedimentation rates and may mark the maximum flooding surface in more proximal settings. We suggested above that transgressive systems tracts have shales with large calcium carbonate concretions, whereas highstand systems tracts are characterized by the out-building of deltas, whose mudstones commonly contain smaller siderite concretions (Fig. 6.22). How universal these patterns

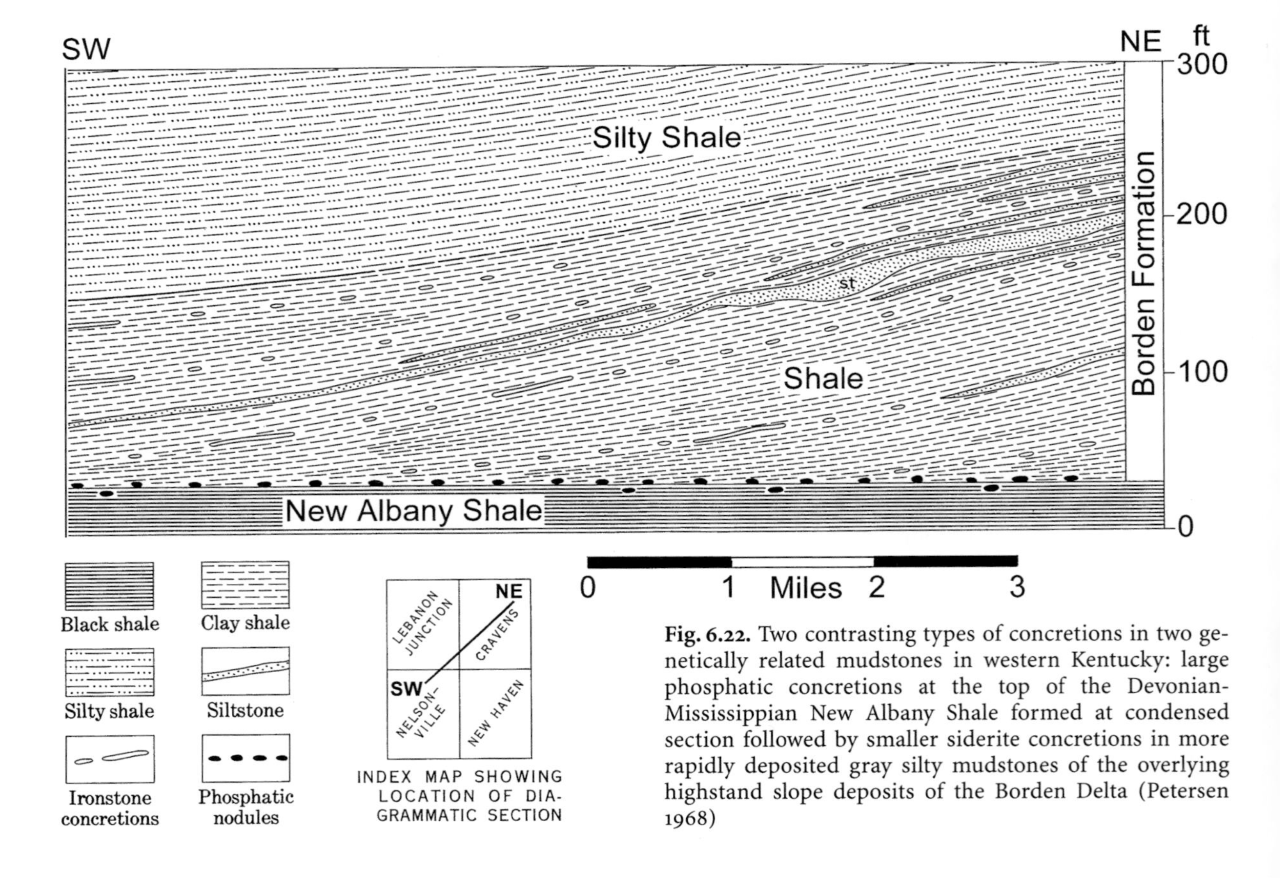

Fig. 6.22. Two contrasting types of concretions in two genetically related mudstones in western Kentucky: large phosphatic concretions at the top of the Devonian-Mississippian New Albany Shale formed at condensed section followed by smaller siderite concretions in more rapidly deposited gray silty mudstones of the overlying highstand slope deposits of the Borden Delta (Petersen 1968)

Table 6.5. Main facts and deductions about concretions and nodules in mudstones

Observation	Interpretation
Found in marine, lacustrine, and fluvial environments and soils	Whatever the processes that produce these segregations, they are widespread
A variety of carbonates, sulfides and phosphates occur, but typically only one dominates (calcite, siderite, pyrite, apatite, etc)	A variety of chemical processes produce a spherical mineralogical segregation
Likely to preserve early, open clay fabrics; more and better preserved fossils and more detrital minerals than mudstone host	Form early, close to the seafloor, before dissolution of fossils and labile minerals by intrastratal solution
Reworked and concentrated along disconformities far from shore	Form at depths sufficiently shallow to be reworked by storms
Occur in specific horizons and vary more in size and shape between horizons than within horizons	Suggests relation to sequence stratigraphy

are is not known, however, and it seems to us that a major future area for fruitful research is tying the morphology, mineralogy, and chemical composition of concretions and nodules to sequence stratigraphy.

6.3.2. Deeper Burial

The mineralogical transformations of deep burial chiefly involve the silicate minerals – clay minerals, feldspars, the different varieties of amorphous and crystalline silica, the zeolites, and, to a lesser degree, the carbonates (Table 6.5). Not only are new minerals formed, but also their crystallinity is enhanced with higher burial temperatures. Understanding these changes helps us unravel the complexities of burial history and how it relates to petroleum generation and mineral deposits. In this section we will focus on typical mudstones, which evolve towards a final assemblage of illite and chlorite clays with quartz, albite, and K-feldspar in the silt fraction. An alternative pathway is present for clays rich in volcanic glass, which may give rise to smectitic clays, to zeolites, or to the sequence: opal-A, opal-CT, and finally quartz.

Clay Minerals. The most important transformation is that of the smectites into illites and chlorites (See Clay Mineral Appendix). When smectite is dioctahedral, its transformation produces illite through an intermediary called *illite-smectite*, I/S, and when trioctahedral, the transformation produces chlorite through an intermediary called *illite/chlorite*, I/C. Because dioctahedral smectites are the most common, illite becomes dominant in older mudstones. The I/S or I/C mixed layer clays consist of intergrown sheets, whose proportions are estimated by X-ray diffraction. These transitions are of prime interest for several reasons: illite is the dominant clay mineral in ancient mudstones and the I/S transition provides insight to the temperature history of a basin. Consequently, this transition has been widely studied in different basins. Although details are still not fully resolved (Melson et al. 1998), the chief controlling factors are temperature and availability of K^+. Contributing factors include permeability (which accelerates virtually all mineralogical transformations), organic matter (which seems to retard clay reactions) and, of course, original detrital composition and the time the mudstone remains at a given temperature. As burial temperature increases, illitic sheets develop in a smectite, making an I/S clay, a process that may be gradual and continuous or occur in distinct steps (Fig. 6.23). The end product of the transition is first a 1 M and finally a 2 M mica at

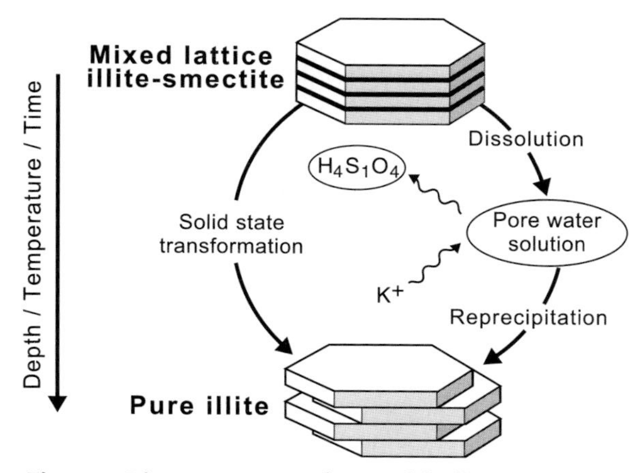

Fig. 6.23. Diverse processes that modify illite morphology with deep burial (Inspired by Meunier 2003)

temperatures above 200 °C in low-grade metamorphism. Either way, irregular, detrital smectite-illite platelets become more illitic, more crystalline and ordered, and thicker and more euhedral. K^+ and Al^{3+} are transferred to the clay fraction and H_4SiO_4 and H_2O are released (Lynch et al. 1997). The dissolved silica reprecipitates as quartz either in the mudstone as new silt-sized grains or in interbedded sandstones as quartz overgrowths. Two chemical reactions of the smectite-to-illite transformation, the first using detrital feldspar as a source of K^+ and the second using introduced K^+ ions from solution are (using idealized clay formulas):

$$\underbrace{Al_2Si_4O_{10}[OH]_2\, n\, H_2O}_{\text{smectite}} + \underbrace{KAlSi_3O_8}_{\text{K-feldspar}} \rightarrow \underbrace{KAl_3Si_3O_{10}[OH]_2}_{\text{illite}} + \underbrace{4SiO_2}_{\text{quartz}} + n\, H_2O \qquad (6.6)$$

and

$$2K^+ + \underbrace{3Al_2Si_4O_{10}[OH]_2\, n\, H_2O}_{\text{smectite}} \rightarrow \underbrace{2KAl_3Si_3O_{10}[OH]_2}_{\text{illite}} + \underbrace{6SiO_2}_{\text{quartz}} + 2H^+ + 3n\, H_2O\,. \qquad (6.7)$$

(Essene and Peacor 1995, pp 549). Similar reactions with Mg^{2+} convert smectite to chlorite. Changes in mineralogy with depth in Gulf Coast mudstones show an increase in illite matched by a decrease in K-feldspar, so reaction (6.6) does occur, but the probability is that reaction (6.7) operates simultaneously. Although other equations have been proposed, the overriding idea is that the detrital mixture is readjusting to higher temperatures and changing pore water chemistry. In so doing, there is a transfer of elements from one mineral to another and new minerals form. The released water is expelled into nearby sands and may carry with it newly-formed petroleum derived from the parallel maturation of organic matter.

In the Gulf of Mexico basin, the illite-smectite transition starts at about 50 °C. By the time the temperature reaches 120 °C there are about four times as much illite as smectite in an I/S clay. But keep in mind that these temperature limits and proportions vary with different initial detrital mixtures and burial histories, so reliable transition temperatures need to be established for each basin. This is done by combining evidence from the crystallinity of illite and chlorite, from the reflectance of organic material and from other mineralogical transformations. Collectively, these transformations provide a broad measure of *reaction progress* for burial and mineralogical change prior to low-grade metamorphism, starting at 200 °C.

Kaolinite decreases with burial depth in the Gulf Coast (Eberl 1993, Table 1) and this is the general rule in other basins. Basinal brines have very high ratios of K^+ to H^+ (Kharaka and Hanor 2003, Tables 1 and 2) and thus convert kaolinite to illite:

$$2K^+ + 3Al_2Si_2O_5[OH]_4 \rightarrow 2KAl_3Si_3O_{10}[OH]_2 + 2H^+ + 3H_2O\,. \qquad (6.8)$$

This reaction likely involves addition of K^+ from outside the system based on analyses of formerly kaolinitic paleosols (Sutton and Maynard 1996). The net of all of these processes is that mudstones older than Carboniferous have mostly lost their smectite, mixed-layer clays and kaolinite, and are dominated by illite and chlorite. This is a key fact for geotechnical engineers, because illite-chlorite compositions are so much more mechanically stable than those of smectite and mixed layer clays.

Silica. The transformations of silica are also significant for understanding the mud to mudstone conversion, because of the role of silica in induration. There are three forms of silica in muds and mudstones – amorphous or opaline silica, *opal-A*, derived from the dissolution of siliceous sponge spicules, radiolarians, diatoms and volcanic glass, which transforms into opal-CT, a dissolved form of low cristobalite, which finally becomes low quartz with increasing temperature and time. There is a progressive increase in crystallinity with each transition (Fig. 6.24). These transformation products have the same chemistry and differ only in their crystalline structure (hence the term *polymorph*), so they lack the complexities of the clay minerals. The transition temperatures are about 45 and 90 °C for these three forms.

Mudstones rich in siliceous diatoms are called diatomaceous mudstones or *diatomites* (the name *porcellanite* is used for their more deeply buried equivalents). Other sources of biogenic silica are radiolarians and sponge spicules; volcanic glass is also a major source. Biogenic textures are preserved in the initial opal-A zone, but dissolution starts at the top of the opal-CT zone and original organic fabrics are totally destroyed in the low quartz zone. Cherts in mudstones or siliceous mudstones of either deep or shallow water origin are, with few exceptions, formed by the above mineralogical transformations and derive their silica by diffusion from the host mud or mudstone. Overgrowths on quartz silt may develop from these processes and possibly some silica in so-

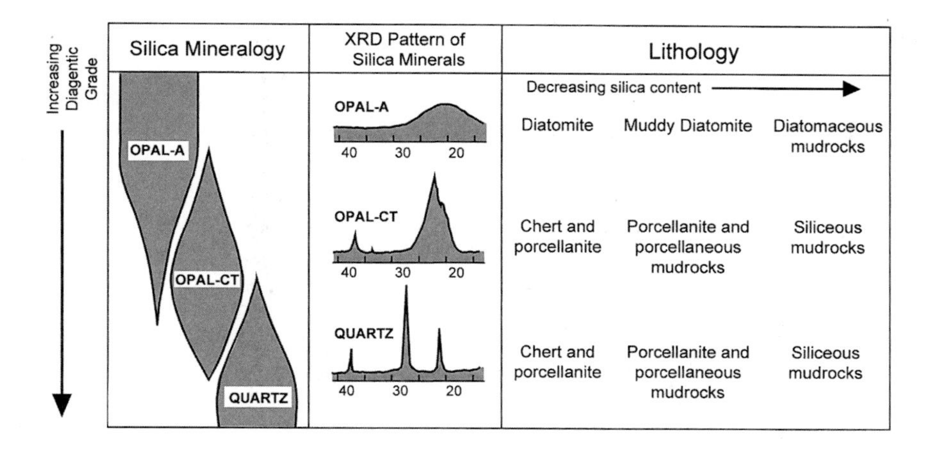

Fig. 6.24. Transformation of silica with burial (after Pisciotto 1981, Fig. 1). Published by permission of the author and the International Association of Sedimentologists

lution even precipitates and fills small pores at depth to form fine authigenic quartz silt.

Zeolites. The zeolite group of minerals has a very clear temperature (depth) dependence. These minerals are most common in sediments that have a volcanic provenance; i. e., those that initially contained some volcanic glass (Fig. 6.25). Although there are many different minerals in the zeolite group, overall, the equation below summarizes key reactions

$$\begin{aligned}\text{Glass} &\rightarrow \text{alkali zeolites(Na + K clinoptilolite} \\ &\quad + \text{mordenite)} \\ &\rightarrow \text{analcite(and locally also heulandite)} \\ &\rightarrow \text{albite + K feldspar}\end{aligned}$$

Because zeolite transitions are strongly temperature dependent, transition depths vary with geothermal gradient and, because reactions are rapid, there are few zeolites in pre-Tertiary mudstones. The contrast between the evolution of zeolites and I/S is striking (Fig. 6.25). Zeolites in mudstones occur as pore fillings in fossils, but principally as small masses of cement between clay platelets. They are always best identified by X-ray diffraction.

Feldspar. Feldspar is only a minor component of most mudstones, but its composition changes with depth. Detrital, fine-grained plagioclase deposited in muds with varied albite-to-anorthite molecular proportions is progressively converted into pure diagenetic albite in ancient mudstones. The more Ca-rich, the more susceptible detrital plagioclase is to diagenetic dissolution and replacement. Ca^{2+} made available by this process may promote calcite cementation in associated sandstones as water is expelled from mudstones during compaction. Additional Ca^{2+} may be supplied by the transformation of smectite into illite and I/S mixed layers, a reaction that is also the source of Na^{+} for albitization. The destruction of de-

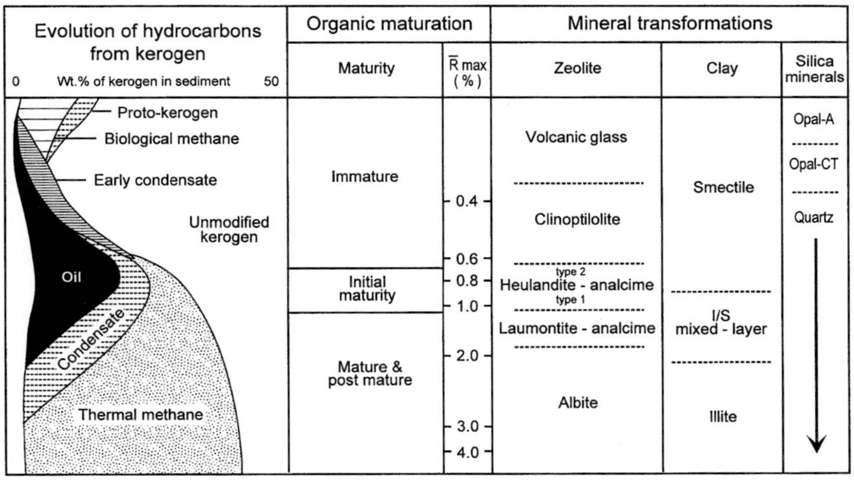

Evolution of hydrocarbons from kerogen (0 – Wt.% of kerogen in sediment – 50)	Organic maturation: Maturity	$\bar{R}$ max (%)	Mineral transformations: Zeolite	Clay	Silica minerals
Proto-kerogen; Biological methane; Early condensate; Unmodified kerogen	Immature	0.4	Volcanic glass; Clinoptilolite	Smectite	Opal-A; Opal-CT; Quartz
Oil	Initial maturity	0.6; 0.8; 1.0	type 2 Heulandite - analcime type 1		
Condensate; Thermal methane	Mature & post mature	2.0; 3.0; 4.0	Laumontite - analcime; Albite	I/S mixed - layer; Illite	

Fig. 6.25. Change of silicate mineralogy with temperature and its relation to vitrinite reflectance (Miki et al. 1991). Published by permission of the authors and the Mineralogical Society

trital plagioclase in mudstones is also enhanced by organic acids generated by the maturation of associated organic matter (Surdam et al. 1984; Pittman and Lewan 1994, Chap. 1).

Organics. In many respects the most important change in deep burial is the maturation of organic matter. Maturation generates the hydrocarbon fuels the world depends on for so much of its energy needs, increases pore pressure in the subsurface, and generates organic acids that attack sandstones and form secondary porosity. The evolutionary path followed by kerogen (the insoluble organic matter in mudstones) depends on its original source (Appendix A.6). Briefly, kerogen has four principal sources: marine, lacustrine, terrestrial, and recycled. Most petroleum is generated from marine and lacustrine source rocks, whereas coal and some natural gas forms from terrestrial organic matter. Recycled kerogen is largely inert. Mudstones vary greatly in the quantity of kerogen they contain (Table 4.4) and in its quality. Quantity is usually expressed as total organic carbon (TOC), which does not distinguish among the kerogen types. Experiments with generating oil from mudstones (Lewan 1987) suggest that greater than about 1.5% TOC is required for a good source rock (Appendix A.6).

Estimates of quality are based on the amount of hydrogen the kerogen contains, expressed as the atomic H/C ratio (see Appendix, Fig. A.21). Marine and lacustrine kerogens have initial H/C ratios of about 1.3 to 2.0, whereas terrestrial organics such as peat have H/C less than 0.9 (Hunt 1996, p 330). With burial, kerogen looses both H and O much faster than C, so the ratios of these elements to C decrease. For marine kerogen, the H/C ratio at the threshold for oil generation is in the range of 1.2 to 1.4, whereas for terrestrial kerogens, the value is about 0.7, too low to generate large amounts of liquid hydrocarbons, although considerable potential for gas generation still remains.

We saw (Chap. 4) the importance of organic matter for muds at the sediment water interface through its control on redox potential, types and abundance of fossils, bioturbation, color and lamination. This organic matter is also important as the source of shallow methane gas of bacterial origin. With depth and higher temperatures, the organics break down to yield oil, then oil mixed with gas and finally dry gas (thermally-generated methane). The part of this sequence in which maximum oil generation occurs is called the *oil window*, which typically lies in the temperature range 60 to 160 °C. Within these limits, individual mudstones behave quite differently. For example, the oil window for the Monterey Shale is quite cool, extending from 50 to 100 °C. At the other extreme, the oil window for Miocene mudstones of the Gulf Coast extends from 130 to 170 °C. More typical values are the Kimmeridge Shale of the North Sea, 110 to 150 °C (Hunt 1996, his chapter 6). Why these big differences? The conversion of kerogen to oil is a *kinetic*, not an equilibrium process. Therefore time (that is rate of burial and the geothermal gradient) is an important factor. Also key is the composition of the kerogen. Surprisingly, it is not so important whether a kerogen is Type I or Type II that matters, but its sulfur content. The kerogen of the Monterey Shale is very high in sulfur, which reduces the activation energy needed to break the carbon bonds, so it generates oil faster than low-S kerogens. Lacustrine kerogens (Type I) are generally lower in sulfur than marine kerogens (Type II), so as a class they are slower-generating, although there is extensive overlap. Type III, the woody kerogens, also overlap the slow end of the Type Is and Type IIs.

Kerogen breakdown also releases large amounts of CO_2 and organic acids, which influence carbonate diagenesis and the transformation of iron compounds in mudstones. In adjacent sandstones, these acidic compounds dissolve framework grains and carbonate cements to produce secondary porosity. Thus the organic matter in mudstones not only generates hydrocarbons, but also generates some of the pore space in the traps where hydrocarbons accumulate. During this maturation process, kerogen becomes denser and darker (think of burnt toast) so that its changes in color and reflectance provide good initial clues to the past temperature history of mudstone (Hunt 1996, 10-38). Notable is the fact that organic matter, in contrast to illite-smectite clays, quickly adjusts to even short periods of high heat flow. Consequently, vitrinite reflectance is regarded as the most reliable paleotemperature indicator for a basin. In a broad way, the illite-smectite conversion and illite crystallinity can be used to confirm temperature history inferred from reflectance and color of organic matter. These assume greater importance in older mudstones that lack vitrinite particles.

6.4. Metamorphic Equivalents

Mudstones that have experienced metamorphism – temperatures above 200 to 250 °C or strong deformation – carry a variety of field names: *argillite*, more indurated than a mudstone, but lacking slaty cleavage; *pelite*, a general term favored by metamorphic petrologists for a metamorphic rock rich in clay-minerals or micas; *slate*, a hard, indurated mudstone that has well developed cleavage, but still recognizable stratification, and *schist*, cleavage so intense that most primary stratification is destroyed. Such cleavage can be either regular or irregular. For example, in some intensely deformed rocks, cleavage has a complex, web-like, anastomosing pattern rather than being parallel. Where metamorphic mudstones are abundant, terms such as slate belt, slate-quartzite association, schist belt, and greenstone belt (when volcanics are present) are used to describe the region geologically. Many of these metamorphic mudstones belong to the lowest realms of metamorphism, the *greenschist facies*. Mudstones of the greenschist facies typically are hard, likely to have cleavage and their original clay minerals are now transformed into chlorites and 2M micas. Perhaps there is also some dispersed, opaque graphite (epidote is likely, if the original mud were tuffaceous). Greenschists remain fine grained in contrast to metamorphosed mudstones of higher metamorphic grade (Fig. 6.26), which are coarser and develop minerals such as garnet and sillimanite.

The transition from diagenesis to metamorphism in mudstones is best exhibited by their illite-smectite clays and by their organic matter. The illitic clays change from 1Md diagenetic illites to 2 M muscovites, potassium content increases, crystallinity improves, and apparent K-Ar ages decrease dramatically (Clauer and Chaudhuri 1995, Fig. 5.4). Organics are now almost entirely carbon, the O and most of the H having been expelled. Carbon-rich argillites are noticeably graphitic, with a greasy, dirty feel in hand sample. In the microscope, all of the particles are black and vitrinite reflectance exceeds 2.5%.

A major question about the processes within the interval spanning high-grade diagenesis and low-grade metamorphism is the degree of chemical exchange between mudstones and the fluids in the basin. Mostly, this is chemical loss as the mudstones compact, but often they gain potassium. Sorting out the relative effects of burial depth and original detrital composition on changes in composition has proved difficult. Study of changes along strike in the Ouachita Basin (Sutton and Land 1996) shows severe loss of Ca, Mg, K, Na, Fe, and Si from higher-grade sections compared to low-grade ones. The volume change that accompanies these losses is 30 to 50%, so the transformations are especially significant. It is not known whether the Ouachitas are unique in this extent of open-system behavior or whether this pattern is broadly applicable to other basins. The shales involved were deposited in an accretionary prism, which is an environment that exhibits large-scale fluid flow in the modern (e. g., Castrec et al. 1996). Perhaps ancient mudstones will prove to have large differences in open- vs. closed-system behavior depending on tectonic setting?

Fig. 6.26. Calcareous laminated argillite – now a garnet-bearing marble – in upper member of Middle Proterozoic Castner Marble on the Trans-Mountain Parkway, Franklin Mountains, El Paso Co., Texas. This rhythmitic facies (originally thin interlaminations of limestone-mudstone) was deposited on a carbonate ramp between two stromatolitic zones. Contact metamorphism produced by proximity to the nearby Red Bluff Granite. See Pittenger et al. (1994) for details

As the grade of thermal metamorphism or the severity of deformation increases, the depositional fabric of mudstones is destroyed before that of the more competent and less reactive interbedded sandstones. Where metamorphism was dominantly thermal, most sedimentary structures in both the mudstones and sandstones will be preserved. It needs to be remembered, however, that "cooking" will coarsen and recrystallize argillaceous beds sooner than sandy interbeds (textural inversion). Moreover, differential shear between mudstones and competent interbeds of sandstone or lava flows quickly destroys sedimentary structures in the mudstones and most body fossils as well. Extremely deformed mapable masses of rock, *mylonites*, commonly have deformed mudstones as their matrix. Thus the argillites are as easy to study as ordinary mudstones, whereas prime reliance for environmental reconstruction shifts to the sandstones at higher metamorphic grades or with more intense deformation. An informative example of environmental analysis of an argillite is that by Schieber (1990) of the Precambrian Belt Supergroup in the western United States. Also see Williams et al. (1982) and Merriman and Frey (1999) for more details of how mudstones evolve in the metamorphic environment.

6.5. Summary

Burial is best thought of as a continuation of settling during sedimentation – down, down, down and still down, deeper and hotter – but at much slower rates and over millions of years instead of hours, days, or weeks. But there is one overriding difference – during burial the minerals themselves change in response to higher temperatures, time, and more concentrated pore waters and thus move toward a new chemical equilibrium. Looking at this in a different way, in the basin of sedimentation and in the first meter or so of a highly porous mud, water dominates the system, but as it is expelled by compaction, solids become dominant and all the minerals and organics transform with temperature and time. Smectite, mixed layer clays, and kaolinite are converted to illite and chlorite in pre-Mesozoic rocks, amorphous silica transforms into quartz, volcanic glass into smectites and zeolites, calcic plagioclase disappears (with its Ca^{2+} probably going mostly into sandstones as carbonate cement) and feldspars become end-member albite and K-feldspar. In addition, kerogen matures to yield first biogenic methane, then oil, then thermogenic methane so that finally all the volatile hydrocarbons are driven off or destroyed. Heat flow, pore water chemistry, subsidence rate, and time are key to this process, with heat being the most important. All this is accompanied by the expulsion of water and a consequent increase in density and seismic velocity with depth, except for local zones of over-pressured mudstones, whose early pore waters could not escape. Where such overpressure develops, the mudstone becomes mechanically unstable and can create complex structures in what otherwise would be an orderly layer-cake, well-behaved (from a stratigrapher's point of view) sedimentary sequence. Significant petroleum traps commonly result from such deformation. Most of the fluid that is expelled from muds and mudstones is from their initial pore water, but some comes at depth from clay mineral transformations and the maturation of organic matter.

Because of their high porosity at the surface, muds creep, liquefy and flow easily on the lowest of slopes (flow slides). Mudstones, because of their dominant platy clay mineral structure, are the most deformable of all rocks (except salt) as demonstrated by load casts, mud volcanoes, small listric faults sourced on bedding planes, markedly thinner mudstones along the axes of folds, and mudstones that form the soles of far-traveled over-thrusts.

Deep burial processes seem to us to still be among the most elusive of all aspects of mudstone development. The processes of deep burial have to be inferred from their products rather than directly observed, in contrast to shallow burial. And unlike sandstones, mudstones do not lend themselves to paragenetic studies of cementation events with the petrographic microscope.

6.6. Challenges

How can euxinic and anoxic facies be distinguished in the field?

> *This facies distinction is based on chemical parameters and thin section petrography. Megascopically, both facies produce dark, organic-rich shales. Is there some distinguishing feature that can be recognized in outcrop or on wireline logs that could be applied more quickly and cheaply?*

What are the relative effects of dilution, productivity, and anoxia on content of organic matter of mudstones in different environments?

> *It seems to us that this problem deserves more attention, especially the role of dilution.*

Are there environmental or early diagenetic factors that are predictive for the quality of petroleum generated by a mudstone?

In general, land-derived (woody) organics are more gas prone, whereas marine-sourced organics are more lipid rich and hence are more oil prone. Beyond this source control, are there environmental overprints? For example, do euxinic facies produce oils with higher H/C ratios? Pratt (1984) documented a strong effect of degree of bioturbation on the Rock-Eval hydrogen index of mudstones. Are there also differences with laminated mudstones?

What is the effect of mineralogy on the release of petroleum generated by a mudstone?

If the original mud were poor in smectite, then the smectite-to-illite transformation will not generate water as a vehicle for driving out generated oil. Are there other mineralogic effects? Amount of silica from diatoms? Presence of thin silt laminae?

How much chemical exchange occurs between mudstones and circulating basinal fluids?

For a rock type with such low permeability, geologic experience suggests considerable open-system behavior for some mudstones, and very little in others. How is this mass transfer accomplished and what accounts for the wide range in the amount of exchange? Amount of fracturing? Degree of bioturbation?

Is there a sequence stratigraphic significance of concretions?

There is some evidence that calcite and pyrite characterize the transgressive systems tract, that siderite characterizes the highstand systems tract, and that phosphate marks the maximum flooding surface. How universal are these patterns and are there others?

What transformations do dispersed carbonate minerals in mudstones undergo with deep burial?

There appears to be virtually no literature on this subject for mudstones outside of the Gulf Coast. There the trend is for substantial loss of $CaCO_3$ with burial depth so it would appear that new carbonate minerals are not growing within the mudstones and that carbonate is not a cementing agent for mudstones in this basin. Some of this exported carbonate goes to interbedded sandstones, but the mass balance is incomplete and the spatial occurrences of carbonate loss from mudstones and carbonate gain by sandstones do not coincide. Does this suggest long-distance transport of mudstone-derived fluids? See Milliken (2003) for a thorough review of mudstone-sandstone interactions during deep burial.

Degree of bioturbation in marine and lacustrine sediments – is there a difference?

There is very little literature on this subject. It could be argued that, on average, lacustrine muds are more likely to be deposited under anoxic conditions and therefore are more likely to be well laminated than marine muds.

References

Almon WR, Dawson WC, Sutton SJ, Ethridge FG, Castelblanco B (2002) Sequence stratigraphy, facies variation and petrophysical properties in deepwater shales, Upper Cretaceous Lewis Shale, south-central Wyoming. Gulf Coast Association of Geological Societies Transactions 52:1041–1053

Angevine CL, Heller PL (1987) Quantitative Basin Modeling. Geol Soc Am, Short Course Notes, 80 p

Azmy K, Bassett MG, Copper P, Veizer J (1998) Oxygen and carbon isotopic composition of Silurian brachiopods; implications for coeval seawater and glaciations. Geol Soc Am Bull 110:1499–1512

Baird GC, Sroka SD, Shabica CW, Beard TL (1985) Mazon Creek-type fossil assemblages in the US midcontinent Pennsylvanian: their recurrent character and palaeoenvironmental significance. Philosophical Transactions of the Royal Soc B311:87–99

Banerjee I, McDonald BC (1975) Nature of esker sedimentation. In: Jopling AV, McDonald BC (eds) Glaciofluvial and Glaciolacustrine Sedimentation. Society Economic Paleontologists and Mineralogists Special Publication 23:32–154

Bloch JD, Schröder-Adams CJ, Leckie DA, Craig J, McIntyre DJ (1999) Sedimentology, micropaleontology, geochemistry, and hydrocarbon potential of shale from the Cretaceous Lower Colorado Group in western Canada. Geol Survey Can Bull 531:185

Brett CE, Allison PA (1998) Paleontological approaches to the environmental interpretation of marine mudrocks. In: Schieber J, Zimmerle W, Sethi PS (eds) Shales and Mudstones. E Schweizerbart'sche, Stuttgart, 1, pp 301–349

Capuano RM (1993) Evidence of fluid flow in shales. American Association Petroleum Geologists Bulletin 77:1303–1314

Carroll AR, Bohacs KM (2001) Lake-type controls on petroleum source rock potential in nonmarine basins. Am Assoc Petrol Geol Bull 85:1033–1053

Castrec M, Dia AN, Boulègue J (1996) Major- and trace-element and Sr isotope constraints on fluid circulation in the Barbados accretionary complex. Part II: circulation rates and fluxes. Earth Planet Sci Lett 142:487–499

Clark SHB, Mosier, EL (1989) Barite nodules in Devonian shale and mudstone of western Virginia. US Geol Survey Bull 1880:30

Coleman JM (1966) Ecological changes in a massive fresh-water clay sequence. Transactions Gulf Coast Association Geological Societies 16:159–174

Collinson JD, Thompson DB (1989) Sedimentary Structures, 2nd edn. Unwin Hyman, Boston, 207 p

Curtis CD (1987) Mineralogical consequences of organic matter degradation in sediments: inorganic/organic diagenesis. In: Leggett JK, Zuffa GG (eds) Marine Clastic Sedimentology:

Concepts and Case Studies. Graham and Trotman, London, pp 108–123
Drever JI, Li Y-H, Maynard JB (1988) Geochemical cycles: the continental crust and the oceans. In: Gregor CB, Garrels RM, MacKenzie FT, Maynard JB (eds) Chemical Cycles in the History of the Earth. Wiley, New York, pp 17–54
Dutta NC (1986) Shale compaction, burial diagenesis, and geopressures, a dynamic model, solution and some results. In: Burrus J (ed) Thermal Modeling in Sedimentary Basins. Gulf, Houston, pp 149–172
Eberl DD (1993) Three zones for illite formation during burial diagenesis and metamorphism. Clays and Clay Minerals 41:26–37
El Albani A, Vachard D, Kuhnt W, Thurows J (2001) The role of diagenetic carbonate concretions in the preservation of the original sedimentary record. Sedimentology 48:875–886
Emery KO, Uchupi E (1984) The Geology of the Atlantic Ocean. Springer, Berlin Heidelberg New York, 1050 p
Essene EJ, Peacor DR (1995) Clay mineral thermometry – a critical perspective. Clays and Clay Minerals 43:540–553
Ettensohn FR, Rast N, Brett CE (eds) (2002) Ancient Seismites. Geol Soc Am, Boulder, Colorado (Special Paper) 359:190
Faust LY (1951) Seismic velocity as a function of depth and geologic time. Geophysics 16:192–206
Fertl WH (1977) Shale density studies and their application. In: Hobson GD (ed) Developments in Petroleum Geology, I. Applied Science Publications, pp 293–327
Fetter CW (1994) Applied Hydrogeology, 3rd edn. MacMillan College, New York, 691 p
Goldhaber MB (2003) Sulfur-rich sediments. In: MacKenzie FT (ed) Treatise of Geochemistry, vol 7. Sediments, Diagenesis, and Sedimentary Rocks. Elsevier, Amsterdam, pp 257–288
Huckriede H, Meischner D (1996) Origin and environment of manganese-rich sediments within black-shale basins. Geochim Cosmochim Acta 60:1399–1413
Hunt JM (1996) Petroleum Geochemistry and Geology. Freeman, New York, 743 p
Kharaka YK, Hanor JS (2003) Deep fluids in the continents: I. Sedimentary basins. In: Drever JI (ed) Treatise on Geochemistry, vol 5. Surface and Ground Water, Weathering, and Soils. Elsevier, Amsterdam, pp 499–540
Kenter JAM, Schlager H (1989) A comparison of shear strength in calcareous and siliciclastic marine sediments. Marine Geol 88:145–152
Kim J-W, Berg RR, Watkins J, Tieh TT (2001) Texture, mineralogy, and petrophysical properties of geopressured shales, Gulf of Mexico. Transactions Gulf Coast Association Geological Societies 51:161–172
Klein G de Vries (1991) Basin-forming processes. In: Force ER, Eidel JJ, Maynard JB (eds) Sedimentary and Diagenetic Mineral Deposits: A Basin Analysis Approach to Exploration. Rev Econ Geol 5:25–41
Kolata DR, Huff WD, Bergström SM (1996) Ordovician K-bentonites of eastern North America. Geol Soc Am (Special Paper) 313:84
Kvenvolden KA (1999) Biogenic methane and gas hydrate. In: Marshall CP, Fairbridge RW (eds) Encyclopedia of Geochemistry. Kluwer Academic, Dordrecht, 30 p
Langmuir D (1997) Aqueous Environmental Geochemistry. Prentice-Hall, New Jersey, 600 p
Lerman A (1978) Lakes: Chemistry, Geology, Physics. Springer, Berlin Heidelberg New York, 363 p
Lewan MD (1987) Petrographic study of primary petroleum migration in the Woodford Shale and related rock units. In: Doligez D (ed) Migration of Hydrocarbons in Sedimentary Basins. Éditions technip, Paris, pp 113–130
Lynch FL, Mack LE, Land LS (1997) Burial diagenesis of illite/smectite in shales and the origins of authigenic quartz and secondary porosity in sandstones. Geochim Cosmochim Acta 61:1995–2006
Mackey SD, Bridge JS (1995) Three dimensional model of alluvial stratigraphy: theory and applications. J Sediment Res B65:7–31
Maynard JB (1980) Sulfur isotopes of iron sulfides in Devonian-Mississippian shales of the Appalachian Basin: control by rate of sedimentation. Am J Sci 280:772–786
Maynard JB (2003) Chapter 15: Manganiferous sediments, rocks, and ores. In: MacKenzie FT (ed) Treatise of Geochemistry, vol 7. Sediments, Diagenesis, and Sedimentary Rocks. Amsterdam, Elsevier, pp 289–308
Melson WG, Haynes JT, O'Hearn TO, Hubbell KE, Locke D, Ross D (1998) K-shales of the Central Appalachian Province: properties and origin. In: Schieber J, Zimmerle W, Sethi P (eds) Shales and Mudstones, vol 2. E Schweizerbart'sche, Stuttgart, pp 143–159
Merewether EA, Gautier DL (2000) Composition and depositional environment of concretionary strata of Early Campanian (early Late Cretaceous) age, Johnson County, Wyoming. US Geol Survey, Bull 1917-U:28
Merriman RJ, Frey M (1999) Patterns of very low grade metamorphism in metapelitic rocks. In: Frey M, Robinson D (eds) Low-Grade Metamorphism. Blackwell, Oxford, pp 61–107
Meunier A (2003) Argiles. Gordon and Breach, London, 433 p
Miki T, Nakamuta Y, Aizawa J (1991) Relationships between authigenic mineral transformation and variation in vitrinite reflectance during diagenesis – an example from the Tertiary of northern Kyushu, Japan. Clay Minerals 26:179–187
Milkov AV (2000) Worldwide distribution of submarine mud volcanoes and associated gas hydrates. Marine Geol 167:29–42
Miller RL, Brosgé WP (1954) Geology and oil resources of the Jonesville District, Lee County, Virginia. US Geological Survey Bulletin 990:240
Milliken KL (2003) Late diagenesis and mass transfer in sandstone-shale sequences. In: MacKenzie FT (ed) Treatise of Geochemistry, vol 7. Sediments, Diagenesis, and Sedimentary Rocks. Elsevier, Amsterdam, pp 159–190
Mozley PS, Wersin P (1992) Isotopic composition of siderite as an indicator of depositional environment. Geology 20:817–820
Pashin JC, Ettensohn FR (1995) Re-evaluation of the Bedford-Berea sequence of Ohio and adjacent states: Forced regression in a foreland basin. Geol Soc Am (Special Paper) 298:68
Paull CK, Buelow WJ, Ussler W, Borowski WS (1995) Increased continental-margin slumping frequency during sea-level lowstands above gas hydrate-bearing sediments. Geology 24:143–146
Peterson WL (1968) Geologic map of the Cravens quadrangle, Bullitt and Nelson Counties, Kentucky. US Geological Survey, Map GQ 737
Pisciotto KA (1981) Diagenetic trends in the siliceous facies of the Monterey Shale in the Santa Maria region, California. Sedimentology 28:547–571
Pittenger MA, Marsaglia KM, Bickford ME (1994) Depositional history of the Middle Proterozoic Castner Marble and basal muddy breccia, Franklin Mountains, West Texas. J Sediment Res B64:282–297
Pittman ED, Lewan MD (1994) Organic Acids in Geological Processes. Springer, Berlin Heidelberg New York, 475 p
Pratt BR (1998) Syneresis cracks: Subaqueous shrinkage in argillaceous sediments caused by earthquake-induced dewatering. Sediment Geol 17:1–10
Pratt LM (1984) Influence of paleoenvironmental factors on preservation of organic matter in Middle Cretaceous Greenhorn Formation, Pueblo, Colorado. Am Assoc Petrol Geol Bull 68:1146–1159
Prior DB, Coleman JM (1980) Active slides and flows in underconsolidated marine sediments on the slopes of the Mississippi Delta. In: Saxov S, Nieuwenhuis JK (eds) Marine Slides and Other Mass Movements. Plenum Press, New York, pp 21–49
Raiswell R, Fisher QJ (2000) Mudrock-hosted carbonate concretions: a review of growth mechanisms and their influence

on chemical and isotopic composition. J Geol Soc London 157:239–251
Retallack GJ (1997) A Color Guide to Paleosols. Wiley, New York, 175 p
Roberts HH (2001) Fluid and gas expulsion on the northern Gulf of Mexico continental slope: mud-prone to mineral-prone responses. American Geophysical Union Monograph 124:145–161
Roberts HH, Stone G, Bentley S (2002) The Mississippi River Delta depositional system. Louisiana State University, Coastal Studies Institute, 123 p
Scasso RA, Kiessling W (2001) Diagenesis of Upper Jurassic concretions from the Antarctic Peninsula. J Sediment Res 71:88–100
Schieber J (1990) Significance of styles of epicontinental shale sedimentation in the Belt Basin, mid-Proterozoic of Montana, USA. Sediment Geol 69:297–312
Spears DA (2003) Bentonites and tonsteins. In: Middleton GV (ed) Encyclopedia of Sediments and Sedimentary Rocks. Kluwer Academic, Dordrecht, pp 61–63
Spencer CW (1987) Hydrocarbon generation as a mechanism for overpressuring in the Rocky Mountain region. Am Assoc Petrol Geol Bull 71:368–388
Sterling GH, Strohbeck EE (1973) The failure of South Pass 70"B" platform in Hurricane Camille. Offshore Technology Conference Paper 1898, pp 719–724
Surdam RC, Boese SW, Crossey LJ (1984) The chemistry of secondary porosity. In: Surdam RC, McDonald DA (eds) Clastic Diagenesis. Am Assoc Petrol Geol Memoir 37:127–149
Sutton SJ, Land LS (1996) Post-depositional chemical alteration of Ouachita shales. Geol Soc Am Bull 108:978–991
Sutton SJ, Maynard JB (1996) Basement unconformity control on alteration, St. Francois Mountains, SE Missouri. J Geol 104:55–70
Van der Leeden F, Troise FL, Todd DK (eds) (1990) The Water Encyclopedia, 2nd edn. Lewis, Chelsea, Michigan, 808 p
Velde B (1996) Compaction trends of clay-rich deep sea sediments. Marine Geol 133:193–201
Watry SM, Ehlig PL (1995) Effect of test method and procedure on measurements of residual shear strength of bentonite from the Portuguese Bend landslide. In: Haneberg WC, Anderson SA (eds) Clay and Shale Slope Instability. Geological Society America, Reviews in Engineering Geology 10:13–25
Whelan T, Coleman JM, Suhayda JN (1975) The geochemistry of recent Mississippi River Delta sediments: Gas concentration and sediment stability. Offshore Technology Conference, Paper 2342:71–77
Wickstrom LH, Gray JD (1989) Geology of the Trenton limestone in northwestern Ohio. In: Keith BD (ed) The Trenton Group (Upper Ordovician Series) of Eastern North America. AAPG Studies In Geology 29:159–172
Wignall PB, Newton R (1998) Pyrite framboid diameter as a measure of oxygen deficiency in ancient mudrocks. Am J Sci 298:537–552
Williams H, Turner FJ, Gilbert CH (1982) Petrography, 2nd edn. Freeman, San Francisco, 626 p
Wiseman WJ, Rabalais NN, Turner RE, Dinnel SP, MacNaughton(1997) Seasonal and interannual variability within the Louisiana coastal current: stratification and hypoxia. J Marine Systems 12:237–248
Wu S-Y, Bally AW (2000) Slope tectonics: comparisons and contrasts of structural styles of salt and shale tectonics of the northern Gulf of Mexico with shale tectonics of offshore Nigeria in Gulf of Guinea. In: Mohriak WU, Talwani M (eds) Atlantic Rifts and Continental Margins. American Geophysical Union, Washington, DC, Monograph 115:151–172

Digging Deeper

Coleman JM, Roberts HH, Stone GW (1998) Mississippi River Delta: An Overview. J Coastal Res 14:698–716.

Thorough, thoughtful overview of work on the delta including much historical insight. Especially valuable for explanation of post-depositional mud mobility, which is a major process moving shallow-water sediments into deeper water.

Domenico PA, Schwartz FW (1997) Physical and Chemical Hydrology. Wiley, New York, p 506.

Chapter 8, "Abnormal Fluid Pressures in Active Depositional Environments", is a clear qualitative appraisal followed by a mathematical approach.

Frey M, Robinson D (1999) Low-Grade Metamorphism. Blackwell Science, Oxford, 313 p.

An outstanding source of information for the student of mudstones, because this text brings together most of the scattered low-temperature literature of diagenesis. Especially interesting is the short discussion of "Very Low-Grade Metamorphism in a Global Setting".

Hedberg H (1936) Gravitational compaction of clays and shales. Am J Sci 31:241–287.

One of the great classics of "burial" far ahead of its time.

Issler DR (1992) A new approach to shale compaction and stratigraphic restoration, Beaufort-Mackenzie Basin and Mackenzie Corridor, Northern Canada. Am Assoc Petrol Geol Bull 76:1170–1189.

Clearly written summary and application contains a formula converting sonic transit times in mudstone to density and much more.

Max MD (ed) (2000) Natural Gas Hydrates. Kluwer Academic, Dordrecht, 414 p.

This technical book for the well-prepared reader has nine parts and a total of 27 short chapters that range from potential promise (a hydrate-gas economy) through geophysical identification and geographic occurrence to history of discovery.

Velde B (1995) Compaction and diagenesis. In: Velde B (ed) Origin and Mineralogy of Clays. Springer, Berlin Heidelberg New York, pp 220–245.

Excellent, well written, concise analysis invoking physical process and wide insight with clear simple diagrams – necessary reading for many of us.

CHAPTER 7

Provenance of Mudstones

Like detective work – fingerprinting samples and profiling suspects

7.1. Introduction

How can we determine the immediate source of the constituents in a mudstone or even its ultimate crystalline rock sources (Table 7.1)? A wide range of techniques exist for this endeavor, and, because the kind of provenance information that is obtained is strongly dependent on which of these techniques is brought to bear, we arrange our discussion according to these methods, as outlined in (Table 7.2).

But before we address the questions that these provenance techniques can answer, let's first ask another, "Why bother?" Recalling that the term provenance comes from the French verb *provenir* meaning to supply or provide for, sedimentary geologists use this term to infer the relief, composition, climate and location of the source area of a terrigenous deposit. Such information helps us establish the broad paleogeographic setting of the source and infer transport paths from source to sink. Achieving these goals also gives us better insights to the tectonic evolution of the source area, perhaps now vanished, or to the original mineralogy of the sedimentary deposit before alteration by burial diagenesis. Moreover, understanding differences in provenance might help explain why two mudstones with similar diagenetic paths have different final compositions.

Historically, these provenance goals were achieved mostly using sandstone petrology and paleocurrent studies, but today there is a growing effort to employ chemical study of the fine fraction of mudstones. Good methods to determine the provenance of both the mudstone and its interbedded siltstones and sandstones would permit us to answer questions such as, "Were both derived from the same source or did the muds come from one source, perhaps far distant, and the sands and silts from another, perhaps nearby?" or "Is there a significant windblown or vol-

Table 7.1. Suggested components of muds and sands from different tectonic settings and weathering intensities

Tectonic Setting	Moderate Weathering	Strong Weathering
Igneous Rocks		
Plateau basalts	Fe oxides, smectite, little sand	Fe oxides, some smectite with kaolinite and gibbsite
Island arcs	Smectite with volcaniclastic sands	Smectite and kaolinite with volcaniclastic sands
Continent-margin arcs	Smectite and illite with quartzo-feldspathic and volcaniclastic sands	Smectite, illite and kaolinite with quartzo-feldspathic and volcaniclastic sands
Basement uplifts	Illite with quartzo-feldspathic sands	Kaolinite with quartzose sands
Sedimentary Rocks		
Fold-thrust belts & strike-slip terranes	Recycled illite, chlorite, kaolinite plus some new smectite; quartzo-feldspathic sands	Recycled illite, chlorite and kaolinite plus abundant new kaolinite; quartzose sands
Craton interiors	Recycled illite, chlorite and kaolinite; quartzo-feldspathic sands	Recycled illite, chlorite and kaolinite plus abundant new kaolinite; quartzose sands
Metamorphic Rocks		
Mountain belts	Recycled chlorite, muscovite, illite; quartzo-feldspathic sands	Recycled chlorite, muscovite, illite with new kaolinite; quartzose sands
Precambrian shields	Recycled muscovite, illite; quartzo-feldspathic sands	Recycled muscovite, illite with new kaolinite; quartzose sands

Table 7.2. Common provenance techniques for mudstones

Techniques	Inference (*italics*) and Commentary
Field properties	
Paleocurrents Color	*Both give direction to source.* Sands and mudstones may have different current systems, however, and color of particles may be either acquired in situ (soil) or be detrital.
Petrographic study	
Quartz content Size gradients of phenocrysts in volcanic ash	*Both indicate proximity and direction to source.* Fortunately, both are independent of diagenesis. Size gradients in phenocrysts in ash falls identify locations of volcanic arcs and rifts (many studies). Quartz content most rapidly determined by x-ray diffraction and less accurately by gamma ray log (few studies).
Thin section study	*Silt-sized grain types.* Special preparation of thin sections required and much petrographic skill. Few studies.
X-Ray Study	All techniques widely used in studies of modern muds, but not used extensively in ancient mudstones, because of diagenesis and low-grade metamorphism. X-ray study also separates polytypes such as IIb chlorites.
Kaolinite	*Weathering intensity*
Smectite	*Volcanic and other sources*
Chlorite	*Mechanical vs. chemical weathering*
Illite crystallinity	*Weathering intensity*
Chemical study	
Major elements CIA[1], Rb/K CIW[2], IC[3]	*Weathering intensity.* Both CIA and Rb/K may be modified by diagenesis such as potassium addition, whereas, the ratios CIW and ICV are less affected by diagenesis. Statistical studies of the whole array of elements are becoming more popular.
Traces Th/Sc Zr/Cr	*Ratio of felsic to mafic rocks.* Selective ratios of both trace elements and the 14 REE's are useful for distinguishing proportions of felsic and mafic rocks – thorium and zirconium indicate a felsic source and scandium and chromium a mafic one.
REE	
Eu/Eu* La/Yb	A strong negative europium anomaly indicates felsic rocks, as does a high LREE/HREE, often approximated by lanthanum/ytterbium (Caution: ratio may be affected by intense weathering). Many studies since 1990, although still fewer than provenance studies of the sand fraction.
Isotopes $^{87}Sr/^{86}Sr$	*Ratio of volcanic to cratonic source rocks.* Volcanic rocks tend to be close to mantle values for this ratio, whereas cratonic sources are much more radiogenic. Values tend to be altered by burial diagenesis.
ε_{Nd}	*Model age of initial source terrain.* Same advantages of REE's, but gives numerical age of separation of igneous rock from mantle.

[1] Chemical index of alteration, [2] chemical index of weathering, and [3] index of chemical variability

canic component in the mudstone compared to the sandstone?" or "Has there been chemical exchange between mudstone and sandstone parts of a turbidite bed?" Below we emphasize methods to determine the provenance of mudstones, building upon the long tradition of provenance studies of sandstones. In so doing we recognize that the clay fraction is more likely to be modified chemically and mineralogically by both weathering in the source and by burial diagenesis than is the sand fraction. We also note that some of the provenance techniques below are based on gradients within the depositional basin whereas others are based on the composition of grains (isotopic or REE values obtained from a detrital mineral) derived far from the basin.

Finally, we note that the long transport distances characteristic of muds favor mixing of particles from different sources. Because of this long transport with potential for mixing, the mineralogy and chemistry of marine mudstones averages provenances from large areas and over extended periods of time compared to sandstones, which tend to reflect more immediate sources. This contrast provides the potential to obtain two somewhat different kinds of provenance information from study of a terrigenous section.

7.2. Stratigraphic Methods: Architecture of the Basin Fill

The distribution of facies in a basin, stacking patterns, and the orientation of its clinoforms tell us much about how mud and silt were transported into and across a basin and thus frequently give us an early indication of the location of the source of its terrigenous detritus. This is well shown by a Cretaceous basin in eastern France near Grenoble

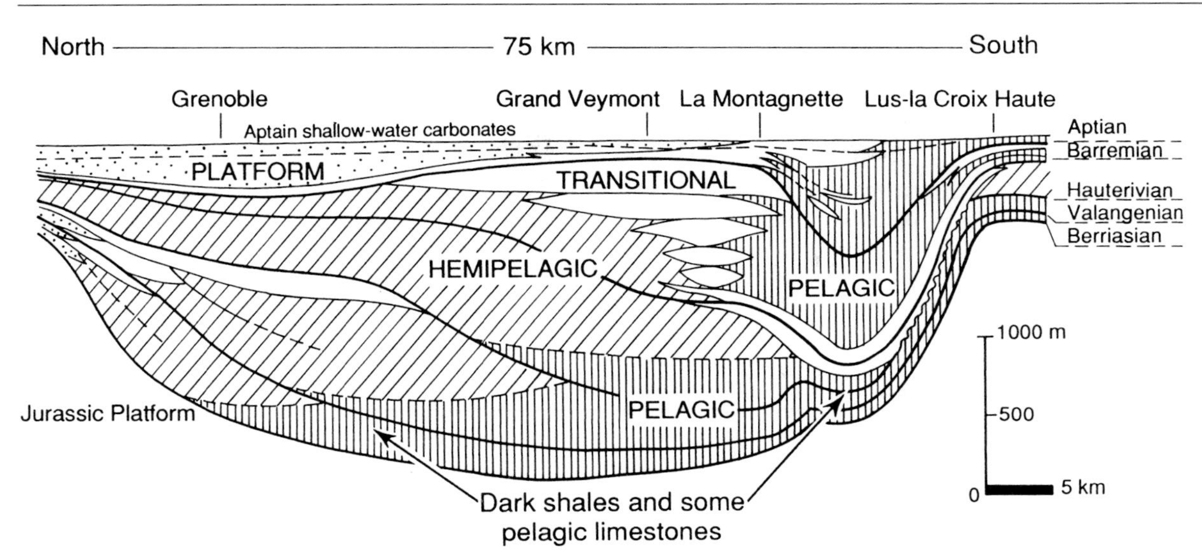

Fig. 7.1. Cross section of a Cretaceous carbonate-rich basin in eastern France has an asymmetrical fill suggesting progradation and influx of mud from north to south (Arnaud et al. 1984, Fig. 6.21)

(Fig. 7.1). Here basin fill is very asymmetrical from north to south – carbonate pelagics dominate the south side, hemipelagics the north side, and stacking pattern progrades from north to south. Thus, as a *first approximation*, we infer that the basin was filled from north to south. But to be sure, of course, we need to know more – did a land mass exist to the north or was there instead a vast carbonate tidal flat? Or is there reason to believe that longitudinal transport of fines from the east was important? Dip direction of clinoforms, which can be read from seismic sections, and sometimes with good outcrops from the field (Chap. 8), would be most telling here.

7.3. Petrographic Methods: Interbedded Sandstones and Siltstones

Much provenance information is carried by the composition of framework grains of interbedded sandstones and siltstones, and there is a long history of efforts to obtain this kind of information (Pettijohn et al. 1987; Zuffa 1985). Insights obtained in this way should always be combined with paleocurrent studies where possible, and ideally every comprehensive provenance study of mudstone should include the paleocurrents of the interbedded siltstones and sandstones. Because this has rarely been done, most sandstone-based provenance studies simply assume that interbedded muds and mudstones were derived from the same source. To see this problem better, think of the Mississippi delta whose muds and sands are all brought to the Gulf of Mexico by the same river and then think of the far-traveled, Andean-derived, coastal muds of Suriname of northeastern South America interbedded with minor sands derived from the nearby Guyana Shield (Fig. 5.33).

Petrography can also sometimes be applied directly to the mudstone itself. Polished thin sections are especially helpful, as in the Eocene-Oligocene Boom Clay of Belgium, which contains 20 to 40% coarse silt and fine sand (Zimmerle 1993). Strong diagenesis obscured the distinctions between primary and secondary clays, but the coarse fraction contained altered volcanic glass and basalt followed by trachyte as the most abundant rock fragments, indicating a volcanic source that was interpreted as being some 200 to 400 km distant. Petrographic study also helped establish a prodelta marine shelf as the most probable environment of deposition (glauconite, large mica flanks, much fine silt).

In addition to the standard petrographic provenance interpretations of detrital grains of sandstone and siltstone, more and more use is made of single-grain analyses, such as SHRIMP (Sensitive High Resolution Ion Micro-Probe) determinations of $^{207}Pb/^{206}Pb$ to determine ages of detrital zircons to match against known ages of possible source rocks (Santos et al. 2002). A similar application of the ion microprobe is measurement of oxygen isotopes on single silt-sized quartz grains. Aléon et al. (2002) used this procedure to trace dust sampled at the Cape Verde Islands back to sources in the Sahara (Fig. 7.2). Perhaps we will soon see instrumental advances that

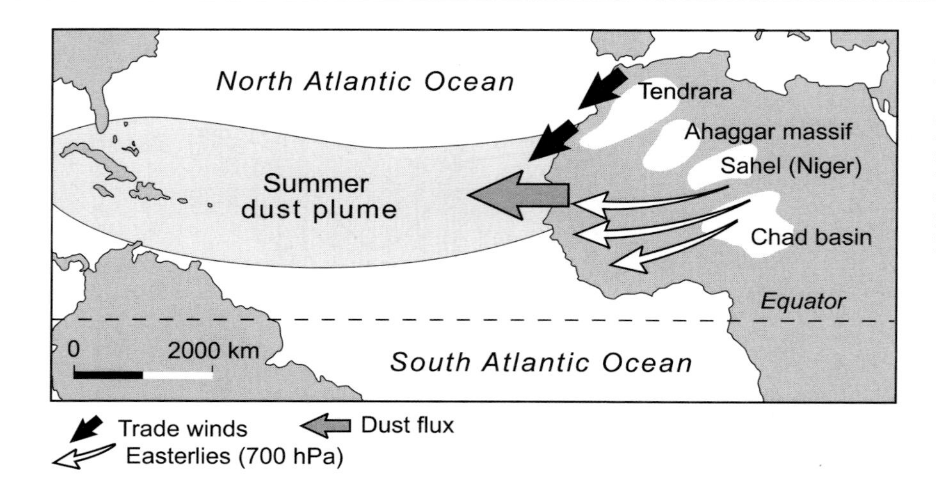

Fig. 7.2. Study of oxygen isotopes of single quartz grains by ion microprobe shows the source for the Saharan dust plume to be mostly from the Niger region (after Aléon et al. 2002). Published by permission of the authors and Pergamon Press

permit this sort of work to be done routinely on fine-grained mudstones.

A very informative component of the silt fraction is the type of particle referred to as a *phytoclast*. These are particles such as wood fragments, pollen grains, or leaf cuticles, and are generally less than 1 mm in size, with the dominant size range 20–50 µm. This material is hydrodynamically equivalent to silt (Tyson 1995, p 215), and commonly is concentrated with the silt fraction, but its lower density makes it more prone to resuspension, so current reworking can lead to organic-rich and organic-poor layers. Mapping of the percent woody fragments can potentially provide a proximal-distal indicator for marine sediments, as was shown in early work by Shepard (1956), but it is more common to use a chemical proxy such as carbon isotopic measurements (see Chemical Methods, below). A phytoclast measurement that is commonly made for modern sediments is the percentage of pollen grains (Fig. 7.3). Pollen falls off rapidly, but more-or-less regularly with distance from the shoreline so it makes a good proximality indicator, except in sub-polar regions where the abundances are vanishingly low. This technique is less used in the ancient and would not apply to rocks older than the Devonian, when land plants were not widely established.

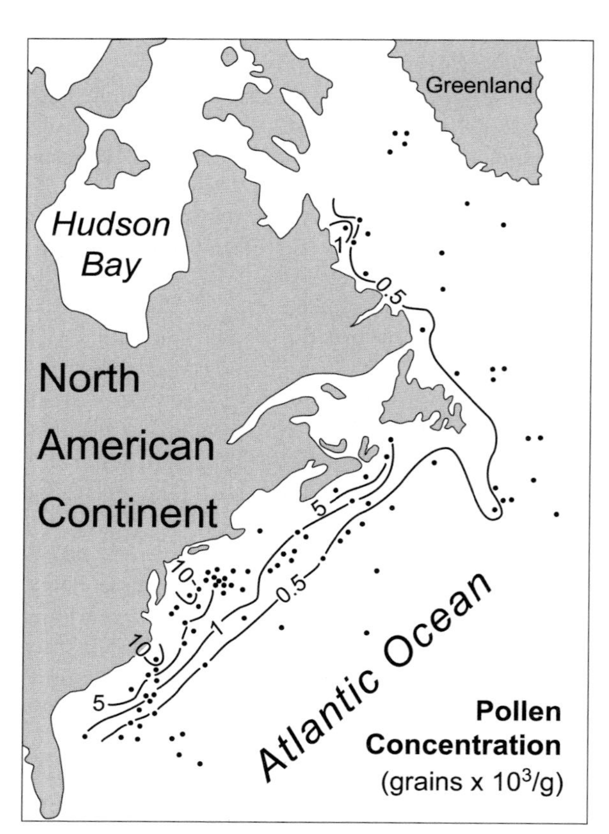

Fig. 7.3. Total pollen grains in sediments from the North Atlantic falls off systematically with distance from the shoreline, except in regions of very cold climate (after Tyson 1995, Fig. 13.13)

7.4. Mineralogical Methods

For muds, clay mineralogy has been the mineralogical procedure most often applied to reconstruct provenance. Clay mineral distribution in the South Atlantic Ocean (Fig. 7.4) illustrates well the utility of systematic study of clay distribution in modern muds and in Tertiary and Mesozoic deposits (before appreciable burial diagenesis converts most of the original metastable components into illite-chlorite). In the modern Atlantic, illite and kaolinite have reciprocal abundances: illite predominates off Antarctica (minimal weathering), whereas kaolinite predominates between the tropical coasts of Africa and Brazil

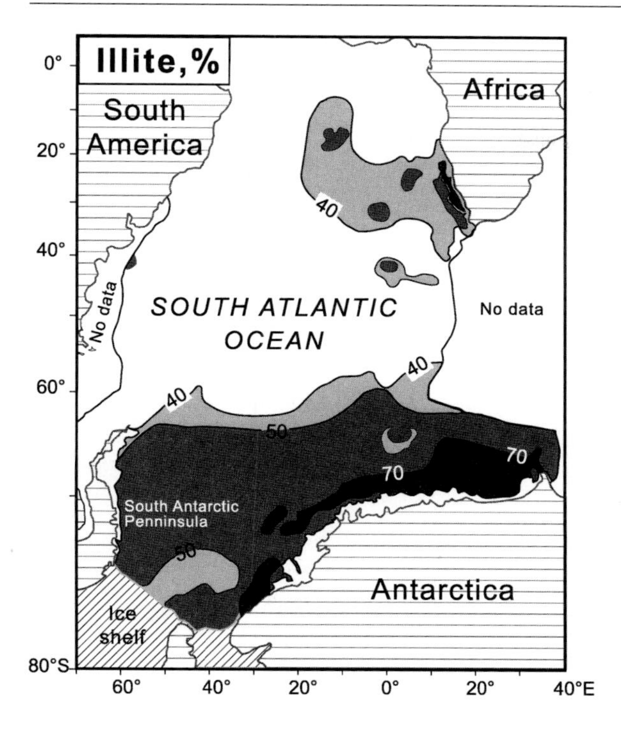

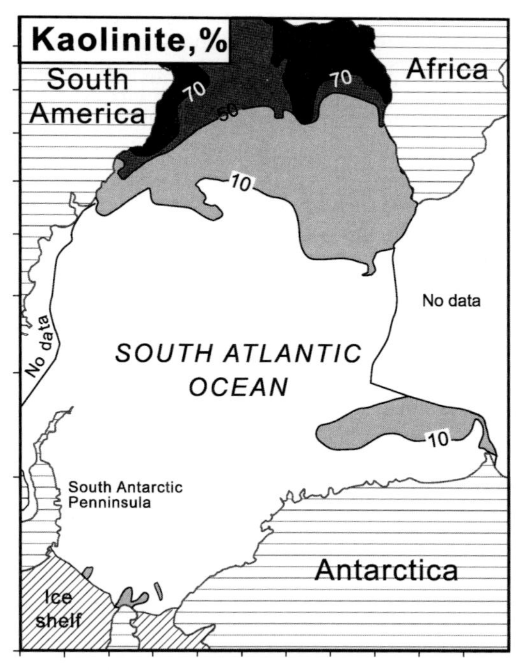

Fig. 7.4. Provenance control of illite and kaolinite in the clay fraction of sediments from the South Atlantic Ocean (after Petschick et al. 1996, Figs. 5 and 8). Illite dominates all the coast of Antarctica and the southwest coast of Africa around the mouth of the Orange River because of minimal chemical weathering (frozen or arid conditions). Kaolinite dominates in the low latitudes because of intense weathering under humid tropical conditions. Map patterns based on over 900 analyses. Published by permission of the authors and Elsevier Science

(maximal leaching in the humid tropics). This distribution is the response to two climate extremes and shows well the importance of considering climate when using clay mineralogy as a provenance indicator; if we could reverse the climates of Antarctica and tropical Brazil and Africa, we would only need to switch the labels of the two maps – nothing more.

A volcanic component in the provenance of a mudstone imparts a strong signal to its clay mineralogy. Tertiary volcanism in the western United States has resulted in a series of alluvial deposits rich in fine and coarse volcanogenic debris. A good example is the White River series of mid- to late Tertiary age in Nebraska and adjacent states (Fig. 7.5), which is composed of massive clay- and silt-rich rhyolitic tuffs, now highly altered. These were deposited by streams and wind over a wide area as a thin non-marine "molasse" cap over fine-grained marine Cretaceous rocks. Alteration products such as smectite, opal, zeolites, and calcite dominate the mineralogy of these deposits (Lander and Hay 1993).

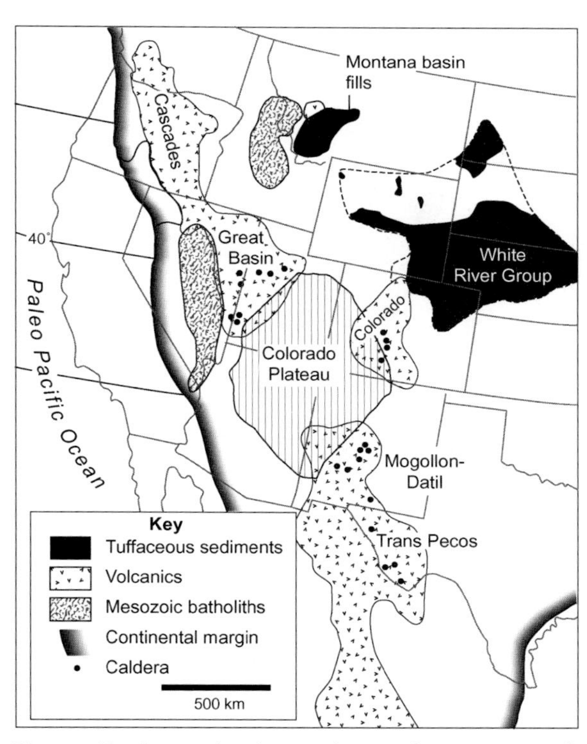

Fig. 7.5. Tertiary volcanic provinces of western North America and their associated blankets of tuffaceous sediments. Erosion of these sediments contributes much of the suspended load of the Missouri River, and thus imparts a strong volcanic signal to sediments of the Mississippi system (Terry et al. 1998, Fig. 5.). Published by permission of the authors and the Geological Society of America

Because these molasse deposits are easily eroded, they form badlands and are major contributors to fines carried to the Gulf of Mexico by the Missouri and subsequently the Mississippi Rivers. The smectite fraction in the western tributaries of the Mississippi typically varies between 45 and 60%, whereas this clay is notably less abundant in the eastern tributaries (Fig. 7.6). This volcanic provenance persists throughout the > 6,000-kilometer length of the Missouri-Mississippi main stem, despite many cycles of reworking, including strong weathering of alluvial deposits in the lower Mississippi Valley. Another example of long-distance transport of fines is the Amazon River, which carries an Andean signature (mica > quartz > smectite = kaolinite > plagioclase > K-feldspar > chlorite and small metamorphic rock fragments) to its mouth and beyond to the coastal muds of Surinam and the Guayanas (Gibbs 1967; Eisma and Van der Marel 1971). Moral: fine-grained suspension load is easily transported from one side of a continent to another, so that a mineralogical signal can be registered by muds very far from their sources.

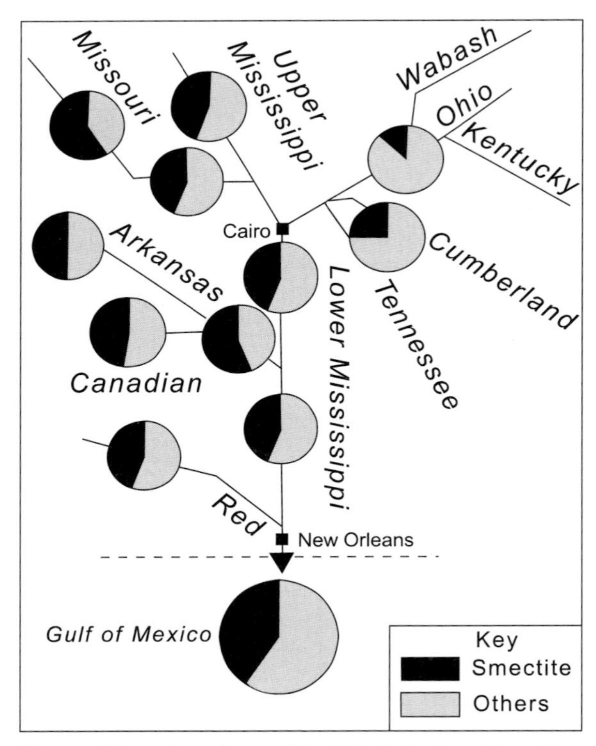

Fig. 7.6. Clay mineralogy of the Mississippi and its tributaries shows abundant smectite in the western tributaries and its lower abundance from the eastern tributaries, with the lower Mississippi and the delta retaining the strong signal of the smectite-rich western sources (after Potter et al. 1975, Fig. 13)

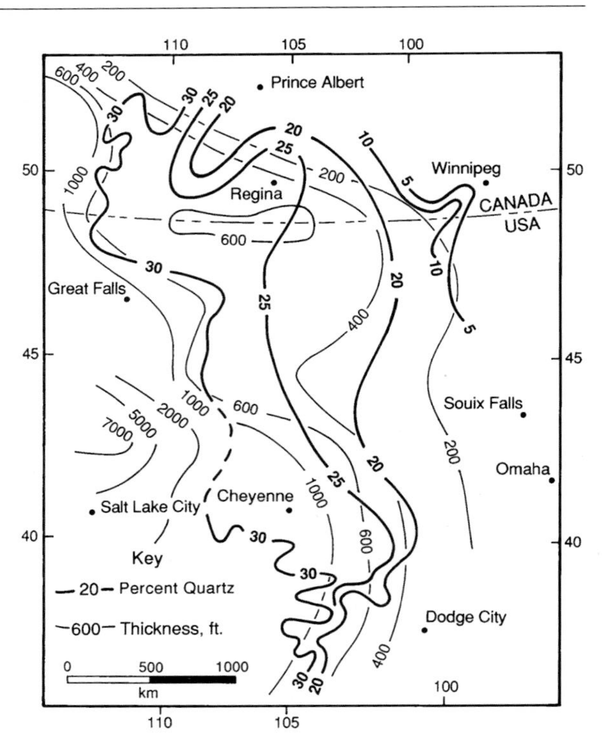

Fig. 7.7. Quartz silt content of Cretaceous Pierre Shale in the High Plains of the United States broadly outlines the source of the mud – the rising Rocky Mountains to the west (Jones and Blatt 1984, Figs. 2 and 3). Published by permission of the authors and the Society for Sedimentary Geology

Measuring the size and amount of quartz grains in the silt fraction of a mud or a mudstone is another approach, although one that has not been widely used because it is so tedious and sampling is not easy in mudstones. In the Cretaceous of the High Plains of United States and southern Canada a map of quartz content clearly identifies the ancestral Rocky Mountains as the source of the mud and silt of the mudstones in the Pierre Shale (Fig. 7.7).

Shoreline indicators such as the systematic mapping of clay mineralogy, palynomorphs, and grain size are all examples of *proximal-distal indicators*. This is the first step to paleogeographic and provenance insight. Below we show that systematic determination of chemical variables in mud and mudstones plays a similar role, but also provides information about the character of the source area.

7.5. Chemical Methods

Most of the provenance studies of mudstones use chemical methods, because separation of individual minerals is difficult, although there are a few studies of heavy minerals in mudstones (e.g. Totten and

Hanan 1998). This section presents various chemical methods of provenance determination, which, as pointed out above, always should be accompanied by petrographic study of the companion sandstones and siltstones.

We divide these chemical methods into groups – major elements, trace elements, and radiogenic and stable isotopes – that correspond to their abundance in common rocks and therefore to the instrumentation needed to measure them. Major element chemistry has been a staple of mudstone studies since at least the early 1900s. Today it is done rapidly and inexpensively by X-ray fluorescence (XRF). Some trace elements can also be measured by this technique, but many of those of greatest interest have concentrations too low for this technique, and more expensive methods such as instrumental neutron activation analysis (INAA) or inductively coupled plasma (ICP) must be employed. The neutron activation techniques have the advantage of using the solid sample directly without dissolving it, but have the disadvantage of missing some key elements. ICP, especially when combined with mass spectrometry (ICP-MS), provides very low detection limits for the greatest number of elements, but is correspondingly expensive and requires the sample to be completely dissolved. Some mineral grains such as zircons, which are very rich in heavy rare-earth elements, are extremely difficult to dissolve and extra care must be taken to ensure complete dissolution. Isotopic measurements are the most powerful ways to study geochemistry, but they involve labor-intensive separations and extremely expensive instrumentation.

7.5.1. Major Element Chemistry

Provenance studies of mudstones that use major elements as their dominant technique are quite rare, perhaps because of the potential for post-depositional alteration. One promising approach is to use discriminant analysis, as shown by Roser and Korsch (1988). They studied turbidite mudstones and sandstones from New Zealand and found that simple bivariate plots did not separate their samples, but that cross-plotting discriminant functions largely eliminated overlap between populations. Mudstone and corresponding sandstones plot in the same fields, which provides strong evidence that both had the same provenance.

Major element chemistry is perhaps best employed to determine the extent of weathering of the source terrain, once the general tectonic setting has been established by other means. Several chemical indices have been proposed to quantify weathering effects. The most popular is the *chemical index of alteration* (CIA) defined as

$$[Al_2O_3/(Al_2O_3 + CaO^* + K_2O + Na_2O)] \times 100$$

where CaO^* is Ca exclusive of carbonates and the values are in molar proportions to emphasize mineralogical relationships (Nesbitt and Young 1982). High values of the index, greater than 90%, indicate extensive conversion of feldspar to clay and hence intense weathering. Lee (2002) used this parameter to characterize Permo-Triassic mudstones from Korea. These rocks show a trend of decreasing CIA values upwards for each formation, which indicates less weathered detritus upwards. Taken by itself, this trend would indicate progressive unroofing of a weathered granitic source area. However, petrographic study of associated sandstones revealed that those formations with high CIA values also have abundant sedimentary rock fragments, an indication that the high CIA values were produced in part by recycling of older sediments rather than by intense weathering of igneous rocks. The vertical change may reflect mixing of first-cycle and recycled sources as much as a change in weathering intensity. Thus this study provides an object lesson in the value of combining petrographic work with geochemistry.

A limitation of the use of the CIA parameter is the prevalence of post-depositional K addition to older clastic rocks (see Chap. 6). Accordingly some workers have preferred a K-free index, the *chemical index of weathering* (CIW), defined as

$$[Al_2O_3/(Al_2O_3 + CaO^* + Na_2O)] \times 100$$

Harnois (1988).

Both the CIA and the CIW indices reflect mostly the amount of feldspar relative to clay minerals, and therefore neither index is strongly sensitive to the type of parent rock. CIA values for unweathered terrestrial basalts are in the range of 40 to 50 whereas for granites the range is 50 to 55. A better discrimination of parent rock types would include Fe and Mg. Cox et al. (1995) proposed a measure called the *index of compositional variability* (ICV):

$$(CaO + K_2O + Na_2O + Fe_2O_3(t) + MgO + MnO + TiO_2)/Al_2O_3 \,.$$

where $Fe_2O_3(t)$ = total iron and CaO includes all sources of Ca. In this index, the weight percents of the oxides are used rather than moles, and the values *decrease* with increasing degree of weathering. Average basalt and average granite Li (2000) yield

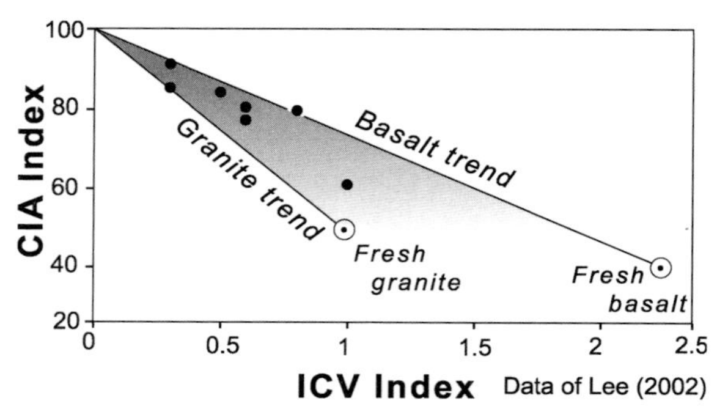

Fig. 7.8. Two indicators of weathering, Chemical Index of Alteration (CIA) and Index of Chemical Variation (ICV) in Late Paleozoic and Early Mesozoic mudstones from Korea (data of Lee 2002)

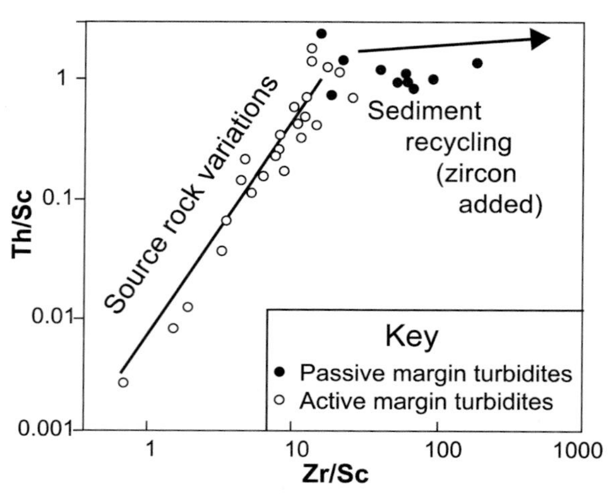

Fig. 7.10. The Th/Sc ratio and the Zr/Sc ratio increase going from mafic to felsic source areas, but passive margin muds show anomalous Zr related to recycling of older sediments (after McLennan et al. 1993, Fig. 3). Published by permission of the authors and Pergamon Press

strongly contrasting ICV values of 2.20 and 0.95. An interesting view of major element chemistry comes from a crossplot of CIA against ICV values. Figure 7.8 shows formation averages from the data of Lee (2002) compared with the rock class averages. Mudstones with the same degree of weathering (the same CIA) commonly have rather different ICV values, indicating differences in the composition of the source area.

7.5.2. Trace Element Chemistry

The ultimate igneous source for a mudstone can usually be categorized easily as relatively mafic or relatively felsic by using trace elements. For example, the elements Ni and Cr are abundant in ultramafic and mafic rocks, but are scarce in rocks of more felsic composition. Thus they are good indicators of ophiolitic sources, and their sudden appearance in a vertical profile may indicate the onset of arc-continent collision in the source area. Also useful is Zr, which is much more abundant in felsic rocks, so plots of Ni/Zr or Cr/Zr are useful to quantify the proportion of mafic material in the source area (Fig. 7.9). Weathering and diagenesis can alter these ratios, however, so it is preferable to use less soluble elements than Ni in provenance studies. Th, Sc, and to a lesser extent Cr, are good indicators of sedimentary provenance, because they are quite insoluble, and thus are transported almost exclusively by terrigenous detritus. Thus they reflect faithfully the chemistry of their sources. They are mostly used in bivariate plots (Fig. 7.10) as aids to detect mixing processes in sediments. For example, Condie and

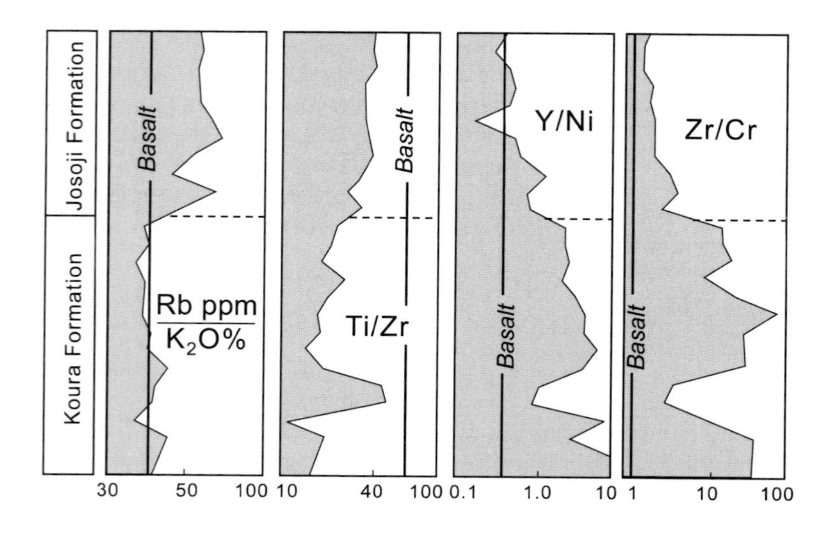

Fig. 7.9. Vertical chemical profile of two Miocene mudstones from the opening of the Sea of Japan shows their changing provenance (after Ishiga et al. 1999, Fig. 7). Interval is about 70 m thick and contains diverse mudstone types with some interbedded sandstones and tuffs. Published by permission of the authors and Pergamon Press

Box 7.1. Rare-earth elements

Although REE are found in very low concentrations in most natural environments, they are widely used as geochemical tracers. Only rarely do they individually reach concentrations of more than a few parts per million (milligrams per kilogram), as when they were discovered in 1787 by CA Arrhenius in a pegmatite at Ytterby, near Stockholm, Sweden. Lanthanum (La, $Z = 57$) and the remaining *lanthanides* (from cerium, Ce, $Z = 58$ to lutetium, Lu, $Z = 71$) are the chemical elements known as rare earths. Most have stable trivalent states. The exceptions are Ce, which can also occur as Ce^{4+} and is used to track oxidation states of water masses and soils, and Eu, which, in igneous processes, can be reduced to $Eu^{+2}$and selectively removed compared to its neighbors. Most other chemical properties of all 14 elements, together with lanthanum, are remarkably similar. The "light rare earths" (LREE, La-Sm) are *incompatible* chemical elements (i. e., are elements which, during magmatic crystallization, are not easily included in the structures of principal minerals and thus become progressively enriched in the diminishing amount of residual liquid). Owing to their smaller ionic radii, the "heavy rare earths" (HREE, Gd-Lu) are more easily incorporated into the crystal structures of rock-forming minerals, and these *selectively compatible* REE are hosted by a number of igneous minerals, like feldspars, hornblende, sphene, pyroxenes, garnet, and zircon. Therefore the light to heavy ratio is a useful tracer of igneous differentiation.

REE are commonly reported normalized either to chondrite values or a composite of mudstone analyses (e. g. NASC) and displayed on spider diagrams. The critical provenance parameters that are extracted from these diagrams are the Eu anomaly (Eu/Eu*) and the ratio of the light to the heavy rare-earth elements, $\sum$LREE/$\sum$HREE. The Eu anomaly is calculated from

$$Eu/Eu^* = Eu_N/[(Sm_N)(Gd_N)]^{1/2}$$

where N refers to the normalized values and

$$La_N/Yb_N$$

is used as a proxy for $\sum$LREE/$\sum$HREE.

Wronkiewicz (1990), McLennan et al. (1990), and McLennan and Taylor (1991) and have favored the use of plots such as Th/Sc against Sc, and Cr/Th against Sc/Th as indicators of the proportions of felsic and mafic source rocks.

Of the trace elements, the most valuable in geochemistry are the *rare earth elements* or REE (Box 7.1), because they are not fractionated from each other by most sedimentary processes. They are largely insoluble under most geological conditions and thus are present in very low concentrations in river water and seawater (typically, from 10×10^{-7} to 10×10^{-2} of the levels found in most rocks). Weathering releases them from primary minerals in the soil, but they are generally retained and concentrated within the weathering profile (Nesbitt 1979) in secondary minerals. Once eroded, these soil minerals then faithfully record the REE signature of the parent material (but watch for Ce variations). Clay minerals typically have much higher total REE concentrations than coarser size-fractions (Cullers et al. 1987), and appear to be the most important mineral fraction in hosting REE in mudstones (Condie 1991). REE concentrations remain unaffected by most geological post-depositional processes; diagenesis, for instance, typically has little influence on the redistribution of the REE because very large water/rock ratios are required to effect any change in their sedimentary chemistry (but see Lev et al. 1998, for an exception).

REE patterns are shown normalized to either a chondrite meteorite standard or to a composite of mudstone analyses to eliminate the strong differences in cosmic abundance of even vs. odd atomic numbers. Mudstone-normalized patterns (Fig. 7.11) are used for sedimentary rock studies because they emphasize subtle differences among sediments and thus provide a method of identifying specific sources of mud and mudstones (McLennan 1989). Comparison to a shale standard quickly shows the amount of departure, if any, that the mudstone under study shows from typical mudstones, and deviations of specific elements identify slight enrichments and deficiencies that shed light on provenance. Owen et al. (1999) supply an example – among many others – of the potential uses of REE normalized patterns as

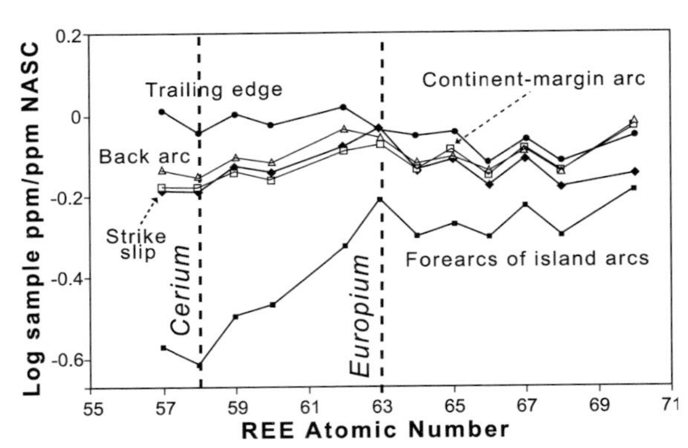

Fig. 7.11. REE profiles for muds from various tectonic settings are similar except for LREE enrichment in Trailing Edge basins and severe LREE depletion in Fore-arc Basins. Back-arc, continental arc, and strike-slip basins are indistinguishable. Data of McLennan et al. (1990) averaged by tectonic setting

provenance indicators. They determined the REE in radiolarian cherts and siliceous mudstones of upper Ordovician age from the Southern Uplands (Scotland) and established that REE signatures can be used as fingerprints to differentiate some of the fault-bounded formations within the study area, and that the provenance for the mudstones had two components: a mature continental block and an andesitic source.

A Precambrian example is provided by Eriksson et al. (1992), who studied the Mt. Isa Basin, an important host of Pb-Zn mineralization that had at least two phases of rift development. The upward trend in this basin is towards lower values of Eu/Eu*, increasing Th/Sc, and decreasing concentrations of Cr, Ni and Co. All of these parameters point to an increasing felsic contribution to the fine-grained detritus in the basin fill as the basin developed. The lower part of the succession, where accommodation space in the basin is generated primarily by mechanical subsidence along rift-boundary faults, has more variability in composition, reflecting variable local source rocks. By contrast, the later sag phase, where broad thermal subsidence dominates, has very uniform REE compositions, reflecting long transport under stable tectonic conditions.

7.5.3. Isotope Chemistry

Radiogenic Isotopes. Isotopes make ideal tracers in the study of sedimentary rock provenance because isotopes of the same element tend to travel together through the sedimentary system, preserving their original source-area signatures. An array of radioactive isotopes has been employed in studies of modern muds, with a somewhat more limited suite used for mudstones (Table 7.1). One that has seen much use in modern sediments is the $^{87}Sr/^{86}Sr$ ratio (Box 7.2). This application was pioneered by Dasch (1969), who demonstrated that even the finest clays in the deep-sea are not in isotopic equilibrium with

Box 7.2. Radioactive Isotopes: Strontium

Strontium isotopes are useful tracers of the sources of modern muds. The method is based on the radioactive decay of ^{87}Rb, which has a half-life of 4.88×10^{10} years, and produces ^{87}Sr. The ratio of the concentration of ^{87}Sr to stable, non-radiogenic ^{86}Sr is used to infer the source of Sr in muds or mudstones as well as in water (Faure 1986:183–199). The Sr isotopic composition and Sr concentration in fine-grained sediment accumulating in the oceans is controlled by, a) detrital minerals, whose Sr isotopic signature is determined by either the weathering products of predominantly young volcanic rocks ($^{87}Sr/^{86}Sr = 0.704 \pm 0.002$), or by those derived from old continental, radiogenic rocks from the crust ($^{87}Sr/^{86}Sr$ generally above 0.730 and commonly as high as 0.800), and b) authigenic minerals (carbonates, sulfides, silicates, etc.), whose $^{87}Sr/^{86}Sr$ is identical to that of the sea water from which they formed. Under advantageous geologic conditions, the noncarbonate fractions can be treated as mixtures of two end-members: the weathering products of young volcanic rocks of basaltic composition at one extreme, and old crustal rocks at the other.

The $^{87}Sr/^{86}Sr$ ratio of sea water has varied in the geological past, increasing from 0.7077 about 50 million years ago (Early Tertiary), to its current value in modern oceans, which is bracketed by 0.709241 ± 0.00032 and 0.709211 ± 0.00037 (Elderfield 1986). The continuous increase of the isotopic ratio since the end of the Eocene epoch is used to date samples of marine carbonates or fossil shells of Tertiary age.

Diagenesis tends to reset the Sr isotopic system. Sr is similar to Ca in its chemistry, and its parent Rb closely parallels K in behavior. Because so many mudstones have experienced post-depositional K addition or Ca loss, changes to Sr isotopes are common.

Box 7.3. Neodymium isotopes

Samarium (Sm) and neodymium (Nd) are rare earth elements linked in a parent-daughter relationship by the alpha decay of ^{147}Sm to stable ^{143}Sm with a half-life of 106×10^9 years. They are especially useful in geochemistry because these two elements undergo very few changes once a rock has formed. The melting of the Earth's mantle, which forms the more evolved plutonic and volcanic rocks of the exposed continental crust, fractionates Sm from Nd and thus allows for measurable isotopic variations over time. Models of the evolution of this reservoir – CHUR (for "chondritic uniform reservoir") or DM ("depleted mantle") – are used to calculate *model ages*, which are a measure of length of time a sample has been separated from the mantle from which it was originally derived.

More useful for mudstones than the model age is ε_{Nd} (De Paolo and Wasserburg 1976), which is a measure of the deviation of the $^{143}Nd/^{144}Nd$ ratio in a sample from the expected ratio value in a uniform reservoir (CHUR's $\varepsilon_{Nd} = 0$). A positive ε_{Nd} implies that the Nd came from depleted sources having a higher Sm/Nd ratio than CHUR (i. e., its source has undergone periods of partial melting or differentiation), whereas negative ε_{Nd} values imply derivation from enriched sources with a lower Sm/Nd ratio than CHUR (i. e., a magma source that earlier separated as a liquid from the primary chondritic uniform reservoir). ε_{Nd} varies widely among possible source rocks, but is unaffected by most sedimentary processes. Hence it provides a useful fingerprint for choosing among likely sources or for distinguishing one mudstone from another.

seawater, but instead reflect the composition of the rocks on adjacent landmasses. This result had compelling implications for understanding burial as well as provenance of clays, because it established that early diagenesis does little to the chemistry or mineralogy of clays, in contrast to the significant changes seen in deeper burial (Chap. 6).

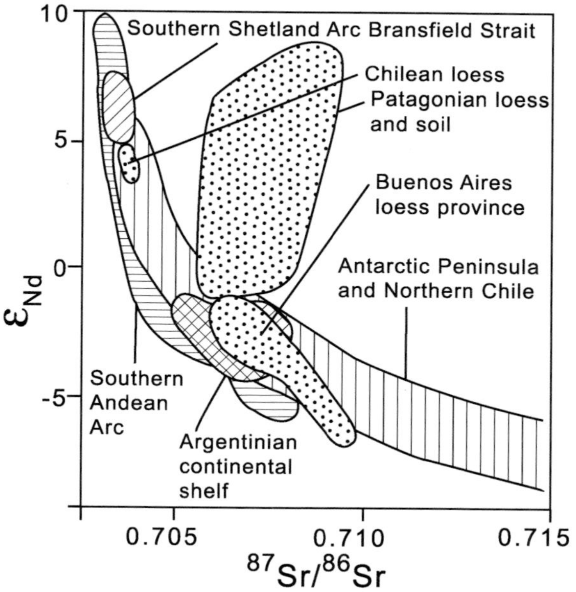

This persistence of $^{87}Sr/^{86}Sr$ values from source to sink has been exploited by a number of workers in provenance studies of modern sediments. For example, Graham et al. (1997) found that oceanic islands in the SW Pacific, which have $^{87}Sr/^{86}Sr$ ratios of 0.7036 to 0.7040, are major sources for clay material only in their immediate vicinity, whereas most of the clays in this region, which have values of 0.7090 or higher, are sourced more remotely from metamorphic rocks on New Zealand.

$^{87}Sr/^{86}Sr$ is often used jointly with $^{143}Nd/^{144}Nd$ in provenance studies, because Sr gives information about the composition of the source terrane, whereas Nd gives information about the age at which the source igneous rocks separated from the mantle. Mahoney et al. (1998) used $^{87}Sr/^{86}Sr$, along with clay mineralogy and $^{143}Nd/^{144}Nd$, to determine the provenance of Plio-Pleistocene hemipelagic muds in the Shikoku Basin of the Philippine Sea. They established that the isotopic signature of sediment sources from the basin margins is not compatible with the sediment in the basin center and identified an episodic eolian influx of materials with a continental signature (Sino-Korean craton), mixing with other marine sources. Walter et al. (2000) also employed both Nd and Sr isotopic evidence to study the sources and transport mechanisms of terrigenous sediments being delivered to the South Atlantic Ocean (Fig. 7.12). They found that sources varied greatly between glacial and interglacial pe-

Fig. 7.12. Nd vs. Sr in the southern cone of South America and the adjacent South Atlantic Ocean (after Walter et al. 2000). The complexity of this diagram probably reflects the proximity of fore-arc, back-arc, and passive margin sources. Published by permission of the authors and Pergamon Press

riods, a possibility that needs to be considered in all provenance studies. Nd isotopic methods are so sensitive that they can trace Greenland dust back to sources in China (Borry et al. 2003).

Sr tends to be reset by burial diagenesis because of the high mobility of both Sr and its radioactive parent, Rb. For example Awwiller (1994) found that Rb-Sr ages of Gulf Coast mudstones decrease with increased burial depth, becoming younger than their depositional age. Therefore older rocks present problems for using Sr isotopes in provenance studies, and Nd isotopes accordingly assume more prominence. Gleason et al. (1994) provide an example of use of Nd isotopes to track contributions to the turbidite mudstones and sandstones of the Ouachita Basin. They found that both lithologies share the same Nd signals: a Lower Paleozoic contribution with ε_{Nd} of −16 to −13, indicating a source from older rocks on the North American craton, which shifts abruptly in the Late Ordovician to an Appalachian orogen source characterized by younger sources with ε_{Nd} of −10 to −6. These changes in source area are consistent with inferences based on the petrography of the interbedded sandstones.

Stable Isotopes. Organic chemical indicators are used to provide information about the proportion of terrestrial- to marine-derived organic matter in a mudstone. A map of the values of this proportion would then provide a good picture of the direction to the shoreline. This is a procedure that has been widely used in the modern, but seldom in the ancient, perhaps because of the difficulty of doing detailed sampling of a narrow stratigraphic interval over a large area.

The quickest way to do this analysis is by stable isotope measurements of carbon (see Tyson 1995, Chapter 23 for a thorough review). Land plants today produce particulate organic matter with $\delta^{13}C$ values of about −27 permil, whereas marine organic carbon is at about −20 permil. Therefore the simple measurement of $\delta^{13}C$ provides a quick estimate of the relative effect of terrestrial-sourced material in the basin (Fig. 7.13a).

Unfortunately, ancient mudstones are more complicated. $\delta^{13}C$ of organic matter in Devonian marine shales of the Appalachian Basin indicates that the terrestrial source to the east has values of about −26 permil, but the marine carbon is lighter still, at about −32 permil (Fig. 7.13b). This difference in behavior of the modern and older systems has prompted considerable debate, but the cause seems to be related to higher atmospheric CO_2 levels in the

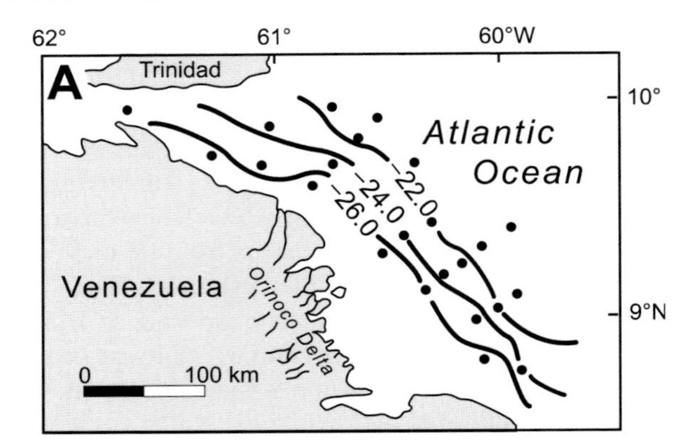

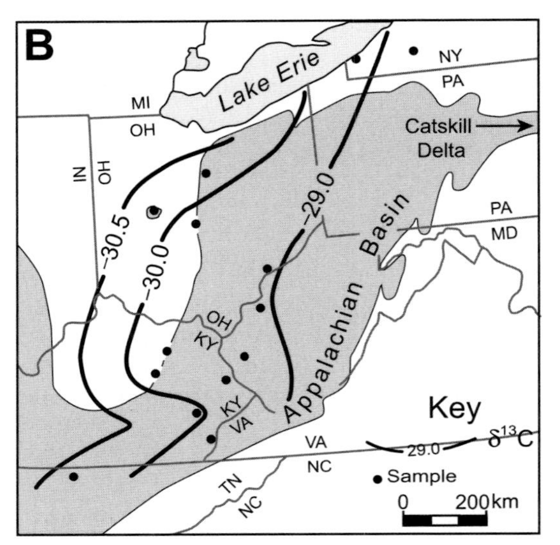

Fig. 7.13A,B. Provenance signal of carbon isotopes in organic matter: (**A**) $\delta^{13}C$ in muds off the Orinoco Delta show clearly the contribution of terrestrial carbon, $\delta^{13}C = -26$ permil, to the marine system with $\delta^{13}C = -20$ permil (after Kennicutt et al. 1987, p 45), and (**B**) Devonian black shales of the Appalachian Basin also show a proximal-distal variation in $\delta^{13}C$, but with the reverse polarity (after Maynard 1981, Fig. 1). Published by permission of the authors, Pergamon Press and the Geological Society of America

Devonian and at most other times in the Phanerozoic (Lewan 1986; Popp et al. 1989). Their data suggest that the change to modern-style carbon isotopic patterns occurred in the Eocene-Miocene, a time of global lowering of CO_2 and accompanying cooling of the climate. Comparable lowered CO_2 at other times in the geologic past should also have resulted in marine carbon that had heavier $\delta^{13}C$ values, but there is very little data from the relevant periods, perhaps because these were times of diminished black shale accumulation. Tyson (1995, his appendix B) lists one Permian occurrence with $\delta^{13}C = -20$ to −18 permil, consistent with this argument. The study of the areal distribution of carbon isotopes in black shales of various ages would seem to be a fruitful area for future research.

7.6. Recycling

An important limitation on the use of mudstones in provenance studies is the extent of recycling, which makes it ambiguous whether the signal observed comes from the immediate source or was inherited from a pre-existing terrane. The problem is more pervasive for muds than for sands because of the greater stability of the clay mineral grains than all but the quartz grains in a sandstone. Does a high CIA value for a mudstone indicate intense weathering in the source area or rather a slightly weathered source made up mostly of older mudstones themselves derived from weathering under wet tropical conditions? Furthermore, how do we know that a particular REE pattern in a mudstone reflects the composition of the parent terrane rather than that of some earlier terrane in the sedimentary recycling history of the material? It turns out that muds and mudstones are commonly the product of the turnover of a pre-existent sedimentary mass because their particles largely consist of stable weathering products, and these particles, like quartz grains in a sandstone, can be recycled through many episodes of burial, uplift and erosion. Veizer and Jansen (1979) estimated that, on a global basis, clastic sediments are 65% recycled. Secondly, most mountainous regions have a high proportion of sedimentary rock exposed, which creates the potential for sedimentary-sourced particles to dominate the suspended load of major rivers. This recycled component of rivers has long been recognized, but it is hard to quantify in the modern and nearly impossible in the ancient for a given deposit.

Gaillardet et al. (1999) provide an initial attempt at estimating the amounts of recycling in individual river basins by comparing the measured suspended load to that predicted from the composition of the dissolved load. The calculations are similar to those developed by Garrels and MacKenzie (1971) for converting average igneous rock into sediment plus ions in solution. For all rivers, there is an excess of suspended load over what would be predicted from the dissolved load if only crystalline rocks are being weathered. The amount of this excess is proportional to the extent of sedimentary rock cover in the source area, and thus provides an estimate of the fraction of recycled material a stream is carrying. The results vary widely for different rivers, with values falling between 45 and 75% recycled material. The overall conclusion from these calculations is that the amount of recycled material in a mudstone is likely to be very large and highly variable depending on the nature of the river bringing the sediment into the basin of deposition. Therefore provenance studies using mudstones need to somehow take into account this potential for extensive recycling.

Recycling requires knowing the composition of the dissolved load, which we cannot recapture for ancient rocks. In our view, the best key to the amount of recycling in mudstones is found in the petrography of the interbedded sandstones. Abundant volcanic rock fragments in the sandstones indicate a dominantly juvenile source, whereas a pure quartz sand generally indicates a recycled source – be on alert for sandstone rock fragments or for quartz grains with old overgrowths. Pure chemical approaches to recycling, on the other hand, are likely to be less compelling in our view. If a mudstone shows very little evidence of weathering (low CIA + high ICV), then it is likely that the parent terrane contains mostly juvenile igneous rocks. The converse is not true, however. If CIA is high and ICV low, then the cause could be either deep weathering of the source or recycling of previously weathered material. Consider for example a Miocene mudstone, encountered in offshore drilling, that has high CIA values and low Th/Sc, indicating a strongly weathered mafic-rich source. Was the immediate source a weathered basalt or a Paleozoic mudstone itself derived from a basaltic terrane? The tectonic reconstruction of the basin would be radically different depending on the answer to this question.

7.7. Ash Fall Mudstones

One component of a mudstone unfailingly reflects primary igneous sources – volcanic ash beds. These come in two main varieties, those deposited in swamps, which become kaolinite-rich *tonsteins*, and those deposited in marine basins, which become smectite-rich *bentonites*. (See Chap. 6 for a discussion of the diagenetic pathways followed by these lithologies, and Chap. 8 for a discussion of their great utility in stratigraphic correlation.) As provenance indicators, these beds are best studied through their trace element chemistry and their phenocryst mineralogy, although some use has been made of tracking the direction to the source by isopach mapping of ash thickness (Huff et al. 1996). A good example of the use of trace elements in tonsteins is Spears and Lyons (1995) who separated sources into rhyodacites and alkali basalts using a plot of Zr/TiO_2 vs. Nb/Y. Ash

Table 7.3. Provenance indicators in volcanic ash beds. Adapted from Bohor and Triplehorn (1993)

Feldspars
Sanidine in rhyolitic ashes, Ca-plagioclase in more mafic ashes, whereas microcline or pure albite indicate a non-volcanic clay or post-depositional alteration; microprobe chemistry of plagioclases a useful fingerprint.

Quartz
Euhedral, water-clear beta-form quartz is an unequivocal indicator of a volcanic origin of the clay bed; glass inclusions in the quartz can be probed to give original magma chemistry

Zircon
Ideal for ion-microprobe Pb-Pb dates

Apatite
Commonly contains glass inclusions useful for magma chemistry

Biotite
Has a large compositional variability that can be useful for fingerprinting each ash bed in a series; typically the most abundant phenocryst

Insoluble trace elements
Zr, Nb, Y, Ti, Th, Ta, and Hf are commonly used on discrimination diagrams that define magma types

Rare-earth elements
REE patterns are good for fingerprinting ashes, and the size of the europium anomaly is an indicator of how felsic the volcano was.

beds tend to be bimodal in grain size, with coarse phenocryst minerals set in a clay matrix. These phenocrysts come in several types (Table 7.3) and contain much about their parent magmas in the form of mineral chemistries and in the chemistries of their melt inclusions (Huff et al. 1998).

These ashes form *event beds* that span several environments, so they are a powerful contributor to understanding the sequence stratigraphy of a mudstone section. See Sarna-Wojcicki and Davis (1991) for examples of widespread Quaternary continental ashfall deposits in lakes, soils, and on river terraces. The largest covers over 2 million km^2. In addition, knowledge of the type of volcano that produced the ash is a great help in reconstructing the tectonic setting of the basin of deposition, always a useful step in mineral and petroleum exploration. Numerous thin ash beds, as are found in the Permian of the Paraná and Karoo Basins of Brazil and South Africa, may indicate that the background clay deposition in the surrounding mudstones also had a high volcanogenic input. Such mudstones might be particularly prone to engineering problems or might be especially effective source beds for petroleum because of their capacity for releasing water during hydrocarbon generation.

7.8. Provenance of Clays in Archeology

The principles developed above also have application to the sourcing of ceramics in archeological work. Pottery contains two main sets of ingredients, *plastic*, comprising mostly clays, and *aplastic*, coarser materials either native to the clay or added to *temper* the clay and make it easier to work and to give strength to the pottery when fired. The plastic component has provenance characters like mudstone, whereas the aplastic component is analogous to interbedded sandstone, and the strategies for characterizing provenance for these two rock types are also applied to the pottery ingredients.

The aplastic component most commonly consists of quartz-rich sand, carbonate particles, either limestone fragments or broken shells, or *grog*, which is the term for ground up old pottery reused in the new, just like reworked sandstone fragments in a second-cycle sandstone. Petrographic techniques are the method of choice for these sand-sized ingredients (e. g. Ferring and Perttula 1987). Microchemical approaches like the electron probe could be used to study the chemistry of these particles separate from that of the clay matrix, but we know of no such attempts recorded in the archeological literature. For example, some material characterized as grog in thin section may actually be mudstone fragments, and these could be distinguished microchemically from broken pottery, which should have a similar composition to the clay matrix.

Chemical techniques are most appropriate for the plastic component, but whole-sherd analyses have the complication of combining both the clay- and the sand-sized ingredients in the same analysis, when they likely have rather different sources. For example, silicon, being an almost universal temper ingredient, will likely not provide useful information about clay sources. Likewise calcium will reflect mostly the amount of temper if shell or limestone was added to the clay. Within this inherent constraint, some useful results have been obtained by careful study of the chemistry of pottery fragments.

Two kinds of questions are addressed. The simpler is, "Are the pottery sherds at two sites (or from two layers in one site) the same or different from each other?" Simple cluster analysis of all of the chemical data is effective in addressing this type of question. Combining major and trace element data, however, presents problems because of the vast difference in the ranges of concentrations. This is usually overcome by using log-transformed data (or by normalizing each data point to a percentage of the total range for this particular variable). Statistical analysis of all chemical constituents obviously combines the clay and the temper, but differences in manufacturing procedure may be just as interesting as differences in the sources of the materials, so this "whole-rock" procedure is beneficial.

The second, more difficult type of question to ask is, "What was the geological source of the clay used in this pottery?" Here we have to exclude the temper components. The procedure to follow is again rather similar to provenance studies of mudstones, in that the trace elements give the best provenance discrimination. The elements Si, Ca, P, and Sr should be dropped from the statistical analysis because they are common ingredients of sand or limestone or shells. One might also wish to exclude Al, K, and Na because they are the elements making up the feldspar component of sands, and a quick check of the temper using the petrographic microscope reveals whether grains of feldspar (or feldspathic rock fragments) are sufficiently abundant to warrant the exclusion of these elements. Soltman (2001) shows good illustrations of thin sections of quartz-rich and feldspar-rich tempers. Excluding these temper elements leaves the trace elements and a few less mobile major elements like Fe and Ti to be included in the statistical treatment. The same array of trace element ratios that is useful as bivariate plots in mudstone provenance appears to work best here as well (see for example Rodriguez-Alegria 2003), or multivariate statistical tests such as principal component analysis can be applied.

An interesting example of this approach comes from a study of Roman pottery of Tunisia (Sherriff et al. 2002). The area contains three likely clay sources, a Miocene marine clay rich in sulfur and gray in color, a Pliocene sandy non-marine brown clay, and a Late Pliocene calcareous non-marine green clay. Principle component analysis defined three distinct fields for the composition of these clays using two linear combinations of chemical variables (Fig. 7.14). The pottery sherds showed a close match to the brown clay, but shifted slightly because of higher Na and Cl, which the authors attributed to seawater added to the mix to harden the pottery.

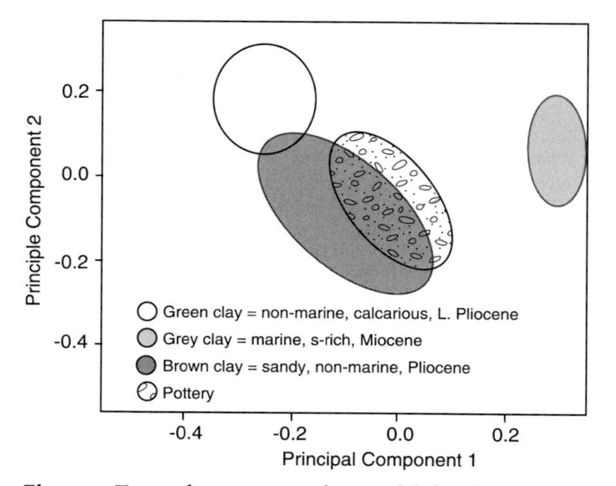

Fig. 7.14. Trace elements are also useful for determining sources of clay used in pottery – an example of Roman ceramics from Tunisia (after Sherriff et al. 2002, Fig. 8). Published by permission of the author and John Wiley & Sons

7.9. Summary

The determination of provenance of muds and mudstones has many similarities to the better-known procedures for sandstones, but differs in two important ways. Because grain size is so small, petrographic techniques are much less useful, and because transport distances can be so much greater, additional care must be taken to separate local from distant sources. In particular, wind-blown material of sand size cannot be transported long distances, but silt- and clay-size material is easily distributed over the area of an entire continent by the wind. Because petrography is less useful, chemical methods assume greater importance for mudstones compared with sandstones. Major elements are most useful for estimating the degree of weathering of the source, whereas trace elements are best suited to estimating the relative proportions of mafic and felsic rocks in the ultimate source. Radiogenic isotopes are best used to "fingerprint" individual sources, and therefore commonly repay the extra investment required for their use. For all techniques, it is better to compare like types of mudstones. Black shales in particular are prone to enrichments in many trace elements compared to red and green mudstones.

A major source of uncertainty in provenance reconstruction is the proportion of recycled material in the basin fill. An apparently mafic source could instead be an older mudstone that itself had a mafic source, perhaps many million years earlier. For this reason, thin clay beds derived from volcanic ash are particularly good indicators of provenance, but bear in mind that the volcanoes may have been only one of several different sources for the basin fill.

7.10. Challenges

Recycling: A particle in a mud or mudstone may be derived from weathering of a crystalline rock, or it may come from weathering of a pre-existing mudstone.

How does one detect recycling and quantify its effects? Another issue is that our hypothetical particle may have languished along its transport path in alluvial deposits where it was subject to additional alteration. How long does such alluvial storage need to be to change the apparent provenance signal of a mud? How long do particles typically spend in this setting? Is there any way to measure this effect?

How much of the fine silt in muds and mudstones is added by wind?

Because wind-blown silt can travel so far, it is quite likely to have a different provenance than water-borne particles in the mud. Is there some combination of chemical variables that can estimate this contribution? The composition of soils on carbonate ocean islands might provide a starting point.

The amount and composition of muds that are produced by a source area are influenced by both climate and relief, yet it is usually difficult to distinguish the relative importance of these two factors from the resulting mudstone alone.

This same dilemma has plagued sandstone provenance studies for years. Would detailed comparisons of interbedded sandstones and mudstones help resolve this?

Many workers have suggested that the Archean weathering environment was much hotter and had much less oxygen than today.

In what ways would weathering have been different then and how would the provenance signals be changed? Tropical lowland rivers today carry little mud and the little that they carry is mostly kaolinite and quartz, neither of which carries much provenance information. Would Archean rivers have been similar?

Geochemists have expended much effort assembling global averages of mudstones through time.

Given that mudstones can be linked geochemically to different tectonic settings, would it not be helpful to separate these analyses by basin type and track the average "rift mudstone" or the average "foreland basin" mudstone through time? Our impression of the "global average shale" may be strongly biased by the vagaries of sampling different plate tectonic settings. Consider also that some settings experience much more recycling than others. Tracking arc-related mudstones through time might provide a better signal than the global average mudstone.

References

Aléon J, Chaussidon M, Marty B, Schütz L, Jaenicke R (2002) Oxygen isotopes in single micrometer-sized quartz grains: Tracing the source of Saharan dust over long-distance atmospheric transport. Geochim Cosmochim Acta 66:3351–3365

Arnaud H, Charollais J, Médioni R (1984) 2.1-Bresse, Jura, Bas-Dauphine Chaînes subalpines septentrionales. In: Debrand-Passard S, Courbouleix S, Leinhardt MJ (eds) Synthèse Géologique du Sud-Est de la France. Mémoire Bureau Recherches Géologiques Miniéres 1, 125:305–313

Awwiller DN (1994) Geochronology and mass transfer in Gulf Coast mudrocks (south-central Texas, U.S.A.): Rb-Sr, Sm-Nd, and REE systematics. Chem Geol 116:61–84

Bohor BF, Triplehorn DM (1993) Tonsteins: altered volcanic ash layers in coal-bearing sequences. Geol Soc Am (Special Paper) 285:44

Bory AJ-M, Biscaye PE, Grousset FE (2003) Two distinct seasonal Asian source regions for mineral dust deposited in Greenland (NorthGRIP). Geophys Res Lett 30:16.1–16.4

Condie KC (1991) Another look at rare earth elements in shales. Geochim Cosmochim Acta 55:2527–2531

Condie KC, Wronkiewicz DS (1990) The Ce/Th ratio of Precambrian pelites from the Kaapvaal Craton as an index of cratonic evolution. Earth Planetary Sci Lett 97:256–267

Cox R, Lowe DR, Cullers RL (1995) The influence of sediment recycling and basement composition of evolution of mudrock chemistry in the southwestern United States. Geochim Cosmochim Acta 59:2919–2940

Cullers RL, Barrett T, Carlson R, Robinson B (1987) Rare-earth element and mineralogic changes in Holocene soil and stream sediment: a case study in the Wet Mountains, Colorado, USA. Chem Geol 63:275–297

Dasch EJ (1969) Strontium isotopes in weathering profiles, deep-sea sediments, and sedimentary rocks. Geochim Cosmochim Acta 33:1521–1552

De Paolo DJ, Wasserburg GJ (1976) Nd isotopic variations and petrogenic models. Geophys Res Lett3:249–252

Eisma D, Van der Marel HW (1971) Marine muds along the Guyana coast and their origin from the Amazon River. Contribut Mineral 31:321–324

Elderfield H (1986) Strontium isotope stratigraphy. Palaeogeography, Palaeoclimatology, Palaeoecology 57:71–90

Eriksson KA, Taylor SR, Korsch RJ (1992) Geochemistry of 1.8–1.67 Ga mudstones and siltstones from the Mount Isa inlier, Queensland, Australia: Provenance and tectonic implications. Geochim Cosmochim Acta 56:899–909

Faure G (1986) Principles of Isotope Geology, 2nd edn. Wiley, New York, 589 p

Ferring CR, Perttula TK (1987) Defining the provenance of red slipped pottery from Texas and Oklahoma by petrographic methods. J Archaeol Sci 14:437–456

Gaillardet J, Dupré B, Allégre CJ (1999) Geochemistry of large river suspended sediments: Silicate weathering or recycling tracer? Geochim Cosmochim Acta 63:4037–4051

Garrels RM, MacKenzie FT (1971) Evolution of Sedimentary Rocks. Norton, New York, 397 p

Gibbs RJ (1967) Geochemistry of the Amazon River System: Part 1. Factors that control the salinity and the composition

and concentration of the suspended load. Geol Soc Am Bull 78:1203–1232
Gleason JD, Patchett JP, Dickinson WR, Ruiz J (1994) Nd isotopes link Ouachita turbidites to Appalachian sources. Geology 22:347–350
Graham IJ, Glasby GP, Churchman GJ (1997) Provenance of the detrital component of deep-sea sediments from the SW Pacific Ocean based on mineralogy, geochemistry and Sr isotopic composition. Marine Geology 140:75–96
Harnois L (1988) The CIW index; a new chemical index of weathering. Sediment Geol 55:319–322
Huff WD, Kolata DR, Zhang Y-S (1996) Large-magnitude Middle Ordovician volcanic ash falls in North America and Europe: dimensions, emplacement and post-emplacement characteristics. J Volcanol Geotherm Res 73:285–301
Huff WD, Bergstrom SM, Kolata DR, Cingolani CA, Astini RA (1998) Ordovician K-bentonites in the Argentine Precordillera: relations to Gondwana margin evolution. In: Pankhurst RJ, Rapela CW (eds) The Proto-Andean Marin of Gondwana. Geological Society, London (Special Publication) 142:107–126
Ishiga H, Dozen K, Sampei Y (1999) Geochemical constraints on marine invasion and provenance change related to the opening of the Japan Sea: an example from the Lower Miocene shales in the Hoda section, Shimane Peninsula, SW Japan. J Asian Earth Sci 17:443–457
Jones RL, Blatt H (1984) Mineral dispersal patterns in the Pierre Shale. Journal Sedimentary Petrology 54:17–28
Kennicutt MC, II, Barker C, Brooks JM, DeFreitas DA, Zhu G (1987) Selected organic matter source indicators in the Orinoco, Nile and Chengjiang deltas. Organic Geochem 11:41–51
Lev SM, Hanson GN, McLennan SM, Meyers WJ (1998) A petrographic approach for evaluating trace-element mobility in a black shale. J Sediment Res 68:970–980
Lander RH, Hay RL (1993) Hydrogeologic control on zeolites diagenesis of the White River Sequence. Geol Soc Am Bull 105:361–376
Lee YI (2002) Provenance derived from the geochemistry of late Paleozoic-early Mesozoic mudrocks of the Pyeongan Supergroup, Korea. Sedimentary Geology 149:219–235
Lewan MD (1986) Stable carbon isotopes of amorphous kerogens from Phanerozoic sedimentary rocks. Geochim Cosmochim Acta 50:1583–1591
Li Y-H (2000) A Compendium of Geochemistry. Princeton University Press, Princeton, NJ, 475 p
Mahoney JB, Hooper RL, Michael G (1998) Resolving compositional variations in fine-grained clastic sediments: A comparison of Sr/Nd isotopic and mineralogical sediment characteristics, Shikoku Basin, Philippine Sea. In: Schieber J, Zimmerle W, Sethi PS (eds) Shales and Mudstones. E. Schweizerbart'sche, Stuttgart, 2, pp 177–194
Maynard JB (1981) Carbon isotopes as indicators of dispersal patterns in Devonian-Mississippian shales of he Appalachian Basin. Geology 9:262–265
McLennan SM (1989) Rare earth element in sedimentary rocks: influence of provenance and sedimentary processes. In: Lipin BR, McKay GA (eds) Geochemistry and Mineralogy of Rare Earth Elements. Rev Mineral 21:169–200
McLennan SM, Taylor SR (1991) Sedimentary rocks and crustal evolution revisited: Tectonic setting and secular trends. J Geol 99:1–21
McLennan SM, Taylor SR, McCulloch MT, Maynard JB (1990) Geochemical and Nd-Sr isotopic composition of deep-sea turbidites: Crustal evolution and plate tectonic associations. Geochim Cosmochim Acta 54:2015–2050
McLennan SM, Hemming S, McDaniel DK, Hanson GN (1993) Geochemical approaches to sedimentation, provenance, and tectonics. In: Johnsson MJ, Basu A (eds) Processes Controlling the Composition of Clastic Sediments. Geol Soc Am (Special Paper) 284, Boulder, pp 21–40
Nesbitt HW (1979) Mobility and fractionation of rare earth elements during weathering of a granodiorite. Nature 279:206–210
Nesbitt HW, Young GM (1982) Prediction of some weathering trends of plutonic and volcanic rocks based on thermodynamic and kinetic considerations. J Geol 48:1523–1534
Owen AW, Armstrong HA, Floyd JD (1999) Rare earth element geochemistry of the upper Ordovician cherts from the Southern Uplands of Scotland. J Geol Soc London 156:191–204
Pettijohn FJ, Potter PE, Siever R (1987) Sand and Sandstone, 2nd edn. Springer, Berlin Heidelberg New York, 553 p
Petschick R, Kuhn G, Givgle F (1996) Clay mineral distribution in surface sediments of the South Atlantic: Sources, transport and relation to oceanography. Marine Geology 130:203–229
Popp BN, Takigiku R, Hayes JM, Louda JW, Baker EW (1989) The post-Paleozoic chronology and mechanism of ^{13}C depletion in primary marine organic matter. Am J Sci 289:436–454
Potter PE, Heling D, Shimp NF, VanWie W (1975) Clay mineralogy of modern alluvial muds of the Mississippi River Basin. Bulletin Centre Recherche Pau-SNPA 9:353–389
Rodriguez-Alegría E (2003) Indígena ware: Spain to Valley of Mexico. In: Glascock MD (ed) Geochemical Evidence for Long-Distance Transport: Westport, CN, Bergin and Garvey, pp 13–31
Roser BP, Korsch RJ (1988) Provenance signatures of sandstone-mudstone suites determined using discriminant function analysis of major-element data. Chem Geol 67:119–139
Santos JOS, Hartmann LA, McNaughton NJ, Easton RM, Rea RG, Potter PE (2002) Sensitive high resolution ion microprobe (SHRIMP) detrital zircon geochronology provides new evidence for a hidden Neoproterozoic foreland basin to the Grenville Orogen in the eastern Midwest; USA. Canad J Earth Sci 39:1505–1515
Sarna-Wojcicki AM, Davis JO (1991) Quaternary tephrochronology. In: Morrison RB (ed) Quaternary Nonglacial Geology: Conterminous U.S. Geological Society America, The Geology of North America K-2:93–116
Shepard FP (1956) Marginal sediments of Mississippi Delta. Am Assoc Petrol Geol Bull 40:2537–2623
Sherriff BL, Court P, Johnston S, Stirling L (2002) The source of raw materials for Roman pottery from Leptiminus, Tunisia. Geoarchaeology 17:835–861
Soltman JB (2001) The role of petrography in the study of archaeological ceramics. In: Goldberg P, Holliday VT, Ferring CR (eds) Earth Sciences and Archeology. Kluwer Academic/Plenum Press, New York, pp 297–326
Spears DA, Lyons PC (1995) An update on British tonsteins. In: Whateley MKG, Spears DA (eds) European Coal Geology. Geological Society, London, Special Paper 82, pp 137–146
Terry DO, LaGarry HE, Hunt RM (1998) Depositional environments, lithostratigraphy, and biostratigraphy of the White River and Arikaree Groups (Late Eocene to Early Miocene, North America). Geol Soc Am (Special Paper) 325, 216 p
Totten MW, Hanan MA (1998) The accessory-mineral fraction of mudrocks and its significance for whole-rock trace element geochemistry. In: Schieber J, Zimmerle W, Sethi P (eds) Shales and Mudstones 2: E. Schweizerbart'sche, Stuttgart, pp 35–53
Tyson RV (1995) Sedimentary Organic Matter. Chapman and Hall, London, 615 p
Veizer J, Jansen SL (1979) Basement and sedimentary recycling and continental evolution. J Geol 87:341–370
Walter HJ, Hegner E, Dickmann B, Kuhn G, Rutgers van der Loeff MM (2000) Provenance and transport of terrigenous sediment in the South Atlantic Ocean and their relations to glacial and interglacial cycles: Nd and Sr isotopic evidence. Geochim Cosmochim Acta 64:3813–3827
Zimmerle W (1993) On the lithology and provenance of the Rupelian Boom Clay in Belgium, a volcaniclastic deposit. Bulletin Societé Belge de Géologie 102:91–103

Zuffa GG (ed) (1985) Provenance of Arenites. Reidel, Dordrecht, 408p

Digging Deeper

Goudie AS, Middleton NJ (2001) Saharan dust storms: nature and consequences. Earth Sci Rev 56:179–204.

Thorough review of information on Saharan dust. Relates clay mineralogy to dust sources. Also gives tables of chemical composition and maps of rate of dust flux to the Atlantic Ocean.

Lorenz S, Zimmerle W (1993) Miozäne Kaolin-Kohlenstonsteine aus dem Braunkohlen-Tagebau von Bełchatów SW von Lódź Polen. Zeitschrift Deutschen Geologischen Gesellschaft 144:187–223.

A little-known, exceptionally detailed and thorough petrographic and chemical study (Rock-Eval, major and trace elements, electron microprobe) of ash falls in the Miocene lignitic brown coals of Poland. Includes an interesting worldwide summary plus an outstanding compilation of individual components and their diagenetic alteration. See especially their Fig. 7 and plates 1 and 2. Excellent reference when beginning your petrographic examination of a suspected tonstein. See also Throw and Von Rad (1992). Has an extended English abstract and the figure captions are also presented in English.

McLennan SM, Hemming S, McDaniel DK, Hanson GN (1993) Geochemical approaches to sedimentation, provenance, and tectonics. In: Johnsson MJ, Basu A (eds) Processes Controlling the Composition of Clastic Sediments. Geol Soc Am (Special Paper) 284, Boulder, pp 21–40.

Excellent review of the use of immobile trace elements and REE in reconstructing provenance, based mostly on deep-sea turbidite associated muds. Table 1 gives a useful summary of the utility of various petrographic and geochemical techniques.

Neff H (1992) Chemical Characterization of Ceramic Pastes in Archeology. Monographs in World Archeology No. 7, Prehistory Press, Madison, Wisconsin, 289 p.

Twenty chapters on chemical analysis of pottery from both the Old and the New World. An interesting format of this book is that three of the chapters are critical commentaries on the contributions in each section. Most of the papers employ Principal Components Analysis to compare pottery with potential sources and, just as with mudstones, the REE prove to be the strongest discriminators.

Taylor SR, McLennan SM (1985) The Continental Crust: Its Composition and Evolution. Blackwell Scientific, Oxford, p 312.

Subtitled "An Examination of the Geochemical Record Preserved in Sedimentary Rocks," this book provides an in depth appraisal of the needed chemical methods, addresses major problems and reviews theories of crustal formation. Principal emphasis on the Archean and Proterozoic. Background needed.

Tyson RV (1995) Sedimentary Organic Matter. Chapman and Hall, London, 615 p.

The premier reference on organic partic Comprehensive, well-written treatment with many fine illustrations along with very useful data compilations in the appendix.

Muddy Basins

To read the history of a basin, understand the language of its mudstones

8.1. Introduction

There are many questions to ask about mudstones and sedimentary basins, in part simply because mudstone is the most abundant sedimentary rock, and thus is almost always present even in small amounts – and even small amounts are important for many engineering problems. Nonetheless, there are some basins that contain mostly all sandstone, all carbonate, or all volcanic rocks, while some have mixtures of all four. One key question is, "Why are some basins mudstone-rich and some not?" Still other questions are, "How do mudstones vary across a basin and what controls this variation?" and, "What do mudstones tell us about the origin of the basin itself?" Here a comparison among muddy, sandy, and carbonate basins is helpful (Table 8.1). Because body fossils, trace fossils, and organic matter are responsive to variations in the oxygen content of bottom waters, mudstones tell us much about the paleo-oceanography of ancient seas and lakes. In addition, dark mudstones concentrate both trace metals and are petroleum source beds. Because of the dominance of mudstone in the stratigraphic record, mudstones are a great source of information, if we can extract it. Recalling that stratigraphy is often described as the "queen" of all of geology's sub-disciplines, our challenge in this chapter is to use the stratigraphy of mudstones in a basin not only to explain the mudstones themselves, but also to amplify our understanding of an entire basin and its place in earth history.

We define a mudstone-rich basin to have at least 50% mudstone (Fig. 8.1). Such basins can be of any size. For example, paleoecologists studying a late

Table 8.1. Muddy, sandy, and carbonate components of basins compared

Muddy Basins
Mostly marine or lacustrine, although small ones occur in all tectonic settings and in all environments. Geometries range from small isolated pods and channel/canyon fills, through widespread thin sheets to large, thick basins (large muddy basins always need a source with ample rainfall and moderate to high relief). Variations of bottom oxygen produce many widespread marker beds in both large and small muddy basins. Late diagenesis more significant than early for mineralogy. Principle source of petroleum and host to Cu and Pb-Zn deposits. Many industrial uses
Sandy Basins
About half as abundant as mudstones, sandstones carry a strong provenance and tectonic signal, and provide information about transport paths and mechanisms within the basin. Typically have less continuity than mudstones and most carbonates; more common around basin margins than in basin centers. Compaction less than mudstones, and mineral diagenesis less if quartz-rich. Major reservoirs for petroleum and water
Carbonate Basins
Approximately half as abundant as mudstone, carbonates are totally intrabasinal, require warm, shallow, mud- and sand-free water. Thus latitude (warm water) and isolation from terrigenous clastics needed. Best developed on cratons and passive margins distant from deltas. Like mudstones, carbonates are mostly marine and lacustrine. Widespread sheets and mounds are the most common geometries. Early diagenesis has major role. Major reservoirs for petroleum and ground water plus host to many Pb-Zn deposits

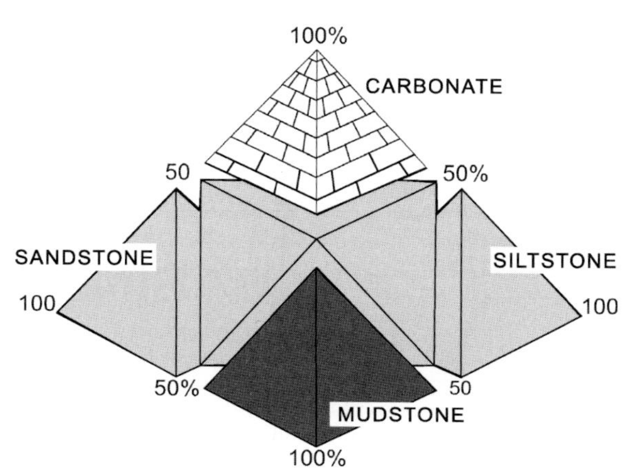

Fig. 8.1. Mudstones in the pyramid of sedimentary lithologies. The hidden fifth corner includes evaporites, coals, ash falls, ignimbrites, and conglomerates

Pleistocene-Holocene abandoned river-channel fill – a clay plug – typically think in terms of a narrow, elongate, curving basin of only a few square kilometers. The geotechnical engineer working in a metropolitan region built on a large modern delta thinks of a muddy basin of perhaps one hundred or more square kilometers. Or, on a still larger scale, a global sequence stratigrapher for a petroleum company routinely thinks of muddy basins as extending over 1,000 s of square kilometers and may even occasionally work with parts of the same basin that now occurs on two separated continents. Thus size is relative to the problem at hand, whereas dominance of fines is all-important, and a minimum of 50% of mudstone seems reasonable.

Beyond the definition of a muddy basin lies another important generalization that needs to be clear in any analysis of muddy basins. With a few exceptions, most mud-rich basins are marine or lacustrine, because only these two environments consistently provide sufficient space to trap extensive fine-grained sediments. Although muds and mudstones can be found in every depositional environment (Chap. 5), where they may be locally important for paleoecology, geotechnical studies, economic geology or correlation, the marine and lacustrine realms are the sites of most mud deposition, past and present, and hence it is they that receive our primary attention.

8.2. Controlling Processes

Controlling processes belong to two groups: those that control the supply and transport of mud *to* the basin and those that control the distribution and thickness of mud *within* the basin (Fig. 8.2).

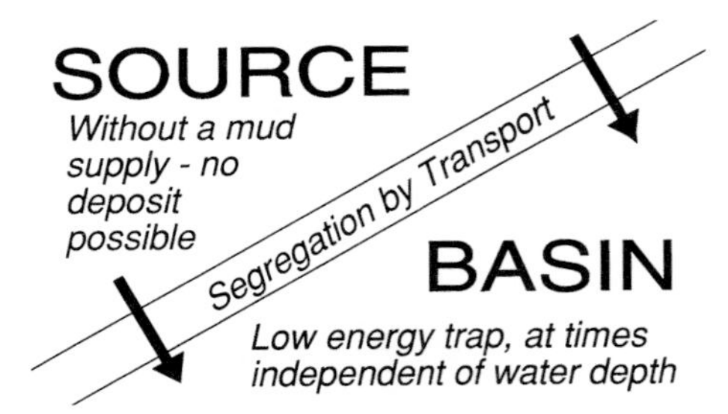

Fig. 8.2. A source area with ample rainfall and moderate to high relief is needed for mudstone-rich basins, whereas an arid, low-relief source area will be a small sediment donor and yield more sand than mud. If bordered by warm waters, neighboring basins will accumulate mostly carbonate rocks (Potter 1998, Fig. 1)

8.2.1. Mud Supply and Transport

In Chap. 2, Mud and Silt Production, we argued that the volume of terrigenous debris is maximized by a combination of high relief and high rainfall and is minimal where relief is low, be it either a wet tropical peneplain with deep residual soils and little sediment runoff or a vast desert with only limited eolian export of fines (Fig. 2.3). In other words, without a supply of terrigenous silt or mud – be it either lack of clay-yielding rocks in the source area or no transport out of it – a basin will accumulate only carbonate or chemical sediment or minor sand. To see this better, imagine a sea transgressing a dry sandy craton whose restricted rivers carry only small volumes of sand. Here, few fines are generated and those that are formed are blown away. Thus terrigenous mud brought to the sea will be minimal and offshore muds, should they exist, will be below the shelf edge and show evidence of lateral, deep-water transport parallel to the slope or platform. Two examples are the Middle Cambrian Burgess Shale, which is distal to carbonates of the Cathedral Escarpment in British Columbia and the fine sediments below the carbonate rim of the Delaware Basin of Texas.

Now let's examine hydraulic segregation in transport, which has several different dimensions. The first and better known is *bypassing*, the winnowing of fines leaving behind a concentrate of sand or pebbles. Bypassing is the way that rivers carry mud to the sea and ocean currents carry mud across a shelf, perhaps even beyond its edge. Modern oceanic examples include the tide-dominated shelf around the

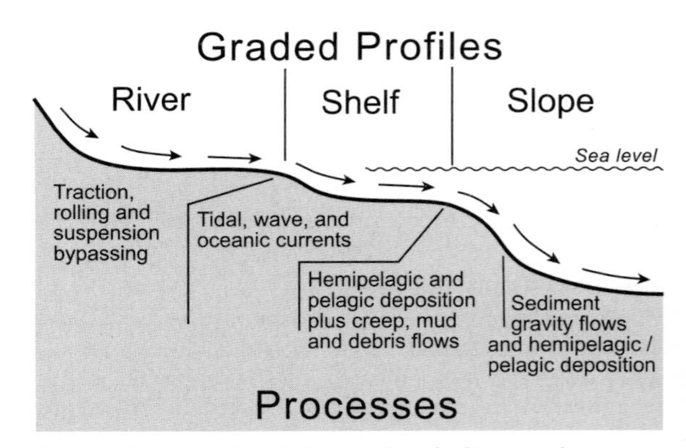

Fig. 8.3. Processes that bring mud and silt to a sedimentary basin (after Ross et al. 1995, Fig. 1). The fines at the "end of the line", be it a large river, a lake, a shelf, or a deep ocean, can be thought of as the end result of "sedimentary runout". See also Fig. 8.4. Published by permission of the authors and the American Association of Petroleum Geologists

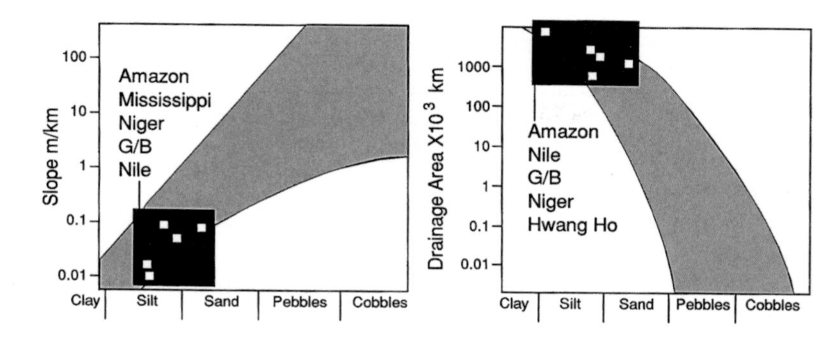

Fig. 8.4. River gradients, drainage basin areas, and proportions of silt and clay of the Amazon, Ganges-Brahmaputra (G/B), Mississippi, Niger, and Nile Rivers (after Orton and Reading 1993, Figs. 3 and 9). The low gradients of these big rivers near their mouths effectively preclude coarse sand, pebbles, and cobbles from entering the ocean. Published by permission of the authors and E. Schweizerbart'sche Verlagsbuchhandlung

British Isles and the Indian Ocean coast of South Africa where the Agulhas Current impinges on the shelf – both examples of extensive sand reworking and export of fines by suspension. Hydraulic segregation also occurs in large river systems, graded shelves and in sub-sea fans (Fig. 8.3). Earlier we saw that the smaller the fall velocity of a particle, the weaker the current needed to transport it and the farther the particle will travel in suspension (Box 3.1). These observations explain the predominance of mud, silt, and fine sands at the mouths of large rivers such as the Amazon, Niger, Ganges-Brahmaputra, Huang-Ho (Yellow), and Mississippi; on many ancient terrigenous shelves (and some modern ones); and in most submarine fan deposits (Fig. 8.4) Two observations are important here. First, the grain size-gradient relationship seen in these large rivers in their lower courses and in distal turbidites is, "part and parcel" so to speak, of the fluvial-turbidity current transport process. Therefore we can almost always think of a distal riverine or distal-turbidite hydraulic concentration of fines. This fine fraction will normally include detrital organic particles and, under anoxic conditions, may also include algal-derived organics from within the basin. On a graded shelf, a combination of storm events, tides and coastal currents has the same effect of segregating fines from inshore coarser detritus. Another way to think of these distal fines is that they represent the end stage of "sedimentary run out". Secondly, all five of the above named rivers drain to passive continental margins. This implies that many of the largest and thickest muddy sections in the geologic column were down slope from large rivers draining to passive margins – with the added requirement of a wet highland somewhere in their watershed (or perhaps vast loess plateaus) to provide the needed fines.

A second major way for size segregation to occur is by trapping coarse debris in an intervening, up dip basin between a wet highland source and the final depositional site. This could be either a rapidly subsiding foreland basin in front of developing thrust sheets that depress the crust (Marañón Basin just east of the Andes in central Peru) or the first and inshore of several fault-bounded borderland basins

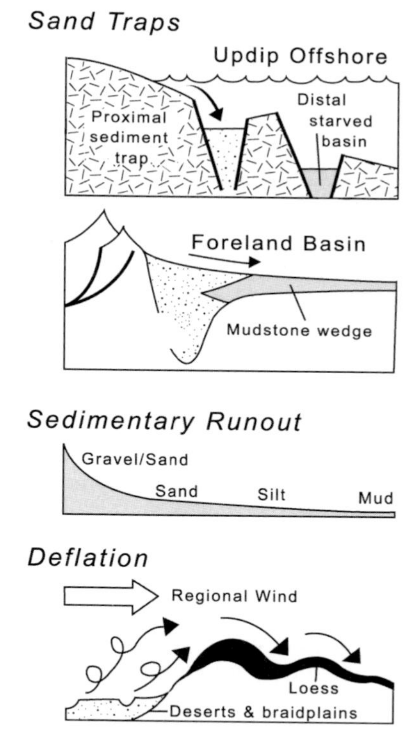

Fig. 8.5. Varieties of basin trapping. The most inshore of successive submerged fault-bounded basins, formed by either pure extension or by obliquely shearing plates along a continental margin, traps most of the coarse detritus, permitting the offshore neighboring basin to accumulate biogenic, organic-rich deposits (diatomites/marls) as well as mudstones. Another possibility is a foreland basin trapping sand while its adjacent cratonic basin accumulates muds, perhaps interbedded with carbonates, given a warm climate. Removal of silt and mud by wind from a closed basin with an arid climate provides still another possibility

on transitional crust marginal to a continent such as offshore southern California basins (Fig. 8.2).

So to sum up, every muddy basin was a quiet water trap and had access to a source that supplied terrigenous mud and silt. Conversely, basins deficient in mudstone result from bypassing or updip trapping or lacked a fine-grained terrigenous source. That is, minimal supply (always favored by little rainfall), bypassing or both produce basin fills with little or no mudstone, whereas the combination of supply and quiet water traps produces basin fills rich in mud. At this point note another important insight from the dependence of mudstone on quiet water, low-energy bottoms: the greater proportion of time a basin is flooded, the more mudstone rich it will be. Additionally, the fewer interbedded sandstones and carbonates a muddy basin has and the less the silt content of its mudstones, the closer it is to an end member "mudstone" basin, one totally composed of claystone. Persistent supply only by weak muddy suspensions (a bottom free of silt-carrying turbidity currents, storm waves, tidal or contour currents) is necessary to deposit a basin filled with claystone, be it a lagoon, bay fill, or a deep tectonic low. Such basins may also have pelagic carbonate and chemical interbeds as well as chert.

8.2.2. Mudstone Distribution and Thickness

Entry points of mud, current regime within the basin, bottom topography and shoreline position (relative sea level) all interact to determine the thickness and distribution of mudstones within a basin.

Entry of mud varies from a single point source, such as a large muddy river, to many small streams draining a well-watered, linear forearc basin, to a large continental glacier calving into the sea on a passive margin. In a large river delta, a regularly thinning sheet of prodelta mud, with some silt and fine sands and slump and storm beds, is deposited. In the forearc case, many small streams feed mud, silt and sand laterally into a foredeep, constructing a fan apron, with final axial transport parallel to the arc. In the glacial example, a thick, texturally varied unit with much mud and silt along with dropstones and irregular isolated masses of till is formed.

Current systems within a basin range from weak and variable (e. g., protected gulfs or long, narrow, deep basins), to open, windswept tidal shelves or even shelves with intruding inshore oceanic currents powered by strong persistent trade winds. The interaction of tidal range, storm tracks, wind regime, and shelf width and depth determines the final mud distribution. In the ancient, paleocurrent study of the mudstone and its interbedded limestone and sandstones, along with facies mapping and insight to the size and shape of the former water bodies, are the keys for reconstructing these current patterns. In addition, at times it may be possible to use knowledge of global wind patterns and former continental positions to estimate ancient trade wind paths.

Table 8.2. Determining paleodip of ancient shelves and ramps

- Orientation of clinoform from log and seismic markers and facies changes
- Darker colored shales in deeper water
- Thinner and fewer storm beds in deeper water plus decrease in sand/mudstone ratio
- Increase of pelagic fauna and bioturbation basinward (up to oxygen restriction)
- Direction of overturning of slump folds
- Shelf channels and canyonheads oriented broadly down dip
- Orientation of cross bedding, ripple marks, sole marks,
- Fossil orientation and gutter casts of interbedded sandstones and carbonates useful when integrated with the above

Bottom topography also varies widely from broad, uniformly sloping shelves and ramps to steep, fault-bounded, small basins that form complex mosaics at the other extreme. The rate of generation of accommodation space is minimal in the first case, but maximum for the case of fault-bounded pull-apart basins. For equal rates of mud supply, the pull-apart will have a far greater mudstone thickness. Mud supply also tends to be greatest in the pull-apart case because of adjacent transpressional uplifts, whereas slowly-subsiding interior basins tend to be supplied from older, deeply-eroded mountain belts or peneplained, sand-rich shields.

The final control is the position of the shoreline – is it far inland at the basin's limits (maximum flooding surface), far seaward (lowstand) or somewhere in between (most of the highstand systems tract)? The stacking patterns of the mudstone and interbedded sandstone and limestone vary with shoreline position and are a major clue to its reconstruction. Shoreline position in turn varies with relative rates of sediment input, basin subsidence, and world sea level (Fig. 8.6). High input, low subsidence rates, and low world sea level cause shorelines to migrate seaward, whereas high worldwide sea level and low input

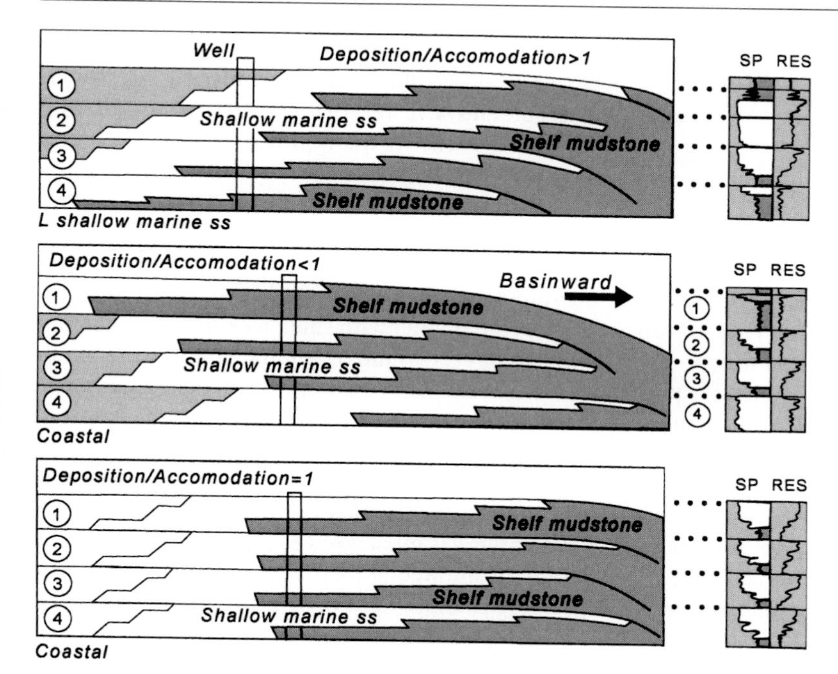

Fig. 8.6. Stacking patterns of interbedded carbonates and sandstones are all-important to understanding the distribution of mudstones and related facies in a basin (after Van Wagoner et al. 1990, Fig. 10). See Appendix A.1 for more about geophysical logging and Krassay (1998) for an excellent application to the Proterozoic of Australia. (SP = self-potential; RES = resistivity). Published by permission of the authors and the American Association of Petroleum Geologists

cause shorelines to advance landward (as do other combinations). Such oscillations shift mud deposition (and thus mudstone distribution, thickness and many mudstone properties as well), back and forth across a basin. Ideally, in such a process, mudstone thickens down-dip away from a shoreline to a maximum (both maximum accommodation and supply) beyond which the mudstone thins as supply diminishes, in spite of ample accommodation. Each such mudstone has a maximum flooding surface identified by having the most clay, most pelagic organisms and least silt. In addition it is likely to be the most organic rich.

The outcomes of the above possibilities vary widely and lend themselves to a deductive analysis. Such an analysis has yet to be made for all the distribution patterns of mudstones that exist in sedimentary basins even though there are some well-established end members. Examples of end members include the graded shelf (more common in the ancient than in the modern), coastal mud belts, high carbonate platforms bounded by abrupt scarps with thick mudstones below, and deep sea muddy fans on basin bottoms. See Ross et al. (1995) for computer modeling of mud distribution on graded shelves and slopes.

And what is known about the thickness of mudstones? In spite of all the many studies of mudstones, very few data exist on mudstone thickness versus environment of deposition or basin type, despite the long history of such analyses for sandstones. One well-documented exception exists for the Oligocene Frio Formation of the Gulf Coast (Fig. 8.7). From the standpoint of modeling fluid flow in either a reservoir or across a basin, the frequency distribution of mudstone thickness and continuity is a critical input, so many more studies such as that of Geehan et al. (1986) on the Prudhoe Bay fields are needed. Could a systematic study of the lithologic logs published by the Ocean Drilling Program provide a better un-

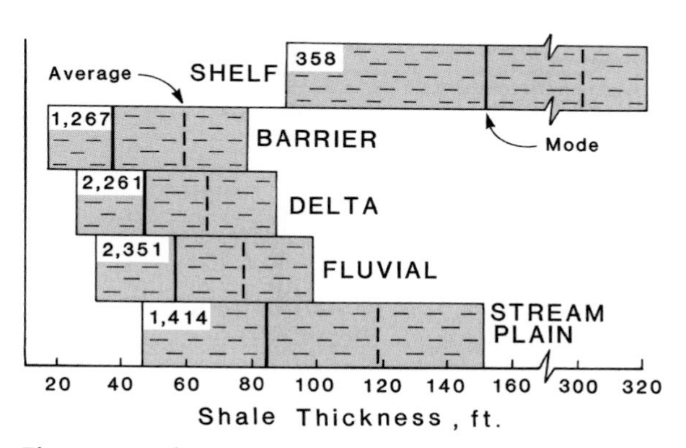

Fig. 8.7. In the Tertiary Frio Formation of the Gulf Coast, shelfal mudstones are the thickest by far (maximum accommodation), those of coastal barriers are the thinnest (minimal accommodation), and all distributions are strongly negatively skewed toward thinner mudstones. Numbers on each bar are numbers of beds (after Galloway and McGara 1982, Fig. 15). More data such as these are needed

derstanding of how mudstone thickness varies with tectonic setting?

8.3. Mudstone Geometry

The geometry of a mudstone tells us much about its origin at all scales from outcrops to basins. This idea, now widely accepted and used, was first fully developed and exploited for all sedimentary rocks with the introduction of seismic stratigraphy by Exxon (Mitchum et al. 1997, Figs. 2.4, 2.5, 2.6, and 2.7).

Mudstones have three common geometries plus several exceptional ones (Fig. 8.8). Three common geometries are thin *sheets*, *wedges* and *drapes*. Sheets have essentially parallel upper and lower boundaries and may be continuous or discontinuous. Sheets range from millimeters (tidalites, varves, ash falls, etc) to tens of meters or more in thickness. The term *mud blanket* is sometimes used for a uniform, thick mudstone with continuous, sheet-like geometry. Thicker, widespread sheets of mudstone represent many events and need an even, broad stable surface of deposition such as exists on cratons, pericratonic basins, subsea fans, and the bottoms of stagnant basins or ocean floors. Such mudstones are widely traceable and have well-defined internal stratigraphy.

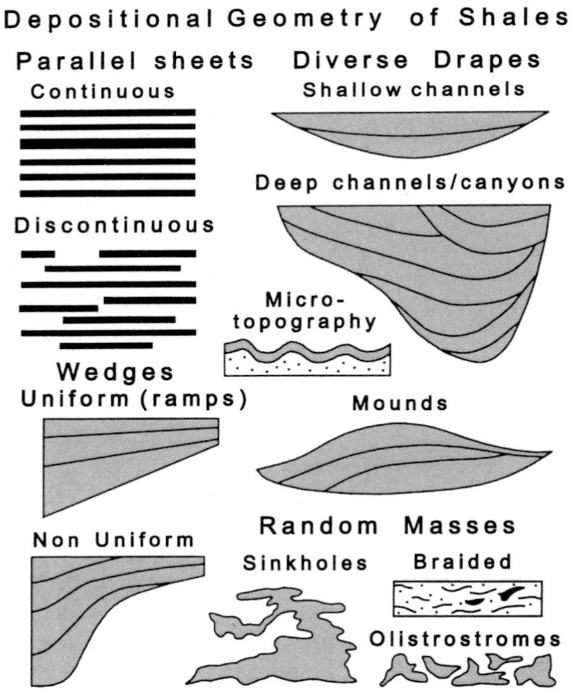

Fig. 8.8. Schematic depositional geometry of mudstones. Most of these patterns form the essential end members of the more complex recurring patterns recognized in seismic stratigraphy in Fig. 8.10. Published by permission of the authors and the American Association of Petroleum Geologists

Wedges form in many different ways on many different scales. A common occurrence is on a carbonate-mudstone ramp, where mudstones thicken downdip and become darker and more organic rich. Downdip there are fewer small-scale discontinuities, benthic fossils become less diverse and trace fossils become smaller with shallower burrows. Another morphology is found in clinoforms, where mudstones first thicken downdip, reach a maximum, and then thin into a distal condensed section. Foreland basins provide still another large-scale recurring example (sometimes referred to as a "clastic wedge"). In this case, thrust loading flexes the crust downward so there is much greater accommodation in the middle of the wedge than at either end with corresponding changes in mud thickness. But wedges can have abrupt transitions too – beyond a carbonate rim or across the edge of a fault-bounded basin.

A clay drape is a sheet that covers an inclined or irregular surface. Small-scale drapes are ubiquitous as laminations only a few millimeters thick blanketing irregular micro-topography such as ripple marks or shallow scours. Larger-scale drapes occur on the sides of river, tidal, and turbidite channels as clay curtains, and within the fill of abandoned small channels or large canyons. Channel or canyon fills, *abandonment facies*, range from shallow, mudstone-filled channels a few meters thick in coal and tidal deposits (Fig. 8.9) to those that are hundreds of meters thick in some shelf edge canyons. Almost all have stratification that dips and thickens toward the axis of the channel. Beds thicken toward channel axes, because deeper water traps more mud at times of low flows (which is most of the time). Other examples are the sinuous paths of turbidity currents that follow the lows at the sides of grouped shale and salt diapirs – the resultant deposit thickens toward the low.

Mound-like geometries also occur. Most common are the distal, transverse, mound-like cross sections of a distal delta or fan. Other mudstone mounds form as clay dunes on mud-swamped coasts, some deep-sea silt-size contourites, and the green mudstone cores of Mississippian Waulsortian carbonate mud mounds. To the above, we should add randomly shaped and distributed mudstone bodies such as might occur in a sinkhole, in a braided stream deposit or turbidite channel, or in an olistostrome. All of these are likely to have random patches or bodies of mudstone with hard-to-generalize geometries: sinkholes (infill of complex, solutional karst geome-

Fig. 8.9A–C. Two channels filled with mudstone and a mudstone "hill": (**A**) small channel at the Pennsylvanian-Mississippian unconformity along U.S. 24 in Schuyler Co., Illinois has a basal coal above which beds dip toward the channel axis; (**B**) sub-sea channel in Mississippian Cowbell Member of the Borden Formation along KY 546 in Lewis County, Kentucky successively truncates underlying beds in direction of car; and (**C**) paleo-hill of green marine, fossil-bearing mudstone over which a Waulsortian carbonate mound of the Mississippian Fort Payne Formation developed about 5 miles north of Burksville, Cumberland Co., Kentucky

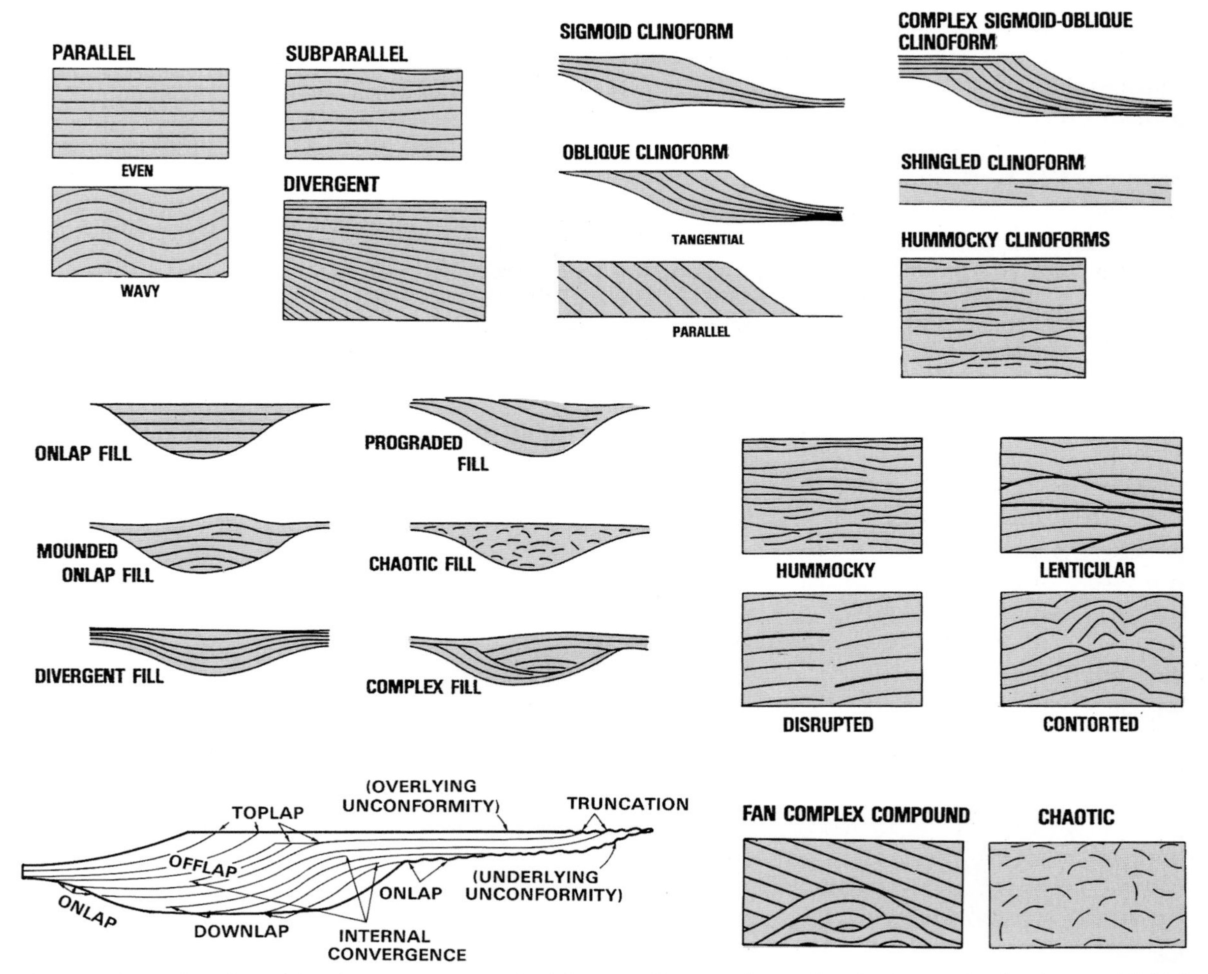

Fig. 8.10. Recurring depositional patterns recognized in seismic stratigraphy (Mitchum et al. 1977, Figs. 6, 9, 11, 13 and 15). Contrasts in lithology and hence density produce the reflections that define these patterns. Thus a thick, *very uniform* mudstone has few internal reflections. Published by permission of the authors and the American Association of Petroleum Geologists

tries); braided channels (an essential random pattern of channel shifting and penecontemporaneous erosion); and olistostromes (complex shearing during tectonic transport).

Seismic stratigraphy provides additional insights to the geometry of mudstones and all the other major sedimentary rocks (Fig. 8.10). With it, the investigator can image the geometry of the entire sedimentary package, possibly thousands of feet thick. Although not always available to the student of mudstones, an awareness of these patterns is everywhere, useful, be it working only with a few outcrops or a thick mudstone-rich basin.

Knowledge of the geometry and continuity of mudstone beds is important for understanding fluid flow at all scales – for both basins and reservoirs of petroleum and water as well as for the paths of mineralizing fluids, because even thin mudstone beds can hinder or block flow. On reservoir scales, there are at least eight commonly occurring sites of mud deposition These include the tops of fining upward alluvial or tidal cycles, the bottoms of coarsening upward coastal cycles, widespread mud blankets deposited from suspension over a turbidite lobe, a thin transgressive mudstone or argillaceous siltstone, back swamp deposits, the clay drapes of point and tidal bars, abandoned clay-filled alluvial, tidal or turbidite channels, and thick basin-bottom muds.

8.4. Lateral Extent and Correlation

Mudstones are the most useful lithology to correlate sedimentary rocks, because they not only have an easily recognized gamma-ray profile that distinguishes them from carbonates, coals, and most sandstones, but also because even thin mudstone beds are more likely to be widespread and continuous than the other lithologies. For example, in turbidite fans, the thickness of the individual mudstone units varies much less across-fan than does that of the interbedded sands. The small fall velocity of the fines permits long lateral transport, the best example being an ash fall tuff followed by marine mudstones (Zeito 1965; Geehan et al. 1986). In addition, because muds record small variations in bottom oxygen and chemistry far better than sandstones and most carbonates, single beds can be distinctly different from neighboring ones and thus much easier to trace laterally.

"Basin-bottom" anoxic events occur at all scales from ocean basins to lagoons to small lakes. High productivity and sluggish circulation are responsible (or surface productivity so high that it overloads the bottom, producing an Oxygen Minimum Zone). In the paleo-Atlantic, high surface productivity in the Mid- to Late Cretaceous (Greenhouse world) created bottom anoxia (Graciansky et al. 1980, Figs. 6 and 7; Hay 1988, Fig. 8). This effect was intensified by the long, narrow geometry of the paleo-Atlantic, similar to the present Red Sea, but much longer.

Rapid flooding of a shelf is considered to have produced anoxia and black shales on gently-dipping cratons. In Europe and North America the names "marine band or zone" are used for the thin marine, laterally extensive, dark to medium gray, fossiliferous mudstones found in dominantly non-marine coal measures like those of the Appalachian Basin (Calver 1968; Bennington 1996). In the Midcontinent and Illinois Basins of the United States, these thin marine black shales, commonly one to two meters or less in thickness, are the most widely traceable stratigraphic units in the section, and are thought to be the maximum flooding surface. See, for example, correlations in the Midcontinent region by Heckel (1991). All of these beds reflect flooding of a broad, even coastal plain coupled to a shallow shelf bordered by an anoxic sea. More recent studies have even established millimeter-scale correlations across several hundreds of kilometers in Kansas in these black shales (Fig. 8.11A). This widespread millimetric-scale lamination is considered to be the result of deposition in an anoxic basin totally lacking in bottom currents with variations in clay influx controlled by climate variations at solar cycle scales.

Two very different environments with unusual correlations are the thin silts in the coastal muds of the German coast (Fig. 8.11B) and the deep-water thin carbonaceous claystones interlaminated with thin siltstones in the Belt-Purcell Super Group of the Idaho, Montana, and British Columbia in the United States and Canada. The latter have been correlated band-for-band across more than 100 km of its basin (Huebschmann 1973). Primitive planktonic algae may have been responsible for these carbonaceous bands, and wind may have deposited the interlaminated silts in a large, below-wave-base stagnant basin during a major flooding event. James (1951) reports a somewhat similar widespread marker, a pyritic slate, within a Precambrian turbidite sequence in Upper Michigan. At the opposite end of the oxygen spectrum are some twenty green and red color zones (Fig. 8.11C) in a 100-m thick interval that have been correlated over 40 km in England in a small part of a 3,700 m thick Triassic red-bed basin containing anhydrite and salt (Wilson 1990). Variations in depth to the water table have been proposed.

Still other examples are thin beds of mud-smothered fossils buried intact (along with some transported ones and possibly some even clay clasts torn from the bottom as the slurry flowed down slope) by storm-induced slurries flowing basinward along a muddy shelf (Brett and Taylor 1996; Frey 1996). These are known to extend as far as 200 km and carry the technical name *obrution beds*. Such deposits tend to have open clay fabrics, the result of rapid deposition, and typically are massive to irregularly bedded (Fig. 3.6). But the real champions are bentonites and tonsteins (Chamley 1989; Spears 2003), thin, distinctive beds of volcanic ash carried hundreds of kilometers downwind and deposited over both water and land. These are best preserved in marine basins (bentonites) and swamps (tonsteins) and lakes. They have distinctive chemical signatures related to differences in their original magmatic compositions. The presence of potassium feldspars and illite (here secondary minerals) give bentonites a strong gamma- ray deflection that makes them easily identifiable, whether hosted by mudstones, sandstones or carbonates. In some mudstone clinothems, bentonites help identify clinoform geometry and aid correlation of individual clinothems (ash accumulates along an omission surface that marks a parasequence boundary). The zircons that are present in many ash falls can be dated by Pb isotopes using ion microprobe techniques to yield absolute ages of the surrounding sediments, while K-rich micas provide age dates of later diagenesis based on K-Ar and Ar-Ar methodologies (Ziegler and Longstaffe 2000, Fig. 3).

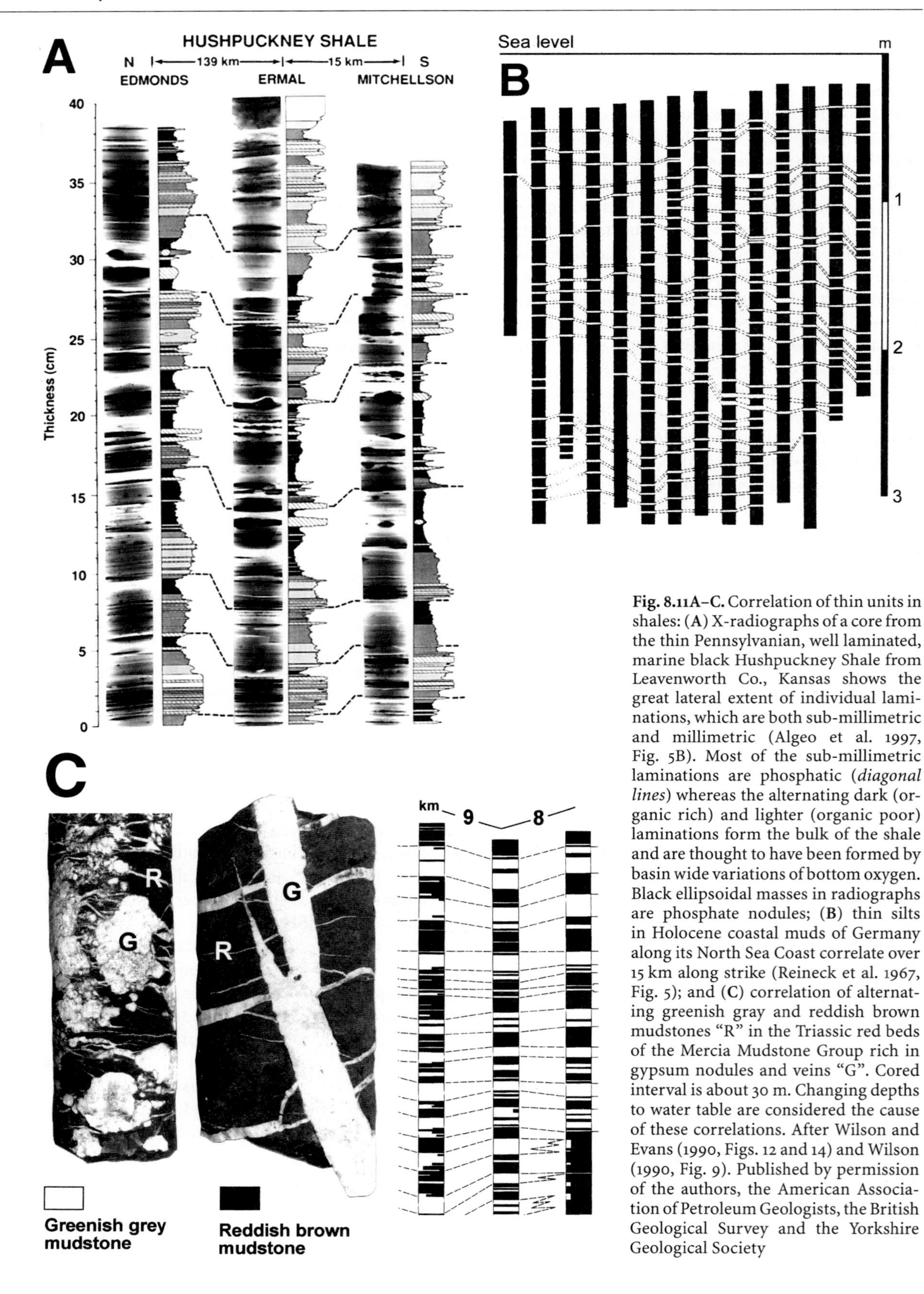

Fig. 8.11A–C. Correlation of thin units in shales: (**A**) X-radiographs of a core from the thin Pennsylvanian, well laminated, marine black Hushpuckney Shale from Leavenworth Co., Kansas shows the great lateral extent of individual laminations, which are both sub-millimetric and millimetric (Algeo et al. 1997, Fig. 5B). Most of the sub-millimetric laminations are phosphatic (*diagonal lines*) whereas the alternating dark (organic rich) and lighter (organic poor) laminations form the bulk of the shale and are thought to have been formed by basin wide variations of bottom oxygen. Black ellipsoidal masses in radiographs are phosphate nodules; (**B**) thin silts in Holocene coastal muds of Germany along its North Sea Coast correlate over 15 km along strike (Reineck et al. 1967, Fig. 5); and (**C**) correlation of alternating greenish gray and reddish brown mudstones "R" in the Triassic red beds of the Mercia Mudstone Group rich in gypsum nodules and veins "G". Cored interval is about 30 m. Changing depths to water table are considered the cause of these correlations. After Wilson and Evans (1990, Figs. 12 and 14) and Wilson (1990, Fig. 9). Published by permission of the authors, the American Association of Petroleum Geologists, the British Geological Survey and the Yorkshire Geological Society

Isotopic profiling of mudstones and limestone-mudstone sections is also becoming increasingly common. This involves two strategies: analysis of the stable isotopes of organic carbon and the 86/87 isotopic ratio of strontium in interbedded limestones or from isolated, interbedded shelly fossils (see Box 7.2 and Appendix A.6). Strontium isotopes are especially useful in the Neoproterozoic for regional correlation in the absence of fossils (Knoll 2000), although great care must be used to select unaltered samples. Sr isotope chemostratigraphy is based on the observation that the strontium isotopic ratio in the sea water of the world ocean has changed systematically through time, and consequently, studies of Sr isotopes of interbedded carbonates permit us to correlate interbedded shales (in the Phanerozoic, strontium isotopes are also used for correlation, although here they are but one of many available tools). The seawater record of the carbon isotopes of organic matter from mudstones and of interbedded limestones can be used to identify global anoxic events. When such an event is suspected, the interval can be closely sampled, and a possible deviation (excursion) from background values of $\delta^{13}C$ goes far to confirm the local presence of the event (Holser 1997). In years to come, as the ease of analysis improves and the needed sample size decreases, we predict that the use of isotopic methods in stratigraphy will expand dramatically.

8.5. Sequence Stratigraphy

Understanding mudstones in basins is closely related to sequence stratigraphy, the heart of which is linked to changes in *relative sea level*. Widespread thin mudstones occur where a craton or super-craton and its margins are flooded during a transgression and early highstand. Here water depths range from a few to 10 s of meters and only rarely exceeded a 100 m or so. Given these fairly shallow depths plus low slopes and low rates of subsidence, small changes of relative sea level shift shorelines and environments long distances. Consequently, lithologies change vertically much more rapidly than laterally – a stratigrapher's paradise.

Because it is these changes in relative sea level that shift depositional environments across a basin and produce different kinds and patterns of mudstones, it is sequence stratigraphy that provides the genetic model for the *stratigraphic consequences* of relative sea level change. The essential element of the genetic model is the close dependence of mudstones on water depth, bottom turbulence, and bottom oxygen. From this point of view it can be argued that mudstones are the *key players* of sequence stratigraphy. Thus we need to understand and freely use it. See Emery and Meyers (1996) and Van Wagoner et al. (1990) for introductions to this rich and important part of stratigraphy and sedimentology, the edited volume by Weimer and Posamentier (1993), and the volume of European examples by Graciansky et al. (1998). A helpful summary for muds and sands is also provided by Catuneanu (2002).

Sequence stratigraphy, like any field, has a specialized terminology (Box 8.1 and Fig. 8.12). Two concepts are needed to use relative sea level and sequence stratigraphy to understand better basinal patterns of mudstones, one new and the other old. These are *accommodation* and the *base level of deposition*. By accommodation is meant, "space available for deposition", a concept early recognized (but not named) by Larry Sloss (1962, p 1051). Accommodation is best thought of as a *dynamic volume* that changes across a basin. This dynamic volume changes in response to six factors. Two important factors are differential tectonic subsidence within the basin and changes in world sea level called *eustasy* (driven by changes in the volumes of ice caps and of mid ocean ridges). In addition, local changes in sediment supply to the basin (induced by climate, uplift or simple switching of a delta distributary) can alter relative sea in a basin, as does compaction and preexisting topography (gradients of a ramp or shelf). And as relative sea level in a basin changes, its shorelines, water depths, and the position of its environments shift. Mud and silt are directly linked to accommodation through their dependence on deposition in quiet water. With greater accommodation, deeper water depths mean less bottom oxygen, reduced fossil diversity, more marine carbon, and enhanced preservation of organic debris.

The second concept is base level of deposition, defined as "the highest level to which a sedimentary deposit can build" (Twenhofel 1939, p 8). Deposition above this level is at best temporary (thus subsequent erosion), while deposition below it is preserved. The role of base level of deposition becomes clearer when we consider some conceptual examples. A few decimeters of siliceous mud deposited over a long time in a deep basin indicate a starved basin (subsidence $\ggg$ supply = large accommodation), whereas a thick, overfilled sand-rich foreland basin is just the opposite (supply $\ggg$ subsidence = minimal accommodation). Or on a smaller scale consider how the internal clinoform geometry (stacking pattern) of a prograding shelf margin changes as

Box 8.1. Sequence stratigraphy – background and definitions

Sequence stratigraphy identifies and maps the natural genetic rock units of strata using bedding surfaces, hiatuses, unconformities, and equivalent conformities to provide lithologically based time lines, or more technically, a *chronostratigraphic framework*. As such, it helps correlate and explain the relations of one facies to another and relates their shifting positions and the shorelines within a basin to changes in *relative base level of deposition*. In marine and lacustrine deposits, base level of deposition is the subaqueous profile of equilibrium (often approximated by sea or lake level) and in fluvial deposits by the graded stream profile, which responds to tectonic tilting plus changes in base level and climate (a rising sea level raises water tables and thus favors water-logged soils, whereas a falling sea level favors better drained, oxidized soils). Thus sequence stratigraphy is applicable to both marine and non-marine deposits. And, because the depositional environments vary with supply, accommodation and preexisting topography, sequence stratigraphy goes far to explain the relationship of one environment to another. The distribution of these linked environments across a basin, called a *depositional system*, varies with supply and accommodation as does their gross geometry or *stacking pattern*. Twenty years after the pioneering paper of Sloss (1962) a group at Exxon (Vail 1987, Van Wagoner et al 1988; Mitchum and Van Wagoner 1991) and many others in industry and the academic world developed sequence stratigraphy as it is today. Based on these studies, Bohacs (1998b) applied it to mudstones and emphasized how useful sequence stratigraphy is for marine sections but also for lacustrine, coastal (paralic) and non-marine rocks

Some useful supplementary definitions:

Accommodation	Space available for deposition
Condensed section	A thin, typically widespread unit of slow or little deposition that is bioturbated and marks a flooding surface; i. e., represents a sediment-starved sea floor, because most sediment is deposited shoreward. Also called an omission surface
Eustatic sea level change	Absolute change in sea level relative to some position in the earth, usually its center
Flooding surface	A surface that separates younger sediments deposited in deeper water from older sediments deposited in shallower water; contact may be abrupt or gradual. The *maximum flooding* surface (downlap surface) is the contact between transgressive and regressive deposits
Highstand	The most landward deposition in a cycle (highest relative sea level)
Lowstand	The most basinward deposition in a cycle (lowest relative sea level)
Offlap	Progressive basinward movement of the shoreline
Offlap point	The break in slope between the topsets and slope deposits of a clinoform; i. e., the shelf break
Onlap	Progressive shoreward movement of the shoreline
Stacking pattern	The geometry of parasequence sets may be progradational (facies of a parasequence evolve or stack seaward), aggradational (stable stacking geometry upward) or retrogradational (stacking evolves landward). Recognition of stacking patterns is key to developing an understanding of basin fill
Systems tract	Originally defined by Brown and Fisher (1977, p 215) as a linkage of contemporaneous depositional systems. Seismically, a systems tract is a three-dimensional unit of deposition defined by toplap, downlap, and onlap (bounded by seismic terminations) with coherent internal geometry; within any one sea level cycle, highstand, transgressive, and lowstand systems tracts are recognized (Emery and Myers 1996, p 26)
Transgressive surface	First flooding surface above lowstand deposits

supply versus subsidence varies (Figs. 8.6 and 8.13). Cessation of deposition with no change in water depth produces a bioturbated, time-rich, condensed section, whereas balanced supply and subsidence results in a balanced, equilibrium graded profile. If water depth decreases in a marine basin, a so-called submarine "deepwater" unconformity may develop on the shelf break or upper slope, which is now exposed to stronger wave energy with bypassing rather than deposition. These concepts serve us well in understanding mudstones, because they apply at all levels – from entire basins to a few meters of section.

Oscillations of relative sea level in a basin produce periods of flooding and marine deposition followed by regression of the sea, erosion in landward areas, and a basinward shift of environments of deposi-

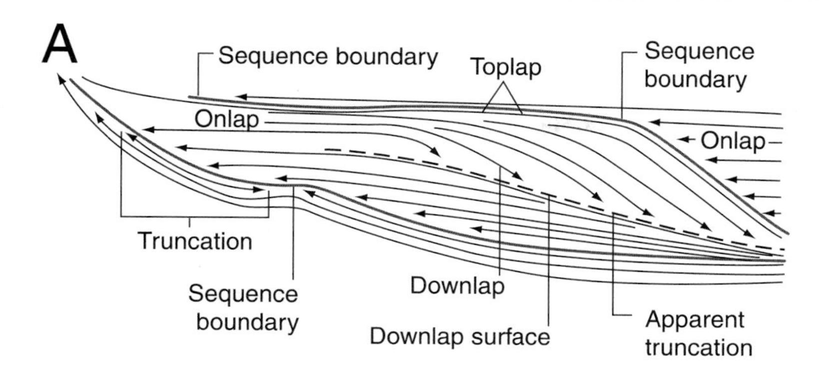

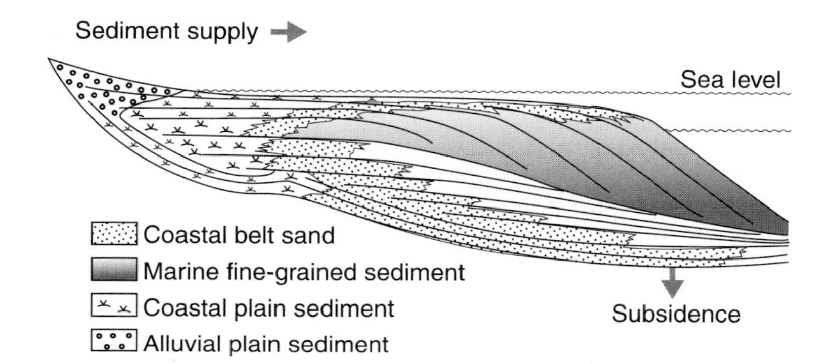

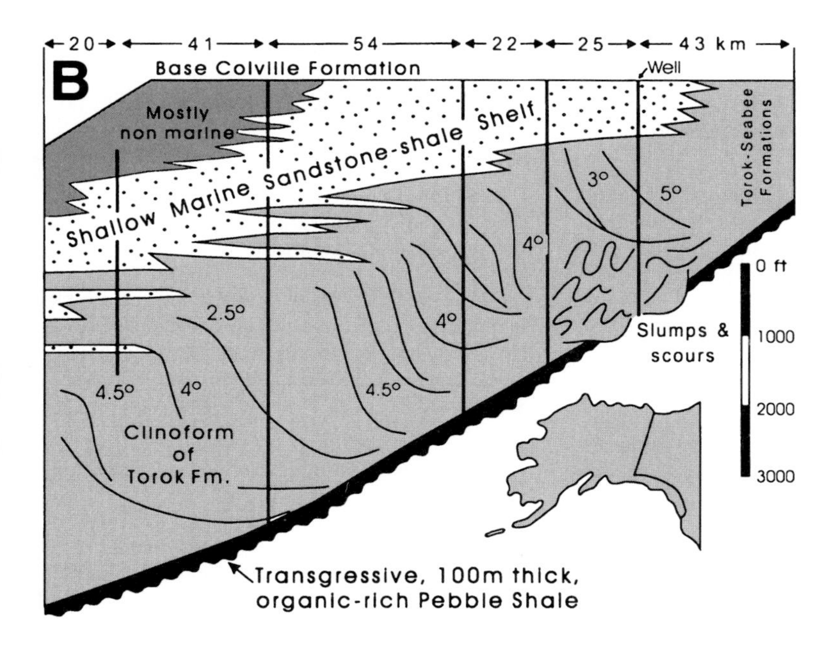

Fig. 8.12A,B. Clinoform structure: (**A**) Definition diagram for key terms of sequence stratigraphy (Vail 1987, Figs. 1 and 3); and (**B**) clinoform of the Cretaceous Torok Formation on the North Slope of Alaska after Molenaar (1985, Fig. 18). Notice how the clinoform dips in the direction of thinning of the entire section. Published by permission of the author and the American Association of Petroleum Geologists

tion. The result is a *sequence* of genetically linked deposits bounded by erosional surfaces (unconformities) or their correlative conformable surfaces such as condensed sections. Criteria to recognize sequence boundaries vary somewhat from the marine to the nonmarine realms in a basin (Table 8.3). Sequences are typically subdivided into smaller units called *parasequences*, which are bounded by *marine flooding surfaces*, or just flooding surfaces for short. Flooding surfaces are marked by deeper marine sediments overlying shallow marine or non-marine strata and by markedly slower terrigenous sedimentation. In distal areas, it is common to see pyritization or phosphatization along these surfaces.

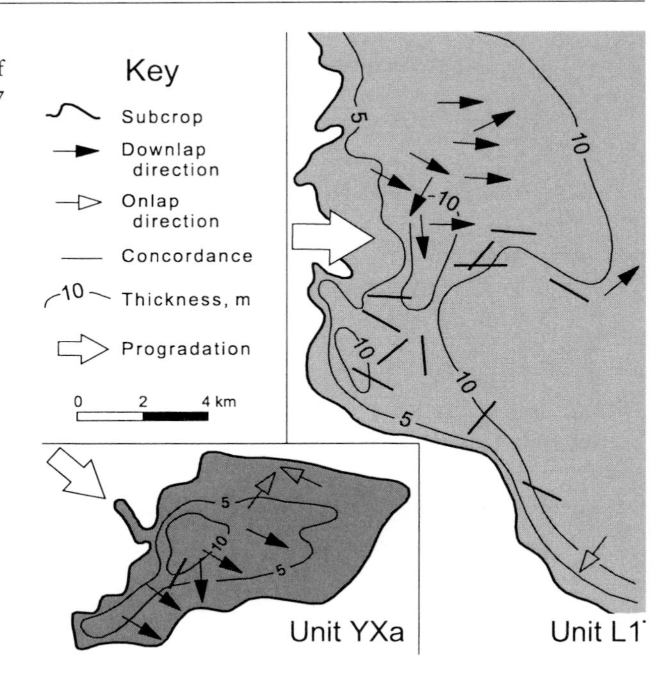

Fig. 8.13. Dip direction of the thin clinoforms in the Lower Eocene Ypresian clay of the English Channel of Belgium (DeBatist et al. 1984, Fig. 9). See also Table 8.7 on how to reconstruct clinoform dip and direction without seismic sections

Table 8.3. Criteria for recognition of sequence boundaries and flooding surfaces in muddy basins (Simplified from Bohacs 1998b, Table 2)

Environment	Sequence boundaries	Flooding surfaces
Marine Basins	More hemipelagics and biogenic debris and lower total gamma-ray counts (GR) *above* rather than *below* low-relief, almost conformable boundary. Little or no shift of basinal environments	Minimal bottom energy and terrigenous input; continuous, even lamination and moderate to high total GR plus more deep water microfossils at or just above surface; phosphatic concretions and lowest terrestrial organic matter above surface
Marine shelf	Greater terrigenous input with more and coarser typically resedimented sandstone, thicker beds, wavy laminations, and lower total GR plus more terrestrial organic matter and bypass of mud *above* than *below* boundary, which has some local scour. Beds below boundary are regionally moderately truncated and onlapped; more shallow-water fossils *above* than *below* boundary. Moderate shift of basinal environments	Minimal bottom energy and terrigenous input plus more concretions, fine-grained pelagics and more marine organic matter at or just above surface; possible concentrates of bones, fish scales, (and exceptionally dropstones) at surface, which is typically widespread and fairly even. Both total GR and deepwater fossils maximal; phosphatic, siliceous or calcareous shales at or above surface
Transitional	Sandstone and siltstone and more pebbles, reworked concretions and body fossils *above* than *below* boundary, which is strongly erosional, laterally extensive and may have rootlets, oxidation, and soil profiles. Organic matter below boundary likely to be oxidized. Strong shift of marginal environments	Reduction in bottom energy results in finer grained, slower sedimentation above surface, which may be bioturbated, rich in concretions and microfossils and have bones, fish scales (and possible dropstones) plus glauconite, phosphate and siderite. High total GR plus possible concentrates of reworked concretions and fossils
Non-marine	Significant increase in sandstone-mudstone ratio *above* extensive, strongly eroded boundary that may have well-drained soils, caliche, desiccation cracks or degraded coals in humid climates; integrated; paleovalley system and low ground water table. Terrestrial organic carbon dominant above boundary. Strong shift of environments above boundary	Less bottom energy at or just above surface as shown by presence of muddy and silty sediments (more flood basins, lagoons, and bay fills) than below surface. Mudstone continuity highest at surface. High ground water tables (wet soils plus coal and peats) mark flooding surfaces; terrestrial organic carbon abundant as in pyrite

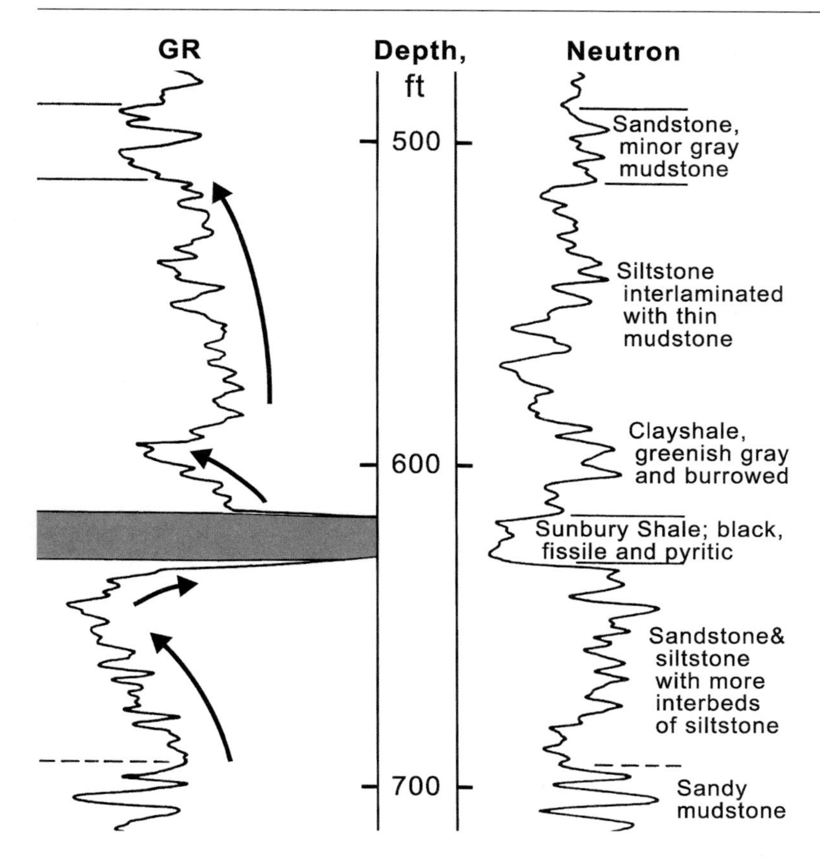

Fig. 8.14. Interpretation of gamma ray-neutron log from the Ashland No.1Wolfe well in 13-Y-77, Lewis County, Kentucky. The maximum flooding surface is in the overlying green shale at the top of the radioactive Lower Mississippian Sunbury Shale. (Note change of GR scale in Sunbury interval.)

An especially important boundary is the *maximum flooding surface*, which represents the greatest landward extent of the sea in a sequence and marks the turnaround from the transgressive to the highstand systems tract. The characteristic stacking patterns of the parasequences changes at this point, which is recognizable on seismic sections. In individual outcrops, it can be difficult to distinguish the maximum flooding surface from a parasequence boundary, and long-distance correlation using both outcrops and wire-line logs are likely to be needed. Clues to recognizing the maximum flooding surface are that its mudstones are the purest – deposited farthest from shore with the least silt and most clay. As a result, they have the most open marine fauna, most marine kerogen, and commonly the highest gamma-ray readings (Fig. 8.14). Exceptions are the most distal parts of a basin beyond clay deposition, where there is a bioturbated carbonate hardground or other type of *condensed section* (Table 8.4). In non-marine sections near a coast the maximum flooding position is still detectable: rising sea level causes water tables to rise, producing water-logged soils. Note that both sequence and parasequence boundaries and the maximum flooding surface are widespread and mapable. Together these two types of surfaces help us

Table 8.4. Condensed sections in mudstones. After Schutter (1998, Table 5)

Lateral continuity
High (time equivalent surfaces)
Colors
Variable (depends mostly on bottom oxygen)
Lithology
Laminated to homogenized and either poor or rich in Pb, Zn and U and trace elements (depends on bottom oxygen); likely to have some chemical sediments (glauconite, if bottom oxygenated), lags of phosphate nodules and pyritized fossils, and volcanic ash possible; thin unit with sharp top
Paleontology and bioturbation
Skeletal lags (bone beds), mixed pelagic and bottom fauna with much bioturbation; oxygen permitting, mostly Nereites and Zoophycos
Organic content
Variable depending on bottom oxygen, but palynomorphs likely to be abraded and weathered
Seal character
More compact than beds above and below and thus reduces cross flow, if not fractured or extensively bioturbated

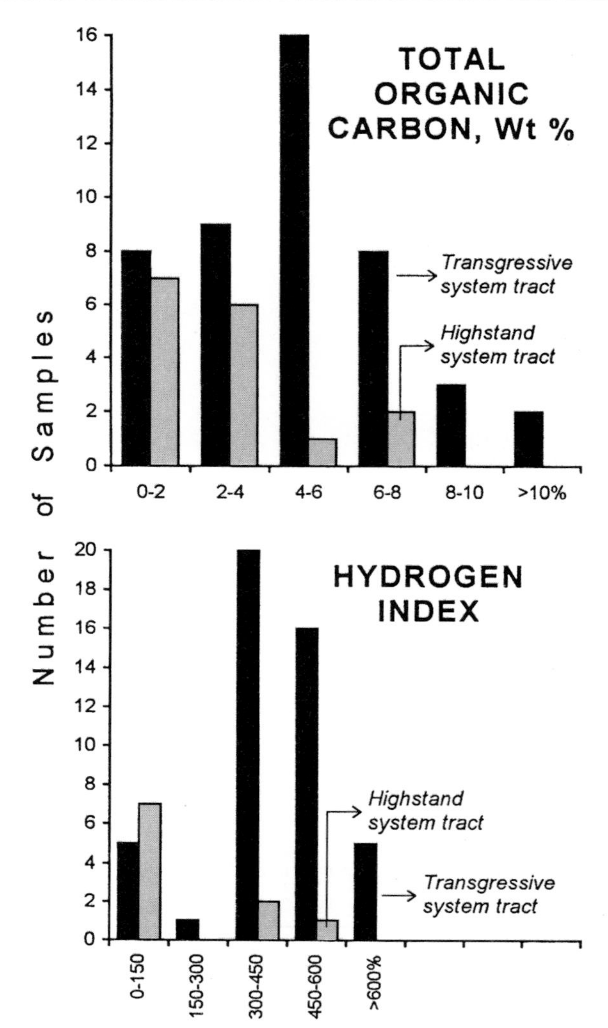

Fig. 8.15. Contrasts in total organic carbon and the hydrogen index between transgressive shales (*black*) and high-stand mudstones (*gray*) in the Late Cretaceous Duwi and Dakhla Formations of Egypt are the result of low versus high mud input (redrawn from Robison and Engel 1993, Figs. 11 and 12). In these two formations the richest oil-prone shales occur in the condensed sections at the top of three transgressive systems tracts (the maximum flooding surface), whereas, most of the organic matter in highstand systems tracts is gas prone. Published by permission of the authors and the American Association of Petroleum Geologists

understand the internal geometry of a sequence and better correlate one facies with another. Hence it is clear that the identification and mapping of these surfaces deserve as much attention as the mudstones and other rocks between them.

In addition to correlation and the recognition of clinoform structure, sequence stratigraphy helps us predict the organic abundance of mudstones. Given low bottom oxygen in a lake or sea, TOC will vary inversely with sedimentation rate (Fig. 8.15) being most abundant at an initial flooding surface (mud mostly trapped far up-dip on the inside of a shelf or in an estuary) and decreases both above and below this surface (faster sedimentation and thus more dilution). Moreover, the kerogen here will be mostly algal-derived and any phytoclasts are likely to be abraded, altered and small (distal derivation equals long transport plus reworking). This reasoning suggests that the amount and the quality of organic matter should be closely linked to the sequence stratigraphy of muds and mudstones. Why? First, both organic carbon and mud settle together to the bottom in quiet water. Second, the initial stages of a transgression trap clastics and woody plant material inshore, leaving only very fine clays, siliceous skeletons, and algal-derived organic matter to accumulate in deeper water. In addition, it is useful to recall that wet climates maximize both mud supply and terrigenous plant debris, whereas dry climates minimize both, so the ideal source bed will be deposited during a rapid transgression in a basin whose hinterland has but moderate rainfall. But suppose the bottom is not anoxic? Under these conditions, there would be more benthic fauna and bioturbation, and glauconite or a hardground with phosphatic nodules will better identify a flooding surface. Furthermore, an increase in the abundance of phosphate indicates a more seaward position along the flooding surface, regardless of bottom oxygen. Consider now the role of accommodation. Where it is high relative to sediment supply, TOC is likely to be high (poor bottom circulation), but where it is low, the bottom may either be well oxygenated (commonly the case) or may be dysoxic or anoxic as is true of some protected lagoons and shallow stratified saline lakes.

The contemporaneous shelf, slope and basinal deposits that accumulated in different water depths in a basin define a *clinoform*. This series of facies has a characteristic geometry that is best recognized in down-dip seismic sections, but was first recognized in outcrop and later in cross sections using geophysical logs. Many organic-rich marine and lacustrine mudstones occur at the coalescence of distal clinothems. Importantly, the maximum dip direction of a clinoform faithfully gives the seaward dip direction of a shelf (Fig. 8.13). Fortunately, this dip direction can be reconstructed in the absence of seismic lines and logs (Table 8.3).

In broad terms, sequence stratigraphy represents as much a revolution for stratigraphy and historical geology as did the introduction of plate tectonics to all of geology – and perhaps even more so, because it is applicable to outcrops, cores and logs as well as seismic sections. Below we use the concepts of sequence stratigraphy to help explain the great diversity of mudstones in the stratigraphic record.

8.6. Mudstones and Basins

Tectonic influences on mudstones include those that affect what is brought into the basin plus those that affect its deposition and burial (Table 8.5). Source area influences are relief (given rainfall, the high relief of orogens generates much mud) and the nature of the delivery system to the sedimentary basin. As we saw in Chap. 2: Mud and Silt Production, there is a fairly good relationship between the discharge of a river and the character of its load. Large river systems on cratons bring mostly fines to the sea, whereas complexly structured, high-relief source areas drained by many small rivers yield both coarse and fine debris to nearby basins. Initial clay mineral compositions are also tied to tectonics through volcanism in Mesozoic and Tertiary Basins, but original compositions are likely to be erased by burial digenesis after about 150 Ma.

Within the basin, accommodation (space available for deposition) determines mudstone geometry. Wide stable cratons and passive margins have thin to moderately thick mudstones that change gradually in thickness, whereas along convergent margins, in rifts, in pull-apart basins or in a salt dome province, abrupt changes in geometry and thickness occur. Basin tectonics also controls rates of subsidence and thus over- pressuring. Faster subsidence favors overpressuring (less time near bottom for muds to dewater) and the development of mudstone diapirs. And finally, organic debris matures more rapidly in basins with high heat flow (young rifts) than in basins with low subsidence rates (old rifts and cratonic basins).

On the other hand, the relationship between bottom oxygen and tectonics is much less direct. Mudstone types are directly linked to bottom oxygen, which depends on water circulation, organic productivity and global sea level as well as the ratio of supply to subsidence (accommodation) within the basin. And to these we need to add climate, because warm water, given nutrients, promotes productivity in the water column. To see this another way, remember that a range of mudstone types may be deposited in a basin as it subsides through time even though the basin retains the same tectonic setting. Consequently, the link between mudstone types and the basin setting are not as close as its link to heat flow, subsidence rate, or perhaps even mudstone geometry. So our final conclusion is that, while tectonics affects many aspects of mudstones, we cannot use mudstone facies and types by themselves to infer tectonic setting, chiefly because local variations in bottom oxygen are largely independent of tectonic setting.

Table 8.5. Tectonic influences on mudstones

Factor	Influence
In source region	
Relief	High relief always yields much mud, given rainfall
Size of river system	Large river systems of cratons and passive margins bring large volumes of fines to the sea, whereas small rivers transport gravel as well as and mud sand either in large or small volumes
Clay mineral composition	Smectites and mixed layer clay have good provenance signals, but if basin older than 150 Ma, the provenance signals blurred by burial digenesis
Climate	Rainfall, a key element of mud production, depends on latitude and orientation of orogenic belts with respect to world circulation (basin on windward or lee side of orogenic belt?)
In Basin	
Mudstone geometry and thickness	Depends mostly on accommodation (differential subsidence in basin) and to lesser degree on supply
Subsidence rate	Determines dewatering/compaction history and thus possibility of overpressure; strongly linked to basin tectonics
Heat flow	Controls depth at which mineral transformations start in mudstones; is closely linked to basin tectonics
Proportion of mudstone in basin	High subsidence rates favor much mud deposition provided supply exists and sand trapped up dip
Bottom oxygen	Largely independent of tectonic setting even though deeper water favors bottom oxygen deficiency; consequently, many of the most important properties of mudstones correlate poorly with tectonic setting and basin types

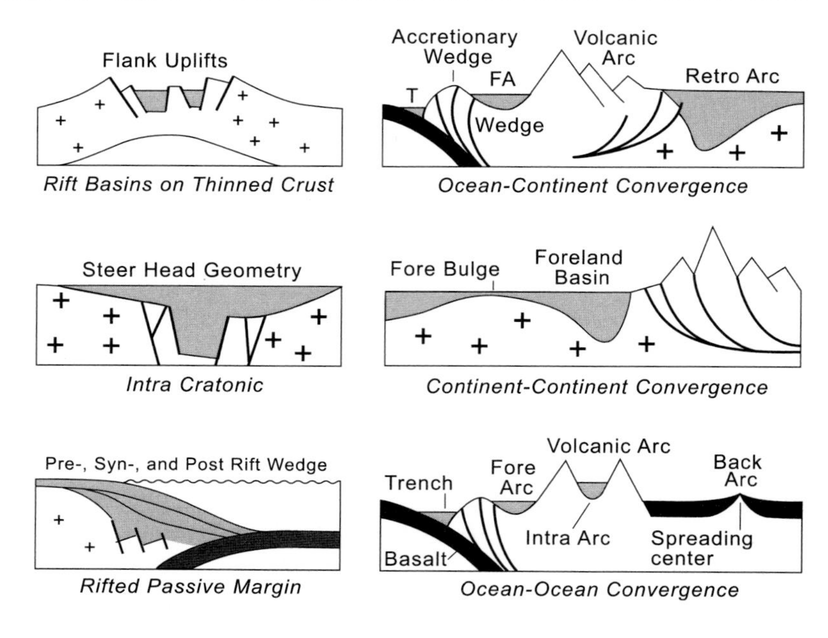

Fig. 8.16. Simplified tectonic settings of end-member sedimentary basins. (T = trench; FA = fore-arc.)

Nonetheless, it is useful to have at hand a straightforward tectonic classification of basins such as that of Fig. 8.16. And indeed at least some weak associations do seem to exist between basin types and abundance of mudstones. Basins that appear to be consistently mudstone-rich include mainly rift basins, foreland basins and pericratonic basins along passive margins all of which provide persistent accommodation for mud derived from either proximal or distal source regions, climate permitting. But mudstones of different types occur in every basin and so the relationship of both abundance and types of mudstones to basin style is one that seems to us to invite much future attention. To see this better, consider the links between the petrographic types of sandstones and tectonic setting.

There is a much stronger link between the sandstone composition of a basin and its tectonic setting (Dickinson 1989) than exists between the types and abundance of mudstones in a basin and their tectonic setting. There are at least three reasons for this. First, the composition of sandstones is a provenance indicator only marginally affected by burial diagenesis, whereas with burial and time, the clay mineralogy of a basin is greatly altered. Thus, in pre-Mesozoic basins, clay mineralogy is not so easily linked to plate tectonic setting. Secondly, the oxygen content of bottom water is not so closely related to tectonics, as mentioned above. And finally, because the fines of mudstones can be very far traveled, possibly sourced on one side of a continent but deposited on the opposite, a mudstone-rich basin can be far removed from its source. Thus it seems clear that the study of the mudstones in a basin provides much more *intra-basinal* information than do its sandstones, chiefly because of their close dependence on bottom oxygen. Carbonates, on the other hand, are totally intrabasinal and, in addition to oxygen, dependent on nutrients and light (Wilson 1975, Chap. 1). Consequently, we think of mudstones as being transitional in this respect between sandstones and carbonates (Table 8.1).

Below we summarize briefly the principal basin types illustrated in Fig. 8.16 and give examples of some of their mudstones starting with intracratonic, pericratonic and foreland basins, where mudstones have received the most attention. These are followed by examples from rifts, crustal shears, back-arc basins, marine glacial deposits, and, as a special case, we summarize the black and red mudstones. See Einsele (2000) and Busby and Ingersoll (1995) in Digging Deeper for background and additional examples and insights to these basins.

8.6.1. Intracratonic and Pericratonic Basins

Cratons are stable portions of the continents consisting of exposed Precambrian rocks or Precambrian rocks covered by large, little deformed, overlying sag basins separated from each other by low arches or uplands (basins and swells). Such basins have, as a rule, stable paleoslopes and persistent areas of greatest subsidence, which localize both marine invasions and the pathways of rivers (pericratonic basins differ only in that one side fronts an ocean). A fill of mostly sandstone occurs if its climate were cold or dry, and a fill of carbonate (possibly with evaporite), if the basin were at low latitudes, flooded by the sea and sheltered from terrigenous debris. Many intracratonic basins have minimal mudstones throughout their history. At times, however, some also have widespread mudstones that even cross the arches that separate one intracratonic basin from another, and exceptionally, cover much of a continent. Widespread flooding of South and North America, first in the Middle Devonian of South America followed by somewhat later flooding of North America in the Late Devonian and Early Mississippian, deposited exceptionally far-ranging black shales on both continents. These Devonian-Mississippian black shales are transgressive deposits formed in foreland and pericratonic basins around the margins of the North American craton and along the western side of the South American craton followed by somewhat later deposition of thinner transgressive black shales inboard on both continents (Ettensohn 1998, Fig. 3; Gohrbrandt 1992).

Bordering highlands, often far distant, are the source of mudstones in intracratonic basins and also for pericratonic basins. Mudstones in both types of basins are likely to be mostly marine rather than lacustrine, range in color from gray to greenish gray to dark gray to black, and consist of thin, far-ranging subunits deposited simultaneously across a basin as relative sea level changed. Condensed sections are notable and likely to be widespread (Table 8.5). Some minor red mudstones may be present and commonly are thought to have been derived from the regolith of bounding arches or shields, although exceptionally, the distal, thin edge of thick late orogenic red beds deposited in a bordering foreland basin may extend far onto the craton. Also present are thin bentonites derived from far distant borderlands (easily recognized on gamma ray logs by higher than normal radioactivity from illites and K-feldspar). Generally absent are lacustrine mudstones, although even these may have a minor presence when a large alluviating river, following the axis of the basin, progrades seaward.

Mudstones in intracratonic and pericratonic basins are much studied, diverse, are rich sources of industrial minerals, and where organic rich and mature, form good source beds. Below are a few examples. Widespread, graptolitic Silurian mudstones that extend across much of Europe, North Africa, and the Middle East provide an excellent example of widespread mudstones. These Lower Silurian mudstones cover a vast region extending in a great band from south of Morocco into Oman and Iran. They form a wide diachronous sheet bordering the north side of the African-Arabian craton (Fig. 8.17). Ages range from Llandoverian to Ludlovian and thicknesses range up to 1,200 m in Arabia, but in wide areas they are much thinner because of lesser subsidence and later erosion. These mudstones carry many local names, but share common characteristics. They are marine with a diverse fauna of which graptolites are foremost, they coarsen upward irregularly from a basal, locally, organic-rich shale with silty mudstones and some marine sandstones; and southward (where not eroded), they pass into sands deposited on an inner shelf/foreshore. Graptolite zonation clearly shows this sequence to have a gentle, north-dipping clinoform structure. From Egypt eastward, carbonates bound the mudstones on the north. Thus, here we have a gigantic, classic graded shelf of sand→mud sheet→limestone.

The origins of this immense sequence began with the termination of the late Ordovician continental glaciations so strongly in evidence in much of Africa and the Middle East. In Africa, ice sheet melting produced a rapid rise in sea level and marine water rapidly flooded a depressed African crust. Poor circulation on this shallowly flooded shelf led to the deposition of initially very dark to black shale, especially in structurally controlled topographic lows as in central Arabia (Jones and Stump 1999). This was followed by regression of the shoreline as the crust returned gradually to isostatic equilibrium starting first in the south where the ice sheet was thickest. Consequently, we have a north dipping clinoform structure defined by successive graptolite zones. Continental glaciation had a vast influence on sedimentation and on mudstones well beyond its limits. It seems likely that these Silurian mudstones, so widespread because of the underlying stable basement, would be an ideal place to test the lateral extent of marine flooding surfaces.

The Upper Devonian Duvernay and Ireton Formations of Alberta illustrate well the value of log markers (here condensed section/omission surfaces, but elsewhere bentonites) in establishing the internal stratigraphy and origin of a mudstone basin bordering a large carbonate bank to its west (Fig. 8.18). The basal thin, dark, bituminous, calcareous, millimetrically laminated shales and limestones of the Duvernay Formation are unfossiliferous and represent the transgressive phase of a second order sequence that includes the Ireton and Nisku Formations, which both prograded eastward from the carbonate platform. The Duvernay is followed by some 55 m of greenish, calcareous to dolomitic mudstones interbedded with a few argillaceous lime-

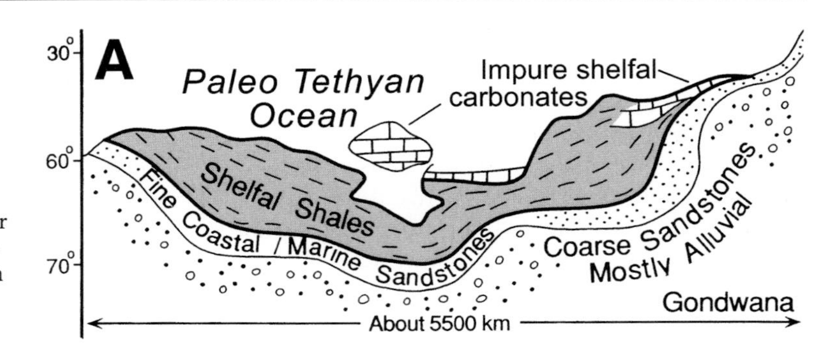

Fig. 8.17A,B. Gigantic pericratonic Silurian mudstone basin of North Africa and Arabia: (**A**) Classic sandstone → mudstone → carbonate shelf bordering the paleo-Tethyan Ocean (after Mahmoud et al. 1992, Fig. 10); and (**B**) the internal clinoform structure of this giant shelf (after Bellini and Massa 1980, Fig. 9). Published by permission of the American Association of Petroleum Geologists and the Geological Society of Lybia

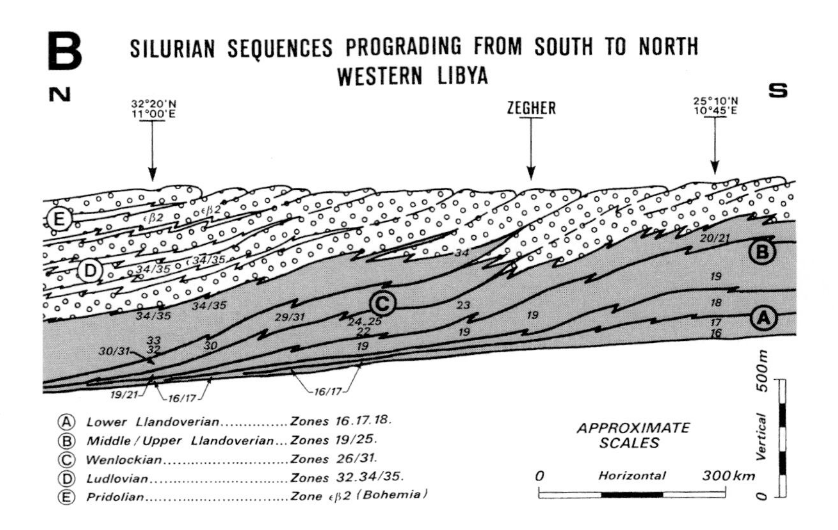

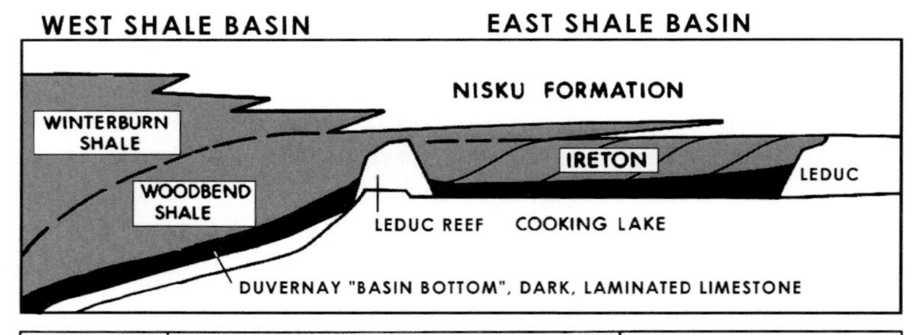

	BASIN	SLOPE: TOE	SLOPE: MIDDLE FORESLOPE	SLOPE: UPPER FORESLOPE	PLATFORM
BIOFACIES	SEA LEVEL; AEROBIC 0-45m; DYSAEROBIC 45-100m; ANAEROBIC > 100m; > 100m	90-100 m	55-90m	35-55m	0-25m; PROFILE FROM ELECTRIC LOGS; CARBONATE PLATFORM BIOHERMAL, INTER-BIOHERMAL AND SUPRATIDAL FACIES
LITHOFACIES	BITUMINOUS LAMINATED LIMESTONE AND SHALE	DENSE MICRITIC LS. AND CALCAR. SHALE	BIOTURBATED CALCAREOUS SHALE; *SKELETAL RUDSTONE*	NODULAR LIMESTONE	SHALE PLATFORM 0-35m; SKELETAL RUDSTONES AND PACKSTONES, INTRACLASTIC PACKSTONES AND RUDSTONES; BIOTURBATED MUDSTONE

Fig. 8.18. West Shale Basin of the Devonian Ireton Shale Formation has well defined clinoform structure above dark, laminated Duvernay Limestone, the basal transgressive unit, which lies directly above a condensed section (after Stoakes 1980, Figs. 33 and 47). Published by permission of the author and the Society of Canadian Petroleum Geologists

stones. The famous Leduc reef, western Canada's first major oil discovery, stands isolated in this mudstone basin. Suspension transport by currents from the north flowed parallel to the front of the bank and supplied the clays, whereas the bank contributed only some carbonate mud.

The log markers defining the clinoform structure of the infilling of this basin are hardgrounds formed during pauses of infilling (that is, they are flooding surfaces). These hardgrounds typically consist of green argillaceous, hard, bioturbated and burrowed limestones with abundant pyritic rims and coats. The hardgrounds have sharp contacts with either the overlying green mudstones or lighter brown grainstones (the grainstones may be clasts of the underlying hardground). Traced down dip, benthic faunas in hard grounds disappear followed by the disappearance of bioturbation as bottom waters became totally anoxic. These hardgrounds define six widespread third order sequences, and one, the maximum flooding surface, extends over as much as 180,000 km^2. Water depths are estimated at about 40 m for the Duvernay and about 100 m for the Ireton.

The Middle and Upper Ordovician of the Central Appalachians and eastern Midwest provide a good example of how far inboard on a craton muds derived from a marginal collision, the Taconic Orogeny, can be transported across a shallow epeiric sea (Fig. 8.19). Here Upper Ordovician mudstones extend almost 1,100 km into Iowa, where some mud may have been derived from the Taconic Orogeny bordering the Appalachians. In this distance, thick eastern red beds (Juniata/Queenston) and turbidites (Martinsburg) pass westward into a carbonate platform, whose mudstone content ranges from little (Galena/Trenton/Black River) to units dominantly composed of mudstone (Kope/Maquoketa). Similar relations can be seen in Upper Ordovician deposits in Ontario to the north. Without knowledge of the Taconic borderland, it would be easy to consider mudstones such as the Kope and Maquoketa as typical marine deposits of cratonic basins when, in fact, they are the distal inboard deposits of a foreland basin.

8.6.2. Foreland Basins

Impingement of a tectonic plate against the margin of another, continental plate generate large thrust sheets inboard of the collision that depress the continental crust and create a new elongated basin, a flexural down-warp (Fig. 8.19). Three subsidence histories exist for foreland basins depending upon the elasticity of the continental plate and the rapidity and magnitude of the thrust stacking, all of which greatly affect the mudstones of the coupled basin. The margin of the continental plate may be deeply depressed. In this case, dark shales and turbidites accumulate and accommodation is much greater than

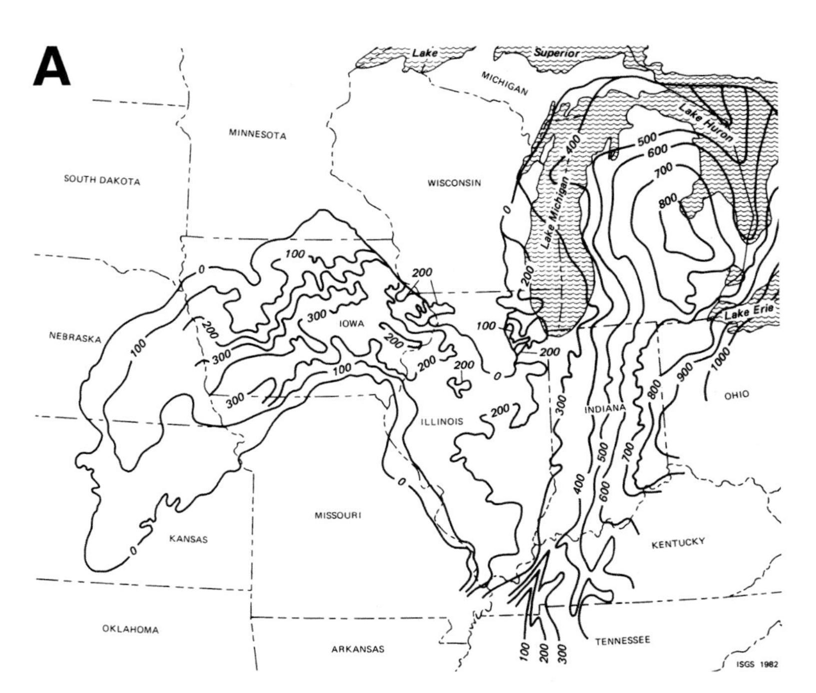

Fig. 8.19A–C. Upper Ordovician Maquoketa Group: (**A**) Thickness map shows it to extend from the Appalachian Basin over 1,100 km westward into eastern Kansas as a thin sheet except in the Michigan Basin (Kolata and Graese 1983, Fig. 1); (**B**) the Maquoketa is notable for its limestone/mudstone interbeds such as seen in the Kope Formation near Cincinnati, Ohio; and (**C**) the mudstone interbeds of the Maquoketa illustrate the "spillover" or runout of a flood of mud onto the craton by the Taconic Orogeny as proximal red beds → thick turbidites → thin, far distal, shelfal mudstones (Keith 1988, Fig. 10) (*continued on next page*)

Fig. 8.19. (*continued*). Published by permission of the authors and the American Association of Petroleum Geologists

sediment supply (*underfilled basin*). If the margin is moderately depressed, a steady state between subsidence and filling may exist (normal marine and some thin, minor continental deposits). If there is little flexure of the depressed plate, the basin may be *overfilled* (accommodation much less than sediment supply) and accumulate a clastic wedge of continental deposits extending far onto the craton. The early Taconic (Ordovician), Acadian (Devonian), and Alleghenian (Pennsylvanian) phases of the Appalachian foreland are familiar examples of these types. Underfilled foreland basins are, as a rule, richer in mudstones (greater accommodation) than overfilled foreland basins (lesser accommodation), but exceptionally may be starved and have only a thin condensed section.

Because thrusting is non-uniform in time and space along the axis of the flexure, there are large changes in accommodation, sometimes abrupt. Consequently, a great variety of mudstones are likely to be deposited. Furthermore, the thickness of a given mudstone unit is likely to vary substantially because of changes in accommodation. Typical geometries for the mudstones and many of the sandstones in foreland basins are widespread elongate wedges: gray and dark gray mudstones thicken and become siltier up dip (and may even pass into reddish brown mudstones near the source) whereas marine black shales thicken down dip into a basin bottom, where they may finally thin again in a starved basin. Basin overfill takes two forms: the first is the progradation of one or more large river systems across the foreland basin (and at times well onto the craton) with coals in either alluvial or low energy deltaic deposits. Here sheet-

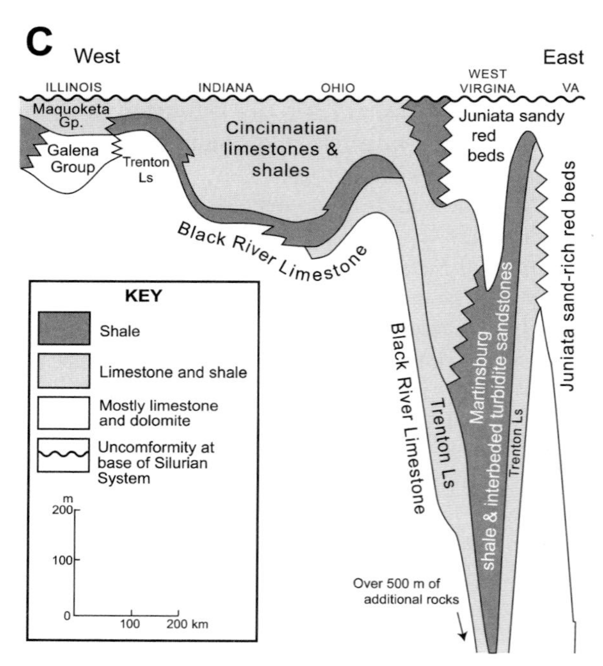

like gray mudstones (marine and nonmarine) and black shales (marine or transitional to non-marine) are numerous and, along with the coals, sandstones and limestones, form well defined cycles. The other possibility is a prograding wedge of red beds of proximal sandstones → siltstones → mudstones that also may extend far cratonward and finally terminate in a red, muddy shoreline bordering a shallow sea, perhaps depositing carbonates. *Because accommodation and supply can change both quickly and often in long lived foreland basins, such basins provide a splendid laboratory for the study of many different*

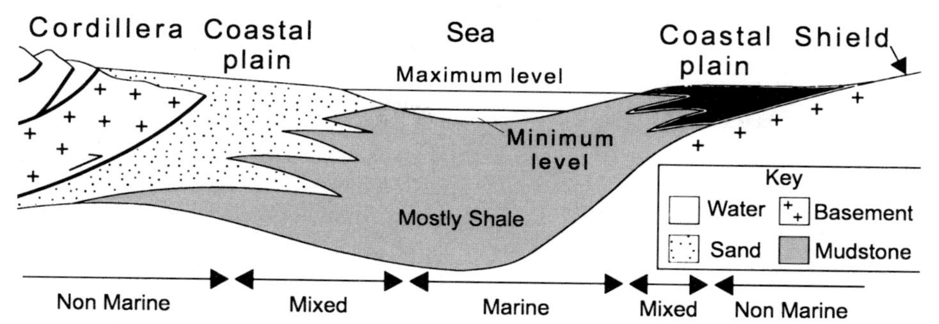

Fig. 8.20. Reconstructed east-west cross section of Tertiary foreland basin from the Magdalena Valley of Colombia eastward into Venezuela, where it onlaps the Guyana Shield (after Cooper et al. 1995b, Fig. 14). Maximum thickness of the Llanos Basin is about 3.5 km and its Paleocene and Lower and Middle Miocene fill is marine and brackish, but its Upper Miocene to Pliocene fill is mostly continental reflecting Andean uplift. Depending on the ratio of thrust-induced basin subsidence to orogenic supply, the mudstones of such a basin may range from black and organic-rich through gray to red, or the basin may be mostly filled with coarse sandstone and marginal conglomerates. Published by permission of the authors and the American Association of Petroleum Geologists

types of mudstones. And, because of deep burial, the organic-rich mudstones of foreland basins are likely to be sufficiently mature to become source rocks for petroleum and to host coals of high quality. Three examples show a few of many possibilities.

An Andean related foreland basin in Colombia and Venezuela began in the Triassic and continues to the present (Cooper et al. 1995b). During these 220 Ma, the limits of this basin have shifted in response to changing thrust loading, development of swells within the foreland basin and changes in relative sea level. Here we focus on the basin's Tertiary history, which records ten third-order cycles – note especially cycles T80 and T90 of the Miocene and Pliocene of the Llanos Basin (Fig. 8.20). In these cycles sand and mud were derived from both the Guyana Shield and the rising Andes and were deposited on alluvial fans, in estuaries, and in and along the borders of a shallow sea. Marine mudstones (Notestein et al. 1944) are typically silty and greenish gray to partially mottled. In contrast, mudstones became brown, red and mottled after the sea withdrew. In fact, the top of cycle T80 is mapped by color change from gray to reddish brown. The marine mudstones of the T80 cycle are the most widespread in the basin (accelerated thrust loading plus high global Miocene sea levels) and make it the best seal in the basin. The overlying T90 cycle, a thick sandstone sequence, deeply buried T80 and caused maturation of its hydrocarbons to form the principal source bed of the Llanos Basin. The present physiography of the Llanos Basin of Venezuela and Colombia – a vast, low, flat region of shifting rivers, lakes and swamps – is not only a good model for much of its earlier alluvial fill, but also for many other foreland basins that had seasonally rainy climates.

The Cretaceous Western Interior Seaway of North America extended all across Canada and the United States and deposited one of the longest (about 5,000 km), most continuous muddy basin fills in the world – from the Arctic into Texas and Northern Mexico (Fig. 8.21). A vast literature exists on the Western Interior Seaway, part of which is summarized by Bohacs (1998b), who emphasizes its sequence stratigraphy. An exceptionally detailed study is that by Bloch et al. (1999) in western Canada, while the collected articles of Caldwell and Kaufman (1993) provide an insightful overview

Most of the mud and silt of the Western Interior Seaway was derived from the ancestral Rockies with lesser amounts from the craton to the east. Thrusting along the western margin formed a series of foreland basins whose sediments onlapped the craton to the east. Over wide areas, bottom waters ranged from dysoxic to anoxic, interrupted only by storm events. Well-known mudstones of this basin include the Pierre, the Mowry, those of the Colorado Group in Canada and the Kaskapau also in Canada. Good outcrops in southern Utah provide insight to how this giant basin was filled (Leithhold 1994).

In southern Utah, the Cenomanian Tropic Shale and its equivalent, the Tununk Shale member of the Mancos Shale, are well exposed down dip for 110 km. They consist chiefly of well bioturbated, fossiliferous, somewhat sandy mudstones and minor sandstones. These mudstones, here about 200 m thick, were deposited in a second-order cycle of 2.5 Ma. Within this time span, there are six third-order cycles and some

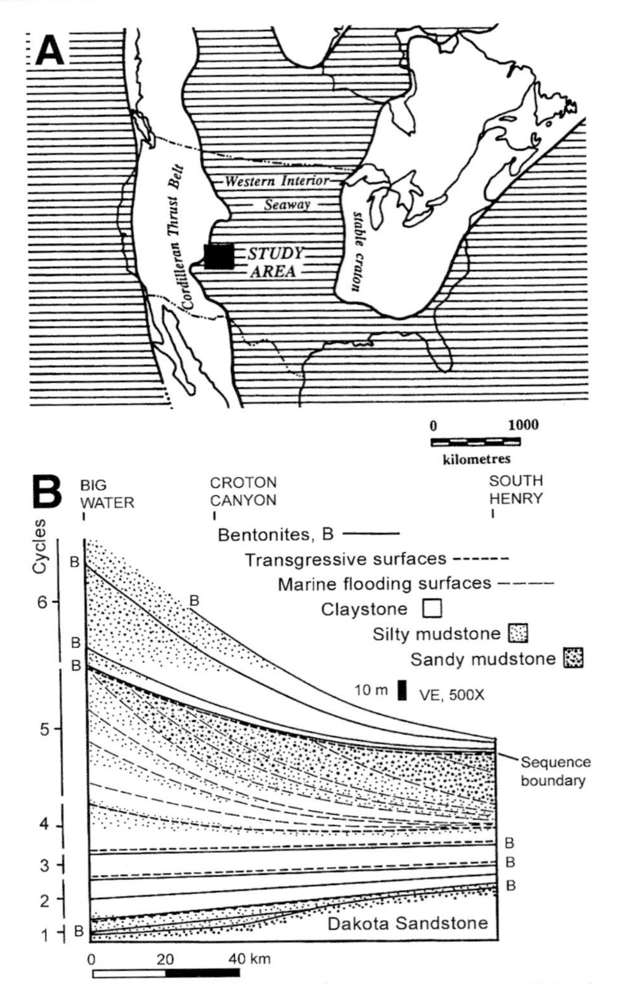

Fig. 8.21A,B. Cretaceous Western Interior Seaway of North American (after Leithold 1994, Figs. 1 and 8): (**A**) Its vast extent; and (**B**) clinoforms near its margin in southern Utah (section is 110 km long). In the center of this interior seaway, clinoforms have lower dips and hemipelagic deposition is more dominant. Published by permission of the author and the International Association of Sedimentologists

37 fourth-order cycles, which have aggrading, retrograding and prograding stacking patterns. These parasequences range from about 2 to 5 m in thickness. Most of the parasequences are more calcareous at their base than at their top (slower terrigenous input on the marine flooding surface). Bentonites derived from the volcanic arc to the west provide excellent, far ranging time lines that help correlate between facies and establish beyond question a clinoform geometry. Bioturbation is widespread and abundant in all but the coarsest sandstones and the darkest mudstones, but is most intense where sedimentation was slowest as at the distal end of clinoforms.

At the base of the Tropic and Tununk Shales is a widespread sequence boundary of bioturbated, thin sandstone rich in oyster shells representing the initial basal transgressive surface of erosion. Five other transgressive surfaces and eight flooding surfaces are recognized in the third- and forth-order cycles of this long section. Thus all the essential elements of sequence stratigraphy are present in this area of superb outcrop. Taken together, these cycles show how the second-order Greenhorn Cycle started with deepening, reached a highstand, and then regressed.

The Cretaceous Ghazij Formation (Johnson et al. 1999) of Pakistan (Fig. 8.22 and Table 8.6) gives us a fascinating insight into a mudstone deposited shortly after the northward-dipping edge of the Indian shelf collided in early Eocene time with a carbonate-rich micro-continent in front of the Asian mainland to the north. This collision formed an intervening foreland basin. In it one can see a parallelism of environmental evidence provided by its mudstones and sandstones. And finally, the role of a weak discordance between its upper and middle

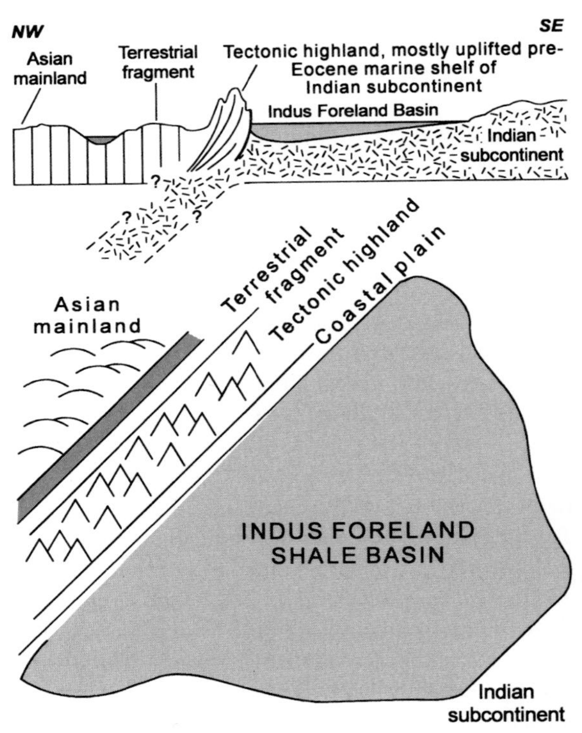

Fig. 8.22. Schematic distribution of mudstone-rich Eocene Ghazij Formation of Pakistan as it onlaps the Indian craton (after Johnson et al. 1999, Fig. 6). These mudstones, derived from the tectonic highlands of the ancestral Himalayas, are carbonate rich, over 1,400 m thick and contain fluvial, coastal and marine facies plus some valuable coals

Table 8.6. Lower Eocene Ghazij Shale, Balochistan, Pakistan. After Johnson et al. (1999)

Basin Setting
Present basin, much deformed, covers an area of 750 × 200 km, thickness locally exceeds 1,000 m; filled with a remarkable 80 to 90% mudstone with minor sandstones, limestones, conglomerates and coals

Characteristics and Interpretation of Mudstones
Upper
Calcareous, non-fossiliferous, light gray to brown to red banded mudstones (possibly paleosols) interbedded with minor alluvial sandstones and a limestone conglomerate → delta plain
Middle
Medium gray, calcareous mudstone with sparse non-marine bivalves, some fine-to medium-grained sandstones plus coals up to one meter thick; interbedded sandstones change from shoreface to tidal to fluvial whereas mudstones change from estuarine to overbank → prodelta to delta fringe
Lower
Dark gray, calcareous mudstones (with foraminifera) interbedded with minor, fine-grained sandstones; organic matter less than 0.5%; gray mudstones grade laterally into green mudstone with limestone nodules and thin limestones → outer shelf and prodelta

members reinforces the depositional model based on its mudstones.

The Ghazij Formation has three internal units: at the base, outer marine shelf and prodelta deposits; in the middle, lower delta plain deposits; and at the top, upper delta plain deposits. Mudstones of the marine shelf and prodelta mudstones are dark gray and calcareous and principal fossils are foraminifera. These mudstones contain a few thin sandstones and form the largest part of the Ghazij. The medial part of the Ghazij conformably overlies the lower part and consists of medium gray, calcareous mudstones with non-marine bivalves, which interbed with three types of sandstones and some mineable coals up to a meter thick. These sandstones change upward from shoreface to tidal to fluvial as shown by sedimentary structures and trace fossils, while the fossil content of interbedded mudstones changes from plant debris (estuarine) to rootlets (overbank). Consequently, a lower delta plain model fits these deposits well. Above a weak discordance are light gray to brown to red banded, non-fossiliferous mudstones (thought to be mostly paleosols) and a few thin, sandstone-filled channels of the upper delta plain. Thus, the Ghazij shows a classic shallowing-upward succession as demonstrated by its mudstones, sandstones, coals, and a weak discordance between upper and lower delta plain deposits.

The calcareous mudstones of the shelf and lower delta plain pose an interesting problem. The sandstones of the Ghazij have abundant framework grains of limestone and there are some limestone clasts and thin limestone conglomerates near the top of the formation. Thus a source rich in carbonate is clear. It is easy to explain the carbonate of the mudstones of the shelf and lower delta plain as having a detrital origin – fine carbonate mud eroded from a carbonate source. But we need to remember that there are many calcareous mudstones where such an explanation is not so easily invoked. So the key question is, "How does one distinguish far-traveled carbonate mud from admixed pelagic carbonate mud in a mudstone?"

8.6.3. Rift Basins

Rifts, extensional basins formed by ruptured lithosphere, form elongate, sharply bounded depressions. Rupture occurs by stretching of either cold, continental lithosphere or by stretching of cooling lithosphere above a hot spot (which may develop a triple junction). Extensional basins range from isolated, small grabens to widespread systems of grabens within rift systems such as underlie the North Sea or occur in southern Argentina along the margins of Gondwana. Some larger rifts on the continents are also the precursors of intracratonic basins as part of the series:

Initial heating → extension and uplift → rifting → cooling and thermal subsidence = an intracratonic basin.

Thus there is an initial phase of rapid mechanical subsidence, followed by a slower, much broader phase of thermal subsidence. Together this pair generates a basin with a "steer head" cross section (broad saucer-like cratonic basin underlain by a rift basin). Rifts are very widespread (as plate motions change so also do areas and kinds of tension) and thus specialists recognize many different types of rifts.

Rifts influence sedimentation and mudstones in important ways. When rifts are young, hot and have high shoulders adjacent to a rapidly subsiding sediment trap, synrift deposits accumulate. At this stage, fill is mostly terrigenous with alluvial fan, Gilbert delta, fan delta, and lacustrine deposits. Both shallow

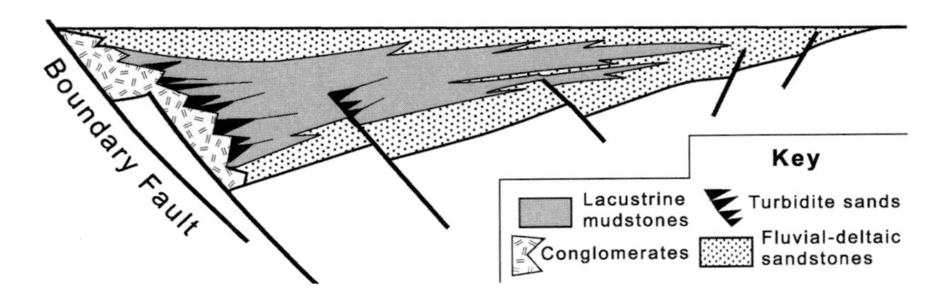

Fig. 8.23. Fault geometries and rift fill (Morley 1999, Figs. 2 and 4). Note the rapid facies change along the main bounding fault, which provides an ideal reservoir-seal pair for hydrocarbons or mineralization. Published by permission of the author and the American Association of Petroleum Geologists

and deep turbidites deposits are typical of synrift fill and thicknesses may reach thousands of meters. Proportions of these lithologies vary widely – some rifts have thick sandstone fills whereas others have thick fills of mudstones perhaps mixed with evaporites and carbonates. Later when the rift is old, cold, and has neither shoulders nor topographic expression (burial by postrift deposits) and is largely inactive, there are two possibilities. An interior rift, although not active, may continue to subside weakly and become the site of an intracratonic basin localizing later marine invasions or a large river. Or should the intracontinental rift form a new ocean, synrift deposits will be overlain first by deep marine slope deposits followed by shallow water shelf deposits (synrift → drift sequence) as the initial small ocean widens to be bounded by passive margins.

The geometry of the rift or graben fill is wedge-like with abrupt borders, length is much greater than width, and thicknesses can reach thousands of meters (Fig. 8.23). This geometry is the result of subsidence in half grabens, mostly formed by deep *listric faulting* (flattening at depth). Because a rift system consists of basins separated by intervening highs (offsets formed by transform faults), the valley of a young rift commonly has many lakes, which may or may not be connected depending on climate.

Such lakes were either open (had an outlet) or closed (insufficient rainfall to overflow a col) and thus likely to have either carbonates or evaporites. Many lacustrine mudstones are organic rich and thus good source beds for petroleum or hosts for strata-from deposits of copper, lead and zinc formed by hot brines reacting with reduced sediments. Two factors seem responsible for the organic richness of lacustrine mudstones in rifts: the greater responsiveness of a smaller rather than a larger body of water to warm climates (enhanced productivity) and the greater likelihood of stratification of the water column in a long narrow and deep basin than in a wide open one. Three examples of marine and non-marine rift mudstones illustrate the diversity of this environment.

Europe's foremost source bed for petroleum is the Kimmeridge Clay Formation of Upper Jurassic-Lower Cretaceous age in the North Sea region with equivalents that extend much farther onto the Russian platform and into the West Siberian Basin. Lithologic variability over such a vast region is great, but almost everywhere its organic-rich claystones are

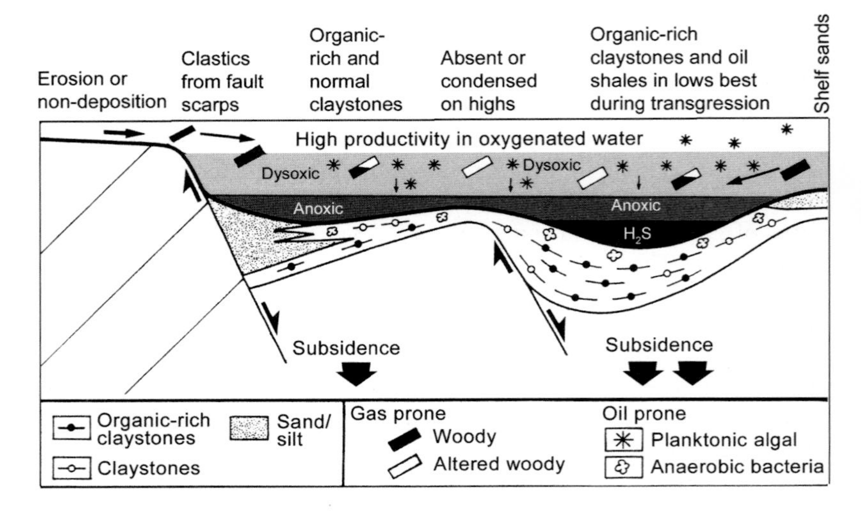

Fig. 8.24. Depositional model of Jurassic Kimmeridge Clay Formation, the source bed of the North Sea, across the rift system of the North Sea, which was active during its deposition (after Cornford 1990, Fig. 11.53). Note how the Kimmeridge thins over highs and thickens into structural lows. Published by permission of the author and Blackwell Scientific

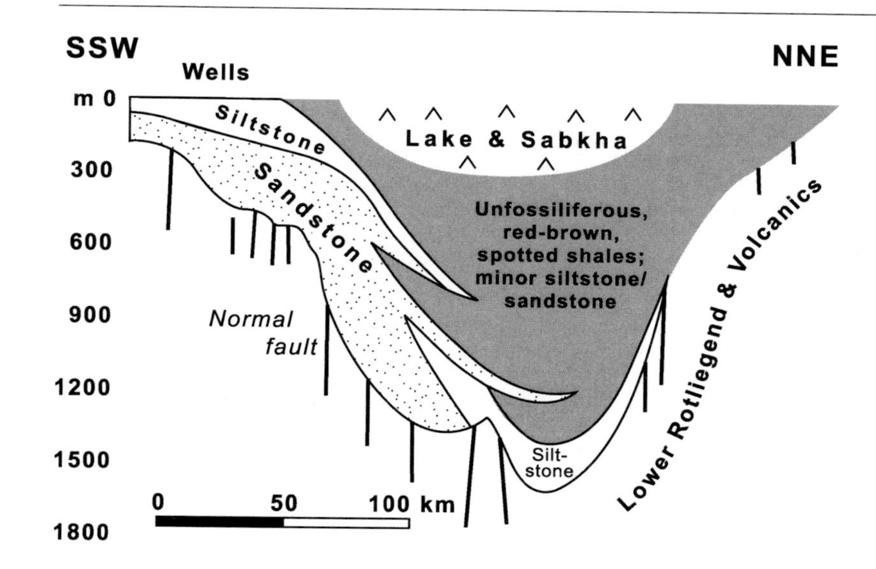

Fig. 8.25. Schematic cross section of thick red bed sequence in North German Basin/southern arm of North Sea Basin culminates in super-saline Haselgeberge desert lake of the Permian Rotliegende Formation (after MacCann 1998, Fig. 4). Omitted for simplicity are the stratigraphic names

source rocks when deeply buried. Here we focus on the North Sea Region, where the Kimmeridge was deposited as an intracratonic basin started to rift during the opening of the North Atlantic (Fig. 8.24).

In the North Sea region, thicknesses range from as little as 50 m on platform highs to as much as 1,250 m in subsiding, anoxic grabens. Lithologies are mostly dark mudstones with turbidite fans against fault scarps, some marginal shelf sandstones, and limestones on topographic highs. Total organic content varies widely, averages some 6% but some oil shales have as much as 40% or more, along with abundant pyrite. (In the cliffs of coastal England these oxidize and give rise to the term "Burning Cliffs"). In outcrop these mudstones are medium to dark gray, blocky, flinty, and calcareous. MacQuaker and Gawthorpe (1993) recognized the following five facies: (1) poorly laminated gray claystones (distal dysaerobic), (2) silty non-laminated gray to black claystones with scare fauna (dysaerobic to anoxic?), (3) gray fissile shales rich in nannofossils and interbedded with limestones (slow sedimentation on highs or in shallow water far from riverine input), (4) brown laminated claystones with pyrite and up to 42% TOC (anoxic bottom interrupted by rare storms), (5) and the much-studied concretionary zones (slow sedimentation far from riverine input).

Most of the Kimmeridge mudstones were deposited in a broad dysaerobic sea that at times was anoxic, particularly in its graben lows during summer. Ager (1975) has suggested that the North Sea in Jurassic time was an ideal basin to accumulate organic-rich source rocks for two reasons. First, it was a protected arm of the ocean, where cool polar waters upwelled into warm surface waters (high productivity) and, secondly, the subsiding graben system provided persistent topographic lows (ample accommodation) favoring anoxia. Sources for much of the above are Cooper et al. (1995a) and Cornford (1990), but like the Western Interior Seaway of North America, a vast literature exists.

The redbeds and evaporites of the Permian upper Rotliegend Group of the North Sea extend from England some 1,500 km eastward into Poland and have a maximum thickness of some 1,500 m. This basin developed in front of the Variscan fold belt above a thick volcanic section, the lower Rotliegend, which depressed the crust to create the accommodation for the upper Rotliegend (Fig. 8.25). The lower volcanic part was deposited in subsiding half grabens during Permian rifting, whereas its upper part represents post-rift fill and caused some differential compaction across graben topography. Because of a dry climate, most of the Rotliegend had inland drainage, much of it a large lake, Lake Haselgeberge, that accumulated mudstones and halite. These deposits, which occur in the southern part of the Rotliegend, were bordered by beaches and sabkhas beyond which were wadis and dunes. It is thought that this large inland lake was as much as 200 to 300 m below sea level.

The mudstones of Lake Haselgeberge and its borders are reddish brown with greenish-gray spots. They are unfossiliferous, contain anhydrite nodules, and are finely interbedded with siltstones that show adhesion ripples. During especially dry periods, widespread beds of halite accumulated in the lake, whereas the reddish-brown mudstones reflect periods of greater moisture. A typical lake cycle is red-brown mudstone → anhydrite-rich mudstone → muddy anhydrite → clear halite. This cycle is a good example of the mudstones clearly representing the maximum flooding surface (pluvials). These red mudstones and evaporites form seals for the many gas fields of the Rotliegend Formation, gas that was sourced from the older coal measures below the volcanics. Sources for much of the above are Almon (1981), Gast (1991), McCann (1998), and Stemmrick et al. (2000).

8.6.4. Pull-apart basins

These are extensional basins formed by oblique rather then perpendicular extension. These develop at the releasing bends of transform faults and along deep crustal shears. Typically, such basins are small, narrow, short lived (usually less than 10 Ma?) and may evolve from early rhomboid geometries into troughs. Young and small pull-aparts are underlain by relatively thin continental crust, whereas larger and older ones by oceanic crust that may even have spreading ridges. Pull-aparts may develop in series separated by pushup blocks, depending on the geometry of the strike-slip fault. Death Valley in California and the Dead Sea in the Middle East are two modern onshore examples of pull-aparts and the Guaymas Basin of the Gulf of California and the Cariaco Basin in Venezuela are examples of pull-aparts flooded by the sea. Mudstones vary greatly in both kind and abundance in pull-apart basins.

In onshore pull-aparts on continental crust, mudstones are both alluvial and lacustrine and likely to be subordinate to sandstone and conglomerate deposited as alluvial fans, Gilbert deltas and valley fills. If the climate were arid, some evaporites may occur. But even here the basin may be flooded from time to time by the sea depending on accommodation and accidents of paleogeography. When offshore, however, a pull-apart basin is prone to have turbidite sands and mudstones plus slumps where it is close to the coast, whereas hemipelagics and pelagics predominate if it is far from the coast or isolated from it by one or more horsts.

The best-known pull-apart basin, one with both continental and marine fill, is the Miocene Ridge Basin of southern California (Crowell 2003). The Ridge Basin lies between two large transcurrent faults, the San Gabriel and San Andreas (Fig. 8.26), is markedly elongate (55 × 20 km), has a sediment section of about 9,000 m and four depositional systems – marine slope aprons and fan deltas, lacustrine fan deltas and alluvial fans and even a few carbonates. Thus at times the Ridge Basin was connected to the sea, but at other times it accumulated thick alluvial and lacustrine deposits with abundant clays. The lacustrine clays include those deposited in both deep and shallow freshwater and deepwater brackish-water lakes. Because the two major boundary faults were active much of the time, there were many changes in accommodation during the short life of the basin and thus many environmental shifts. The rapid shear of the boundary faults is also responsible for the "side by side," shingled fill. Consequently, the Ridge Basin is a remarkable end member of terrigenous sedimentation with much to teach us about clays.

Totally different is the nearby Miocene Monterey Formation most of which was deposited in pull-aparts flooded by the sea. The Monterey Formation of California extends southward from San Francisco along the coast almost to San Diego and has many different facies ranging from thick turbidites to far distal, largely sand- and silt-free, anoxic to dysoxic starved basins rich in biochemical rocks and also some bentonites. Additionally, the Monterey is of interest to us because it is both a prime source bed and a petroleum reservoir (its siliceous facies develops good fractures during the opal-CT to quartz transition).

An unusual combination of world events seems to be responsible for the deposition of the Monterey. Middle Miocene cooling started glaciation in polar regions and reduced world sea levels. At the same time, the Isthmus of Panama closed the connection between the Atlantic and Pacific Oceans increasing silica-rich bottom waters in the Pacific. Together these two events introduced cold, nutrient-rich water to the California coast. In California, the San Andreas Fault System, starting in the Oligocene, created many deep, offshore, pull-apart, extensional basins isolated by horsts from mainland detritus. These three events combined to produce the depositional setting of the Monterey in the Santa Maria Basin near Santa Barbara – a deep, elongate, sediment-starved basin with oxygen deficient water overloaded by organic matter as upwelling of cold, nutrient-rich water mixed with warm surface water. In sum, the Monterey is the result of both worldwide and local events that occurred between about 17.5 and 6.3 Ma.

In outcrop at Shell Beach (Fig. 8.27) near Santa Barbara and in adjacent cores, one can recognize two sequence boundaries in the Monterey, plus two sets of lowstand, transgressive, and highstand deposits as well as two down lap surfaces. Changing lithologic proportions, low-relief scour

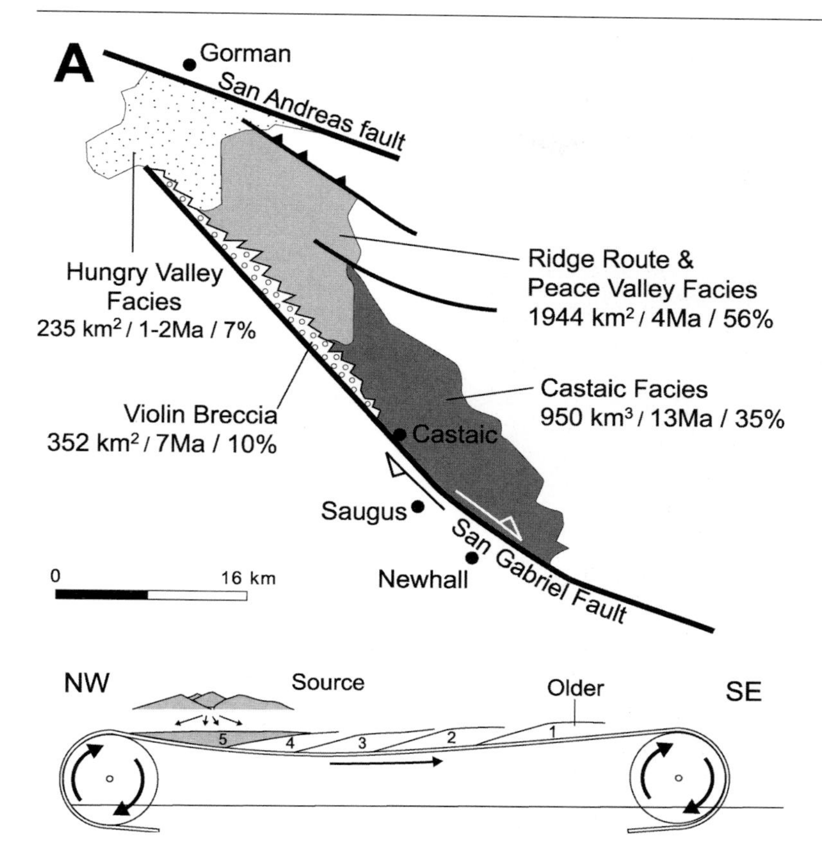

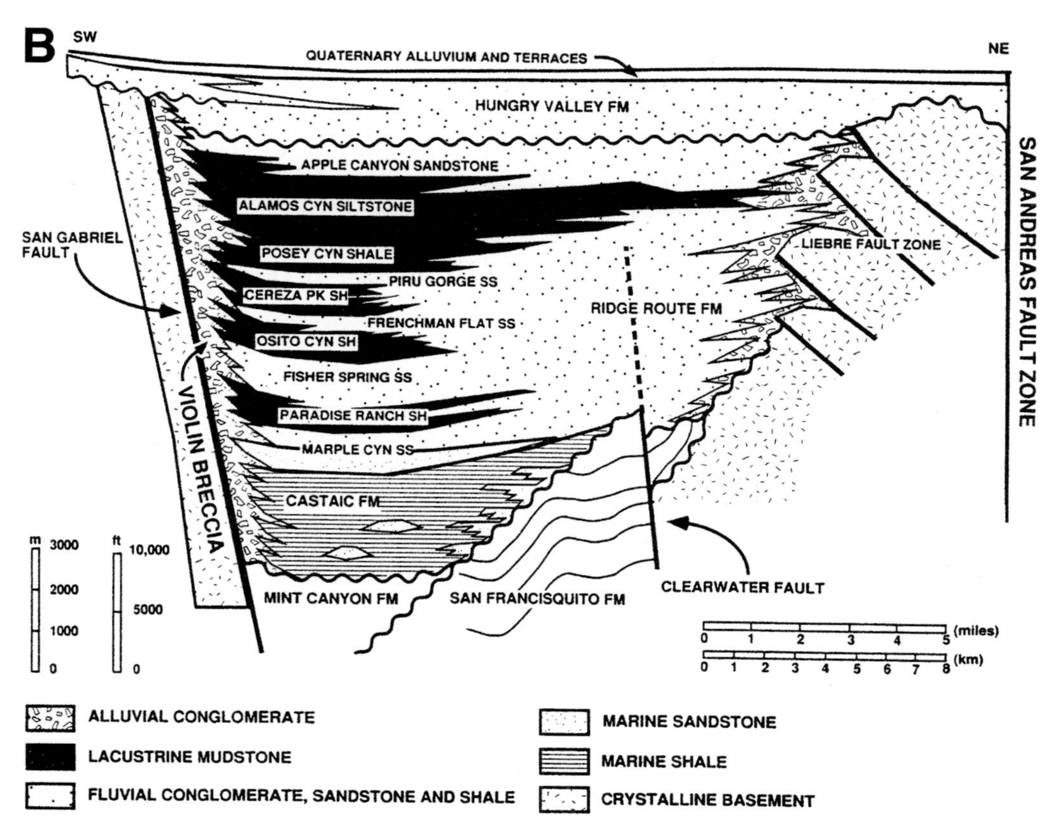

Fig. 8.26A,B. Ridge Basin of southern California, (**A**) Ridge Basin of Miocene age has four depositional systems that are shingled from northwest to southeast as the basin floor was carried to the southeast away from the source. *Numbers* below depositional systems are volumes, durations and percent of basin fill (after Link 2003, Fig. 4 and Crowell 1982, Fig. 6); (**B**) cross section of Ridge Basin shows the variability of its fill (Crowell 1982, Fig. 4). Published by permission of the authors and the Geological Society of America

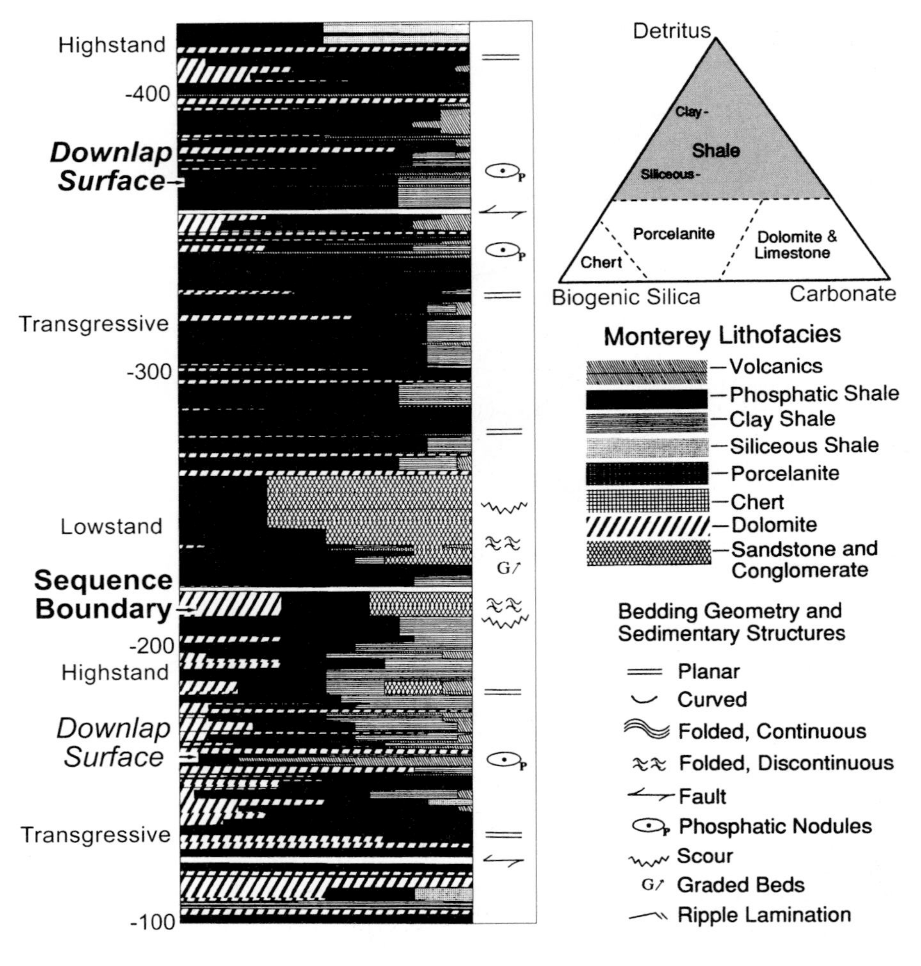

Fig. 8.27. Long section of Monterey Formation measured in the field at Shell Beach, San Luis Obispo Co. California (Bohacs and Schwalbach 1992b, Fig. 5). Published by permission of the authors and the Pacific Section of the Society for Sedimentary Geology

surfaces with phosphate nodules, and chemistry as measured by a three-channel scintillometer (Appendix A.1) provided the information on which these interpretations are based. This section establishes how well the concepts of sequence stratigraphy apply even in basin centers and in isolated starved basins.

The lithologies of the Monterey in the Santa Barbara Basin are dominantly claystones with only a few interbedded siltstones, sandstones, conglomerates and dolomite (Fig. 8.28). The carbonates and the many thin phosphatic lenses represent periods of very slow sedimentation. The claystones are phosphatic, siliceous and calcareous, dark gray to black and organic rich. Overall the Monterey has 6 to 8% TOC. The upper Monterey (the least diluted by detritus) has as much as 42% biogenic silica followed by 33% biogenic calcite, 20% clay, and 7% TOC. Persistent periods of bottom anoxia related to intensification of the oxygen minimum zone preserved this organic material. Sources for the above include Pisciotto and Garrison (1981), Schwalbach and Bohacs (1992a), Dunham and Cotton-Thornton (1990), and Issacs (1984 and 2000). See also Bramlette (1946) in Digging Deeper.

8.6.5. Forearc, Intra-arc, and Backarc Basins

Most of the studies of these basins are focused on their tectonic evolution and significance so their mudstones are less well known than those of basins underlain by continental crust. In part this is the result of lesser economic significance (fewer wells and seismic lines), complex structure and their hard-to-grasp stratigraphy – all the result of their location on a convergent margin. Other complications are that these basins are short lived and may be incorporated into a suture belt, the convergent margin may become a transform margin, or the direction of convergence may reverse. All of these possibilities affect accommodation, provenance, and supply and thus the basin fill is not so easy to generalize. Forearc and backarc basins are, however, mostly marine with fills ranging from deep to shallow as accommodation

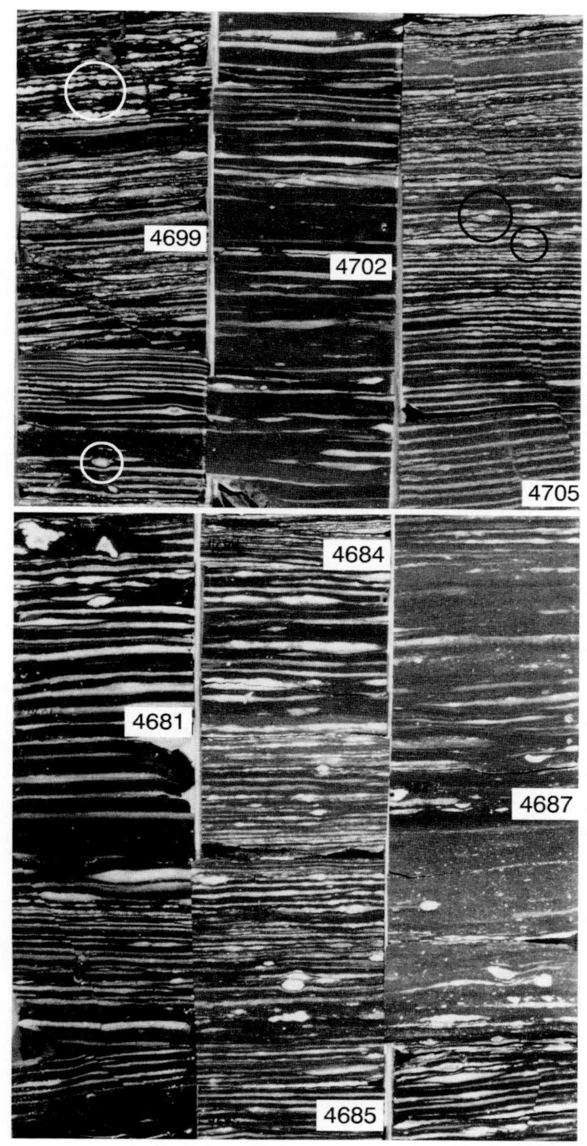

Fig. 8.28. Core from the Monterey Formation from the Leroy 51–18 well in the Santa Maria Valley Field, Santa Barbara County, California (Dunham and Cotton-Thorton 1990). Most of this long core has abundant thin laminations of either dolomite or siliceous material (originally diatoms with some now altered to chert and porcelanite) interlaminated with dark to black organic-rich shale. Also present are scattered phosphatic nodules and some thin dolomite beds. Present thickness of lamination is less than the original, because of both burial compaction and the transformation of opal A to opal CT, which reduces volume. Published by permission of the authors and the Society for Sedimentary Geology

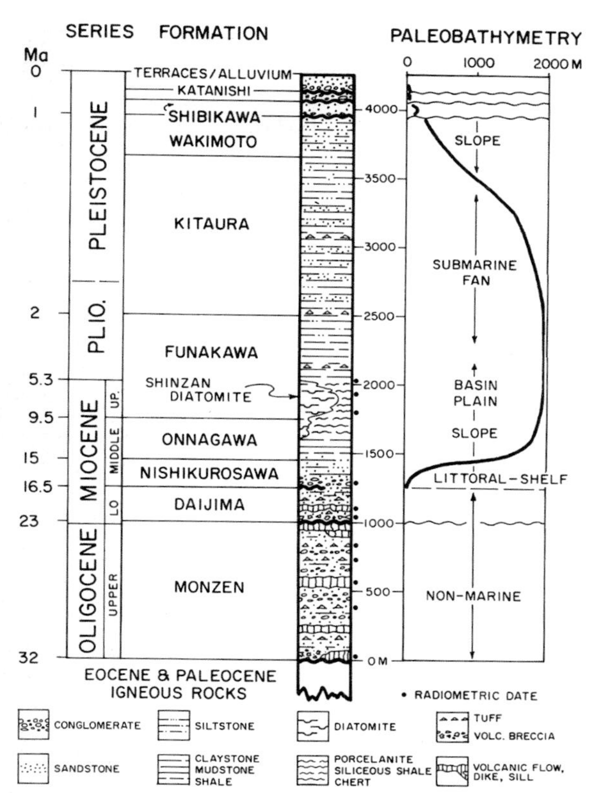

Fig. 8.29. Fill of a back-arc basin: Four-thousand meter section of back-arc basin sediments on the Oga Peninsula, Akita Prefecture, North Honshu Island, Japan (Ingle 1992, Fig. 4). This long section on the east side of the Sea of Japan starts with non-marine coarse clastics with abundant volcanics. The basin then deepened rapidly, accumulating diatomites, then passed through a complete cycle of mostly deep-water mudstones and turbidites, and ended once more in shallow-water sediments – a rare complete cycle

changes (Fig. 8.29), while intra-arc basins have both continental and marine fills. Below we briefly report on these basins largely drawing upon summaries by Dickinson (1995), Smith and Landis (1995), Marsaglia (1995), and Einsele (2000) in Digging Deeper.

Forearc basins fills are elongate, have thicknesses in the thousands of meters, and are bordered by a calc-alkaline magmatic arc on one side and a complexly deformed accretionary wedge of blue-schists and mélanges on the other. Mesozoic and Cenozoic forearc basins occur both along the margins of continents and in the open ocean far from land. Forearc basins along continental margins have thicker fills (because of proximity to an eroding landmass) than those in the ocean, which are underfilled and contain more pelagic deposits. Continental margin arcs generate thick fills of thousands of meters of volcaniclastic and arkosic turbidite sands and muds that may either overlie or be capped by shallow water deltas, shelfal sands and muds or even reefal carbonates as accommodation rapidly changed in response to changes in subsidence, uplift and plate motions. Thus mudstones range from deep-sea red muds to muds associated with deltas and fan deltas, but most of the fill is marine, much of it deep marine. Smectitic clays, fine volcanic rock fragments and ash fall interbeds are common as are siliceous mudstones. The association of diatom- and radiolarian-rich mudstones with volcanic ash has been noted for many years and has been attributed to leaching of silica from devitrifying ash, but upwelling seems to be the dominant control (Barron 1987, p 173). But this only removes the issue one level – does our familiarity with the western North America bias our assessment of this association or is it in fact universal, and, if so, why? Reverse grading in laminations or beds may be the rule, because even waterlogged pumice is still less dense than grains of quartz, feldspar or rock fragments. See Dickinson (1995, his Tables 6.1 and 6.2) for an inventory of forearc basins, which illustrates well their complexity.

Intra-arc basins occur within the magmatic arc. Intra-arc basins may be bounded by faults or by volcanic piles and are thinner as a rule than forearc basins. Intra-arc basins are almost entirely dominated by volcanic debris ranging widely in composition (rhyolites to basalts) and may be either continental or flooded by the sea. Where continental, there are debris flows and ignimbrites interbedded with alluvial fan and braided stream deposits, all with tephra. Alluvial, tuffaceous muds are abundant only in the distal part of the system. Clay minerals are dominated by smectites. Where flooded by the sea and the basin is wide, there are pillow lavas and submarine ignimbrites interbedded with turbidites, tuffs and siliceous and calcareous mudstones (both pelagic in origin). Hydrothermal vent structures may also be present. There may also be bordering reefs around volcanoes (if in low latitudes). The site of the initial intra-arc basin may evolve into a backarc basin through back-arc spreading.

Backarc basins occur behind volcanic arcs. Most modern ones occur along the margins of the Pacific Ocean and are of several types (Marsaglia 1995), divided broadly into those that form on oceanic crust and those that form on continental crust (retro-arc basins). Volcaniclastic debris (tuffs and pillow lavas), slumps and turbidites plus hemipelagic clays and siliceous and carbonate oozes are the typical facies of both types, and the tuffaceous sandstones form petroleum reservoirs in a few cases, such as in the Miocene of Japan (Aoyagi and Iijima 1987). Klein (1985) gives a careful description of the sedimentary fill of backarc basins based on the description of cores from 28 deep-sea sites (Table 8.7). These proportions will vary with distance from a magmatic arc; if far from the arc, turbidites will be distal and hemipelagics and biogenic pelagics will be most abundant.

Table 8.7. Sediments of back-arc basins. After Klein (1985, pp 2–10)

Sediment Types	%
Debris flow conglomerates	1.2
Submarine fan depositional systems	20.0
Silty basinal turbidites	5.7
Hemipelagic clays	21.8
Pelagic red clays	4.2
Siliceous oozes and cherts*	4.3
Carbonate oozes	23.8
Resedimented carbonates	9.5
Pyroclastics	9.5

*Rich in radiolarians and silicoflagellates

To sum up, recognizing forearc, intra-arc, and backarc basins in convergent zones and distinguishing one from the others is a major task in pre-Mesozoic deposits. To see this better, one needs only to read the description by Taira (1985) of the thousands of meters of muddy rocks in the Shimanto Belt and Nankai Trough of Japan. It seems certain that better understanding the mudstones of such active margins would contribute much to our understanding of their evolution (Fig. 5.13A).

8.6.6. Glacial Basins

Earlier we examined Pleistocene muddy sediments produced by melting ice sheets well inboard on a continent (Chap. 5). The study of Paleozoic and Precambrian glacial deposits adds a new dimension to what we can learn about glacially- related ancient mudstones, some of which are even source rocks for petroleum such as the Silurian Qusaiba Member of Arabia (Mahmoud et al. 1992). Identification in the ancient is most secure where boulder-rich fine-grained deposits (diamictites) and rhythmites such as varves are intercalated. Many, perhaps most of these older deposits are glacial marine because continental glacial deposits have little likelihood of long-term preservation. See Eyles (1993) for a global summary of the tectonic setting of pre-Pleistocene glaciations (one that stresses mountain building and high plateaus to reduce world air temperatures so

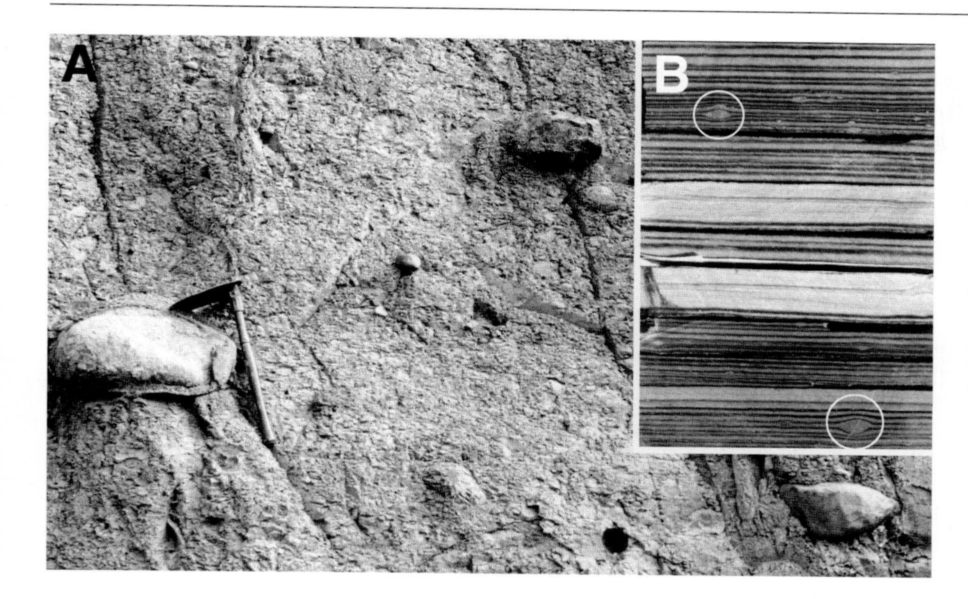

Fig. 8.30A,B. Glacial deposits of Carboniferous-Permian Itararé Group of the Paraná Basin in Brazil: (**A**) Isolated, subhorizontal clasts ranging widely in size contained in massive, medium gray, gritty structureless mudstone with weak horizontal lamination, possibly a marine rainout till; and (**B**) laminites with small dropstones

ice sheets can grow) and Eyles et al. (1992) for the response of glacial facies to sea level change.

Consider a continental ice sheet invading an epeiric sea or broad shelf at a lowstand (sea water locked in the ice cap) well inboard on a large continent such as Gondwana or Pangea, where the weight of the ice isostatically depresses the crust. As the ice melts, sea level rapidly rises, faster than the crust can rebound, and the former ice basin is flooded by a muddy sea well supplied by floating ice blocks. These ice blocks melt and deposit dropstones and frozen till on muddy and silty bottoms (Fig. 8.30). The resultant deposit is a widespread mud blanket with dropstones that overlies sub-glacial deposits. Thus the vertical sequence is likely to be massive diamictite → stratified diamictite + slump deposits → turbidite sands → marine mud (possibly even a black shale)+dropstones and masses of diamictite dropped from floating ice → normal marine mud (Fig. 8.31). This cycle represents cooling (ice invades recently exposed shelf) followed by warming (ice melts, sea level rises and shoreline moves inland) and finally regression with deltaic and alluvial deposits as the crust rebounds. This cycle may be repeated as in southwest Africa, where there is a sequence break at the base of each of three massive diamictites (Visser 1997). Where rebound is great, such a glacial marine sequence will be capped by prograding fluvial, beach or deltaic sands as is true in much of the area around Hudson Bay (Fig. 5.15). Where a continental ice sheet retreats down regional slope inboard on a continent, a lake is created at the glacial front and here glacial freshwater clays directly overlie tills and ablation deposits.

The Miocene to Late Pleistocene Yakataga Formation of Alaska (Armentrout et al. 1983; Eyles et al. 1991) comprises still another type of glacial marine deposit. Here very active mountain glaciers extended to tidewater bounded by a narrow shelf terminating in the steep slope and deep water of the Pacific Ocean. This formation, up to 5 km thick, contains turbidites, slump deposits, "rainouts" of diamictites as well as shallow water deposits all interbedded with mudstones many of which have dropstones and irregular masses of diamictite. Interbedded marine mudstones and coquinas indicate recession of the ice from tidewater and boulder pavements its readvance. The great thickness of the Yakataga is the result of high snowfall on nearby high-relief mountains in British Columbia coupled with ample accommodation on a deep ramp along an active margin. Although well known in the modern as glacial marine basins with anoxic, muddy bottoms, flooded paleovalleys or *fjords*, are seldom reported in the ancient. An exception is in southwest Africa, where deeply-silled, glacially-scoured basins accumulated black, organic-rich muds as part of the Permian Dwyka Group (Cole and Christie 1994).

8.7. Stratigraphy of Black and Red Mudstones

Of all the mudstones, the black and red mudstones have attracted the most attention and generated a vast literature with much diversity of opinion. Here we concentrate on the black shales (Fig. 5.31) and include the red mudstones only for contrast (See Chap. 4: Oxygen for a full treatment of color of mudstones).

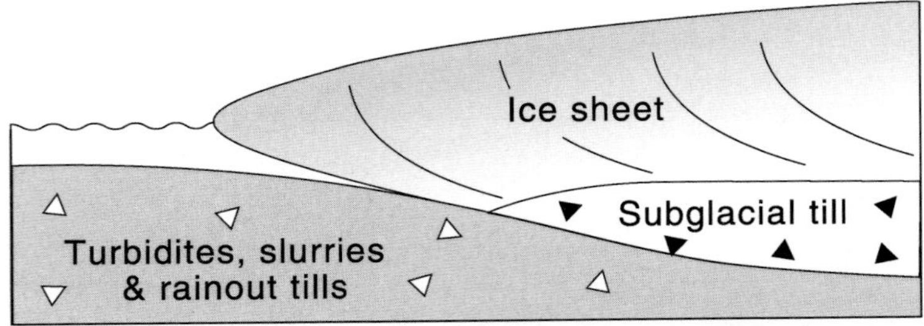

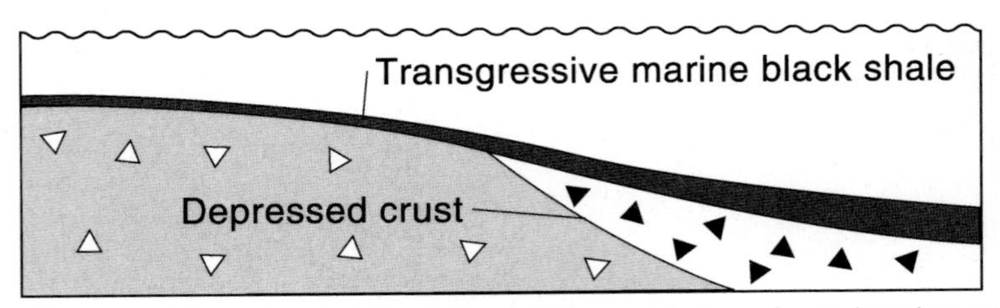

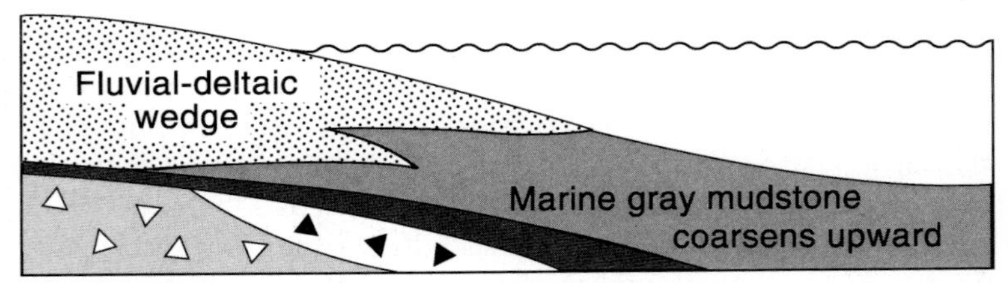

Fig. 8.31. A first approximation to the stratigraphic sequence deposited when an ice sheet advances over a shallow shelf

Black shales have been studied intensively by instrumental techniques. Why is this? First, from the very earliest, their striking color made them stand out and invited attention. Secondly, most of the very widespread ancient black shales have no direct modern equivalents and thus it was necessary to substitute measurement, deductive reasoning and speculation to replace modern analogs (to see this better, think how well modern and ancient alluvial deposits and tidal carbonates compare, or imagine trying to understand glacial deposits without modern analogs). And finally, black shales invited attention because of their economic importance. They are sources of sulfides and heavy trace metals, have been mined for copper, some are sufficiently organic rich to be retorted for oil, and, where sufficiently deeply buried, they are the prime source rocks for petroleum. And exceptionally, where metamorphosed, black shales are even hosts for emeralds as in Colombia.

So then, what is a black shale? Agreement on definition is not universal, because with so much to study and with such broad economic importance, different properties have been given more weight than others. Consequently, definitions vary. Recognizing this, here a black shale is considered to range in color from dark gray to brownish to greenish black to black, is typically but not always well laminated, and generally is organic rich with more than one percent organic carbon, which imparts its dark color. Black shales are commonly interstratified either regularly or randomly with thin laminations or beds of gray mudstone, siltstone, sandstone or carbonate. These interbeds may be calcareous, less commonly siliceous, and exceptionally phosphatic or even tuffaceous. This broad definition in reality defines a *black or organic-rich shale series* that extends from the mid-Precambrian up to and including dark-colored, organic-rich muds in the Holocene. The above definition is that of a field geologist and hence easy and practical to use. But be aware that there is much more to add to the characterization of black shales (Huyck 1990) and their origins.

First, consider their total content of organic carbon, which, while generally only a few percent, can range up to as much as 30 or more percent in some oil

Fig. 8.32. Rhythmic, well defined stratification in the Devonian Ohio Shale at Copperas Mountain, Ross County, Ohio, USA. Gray beds (less resistant layers) interpreted as deposited from dysaerobic suspensions transported by weak distal turbidites into an anoxic basin

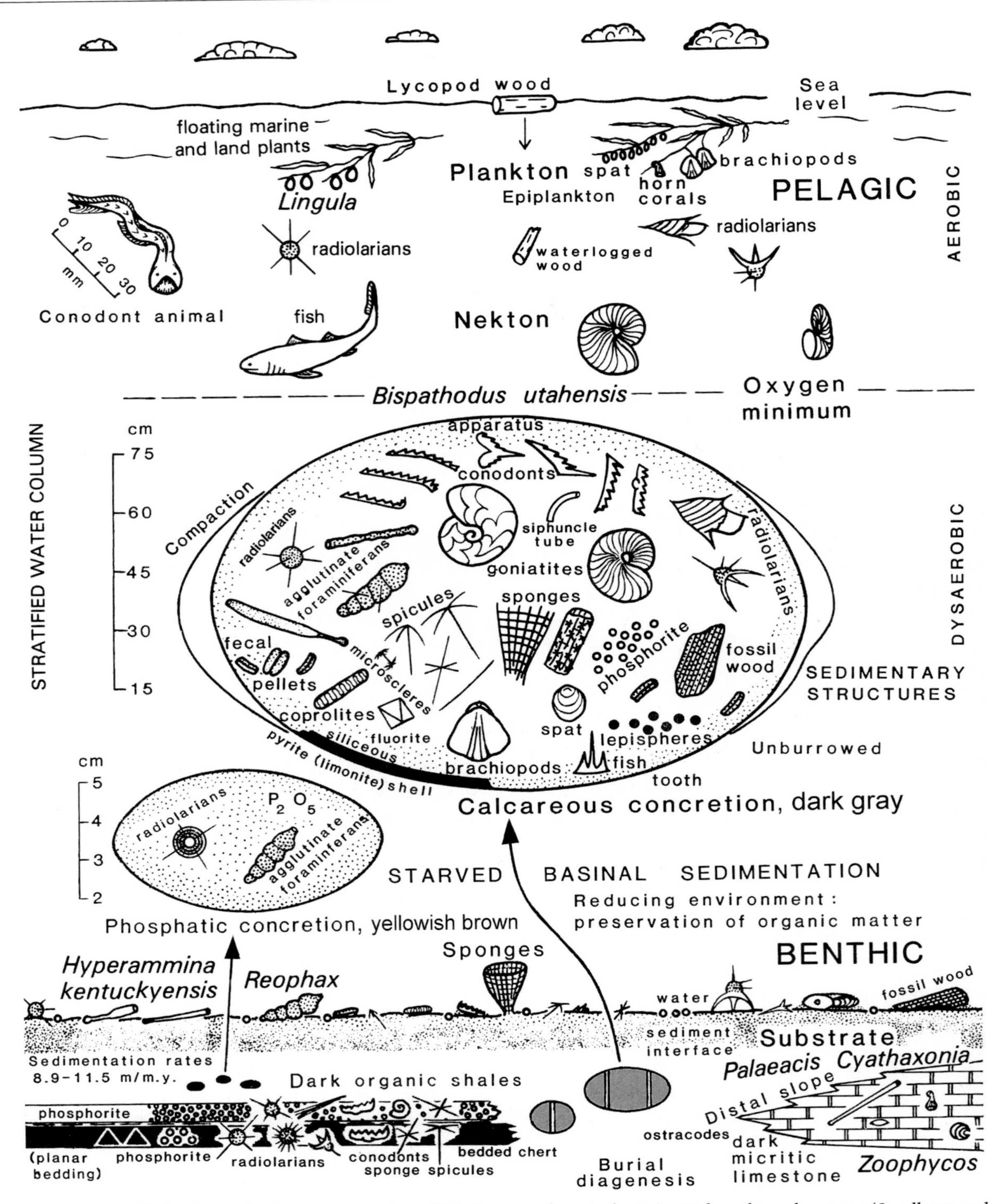

Fig. 8.33. Detailed paleoecologic reconstruction of life in a mid-Paleozoic Sea with anoxic bottom waters in the Devonian Delle Phosphate Member of the Woodman Formation and equivalents in Utah and nearby states (Sandberg and Gutschick 1984, Fig. 13)

shales. This implies that the bottoms of black shale basins were overloaded with organic material – that oxygen demand far exceeded oxygen supply – and that terrigenous dilution was small. Organic-rich and clay-rich laminations in black shales are finely interlaminated (Fig. 8.32), and weather differentially, with the organic-rich layers being significantly more resistant. Terrigenous organic matter may be abundant even in marine black shales, because wood, spores and palynomorphs may float far from shore before settling to the bottom when finally waterlogged. Thus the sedimentologic study of this fine-grained plant material and its kerogen types can help estimate distance to shoreline and the environment (Tyson 1995, Table 1.6). Pyrite is almost universal in black shales and occurs as fine cubes, framboids, or irregular masses. It may also replace fossils, may be present as thin laminations and exceptionally is present as granule lags. Such lags provide evidence of erosional breaks. Exceptionally, black shales can even be interbedded with thin beds of chert or evaporites.

Fossil content of Phanerozoic black shales varies widely depending upon dissolved oxygen and the age of the deposit, but benthic fauna is always very specialized with low diversity but can be exceptionally abundant on single bedding planes. The quality of preservation of floating invertebrates and vertebrates can be exceptional, because bottom anoxia minimizes predators (Fig. 8.33). The pelagic fauna of black shales changes with their age. Graptolites are most abundant in Cambrian through Silurian black shales. In many Devonian and later black shales, however, woody material, pollen, spores, ostracodes, and conodonts are present as well as some algae. Beginning in the Cretaceous and later, a pelagic microfauna of dinoflagellates, phytoplankton, coccoliths, and pelagic foraminifera and algae, especially diatoms, became abundant. In almost all black shales, benthic organisms are rare except for trace fossils made by small, soft-bodied organisms, which lived in temporarily oxygenated bottom waters. Such oxygenated waters, introduced by weak distal turbidity currents or by influx of colder or more saline surface waters to the bottom, deposited gray mudstone interbeds. These interbeds have a specialized bottom fauna of opportunists, the so-called "doomed pioneers," that temporarily lived on the bottom and burrowed into the underlying nutrient-rich, fetid black mud trying not to suffocate.

Many black shales have both laminated and non-laminated beds and X-ray radiography nearly always shows much bioturbation in seemingly massive beds of black shale or mud (and less commonly evidence of flowage). Hence most black shales in reality represent alternations of anoxic with dysoxic bottom waters, which were by no means stagnant as shown by the micro cross-lamination and sole marks of many of its thin siltstones, by their oriented wood, graptolites, and ostracodes and by lags of pyrite and phosphate nodules. Typically, all these features reveal consistent rather than random current patterns (see Chap. 3). Add to this the low- relief, erosional, shale-on-shale scours that are present even in the centers of muddy basins far from shore during low stands (see Table 8.1) and it is clear that the anoxic bottom waters of black shale basins were anything but stagnant. Such anoxic environments led to rich concentrations of metals so that some very mineralized black shales are called *metalliferous black shales*. In addition, black shales provide a treasure trove for both inorganic and organic biogeochemists (See Peters and Moldowan 1993, Chap. 1). Consequently, they have a vast geochemical literature. They have also been studied in great detail paleontologically (Fig. 8.33).

Black shales form at all water depths and have been deposited on stable continental crust (where some are exceptionally widespread), on continental borderlands, and on oceanic crust. Thickness varies from less than a meter to 100 s of meters or more depending on tectonic setting. The broad factors responsible for the deposition of organic-rich shales are now widely agreed upon (Table 8.8) and five major models exist. The model of restricted circulation on a shallow shelf is probably the most common, because shelfal black shales seem to be the most abundant in the geological record. When such a shelf is initially flooded, most sand and silt are trapped shoreward in paleovalley estuaries but the clay and the organic matter are concentrated offshore, especially early in the transgression. Thus the lowest part of a transgressive black shale is the most clay rich and commonly has the highest gamma-ray signature (Fig. 8.14). Because the basin is so sediment starved, thicknesses range from a meter or less to a few tens of meters. Additionally, on a wide shallow shelf, bottom friction progressively reduces inshore wave power favoring stratification of the water column. Thus the wider and shallower a shelf, or the longer and narrower an embayment, the more likely bottom anoxia, which develops first in closed topographic lows or in canyons and paleovalleys cut during the prior lowstand. From such lows – be they a basin center, a small subsiding syncline in an intracratonic basin, or a rapidly subsiding graben in a flooded rift system – anoxia spreads outward. Here bottom currents are minimal in these specially protected environments

Table 8.8. Dark, organic-rich shales

Part A – Responsible Factors
Oceanographic and Biologic
Stratified water mass in silled basin effectively isolated from turbulent, well-oxygenated surface water (*preservation model*)
Excess organic productivity in surface water (upwelling) causes organic overloading on lake or sea bottom even in well mixed waters *(productivity model)*
Excess terrestrial organic matter in lakes/lagoons *(productivity model)*
Geologic/Geomorphic/Climatic
Deep basins below storm wave base (tectonics/global sea level highs/minimal sediment input)
Shallow, stagnant, stratified lagoons, swamps, oxbows and lakes
Globally warm climates
Sedimentologic
Clay, fine silt and organic debris all have low settling velocities and accumulate together in topographic lows in either deep or shallow water

Part B, End Member Models
Silled Basins
Most common (senso lato includes all lakes), range widely in size, and all have barriers (tectonic ridges, reefs, glacial moraines/scoured bed rock, etc.) that restrict circulation favoring stratification of the water column. Examples include Black Sea, Cariacao Basin (Venezuela), Norwegian fjords, borderland basins of California, and lakes in rift systems
Shelf/Slope Basins
Oxygen minimum, topographic lows such as shelf canyons and channels plus initial flooding of shelf (creates temporary "starved basin") all favored by enhanced summer time and Greenhouse, anoxia. Examples include Louisiana Texas shelf of Gulf Coast, Arabian Sea, Pacific coast of Peru, etc.
Deep Water
Overloading of bottom by enhanced productivity favored by mixing of polar and warm surface water (Greenhouse climate). Many modern examples in Stow et al. (2001) and best ancient examples are Cenomanian-Turonian black shales of ancestral North Atlantic

(this is a variant of the silled basin model of Fig. 4.7). Warm global climates also help – the flow of cold bottom polar waters toward the equator decreases at such times so that medium and low latitude ocean water is warmed and thus contains less dissolved oxygen than does cold water.

Interbedded lithologies include thin, distal turbidite siltstones and sandstones plus possible pelagic carbonates. Traced over a regional arch, black shales thin or lap out and, conversely, they thicken into cratonic and pericratonic basins and especially into bordering foreland basins. Updip toward a terrigenous source, black shales typically split and pass into gray mudstones and siltstones. Ettensohn (1998) has emphasized the importance of subsidence for providing the needed accommodation in foreland and rift basins to become repositories for thick, organic-rich sedimentation.

Paleogeography is still another factor, and perhaps the most critical one, in the formation of black shales. Consider, for example, the Jurassic black shales of the North Sea, the Cretaceous black shales of the deep Atlantic, those of the western Interior Seaway of North America and the Jurassic and Cretaceous black shales of the Western Siberian Basin, the great Mesozoic and Tertiary basin that underlies much of the watershed of the Ob River. What all four have in common is north-south elongation that extends from high to mid- or almost low latitudes. This permits cold polar, nutrient-rich water to enter, rise and warm at the surface, and produce high surface productivity that created bottom anoxia in long narrow arms of the sea.

Surely there are many more different examples of paleogeography affecting the formation of black shales and other mudstones. For example, consider the onset of a sill at the entrance of a long arm of the sea. Sills restrict circulation in both shallow and deep-water basins and may result from reefs or tectonism such as faulting or the development of broad, regional arches. Where the resulting basin is shallow and the climate arid, evaporites, organic carbonates, and black shales can all be deposited in the same basin as in the Paradox Basin of Utah. Where the sill creates a deep basin – a fjord or a tectonic basin along an active margin or in a continental borderland, deep water, organic-rich facies will predominate with thin interbedded distal turbidites and thicknesses may be several 100 s to several 1,000 s of meters, because accommodation is great. In this setting chert interbeds result from rainouts of siliceous pelagic microfauna such as diatoms and radiolarians (and here the black shales will be siliceous). Upwelling occurs where deep, nutrient-rich, cold water is brought to the surface photic zone creating a productivity bloom. As this pelagic debris settles to the bottom in great abundance, it overwhelms oxygen supply and either dysoxia or anoxia results.

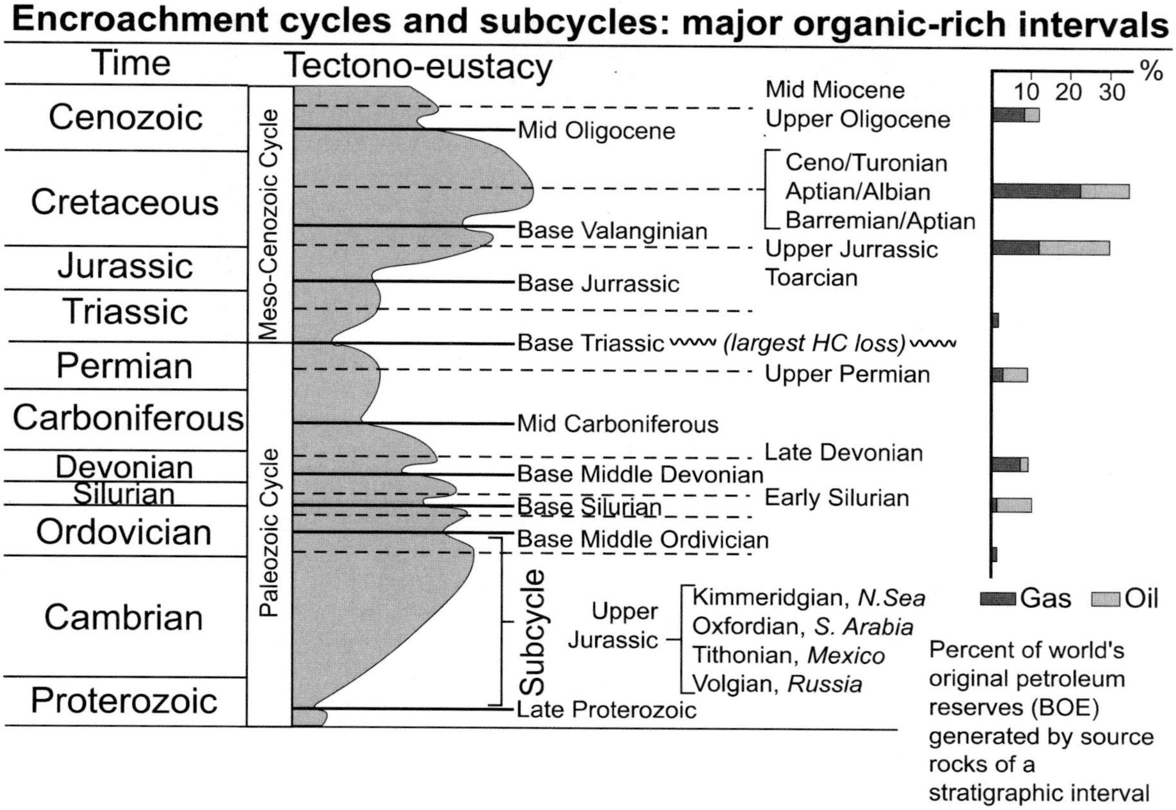

Fig. 8.34. Major times of black shale deposition seem to have occurred in second order cycles, when world sea levels were rising and shorelines retreating (Duval et al. 1998, Fig. 4). Published by permission of the authors and the Society for Sedimentary Geology

Ancient examples abound. The thin black marine shales of the Pennsylvanian coal measures of the Illinois Basin and those of Mid-continent Basin (Heckel 1991) in the United States are classic examples as are the Devonian-Mississippian black shales of North America (Ettensohn 1997 and 1998) and South America (Moretti et al. 1995). The Cambrian Alum Shale of Scandinavia (Thickpenny and Leggett 1987) and the Jurassic Kimmeridge Shale of the North Sea (Tyson 1996; Cooper et al. 1995a) are others that differ in important ways. The Cretaceous La Luna Formation, the source bed of much of Venezuela's oil, is summarized by Palaios (2003) in Digging Deeper. Although not all are organic rich, many of mudstones of the interior Cretaceous sea of North America such as the Mowry Shale provide additional insights to the great variety of black shales. See also the review by Arthur and Sageman (1994).

Above we have discussed black shales from the viewpoint of "day in day out" processes that form black shales. But what about possible special events in earth history that may have formed widespread black shales? There seem to be at least six stratigraphic intervals covering about 30% of Phanerozoic time that are rich in black shales (Klemme and Ulmishek 1991; Duval et al. 1998). These times appear to be closely related to worldwide intervals when shorelines were displaced far inland by worldwide high stands (Fig. 8.34). Two notable but short intervals in the Cretaceous provide another example. First in the Aptian and Albian and later in the Cenomanian and Turonian black shales were deposited in the Atlantic Ocean at mid depths (See Wignall 1994, pp 76–79 in Digging Deeper), where today the Atlantic is mildly oxidizing (Oceanic Anoxic Event, OAE). A positive carbon isotope excursion occurred when this black shale deposition started. This was also a time of worldwide high temperatures (the term "greenhouse state" is popular for such periods).

The red mudstones of red bed sequences represent the other extreme, and their interpretation has been just as controversial as that of black shales. Formerly, deserts were invoked, whereas today most workers view red mudstone color as the result of prolonged exposure to oxidizing soil- and groundwater that converts Fe^{++} to Fe^{+++}. (But remember that in the deep ocean, it is mildly oxidizing bottom water and lack of organic matter that forms the red clay deposits that accumulated slowly on the bottom far from a terrigenous source. See Color in Chap. 4). Oxygenated ground waters are likely when the continents were well above sea level (Triassic/Tertiary) or late in an orogenic cycle when uplift produces Himalayan style highlands with many high intermontane basins or along the borders of some rift basins, especially, when such events occur under a semiarid or arid climate. Well-studied examples of red bed sequences emphasizing their mudstones include the Ordovician Queenston delta that terminates in an arid muddy shoreline that grades into offshore carbonates (Brogly et al. 1998), the 3,000 m thick Triassic mudstones of the East Irish Sea Basin Wales (Wilson 1990), and the Dunkard Group of Pennsylvanian to Permian age in West Virginia and Ohio (Martin 1998). The famous Old and New Red Sandstones and the widespread Triassic redbeds of Europe also contain much reddish brown mudstone. In broad perspective, the contrast of black versus red colors goes back to tectonics and climate – poorly oxygenated sediment traps in topographic lows (many of which are directly related to subsidence tectonics) for black shales versus oxygenated groundwater in continental basins with arid and semiarid climates for red mudstones.

Box 8.2. Milankovitch cycles

Variations in the Earth's distance and inclination to the Sun, Moon and nearby planets were proposed by Milutin Milankovitch in 1941 to explain the alternating ice ages of the Pleistocene. He reasoned that these orbital variations caused changes in the solar energy (insolation) intercepted by the Earth and thus in its global temperature. Today, these orbital variations are known as Milankovitch cycles of which there are three – eccentricity (404, 128, and 98 ka), obliquity (54, 41 and 29 ka) and precession (24, 22 and 19 ka). These frequencies are collectively known as the *Milankovitch parameters*. The term "pace maker" has been proposed for them, because they appear to modulate or pulse much fine-grained sedimentation (both terrigenous and carbonate) below wave base. Thus we should think of the Milankovitch parameters as exterior modulators of sedimentation.

How are Milankovitch cycles identified? You can simply use your eye to count cycle thickness based on grain size, color, etc. and, knowing the limiting ages of the section, try to identify the Milankovitch parameters (Gilbert did this successfully in the Green River oil shale over 100 years ago). A *time series analysis* can be applied to such datasets as well. Among the many variables studied are Si/Al (a proxy for grain size), $\delta^{18}O$, bioturbation, clay mineralogy, gamma radiation, magnetic susceptibility, etc. Once this is done, a *power spectrum* calculated from the time series analysis is obtained for all the frequencies. The power spectrum shows the relative importance of each frequency (the sum of the areas under all the peaks equals the total variance of all the frequencies). The great advantage of the power spectrum is that it provides a graphic display of the relative importance of the Milankovitch frequencies for each variable. When multiple variables are studied in a section, a comparison of their spectra shows their differential response to the Milankovitch parameters. This facilitates a sedimentologic explanation of the bedding cyclicity. For a successful analysis, a section free of breaks is required, and the more points and the more closely they are spaced, the better the resolution of the power spectrum. Because time series analysis requires a fixed sampling interval, it is customary to smooth the data and record values at fixed intervals.

In the Plio-Pleistocene a combination of radiometric dating, magnetic stratigraphy and micropaleontology permit us to use an absolute time scale. But in older sediments, absolute dating is much less precise and the relative proportions of the peaks of the power spectrum are used. When these have ratios close to those of today, 1:5:20, orbital control can be inferred. The word "tuning" is commonly encountered in the Milankovitch studies and refers to the comparison or matching of an observed frequency to an astronomically calculated frequency.

See Hinnov (2003) and Weedon (1993) for good introductions to Milankovitch cycles, Fisher (1993) for a detailed analysis in rocks of the Cretaceous seaway of North America, and Schwarzacher (1999) for a thoughtful comparison of Milankovitch cycles to the parasequences of sequence stratigraphy.

8.8. Milankovitch Cyclicity and Mudstone Deposition

There is a large literature about Milankovitch cyclicity in sedimentation and sedimentary rocks. These are orbital cycles of 19 through 404 ka that cause recurring changes in solar energy intercepted by the Earth (Box 8.2). Milankovitch cycles have been called the "pacemaker" of cyclic sedimentation and stratigraphy, because they are widely considered to be the principal causes of the rhythmic deposition seen in many fine-grained deposits (excluding the annual cycle of glacial varves and the shorter cycles in tidal rhythmites). Examples of such rhythmic couplets include claystones-marlstones, marlstones-limestones, shales-siltstones, claystones-diatomites, and claystones-evaporites. A changing clay flux is the most common denominator in these pairs. But how could a changing clay flux be related to cycles of solar energy intercepted by the earth? Although much remains to be learned, the broad argument is as follows: changes in intercepted solar energy alter average global temperatures (bottom water temperatures and the position of the jet stream), changing atmospheric weather patterns and dust flux. Moreover, the positions and intensities of major ocean currents change, and consequently, worldwide patterns of seasonality and rainfall. It is seasonality and rainfall that directly alter mud production (weathering rates) and the erosion and transport of mud to a quiet water trap.

We also remember that higher surface water temperatures promote higher organic productivity in the photic zone – more pelagic rainouts – and thus enhancement of bottom dysoxia and anoxia. So there are also changing fluxes of organic matter and carbonate (estimated from Ca/Fe) to the bottom. Thus there are many parameters measurable in mudstones that should show Milankovitch patterns, especially through the Mesozoic and Cenozoic (Fig. 8.35). Notable in Fig. 8.35 are: (1) the presence of Milankovitch

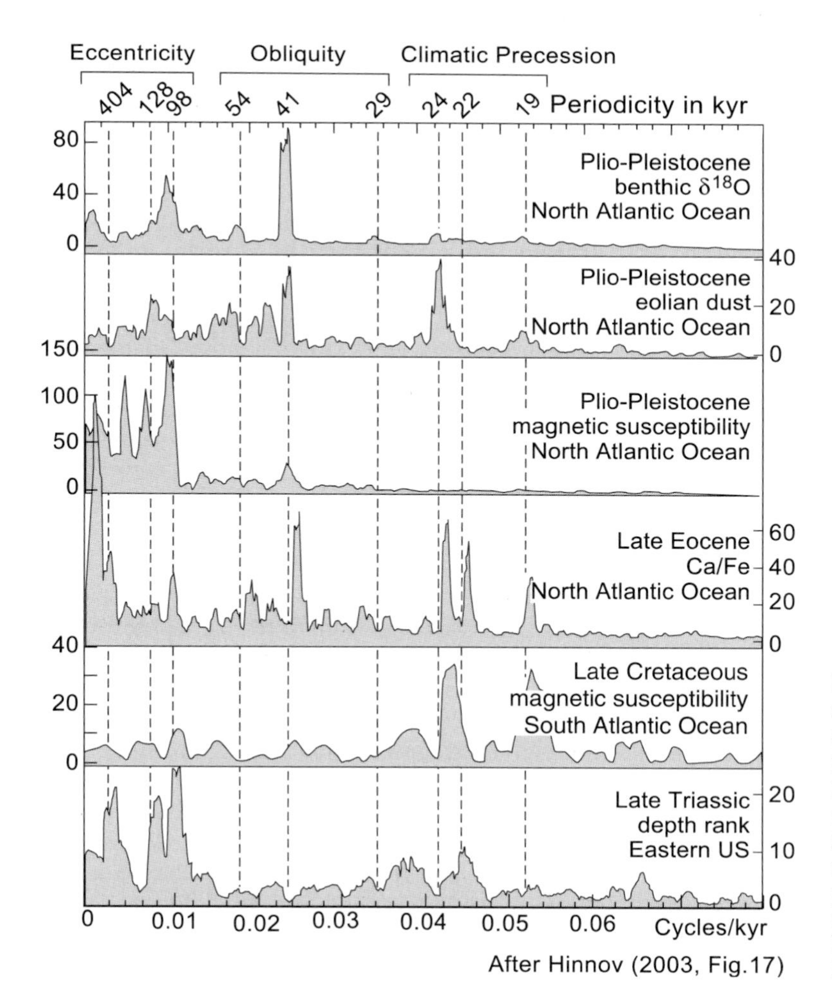

Fig. 8.35. Milankovitch signals recorded in sediments ranging in age from Late Triassic to the Plio-Pleistocene (after Hinnov 2003, Fig. M17). These spectra, although unevenly developed because of different variables and different ages, nonetheless show a remarkable expression of orbital frequencies through 200 Ma of geologic history. Published by permission of the author and Kluwer/Academic Publishers

Fig. 8.36. One complete Milankovitch cycle (van Houten Cycle) well displayed in the Triassic New Berlin Formation at New Berlin, Conn., USA. (Photo courtesy of Paul Olsen). *Arrow* represents one complete cycle (*R* is red sandstone, *G* is gray sandstone, and *B* is grayish black mudstone)

frequencies in a variety of environments from lakes to the deep sea; (2) many kinds of measurements yield consistent results, reinforcing our confidence in orbital mechanisms for the patterns seen; and (3) different parameters seem to be better at revealing different frequencies of the orbital record. Hence, the term pacemaker for Milankovitch cycles seems most appropriate.

Judged by their timing and thickness, most Milankovitch cycles are longer in duration and thicker than those of varves and tidalites, but briefer and thinner than those of most parasequences. Thus they have a definite place in studies of cyclicity and sedimentation – and especially for students of mudstones. Below are two examples of such studies.

Eastern North America in the Late Triassic (Olsen and Kent 1999) had a series of half-graben rifts that accumulated thick successions of lake deposits (Fig. 5.23). Cores through these lake deposits have provided an unsurpassed record of climate signals for this period in Earth history (Fig. 8.36). The lake facies vary from deep, perennial lacustrine to playa. They are interpreted to reflect flooding and drying in response to wet-dry climate cycles. Apparently these Triassic lakes were poised at the ideal combination of accommodation, mud flux, and water flux to make them faithful recorders of the orbital climate signal. All of the Milankovitch cycles have been identified in these lake beds, plus a 3 million year signal. Significant for the study of mudstones is the response of color: the wet-climate lakes deposited black and gray mudstones, whereas under dry climate conditions, red and purple colors predominated. Thus simple color provides a quick proxy for the Milankovitch signals.

The Jurassic Pliensbachian Banded Shales of Yorkshire are about 30 m thick with many couplets of silt-rich (lighter) and clay-rich (darker) evenly laminated shales (van Buchem et al. 1994). These shales contain marine fossils, are bioturbated and were deposited on a dysaerobic shelf. A 10 m subset of the section was studied in detail. In this interval there are 64 couplets that vary from 10 to 70 cm in thickness. Data on remnant magnetism, total organic context, the Si/Al ratio (a measure of silt-clay proportions) and gamma radiation were collected at 5 cm intervals, smoothed and analyzed using time series. The Si/Al ratio, grain size and sedimentary structures and organic matter appear to be related to the precession and obliquity cycles (stormy periods), whereas clay mineralogy appears to be related to the longer eccentricity cycle (gradual changes in the weathering of the source rocks that supplied clay to the basin). To sum up, different parameters in a mudstone section provide different insights into controlling variables. Thus the careful pursuit of a whole battery of measurements through the length of a good core or outcrop pays great dividends in understanding the geological history of its basin of deposition and its source area.

8.9. Mudstone Through Time

Long-term changes in mudstones through time have received some attention, but not the systematic study that they deserve. These changes seem to be of two types – gradual and abrupt. Below we summarize what is known, suggest some additional possible long-term changes, and briefly comment on some of the controlling factors. We think these are of three kinds: (1) those related to earth surface processes such as climate, atmospheric oxygen and oceanic circulation; (2) those related to crustal processes such as burial and plate migration; and (3), perhaps most important of all, changes related to organic evolution. See, for example, Seilacher (1994) in Digging Deeper for the change from biomats to bioturbation.

The best-defined and understood trends are mineralogical, and to a lesser degree, chemical. The mineralogical changes of mudstones are gradual and are driven by increasing burial temperatures, which transform a disequilibrium mixture of unweathered and weathered detritus at the earth's surface into a new mixture at equilibrium with the much higher (> 100 °C) temperatures of deep burial (diverse clay minerals → illite-chlorite, enhanced illite crystallinity; amorphous silica → opal CT → quartz; and Ca-plagioclase → albite). A chemical change that parallels these mineralogical changes is an increase in the potassium content of older mudstones, which suggests that post-depositional addition of K is needed to drive some of the mineralogic changes. See Ziegler and Longstaffe (2000) for an example of significant potassium addition, an event that seems to occur in most older basins. The biggest chemical change in mudstones occurs across the Archean-Proterozoic boundary, which is marked by a dramatic shift from mafic-sourced muds with high Ni and Cr and virtually no Eu anomaly (see Box 7.1) to felsic sourced muds much lower in Ni and Cr and with a pronounced negative anomaly (Taylor and McLennan 1985, Chapter 5).

Organic evolution has also had important effects on mudstones. The most dramatic change is the appearance of free oxygen in the atmosphere at about 2.3 Ga (Holland 2002), which is surely related to organisms, perhaps through changes in oceanic phosphate (Bjerrum and Canfield 2002). On land, this change is marked by the appearance of thick, well-developed red-beds, so some oxidized mud would have been transported into the marine realm beginning at this time. In the younger record, the kinds, abundances and depths of bioturbation have changed since the first appearance of benthic organisms in the Neoproterozoic (Ausich and Bottjer 2001; Droser et al. 2001). The calcium content of mudstones increases sharply in the Paleozoic (Garrels and MacKenzie 1971, Fig. 9.5), presumably because of the appearance of shelled organisms. And starting in the late Mesozoic, the main locus of carbonate deposition shifted from shelves to the deep sea owing to the appearance of pelagic organisms with calcite shells like *Globigerina*. The result is the appearance of abundant deep-water marlstones in the geologic record, and a small shift upwards in the calcium content of shales. Silica increases steadily through the Phanerozoic, possibly in response to the successive evolution of more efficient siliceous organisms in the sequence sponges → radiolarians → diatoms. A major influence on mud production is likely to have been the evolution of large land plants in the Devonian (Algeo et al. 2001). The appearance of trees had a similar consequence to the development of deeper bioturbation in the marine realm – trees growing on soils mixed and aerated it to greater depths. Thus weathering rates increased at least until a new equilibrium soil profile was developed. Consequently, probably more mud was delivered to the ocean and there was a spike in nutrient delivery that led to widespread anoxia in the Late Devonian.

Important controls on the above changes are those that affected the earth surface processes of weathering and carbonate and organic production in the water column (organic evolution). Other events that might initiate such changes are oscillations between Greenhouse and Icehouse climate regimes (wet and warm climates and stagnant oceans versus cold and dry climates and energetic oceans); the distribution and collision of major continental plates as they migrate and open or close pathways for warm water to move to the poles; the evolution of grasses; the appearance of strongly bioturbating organisms like termites; and many more. Such events and processes should have provoked changes in the record of mudstones, but we are aware of no convincing documentation of any effects. Perhaps it is time to repeat Ronov's work (1983 and 1992) on the composition of the rock record using new techniques and a broader sampling base.

8.10. Summary

We argue that the study of mudstones in and across a basin has parallel importance to that of their companion limestones, dolostones and sandstones. Moreover, for a number of key questions, only the mudstones supply the answers. Suppose for a moment that we are presented with the task of exploring a totally unknown sedimentary basin, but have no knowledge of the geology, geochemistry or geophysics of mudstones. Then it seems to us that the following important questions could not be usefully answered or, in some cases, even addressed. Think here of an astronaut landing on a new earth-like planet...

- *Bottom water oxygenation* – Interpretation of degree of anoxia of the bottom water in a basin depends totally on examination of its mudstones – their lamination, color, TOC, bottom fauna, iron and manganese minerals.
- *Compaction* – Compaction could not be predicted and hence burial history curves would not be reliable; overpressuring would be a total mystery (think of the rig blowouts!); and all deformation would be attributed to deep crustal movements.
- *Correlation* – Mudstones usually have much greater lateral continuity than carbonates or sandstones and so form the best correlation framework. Imagine attempting to unravel a coal-measure section like that of the Appalachians with no knowledge of marine mudstone bands or of ash beds.
- *Thermal history* – The best thermal indicators are concentrated in the mudstones, and any attempt to locate the oil window from the other lithologies seems doomed to failure.
- *Seals* – Evaluation of reservoir quality for sandstones would be made even more difficult without an ability to evaluate the quality (extent, thickness, fracturing, bioturbation) of the mudstone seal. The same would be true of landfills. Suppose the relationship of swelling properties to clay mineralogy were unknown!
- *Mudstone-hosted ores* – The largest deposits of Pb and Zn and some of the most important Cu deposits are hosted by mudstones. Many have surface expressions that can be stumbled upon (Red Dog, Roan Antelope), but more often they are blind and cannot be located without detailed knowledge of the stratigraphy and structure of the surrounding mudstones.
- *Industrial minerals* – Never glamorous, but so essential for a pioneering colony on a distant planet taking the first steps to self-sufficiency.

8.11. Challenges

The mudstones of fore-, intra-, and backarc basins

Here is truly an overlooked area of study for the ambitious, because to date most research on these basins has been directed to their tectonic evolution and structure rather than their fill (with mudstones far behind as usual ...).

We have suggested in this chapter that mudstone types, unlike sandstone types, are largely independent of the tectonic setting of the basin in which they were deposited.

Would research really support this idea?

How does mudstone mineralogy and chemistry vary with age?

A renewed effort here, one building on earlier studies, seems most rewarding to us and should begin with a systematic review of the ever growing literature as more and more new instrumental techniques are applied to sedimentary rocks. As the most abundant sedimentary rock and the end product of weathering, mudstones should be the principal repository of information about an evolving earth.

Correlation of thin and ultra thin layers in mudstones.

The lateral persistence of a sedimentary unit, be it great or small, is one of its more important characteristics and also one with great value for stratigraphy. The examples cited in this chapter suggest much more could be learned about basin filling, if we knew how far thin units in mudstones can extend and how common they are.

Mudstones as baffles and seals to fluid flow.

Ten years ago there was very little published on this topic and today we have a good start, but we think there are many more opportunities that combine sequence stratigraphy, geometry, petrology, and a few key engineering measurements. To this list should be added the seismic assessment of seals at the scales of both basins and individual reservoirs.

Thicknesses of mudstones in different environments.

The only systematic compilation we are aware of is that of Galloway and McGara (1982, Fig. 2) for the Tertiary of the Gulf Coast. This information is easy to obtain from logs and surely is a fundamental characteristic of mudstones that deserves far more attention.

Radiometric dating of mudstones.

Good radiometric dating of mudstones is one of the greatest needs for better understanding their role in basin filling. Think of how direct dating of a mudstone would help correlation and thus sequence stratigraphy and tectonic subsidence analysis. Or think of dating fluid events with authigenic overgrowths? Imagine what the routine application of a nano SHRIMP would do for mudstones.

References

Ager DV (1975) The Jurassic World Ocean. In: Finstad KG (ed) Jurassic Northern North Sea Symposium Proceedings. Norsk Petroleumsforening (NPF), p 1.1–1.43

Algeo TJ, Hoffmann DL, Maynard JB, Joachimski MM, Hower JC, Watney WL (1997) Environmental reconstruction of anoxic marine core black shales of Upper Pennsylvanian Midcontinent cyclothems. In: Algeo TJ, Maynard J Barry (Co-Leaders) Cyclic Sedimentation of Appalachian Devonian and Midcontinent Pennsylvanian Black Shale: Analysis of Ancient Marine Anoxic Systems. American Association Petroleum Geologists, Eastern Section/Society of Organic Petrology, Lexington, KY, Sept 27–28

Algeo TJ, Scheckler SE, Maynard JB (2001) Effects of early vascular land plants on weathering processes and global chemical fluxes during the Middle and Late Devonian. In: Gensel P, Edwards D (eds) Plants Invade the Land: Evolutionary and Environmental Perspectives. Columbia University Press, New York, pp 213–236

Almon WR (1981) Depositional environment and diagenesis of the Permian Rotliegend in the Dutch Sector of the North Sea. In: Longstaffe FJ (ed) Clays and the Resource Geologist, Short Course Handbook. Mineralogists Association Canada, 7, pp 119–147

Aoyagi K, Iijima (1987) Petroleum occurrence, generation, and accumulation in the Miocene siliceous rocks of Japan. In: Hein JR (ed) Siliceous Sedimentary Rock-hosted Ores and Petroleum. Van Nostrand Reinhold, New York, pp 117–137

Armentrout JM (1983) Glacial lithofacies of the Neogene Yakataga Formation, Robinson Mountains of southern Alaska Coast Range, Alaska. In: Molina BF (ed) Glacial-Marine Sedimentation. Plenum Press, New York, pp 421–449

Arthur MA, Sageman BB (1994) Marine black shales. Annual Review Earth Planetary Sciences 22, pp 499–551

Ausich WI, Bottjer DJ (2001) Sessile invertebrates. In: Briggs DEG, Crowther PR (eds) Paleobiology II. Blackwell, Oxford, pp 384–386

Barron JA (1987) Diatomite: environmental and geologic factors affecting its distribution. In: Hein JR (ed) Siliceous Sedimentary Rock-hosted Ores and Petroleum. Van Nostrand Reinhold, New York, pp 164–180

Bellini E, Massa D (1980) A stratigraphic contribution to the Paleozoic of the southern basins of Libya. In: Salem MJ, Busrewil MT (eds) Symposium on the Geology of the Libya 1. Academic Press, London, pp 3–56

Bennington JB (1996) Stratigraphic and biofacies patterns in the Middle Pennsylvanian Magoffin Marine unit in the Appalachian Basin, USA. Int J Coal Geol 31:169–193

Bjerrum CJ, Canfield DE (2002) Ocean productivity before about 1.9 Gyr ago limited by phosphorus adsorption onto iron oxides. Nature 417:159–162

Bloch JD, Schroeder-Adams CJ, Leicke DA, Craig J, McIntyre DJ (1999) Sedimentology, micropaleontology, geochemistry and hydrocarbon potential of shale from the Cretaceous Lower Colorado Group in western Canada. Geol Survey Can Bull 531:185

Bohacs KM (1998a) Introduction: mudrock sedimentology and stratigraphy – challenges at the basin to local scales. In: Schieber J, Zimmerle W, Sethi PS (eds) Shales and Mudstones. E. Schweizerbart'sche, Stuttgart, 1, pp 13–20

Bohacs KM (1998b) Contrasting expressions of depositional sequences in mudrocks from marine to non-marine environs. In: Schieber J, Zimmerle W, Sethi PS (eds) Shales and Mudstones. E. Schweizerbart'sche, Stuttgart, 1, pp 33–78

Bohacs KM, Schwalbach JR (1992) Sequence stratigraphy of fine-grained rocks with special reference to the Monterey Formation. In: Schwalbach JR, Bohacs KM (eds) Sequence Stratigraphy in Fine-grained Rocks: Examples from the Monterey Formation. SEPM Pacific Section Guidebook, Bakersfield 70:7–20

Brett CE, Taylor WL (1996) The Homocrinus Beds: Silurian Crinoid Lagersätten of western New York and southern Ontario. In: Brett CE, Baird GC (eds) Paleontological Events: Stratigraphic, Ecological, and Evolutionary Implications. Columbia University Press, New York, pp 181–223

Brogly PJ, Martini IP, Middleton GV (1998) The Queenston Formation: shale- dominated, mixed terrigenous-carbonate deposits of Upper Ordovician, semi-arid muddy shores in Ontario, Canada. Can J Earth Sci 35:702–719

Brown LF, Fischer WL (1977) Seismic-stratigraphic interpretation of depositional systems: examples from Brazilian rift and pull-apart basins. In: Payton CE (ed) Seismic Stratigraphy – Applications to Hydrocarbon Exploration. Am Assoc Petrol Geol Memoir 26:213–248

Caldwell WGE, Kauffman EG (eds) (1993) Evolution of the Western Interior-Basin. Geological Survey Canada (Special Paper) 39:680

Calver MA (1968) Distribution of Westphalian faunas in northern England and adjacent areas. Proc Yorkshire Geol Soc 37:1–72

Catuneanu O (2002) Sequence stratigraphy of clastic systems: concepts, merits, and pitfalls. J African Earth Sci 35:1–43

Chamley H (1989) Clay Sedimentology. Springer, Berlin Heidelberg New York, 623 p

Cole DI, Christie ADM (1994) A paleoenvironmental study of black mudrock in the glacigenic Dwyka Group from the Boshof-Hertzogville regions, northern part of the Karoo Basin, South Africa. In: Deynoux M, Miller JMG, Domack EW, Eyles N, Fairchild IJ, Young GM (eds) Earth's Glacial Record. Cambridge University Press, Cambridge, pp 204–214

Cooper BS, Barnard PC, Telnaes N (1995a) The Kimmeridge Clay Formation of the North Sea. In: Katz BJ (ed) Petroleum Source Rocks. Springer, Berlin Heidelberg New York

Cooper MA, Addison FT, Alvarez R, Coral M, Graham RH, Hayward AB, Howe S, Martinez J, Naar J, Peñas R, Pulham AJ, Taborda A (1995b) Basin development and tectonic history of the Llanos Basin, Eastern Cordillera, and Middle Magdalena Valley, Colombia. Am Assoc Petrol Geol Bull 79:1421–1443

Cornford C (1990) Source rocks and hydrocarbons of the North Sea. In: Glennie KW (ed) Introduction to the Petroleum Geology of the North Sea; 3rd edn. Blackwell, Oxford, pp 294–361

Crowell JC (1982) The tectonics of Ridge Basin, southern California. In: Crowell JC, Link MH (eds) Geologic History of Ridge Basin Southern California. Society Economic Paleontologists Mineralogists, Pacific Section, Book 22, pp 21–41

Crowell JC (ed) (2003) Evolution of Ridge Basin, Southern California: An Interplay of Sedimentation and Tectonics. Geological Society America (Special Paper) 367:247

de Batist M, de Bruyne H, Henriet JP, Mostaert F (1984) Stratigraphic analysis of the Ypresian off the Belgian coast. In: Henriet JP, de Moor G, de Batist M (eds) The Quaternary and Tertiary Geology of the Southern Bight, North Sea. Belgian Geological Survey Free University Brussels (Earth Sciences), pp 75–88

De Graciansky P-C, Deroo JP, Herbin JP, Jacquin T, Magniez F, Montadert L, Müller C, Ponsot C, Schaaf A, Sigal J (1986) Ocean-wide stagnation episodes in the Late Cretaceous. Geologische Rundschau 75:17–41

De Graciansky P-C, Hardenbol J, Jacquim T, Vail PR (eds) (1998) Mesozoic and Cenozoic Sequence Stratigraphy of European Basins. Soc Sediment Geol (SEPM) Special Publication 60:781

Dickinson WR (1989) Provenance and sediment dispersal in relation to paleotectonics and paleogeograpy of sedimentary basins. In: Kleinspehn KI, Paola C (eds) New Perspectives in Basin Analysis. Springer, Berlin Heidelberg New York, pp 3–25

Dickinson WR (1995) Forearc basins. In: Busby CJ, Ingersoll RV (eds) Tectonics of Sedimentary Basins. Blackwell, Oxford, pp 221–261

Dunham JB, Cotton-Thornton ML (1990) Lithology of the Monterey Formation in the Santa Maria Valley Field, Santa Maria Basin, California. In: Keller MA, McGowen MK (eds) Miocene and Oligocene Petroleum Reservoirs of the Santa Maria and Santa Barbara-Ventura Basins, California. SEPM Core Workshop No. 14, San Francisco, June 3 1990, pp 203–222

Duval BC, Cramez C, Vail PR (1998) Stratigraphic cycles and major marine source rocks. In: de Graciansky P-C, Hardenbol J, Jacquin T, Vail PR (eds) Mesozoic and Cenozoic Sequence Stratigraphy of European Basins. Soc Sediment Geol (Special Publication) 60:43–51

Emery D, Meyers KJ (eds) (1996) Sequence Stratigraphy. Blackwell Science, Cambridge, 297 p

Ettensohn FR (1997) Assembly and dispersal of Pangea: large-scale tectonic effects on coeval deposition of North American, marine epicontinental black shales. J Geodynamics 23:287–309

Ettensohn FR (1998) Compressional tectonic controls on epicontinental black shale deposition: Devonian-Mississippian examples from North America. In: Schieber J, Zimmerle W, Sethi P (eds) Shales and Mudstones, vol 1. E Schweizerbart'sche, Stuttgart, pp 109–128

Eyles N (1993) Earth's glacial record and its tectonic setting. Earth Science Reviews 35:1–248

Eyles CH, Eyles N, Lagoe MB (1991) The Yakataga Formation: A late Miocene to Pleistocene record of temperate glacial marine sedimentation in the Gulf of Alaska. In: Eyles CH, Ashley GM (eds) Glacial marine sedimentation; paleoclimatic significance. Geological Society of America (Special Paper) 261:159–180

Eyles N, Eyles CH, James NP (1992) Glacial depositional systems. In: Walker RG (ed) Facies Models: Response to Sea Level Change. Geological Association Canada, St. Johns, Newfoundland, pp 73–100

Fisher AG (1993) Cyclostratigraphy of Cretaceous chalk-marl sequences. In: Caldwell WGE, Kauffman EG (eds) Evolution of the Western Interior Basin. Geol Assoc Can (Special Paper) 39:283–295

Frey RC (1996) The utility of epiboles in the regional correlation of Paleozoic epeiric sea strata: An example of the Upper Ordovician of Ohio and Indiana. In: Brett CE, Baird GC (eds) Paleontological Events. Columbia University Press, New York, pp 334–368

Galloway DK, McGara K (1982) Frio Formation of Texas Gulf Coast Basin – depositional systems, structural framework and hydrocarbon origin, migration, distribution, and exploration potential. Texas Bureau Economic Geology, Report Investigations 122:78

Garrels RM, MacKenzie FT (1971) Evolution of Sedimentary Rocks. Norton, New York WW, 397 p

Gast RE (1991) The perennial Rotliegend Saline Lake in NW Germany. Geologisches Jahrbuch 119:25–59

Geehan GW, Lawton TF, Sakural S, Klob H, Clifton TR, Inman KF, Nitzbery (1986) Geological prediction of shale continuity, Prudhoe Bay Field. In: Lake KW, Lawton TF, Sakura S, Klob H, Clifton TR, Inman KF, Nitzberg KE (eds) Reservoir Characterization. Academic Press, Orlando, pp 63–82

Gohrbandt KHA (1992) Paleozoic paleogeographic depositional developments on the central proto-Pacific Margin of Gondwana: Their importance for hydrocarbon accumulation. J South Am Earth Sci 6:267–287

Hay WH (1988) Paleoceanography: a review for the GSA Centennial. Geol Soc Am Bull 100:1934–1956

Heckel PH (1991) Thin, widespread Pennsylvanian black shales of Midcontinent North America: a record a cyclic succession of widespread pycnoclines in a fluctuating epeiric sea. In: Dennison JM, Ettensohn FR (eds) Tectonic and Eustatic Controls on Sedimentary Cycles (Concepts in Sedimentology and Paleontology, vol 4), Society Sedimentary Geology, pp 65–87

Hinnov, Linda (2003) Milankovitch cycles. In: Middleton GV (ed) Encyclopedia of Sediments and Sedimentary Rocks. Kluwer Academic, Dordrecht, pp 441–443

Holland HD (2002) Volcanic gasses, black smokers, and the Great Oxidation Event. Geochim Cosmochim Acta 66:3811–3826

Holser WT (1997) Geochemical events documented in inorganic carbon isotopes. Palaeogeogr, Palaeoclimatol, Palaeoecol 132:173–182

Huebschman RP (1973) Correlation of fine carbonaceous bands across a Precambrian stagnant basin. J Sediment Petrol 43:688–699

Huyck HLO (1990) When is a metalliferous black shale not a black shale? In: Grauch RI, Huyck HLO (eds) Metalliferous Black Shales and Related Ore Deposits – Proceedings 1989, United States Working Group Meeting, International Geological Correlation Program Project 254. US Geological Circular 1058:42–56

Ingle JC Jr (1992) Chapter 7b: Subsidence of the Sea of Japan: stratigraphic evidence from ODP sites and onshore sections. In: Tamaki K, Suyehiro K, Allan J, McWilliams M et al. (eds) Proceedings of the Ocean Drilling Program, Scientific Results, College Station TX 127/128, part 2:1197–1218

Issacs CM (2000) Depositional framework of the Monterey Formation, California. In: Issacs C, Rullkotter J (eds) The Monterey Formation from Rocks to Molecules. Columbia University Press, New York, pp 1–30

Issacs CM (1984) Hemipelagic deposits in Miocene Basin, California: Toward a model of lithologic variation and sequence. In: Stow DA, Piper DJM (eds) Fine-Grained Sediments: Deep-water Processes and Facies. Geological Society London Special Paper 15:481–496

James HJ (1951) Iron formation and associated rocks in the Iron River District, Michigan. Geol Soc Am Bull 62:251–266

Johnson EA, Warwick PD, Roberts SB, Khan IH (1999) Lithofacies, depositional environments, and regional stratigraphy of the Lower Eocene Ghazij Formation, Balochistan, Pakistan. US Geological Survey (Professional Paper) 1599:76

Jones PJ, Stump TE (1999) Depositional and tectonic setting of the lower Silurian hydrocarbon source rock facies, central Saudi Arabia. Am Assoc Petrol Geol Bull 83:314–332

Joseph P, Eschard R, Granjeon DH, Lerat O, Ravenne C, Souza OG (1998) Characterization of the 3-D architecture of deep-water reservoirs from outcrop analogues in the French Alps and application to the Namorando Field (offshore Brazil). Am Assoc Petrol Geol Bull 82:1928

Keith BD (1988) Regional facies of Upper Ordovician Series of eastern North America. In: Keith Brian D (ed) The Trenton Group (Upper Ordovician Series) of Eastern North America. AAPG Studies in Geology 29:1–19

Klein G, de Vries (1985) The control of depositional depth, tectonic uplift and volcanism on sedimentary processes in the back-arc basins of the western Pacific Ocean. J Geol 93:1–25

Klemme HD, Ulmishek GF (1991) Effective petroleum source rocks of the world: Stratigraphic distribution and controlling factors. Am Assoc Petrol Geol Bull 75:1809–1851

Knoll AH (2000) Learning to tell Neoproterozoic time. Precambrian Res 100:3–20

Kolata DR, Graese AM (1983) Lithostratigraphy and depositional environments of the Maquoketa Group (Ordovician) in northern Illinois. Illinois State Geological Survey Circular 528:49

Krassay AA (1998) Outcrop and drill core gamma-ray logging integrated with sequence stratigraphy: examples from Proterozoic sedimentary successions of northern Australia. AGSO J Geol Geophys 17:285–299

Leithhold EA (1994) Stratigraphical architecture at the muddy margin of the Cretaceous Western Interior Seaway, southern Utah. Sedimentology 41:521–542

Link MH (2003) Depositional systems and sedimentary facies of the Miocene-Pliocene Ridge Basin Group Ridge Basin, Southern California. In: Crowell JC (ed) Evolution of Ridge Basin: An Interplay of Sedimentation and Tectonics. Geol Soc Am (Special Publication) 357:17–88

MacQuaker JHS, Gawthorpe RL (1993) Mudstone lithofacies in the Kimmeridge Clay Formation, Wessex Basin, southern England: implications for the origin and control of mudstones. J Sediment Petrol 53:1129–1143

Mahmoud MD, Vaslet D, Husseini MI (1992) The Lower Silurian Qalibah Formation of Saudi Arabia: An important hydrocarbon source rock. American Association Petroleum Geologists Bulletin 76:1491–1508

Marsaglia KM (1995) Inter-arc and back-arc basins. In: Busby CJ, Ingersoll RV (eds) Tectonics of Sedimentary Basins. Blackwell, Oxford, pp 299–329

Martin WD (1998) Geology of the Dunkard Group (Upper Pennsylvanian-Lower Permian) in Ohio, West Virginia, and Pennsylvania. Ohio Div Geol, Bull 73:49

McCann T (1998) Sandstone composition and provenance of the Rotliegend of the NE German Basin. Sedimentary Geol 116:177–198

Mitchum RM Jr, Vail PR, Thompson S (1977) Seismic stratigraphy and global changes of sea level, Part 6: Stratigraphic interpretation of seismic reflection patterns in depositional sequences. In: Payton CE (ed) Seismic Stratigraphy – Applications to Hydrocarbon Exploration. Am Assoc Petrol Geol Memoir 26:117–133

Mitchum RM Jr, Van Wagoner JC (1991) High-frequency sequences and their stacking patterns: sequence-stratigraphic evidence of high-frequency eustatic cycles. Sedimentary Geol 70:131–160

Moretti I, Martinez ED, Montemurro G, Aguilera E, Pérez M (1995) The Bolivian source rocks. Revue l'Institut Francais du Petrole 50:753–775

Molenaar CM (1985) Subsurface correlations and depositional history of the Nanushuk Group and related strata North Slope, Alaska. In: Huffman AC Jr (ed) Geology and Exploration of the National Petroleum Reserve in Alaska. USGS Bull 1614:37–59

Morley CK (1999) Comparison of hydrocarbon prospectivity in rift systems. In: Morley CD (ed) Geoscience of Rift Systems – Evolution of East Africa. AAPG Studies in Geology 44:233–242

Notestein FB, Hubman CW, Bowler JW (1944) Geology of the Barco Concession, Republic of Columbia, South America. Geol Soc Am 55:1165–1216

Olsen PE, Kent DV (1999) Long-period Milankovitch cycles from the late Triassic and Early Triassic of eastern North America and their implications for the calibration of the early Mesozoic time scale and long term behavior of the planets. Philosophical Transactions Royal Society, Mathematical Physical Engineering Sciences 357:1761–1786

Orton GJ, Reading HG (1993) Variability of deltaic processes in terms of sediment supply, with particular emphasis on grain size. Sedimentology 40:475–512

Peters KE, Moldowan JM (1993) The Biomarker Guide. Prentice Hall, Englewood Cliffs, 363 p

Pisciotto KA, Garrison RE (1981) Lithofacies and depositional environments of the Monterey Formation, California. In: Garrison RE, Douglas RG (eds) The Monterey Formation and Related Siliceous Rocks of California. Society Economic Paleontologists Mineralogists Pacific Section, Special Publication, pp 97–122

Potter PE (1998) Shale-rich basins: controls and origin. In: Schieber J, Zimmerle W, Sethi PS (eds) Shales and Mudstones. E. Schweizerbart'sche, Stuttgart, 1, pp 21–32

Reineck H-E, Gutmann WF, Hertweck G (1967) Das Schlickgebiet südlich Helgoland als Beispiel rezenter Schelfablagerungen. Senckenbergiana Lethaea 48:219–261

Robison VC, Engel M (1993) Characterization of the source horizons within the Late Cretaceous Transgressive sequence in Egypt. In: Katz BJ, Pratt LM (eds) Source Rocks in a Sequence Stratigraphic Framework. AAPG Studies in Geology 37:101–117

Ronov AB (1983) The Earth's sedimentary shell (quantitative patterns of its structure, composition, and evolution). International Geology Review 24:1313–1363 and 1365–1388 (originally published as Osad. obolachka Zemli Ckolichestvennyye zakonomernosti stroyeniya, sostavai evol.: Izvd. Nauk Moscow 1980)

Ronov AB, Bredanova NV, Migdisov AA (1992) General trends in the evolution of the chemical composition of sedimentary and magmatic rocks of the continental Earth crust. Soviet Scientific Rev, Sect G, Geol Rev 1:1–37

Ross WC, Watts DE, May JA (1995) Insights from stratigraphic modeling: Mud-limited versus sand-limited depositional systems. Am Assoc Petrol Geol Bull 79:231–258

Sandberg CA, Gutschick RC (1984) Distribution, microfauna, and source-rock potential of Mississippian Delle Phosphate Member of Woodman Formation and equivalents, Utah and adjacent states. In: Woodward et al. (eds) Hydrocarbon Source Rocks of the Greater Rocky Mountain Region. Rocky Mountain Association Geologists, Denver, CO, pp 135–178

Schutter SR (1998) Characteristics of shale deposition in relation to stratigraphic sequence systems tracts. In: Schieber J, Zimmerle W, Sethi PS (eds) Shales and Mudstones: E. Schweizerbart'sche, Stuttgart, 1, pp 79–108

Schwalbach JR, Bohacs KM (eds) (1992a) Sequence stratigraphy in fine-grained rocks: Examples from the Monterey Formation. Society Sedimentary Geology, Pacific Section (Los Angeles), 80 p

Schwalbach JR, Bohacs KM (eds) (1992b) Field investigation techniques for analysis of the Monterey Formation. In: Schwalbach JR, Bohacs KM (eds) Sequence Stratigraphy in Fine-grained Rocks: Examples from the Monterey Formation. SEPM Pacific Section, Bakersfield 70:21–30

Schwarzacher W (1999) Milankovitch cycles and sequences: two different stratigraphic tools. In: Harff J (ed) Computerized Modeling of Sedimentary Systems. Springer, Berlin Heidelberg New York, pp 247–261

Sloss LL (1962) Stratigraphic models in exploration. Am Assoc Petrol Geol Bull 46:1050–1057

Smith GA, Landis CA (1995) Intra-arc basins. In: Busby CJ, Ingersoll RV (eds) Tectonics of Sedimentary Basins. Blackwell, Oxford, pp 263–298

Spears DA (2003) Bentonites and tonsteins. In: Middleton GV (ed) Encyclopedia of Sediments and Sedimentary Rocks. Kluwer Academic Press, Dordrecht, pp 61–66

Stemmerik L, Ineson JR, Mitchell JG (2000) Stratigraphy of the Rotliegend Group in the Danish part of the North Sea. J Geol Soc 157:1127–1136

Stoakes FA (1980) Nature and control of shale basin fill and its effect on reef growth and termination: Upper Devonian Duvernay and Ireton Formations of Alberta, Canada. Bull Can Petrol Geol 28:345–410

Stow DAV, Huc A-Y, Bertrand P (2001) Depositional processes of black shales in deep water. Marine and Petroleum Geology 18:491–498
Taira A (1985) Sedimentary evolution of Shikoku Subduction Zone: The Shimanto Belt and Nankai Trough. In: Nasu N, Kobayashi K, Uyeda S, Kushiro I, Kagami H (eds) Formation of Active Ocean Margins. Terra/D. Reidel, Tokyo, Dordrecht, pp 835–851
Taylor SR, McLennan SM (1985) The Continental Crust: its Composition and Evolution. Blackwell Scientific Publications, Oxford, 312 p
Thickpenny A, Leggett JK (1987) Stratigraphic distribution and palaeo-oceanographic significance of European early Paleozoic organic-rich sediments. In: Brocks J, Fleet AJ (eds) Marine Petroleum Source Rocks. Geological Society, Special Publication 26:232–247
Twenhofel WH (1938) Principles of Sedimentation. McGraw-Hill Book, New York, 610 p
Tyson RV (1995) Sedimentary Organic Matter: Organic Facies and Palynofacies. Chapman and Hall, London, 615 p
Tyson RV (1996) Sequence-stratigraphical interpretation of organic facies variations in marine siliciclastic systems: general principles and application to the onshore Kimmeridge Clay Formation, UK. In: Hesselbo SP, Parkinson DN (eds) Sequence Stratigraphy in British Geology. Geological Society Special Publication 103:75–96
Vail PR (1987) Seismic stratigraphy interpretation using sequence stratigraphy, part 1: seismic stratigraphy interpretation procedure. In: Bally AW (ed) Atlas of Seismic Stratigraphy. American Association Petroleum Geologists, Studies in Geology 27 1:1–10
van Buchem FSP, McCave IN, Weedon GP (1994) Orbitally induced small-scale cyclicity in a siliciclastic epicontinental setting (Lower Lias, Yorkshire, UK). Special Publications International Association Sedimentologists 19:345–366
Van Wagoner JC, Posamentier HW, Mitchum RM Jr, Vail PR, Sarg JF, Loutit TS, Hardenbol J (1988) A overview of the fundamentals of sequence stratigraphy and key definitions. In: Wilgus CK, Hasting BS, Kendall CGSC, Posamentier HW, Ross CA, Van Wagoner JC (eds) Sea-level Changes – an Integrated Approach. Society Economic Paleontologists Mineralogists Special Publication 42:39–46
Van Wagoner JC, Mitchum RM, Campion KM, Rahmanian VD (1990) Siliciclastic Sequence Stratigraphy in Well Logs, Cores, and Outcrops: Concepts for High-Resolution Correlation of Time and Facies. AAPG Methods in Exploration Series 7:55
Visser JNJ (1997) Deglaciation sequences in the Permo-Carboniferous Karoo and Kalahari basins of Southern Africa: a tool in the analysis of glacial-marine fills. Sedimentology 44:507–522
Weedon GP (1993) The recognition and stratigraphic implications of orbital-forcing of climate and sedimentary cycles. In: Wright VP (ed) Sedimentology Review. Blackwell, Oxford, pp 31–50
Weimer P, Posamentier H (1993) Siliciclastic Sequence Stratigraphy. American Association Petroleum Geologists Memoir 58:492
Wilson JL (1975) Carbonate Facies in Geologic History. Springer, Berlin Heidelberg New York, 471 p
Wilson AA (1990) The Mercia Mudstone Group (Trias) of the East Irish Sea Basin. Proceedings Yorkshire Geological Society 48:1–22
Wilson AA, Evans WB (1990) Geology of the country around Blackpool. Memoir for 1:50,000 geologic sheet no. 66, British Geological Survey, London, 82 p
Zeito GA (1965) Interbedding of shale breaks and reservoir inhomogeneities. J Petrol Technol 17:1223–1228
Ziegler K, Longstaffe FJ (2000) Clay mineral authigenesis along a mid-continental scale fluid conduit in Palaeozoic sedimentary rocks from southern Ontario, Canada. Clay Minerals 35:239–260

Digging Deeper

Bramlette MN (1946) The Monterey Formation of California and the origin of its siliceous rocks. U.S. Geological Survey Professional Paper 212, 57 p.

This classic gives a well illustrated overview of the entire Monterey over all its outcrop areas. Thus more proximal as well as distal facies are described. Well worth examination over 50 years later.

Briggs DEG, Crowther PR Eds (2001) Paleobiology II. Blackwell Science, Oxford, 583 p.

Short, informative, well illustrated articles that provide excellent introductions and summaries of a multitude of fossil-related topics relevant to mudstones. Examples: sequence stratigraphy and fossils, high resolution biostratigraphy, fossils as environmental indicators (8 articles), laggerstätten (beds ultra rich in fossils), oxygen in the ocean, and more.

Busby CJ, Ingersoll RV (eds) (1995) Tectonics of Sedimentary Basins. Blackwell, Cambridge, 579 p.

Thirteen informative chapters each with many illustrations. We frequently consulted this book and found its section on rifts most helpful.

Cluff RM, Reinbold ML, Lineback JA (1981) The New Albany Shale Group of Illinois. Illinois Geological Survey Circular 518, 83 p.

Careful, systematic, and complete description includes many photographs and radiographs of slabbed cores of the Devonian-Mississippian New Albany Shale of the intracratonic Illinois Basin. Its shales range from greenish-gray and bioturbated to dusty yellow to olive and brownish black, but also include thin micritic interbedded limestones and a thin transgressive, basal sandstone. This circular is one of the best sources to see the lithology and stratigraphic subdivisions of an important source bed of an intracratonic basin.

Coe AL (ed) (2003) The Sedimentary Record of Sealevel Change. Open University/Cambridge University Press, 287 p.

A clear, abundantly color-illustrated presentation of the most important topic in stratigraphy today. Thirteen chapters, six authors. Excellent book to begin your study or to refresh yourself.

Dorobeck SL, Ross GM (eds) (1995) Stratigraphic Evolution of Foreland Basins. SEPM Special Publications 52 (Society for Sedimentary Geology), Tulsa OK, 310 p.

The 15 papers of this volume start with two based on computer modeling, include three provenance studies (one of mudstones) and conclude with 10 case histories. Although mudstones are not the main theme of this volume, much can be learned about foreland basins that is relevant to understanding mudstone.

Einsele, Gerhard (2000) Sedimentary Basins, 2nd ed. Springer, Berlin Heidelberg New York, 792 p.

This most useful, well referenced (over 4,000!), and well illustrated book includes good discussion of depositional systems, has a long section of subsidence and sedimentary budgets, an excellent section on basin evolution, and a final section on diagenesis and fluid flow. Highly recommended for students of mudstones.

Kemper E (Coordinator) () Das späte Apt und frühe Alb, Nordwestdeutschlands. Geologische Jahrbuch, Reihe A, Vol. 65, 703 p.

Super detailed description and interpretation at a small late Aptian-early Albian sub-basin in central Germany has four principal parts and 32 short papers ranging from sedimentary components to thin section petrology to micropaleontology and organic chemistry of dark shales bordered by a oxygenated facies with glauconite shelf sands. Mudstones contain an appreciable volcanic debris and some tuffs also. During the late Aptian to early Albian, a partially isolated epeiric sea became much better connected to the open ocean of the ancestral North Sea. Rich source of data and fine example of German attention to detail.

Gonzales S, Johnson KS (1985) Shales and Other Argillaceous Strata in the United States. Oak Ridge National Laboratory, ORNL Sub/84-64794-1, 596 p.

Unusual, one-of-its-kind effort to compile a stratigraphic atlas of mudstones in twelve chapters – one on properties of mudstone and their significance for waste disposal, and 11 on the mudstones of different regions in the United States (includes origin, thickness, burial, and hydrology). Important source document.

Hallam A (1987) Mesozoic marine organic-rich shales. In: Brooks J, Fleet AJ (eds) Marine Petroleum Source Rocks. Geological Society, Special Publication 26, pp 232–247.

Short but telling summary with discussion of environments of deposition. Excellent.

Kolata DR, Huff WD, Bergstrom SM (1996) Ordovician K-bentonites of Eastern North America. Geological Society of America Special Paper 313, 84 p.

Over 60 Ordovician bentonites identified in eastern North America, traceable more than 2,100 km using geophysical logs and outcrops, and their thickness, coarseness and number indicate a source along the southeastern collisional margin of Laurentia. Model study.

Palaios (2003) La Luna Special Issue Research Reports. Palaios 18:302–378.

The Cretaceous La Luna Formation and related anoxic equivalents consist of dark, calcareous and carbonaceous mudstones interbedded with marlstones, limestones, and some sandstones in much of northern South America. Here it is wide spread and both a source rock and reservoir. The La Luna is especially well developed in the Maracaibo Basin, one of the richest petroleum basins in the world. Deposition extended over some 20 Ma. The six papers of this special issue provide an integrated basin analysis ranging from paleogeography and stratigraphy to microfossils and paleoecology through bio- and chemostratigraphy to carbon isotopes. Explanations invoke paleo-climate, paleo-geography and paleo-oceanography (upwelling) plus plate tectonics. Collective references sum to over 300. A Cadillac-Lexus-Mercedes/Benz study by 18 specialists.

Piecha M (1999) Stratigraphie, Fazies und Sedimentpetrographie der rhythmisch und zyklisch abgelagerten, tiefoberdevonischen Beckensedimente im Rechts-rheinischen Schiefergebirge (Adorf-Bänderschiefer). Courier Forschungsinstitut Senckenberg (Frankfort a.m) 163, 151 p.

An exceptionally well documented (75 photographs, 22 plates and 6 tables) "close up" study of thinly laminated Upper Devonian black and dark gray mudstones interbedded with thin carbonates and some siltstones all well exposed in a small part of the Rhenish Schiefergebirge of southwestern Germany. Many polished and thin sections show complete details of microstructures and lamination much of which is similar to that of the Belt Series of Montana. Careful paleontology, some chemical and microprobe analyses and many meters of described section suggest that thin 20 to 40 ka stratification cycles are climate related. End member model study?

Seilacher, A (1998) Patterns of macro evolution: how to be prepared for extinction. C.R. Acad. Sci. Paris, Sci. terre planètes 327, 431–440.

This is a far-ranging, thoughtful, big picture, easy-to-understand view of the macro evolution of life, one that considers movements of tectonic plates and long term atmospheric and oceanic changes. It seems to us that this and other related papers are essential reading for those who wish to trace mudstones through time. Why not begin a seminar on this subject by assigning Seilacher's paper for the second meeting?

Seilacher A, Plüger F (1994) From biomats to benthic agriculture: a biohistoric revolution. In: Krumbein W, Paterson DM, Stal LJ (eds.) Biostabilization of Sediments. Bibliothec and Information Systems der Universität Oldenburg, pp 97–105

This short incisive paper calls attention to the contrast between "matted" mudstones in the Precambrian and burrowed and stirred mudstones in the Phanerozoic.

Wignall PB (1994) Black Shales. Oxford Science, Oxford, 127 p.

Nine chapters with over 600 references cover almost all key ideas. Excellent place to start your study of black shales.

Zangerl R, Richardson ESJ (1963) The Paleoecological History of Two Pennsylvanian Black Shales. Fieldiana 4:1–239.

A remarkable, centimeter-by-centimeter study of two, thin Middle Pennsylvanian laminated black shales in two small quarries. Many illustrations and plates support an interesting and original interpretation. Time to revisit this effort?

CHAPTER 9

Practicalities

Mudstones can be great friends or dangerous foes

9.1. Introduction

The economic and environmental aspects of muds, clays, and mudstones are very diverse and have great importance to modern society, a process that began with the making of sun dried bricks about 6,000 years ago on the alluvial plains of the Tigris and Euphrates River valleys of the Near East (Fig. 9.1).

Muds, clays and mudstones are used today in many diverse industrial products and processes. They host metals such as lead and zinc, are source beds for petroleum, play an essential role in the drilling of deep wells, prevent ponds from leaking, are the main ingredient in pottery, and are important in construction (where they often cause unwanted problems). Many common everyday products such as toothpaste, face powder, desiccants, adhesives, spill clean-up products, cat box filler and others contain specialized clays. Clays and mudstones also serve as both natural and artificial barriers to the flow of fluids. This great range of uses and the special problems associated with clays and mudstones is a direct consequence of the wide range of physical and chemical properties of the clay minerals that make them stand apart from all other rocks. First, it is the plasticity of the clay minerals that makes clays and mudstones so different from all other rocks and gives them so many different uses. Secondly, the clay mineral groups have markedly different particle sizes and shapes, cation exchange capacities, surface areas, colors and many other contrasts – much greater variability than any other mineral family. There is still another factor to consider, their variable organic content. Organic contents ranging from zero to more than 25% also make clays and mudstones very different. For many industrial uses the less organic matter, the better, whereas the contrary is true for mudstones to serve as hosts for zinc, lead, or copper, to be oil shales or source rocks for petroleum. This characteristic – deficiency or richness of organic matter – goes back to their deposition in quiet water, which permits low density organic matter to accumulate with slowly settling clay particles *provided it was available and not oxygenated prior to burial* (see Oxygen Chap. 4).

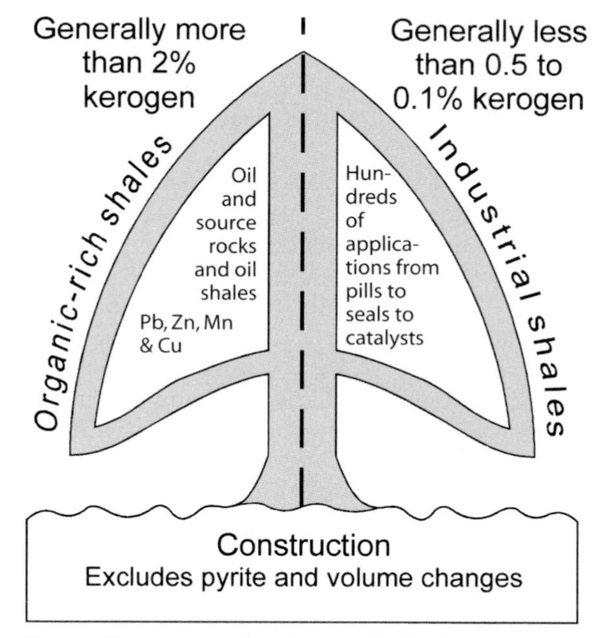

Fig. 9.1. Human uses of mudstones depicted as a tree

The literature of resources, uses and problems associated with muds and mudstones is vast and perhaps includes more disciplines than any other rock family, because it spans civil-, environmental-, chemical-, mining- and ceramic engineering as well as material sciences and geology. Fortunately, an excellent, broad, recent overview is provided by Sethi and Schieber (1998).

9.2. Mudstones as Barriers to Flow

It can be argued that it is the low permeability of mudstones that gives them their greatest social significance. To see this, think as follows. It is the mudstones that provide barriers to leakage of fluids from landfills around cities or from depositories of hazardous waste; mudstones also are the hosts for

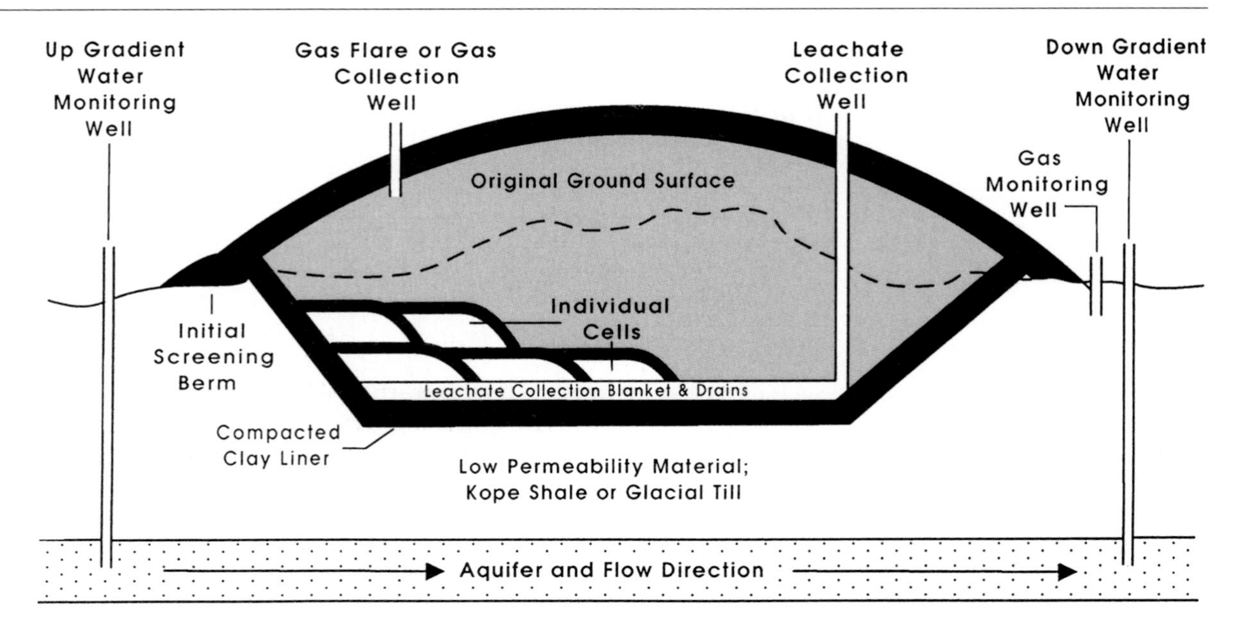

Fig. 9.2. Diagrammatic representation of a landfill (Aughenbaugh 1990, Fig. 1)

mineralization by solutions precipitating lead, zinc, copper and manganese, and mudstones form the confining layer that prevents groundwater escaping from a sandstone or carbonate aquifer. And perhaps most important, mudstones prevent the migration of oil and gas to the surface after it was formed in an underlying organic-rich shale. Although no one appears to have made the calculation, could it be that these benefits to society far outweigh even those from the making of bricks from mud and mudstone for the last 6,000 years? Because the sealing properties of mudstones play such an important role, we start practicalities with mudstones as seals beginning with landfills and hazardous waste (Fig. 9.2).

Clays and mudstones, along with bentonites and clayey tills, are the best rocks to use as seals for dams, landfills, lining leaky ponds, for creating barriers to flow in a surface aquifer, and are used for the long-term storage of both low- and high-level chemical and nuclear waste. There are many reasons for this. All four of these clay-rich materials have low permeability, the clays and mudstones are likely to be available over wide areas, all are easy to emplace, all are plastic so that rather than develop brittle fractures when stressed, they deform and still block fluid flow. There is still another good geological reason to consider. Because most of the clay minerals of mudstones were formed by weathering at the Earths surface, they are already chemically stable. Below are some generalizations about seals, their use and care.

First, when a clay, mudstone, or till is used for a man-made seal, it should be first ploughed and recompacted to eliminate possible fractures, which may have had either a tectonic origin or formed early as it dewatered and compacted. Typically, such fracture permeability (Neuzil 1994), be it large or small, is much greater than intrinsic permeability and, in fact, it is such fracture systems that locally make some mudstones aquifers (Neuzil and Belitz 1998). A layer of swelling bentonite provides additional protection and also helps absorb possible harmful organic compounds, radionuclides, or metals. (As a rule, the greater the cation capacity of the clay mineral, the more efficient its trapping capacity). When serving as a barrier, all seals are only as good as their most permeable portion such as a fracture or where it is very thin. Thus much care is needed in both the construction and in the analyses of seals. The best seals are clay rich, plastic (deform rather than fracture), and are homogeneous and widespread. The clayshales, organic rich or not, best satisfy these criteria, because they are both rich in clay and laminated. For natural seals, petrographic examination is always recommended to insure against surprises such as soluble minerals, the presence of fossils with large open voids, open clay fabrics, well dispersed silt, etc. in addition to the standard tests used in petroleum and civil engineering. And finally, it should be kept in mind that different fluids have different interfacial tensions so that a mudstone that forms a seal for one

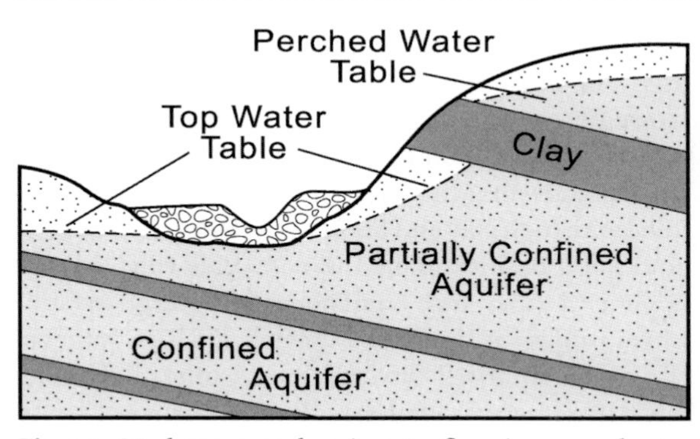

Fig. 9.3. Mudstones as barriers to flow in groundwater systems

Fig. 9.4. Seals can be thought of as mud blankets as shown by this two-meter thick turbiditic mudstone in a tight fold in the Cretaceous Carocal Formation at Rancho de los Guzman near San Luis Potosi, Mexico

fluid, say oil, may at the same time be permeable to another, say, water. Either overpressure (unvented gas in a landfill, excess gas pressure in an man made storage reservoir, or the natural overpressure of fluids) or tectonic deformation may rupture seals.

Mudstones have been proposed as underground hosts for nuclear waste. Many more surface and underground repositories on and in mudstones seem likely to be constructed in years to come and fortunate is the region that has suitable mudstones. To appreciate this fully, think of a granitic terrain exposed to a tropical climate, one that is capped by a permeable regolith of reddish brown oxisols. Here leachate from the landfill can drain directly into the fracture system of the granite terrain, the principal aquifer system of the region, unless special precautions are taken.

Mudstones as barriers to groundwater flow, *aquicludes,* vary greatly in thickness, lateral extent, and how they interact with topography to form, confined, unconfined and perched aquifers (Fig. 9.3). Only a few tens of decimeters of stiff plastic clay in unconsolidated sands and gravels can be a most effective barrier to groundwater flow. This is also true of petroleum reservoirs, where a thin, laterally extensive mudstone can effectively compartmentalize a seemingly homogeneous sandstone or carbonate reservoir (think of a thin mudstone at a parasequence boundary). Such seals, many of which are only as thick as a meter or less, can be thought of as *mud blankets* (Fig. 9.4). These occur in marine and lacustrine mudstones, especially in the transgressive basal part of a parasequence and along its downlap surface (Chap. 8). Conversely, many thin, discontinuous mudstones in a reservoir may have minimal effect on flow unless they seal faults.

On a basin scale, mudstone aquicludes may either have outlets or have down dip plugs (Fig. 9.5). Such mudstones can either be tight or leaky seals. The consequences of these different possibilities are great for the water chemistry of the confined aquifer and the aerial extent of its porosity. If the mudstone seal is tight and the underlying sandstone aquifer has an outlet, fresh water may travel though it for hundreds of kilometers to outcrop on the other side of the basin or to a major down dip fault. Under these conditions,

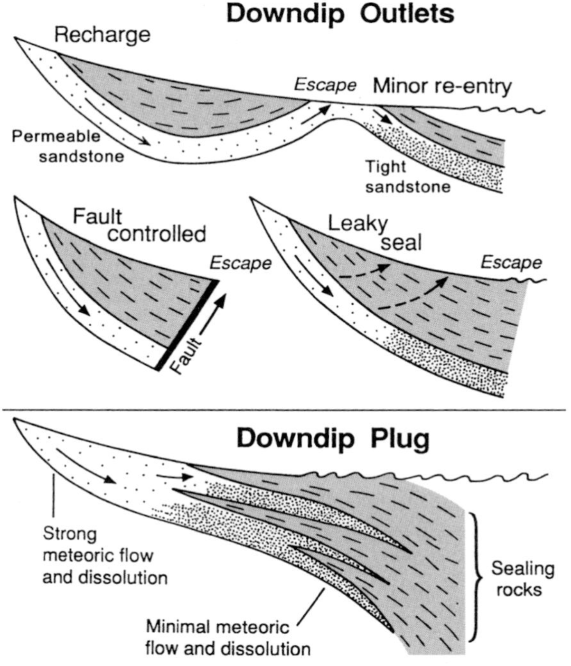

Fig. 9.5. Mudstones as basin wide barriers to flow. Here mudstones extend blanket-like over much of a basin, commonly hundreds of kilometers, and they may be either very effective or leaky seals (dense fractures, many small faults or very silt-and sand rich)

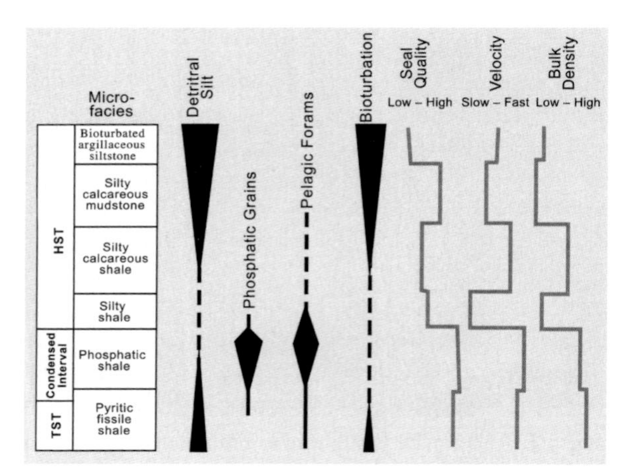

Fig. 9.6. Correlation of mudstone microfacies with seal quality, compressional seismic velocity, and bulk density (Almon et al. 2002, Fig. 9). Published by permission of the authors and the Association of Gulf Coast Geological Societies

slightly acid ground water can leach calcite cements and feldspar and transform many of its clay minerals to kaolinite (Franca et al. 2003). Thus an initially low permeability sandstone becomes a good source of ground water. But, if the seal is leaky and there is no down dip outlet, fresh water inflow into the aquifer will be much less and its down dip permeability will also be much less, because its pore waters are not continually recharged by incoming fresh water. Or aquifers may simply 'finger out' into mudstones as is true of the Gulf Coast of the United States. For more about basin-wide ground water flow see Garven (1992).

Finally, we suggest that perhaps the greatest economic impact of mudstones is their ability to form the seals above traps for hydrocarbons. Three informative studies of petroleum seals are those by Dawson and Almon (1999), Almon et al. (2002), and by Dawson and Almon et al. (2002). Collectively, these papers show that small pore size, low silt content coupled with high clay and organic contents, high density and good fissility make the best seals (Fig. 9.6). As a general rule, such mudstones have high seismic shear velocities, high shear moduli and high values of Young's modulus. Establishing mudstone microfacies via petrography is vital here. From a sequence stratigraphic point of view, the mudstones from transgressive systems tracts and condensed sections make the best seals (likely to be slowly-deposited, organic-rich claystones). See OCED (1996) for a wide range of studies of mudstones both as barriers and conduits to flow including their response to deformation.

9.3. Ores in Mudstones

Organic-rich mudstones plus interbedded carbonates and siltstones contain large reserves of mineable sphalerite (ZnS) and galena (PbS) as well as accessory traces of gold, silver and platinum plus much pyrite (FeS_2), barite ($BaSO_4$) and many other commodities in minor amounts. More than 50% of the world's zinc reserves and about 60% of its lead reserves are found in such deposits whose host sediments include not only dark, non-bioturbated mudstones but also cherts, fine-grained carbonates and graded and cross laminated siltstones plus sandstones. These ores occur in three ways: as beds and lenses parallel to bedding – thus the term *stratiform* – and as domes, and as breccias along faults (Fig. 9.7). Stratiform ore beds and laminations have typical detrital

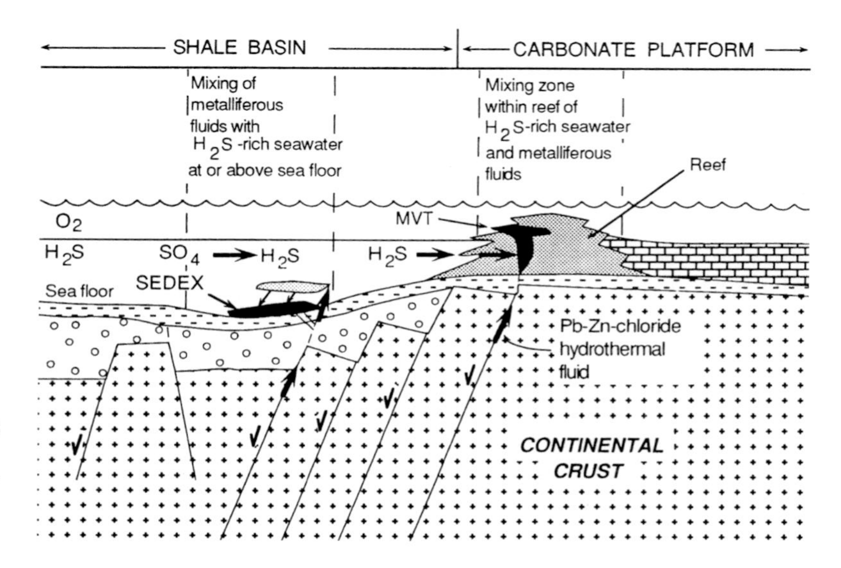

Fig. 9.7. Idealized model of origin of lead and zinc deposits in mudstone-rich basins that emphasizes the mixing of hot basinal brines with sulfur-rich seawater. (Goodfellow et al. 1993, Fig. 29). Published by permission of the authors and the Geological Association of Canada

textures such as grading and inclined stratification clearly indicating precipitation and some local transport of the ore on the bottom. These deposits range in age mostly from middle Proterozoic to Carboniferous and occur in large reactivated rifts above thick sandstone sections typically near deep seated faults. Most properly referred to by the descriptive term "shale hosted massive sulfides" or SHMS, analogous to their volcanic cousins VHMS, they are often called by the genetic term "sedimentary exhalative" or SEDEX deposits. These are believed to have been mostly formed by hydrothermal fluids that reached the sea bottom along fractures, as in the modern Red Sea, although some precipitation of ore minerals in pores beneath the seafloor, as in the modern Guaymas Basin, also occurs.

While organic-rich shales are not always the host rock, they are almost always present and of key importance. Why? Because they provided seals which confined the pore water of the underlying sandstones permitting it to become hot and supersaturated in cations such as Zn, Cu, Pb and Fe (Hemley and Hunt 1992, Fig. 2). When some of this water escapes in a vent to the sea floor as a hot brine, it cools and metals in solution combine with HS^{-1} to form lead and zinc sulfides. When these solutions combine with SO_4^{-2} barite is precipitated. Together galena, sphalerite, pyrite, and barite are the most common minerals found in these deposits. Although seldom of economic significance, many other metal elements may be adsorbed or complexed by organic matter and clay minerals in this process.

When such a brine is less dense than bottom waters, the vented fluid mixes upwards into the overlying seawater, and a layered mineralized dome near the vent is formed. When the brine is more dense than seawater, a hot, super saline density current forms and an elongate sheet-like bed is deposited in a topographic low some distance from the vent. Note that for the first case bottom water anoxia is crucial for the preservation of the sulfides on the seafloor, whereas in the dense brine case, the brine itself can provide protection from mixing and oxygenation. A third scenario is that the vent fluids precipitate their metals within the sediment pile immediately beneath the seafloor, as appears to have happened at the world's largest SEDEX deposit, Red Dog in Alaska (Moore et al. 1986), a process that again protects the deposit from destruction by the oxygen in the overlying seawater. The common theme for all three of these scenarios is a large, steep-sided, under filled (flooded by an arm of the sea) reactivated rift basin with high thermal gradients. This is an environment that strongly favors anoxic bottom waters and hence the common association of these deposits with organic-rich shales and limestones. See Maynard (1991) and Goodfellow et al. (1993) for useful overviews of SEDEX deposits from which much of the above is summarized.

Organic-rich shales also concentrate a large group of heavy metals even where not associated with hydrothermal brines. These can be thought of as the *cold-water mineralized black shales* and are well summarized by Leventhal (1998) and Sethi and Schieber (1998). More rigorously, a metal-rich black shale, MRBS, has been defined (Huyck 1990, p.46 and Table 4) as one that is enriched two times in any given metal relative to the standard SDO-1 shale (Huyck 1990, p 53) of the U.S. Geological Survey (beryllium, cobalt, molybdenum and uranium only need unitary equivalence). These high trace element concentrations and their minor associated sulfides come from the marine water column itself. Three factors are needed for this to occur: a rich source for the metals in the seawater (water derived from acid, fine-grained ash falls is excellent as is water from the weathering and leaching of crystalline rocks), anoxic bottom waters to provide HS^{-1} and slow terrigenous sedimentation to permit long residence time on the sea floor before burial (think of condensed sections). Some metals in dilute solution in seawater, especially Mo, combine with HS^{-1} to form minor fine sulfides, but most are adsorbed on or complexed by organic matter or by the broken bonds of clay minerals. Concentrations of V and Zn reach thousands of parts per million, whereas those of Cr, U, Mo, Cu and As reach hundreds of parts per million. With the possible exception of some very fine pyrite and perhaps sphalerite, none of these are visible with a hand lens or in the microscope. Today, few metal-rich shales are mined, although some have been mined for their oil content such as the famous Alum Shale of Scandinavia. Most of these metal-rich black shales occur in thin, condensed zones so understanding the sequence stratigraphy of the interval is vital. Such condensed sections occur in either transgressive deposits or at the downlap surface toward basin centers far from shore. Slow sedimentation provides time for metals to combine with HS^{-1} or for ion exchange with clay minerals or organic complexing to remove them from seawater. Pasava (1996) provides a broad overview (nine papers) on the mineralization of black shales.

What are the similarities and differences between metal-rich black shales and petroleum source rocks? What they have in common, of course, is that both are organic rich – five percent TOC is thought to be sufficient for a metal rich black shale to form (Leventhal 1998, p 257), although less is needed for a petroleum source. Contrasts, on the other hand, include thickness and persistence. Most metal-rich black shales are thin, commonly only one or two meters or less, whereas the thicker the hydrocarbon source rock, the better. Longevity is also different. Once a metal-rich black shale is formed, it remains as such through even medium grade metamorphism (Arkimsa et al. 1999). Hydrocarbons in a source bed, on the hand, are all driven off at temperatures as low as 200 °C and thus are easily volatilized during burial, especially where geothermal gradients are high. And, unlike a petroleum trap, subsequent faulting or tilting does not destroy a mineralized mudstone. Weathering of metal-rich black shales in humid climates, either in outcrops or in spoil piles, releases high concentrations of heavy metals to surface or ground water that may be potentially harmful (see Construction and the Environment later in this chapter).

Barite, noted above in SEDEX deposits, deserves additional comment. Its largest reserves occur in Paleozoic black shales along with sulfides, cherts, and phosphates, but it is also widely found as minor nodules, rosettes, and in many concretions (Chap. 6). $^{87}Sr/^{86}Sr$ ratios of the Sr contained in the barium of SEDEX Pb-Zn deposits typically have radiogenic compositions. These indicate derivation of the hydrothermal fluids from underlying continental crust (Maynard et al. 1995). Massively bedded barite also occurs without Pb and Zn in economically important deposits in Arkansas, China, and Nevada. These have seawater $^{87}Sr/^{86}Sr$ ratios and may have been precipitated from upwelling seawater (Jewell 2000).

9.4. Industrial Minerals

There are six commercial names for the industrial uses of clays and mudstones (Table 9.1), which are used from thousands of tons each year (bricks and tile, which demand little more than plasticity) to small quantities of great purity (pharmaceuticals, adhesives, and molecular sieves). Two excellent recent reviews of these uses include Virta (1998) and Murray (2000). In broad terms, uses include agriculture, construction, engineering, environmental applications, as well as many uses in the process industries of refining, paper and paint production, foundry bondants, decolorization, pharmaceuticals, and more.

All of these clays and mudstones consist of one or more of the common clay minerals as well as some minor quartz, feldspar, pyrite and marcasite, siderite,

Table 9.1. Industrial categories (after Ampian 1985)

Kaolin (China clay)
A white clay rich in kaolin, halloysite or dickite, which fuses at about 1,785 °C and excels for high temperature applications. Also has many uses as a minor additive to industrial products, because it is inert and plastic. May be either residual or transported

Bentonite
Mostly consists of smectite clays with exchangeable Na^{+}, Ca^{++}, or Mg^{++}; the sodium type swells (western or Wyoming bentonite) whereas the calcium type is non-swelling (southern type). Forms by diagenesis from volcanic ash

Ball clay
A highly plastic, white-firing clay used for bonding ceramic ware; may contain some organic matter and is commonly finer grained than the kaolinitic clays. A "dirty kaolin"

Fire clay
May be plastic or indurated and commonly contains a wide range of impurities that produce buff to gray fired colors. Typically occurs below coal seams

Fuller's earth
Any clay that cleans fibers; thus variable mineralogy, although lath-like minerals such as attapulgite or smectites are most effective. Opal commonly present

Common clay
Any clay that is sufficiently plastic to remold and vitrifies at about 1,000 °C; used primarily for heavy clay products such as bricks and sewer and decorative tile. No mineralogical restrictions, but illite and chlorite predominate

calcite and dolomite plus gypsum. Depending on the intended use, the industrial value of the clay or mudstone is related to understanding not only the precise properties of the clay minerals themselves, but also the absolute abundance, kinds and properties of the non-clay minerals. The deposits used include alluvium, soil and colluvium, till, loess, a wide range of mudstones and hydrothermally altered host rocks (argillization). Almost all of these are mined in open pits and, and there are some environmental impacts, particularly if any pyrite-bearing strata are exposed. In addition, most are mined alone, although hopefully with more care and planning in the future, more will be mined as by-products; for example, using the underclay of a coal seam. Today, many industrial clay deposits are lost to suburban development because they are considered valueless.

Perhaps there are as many as several thousand different industrial process and consumer uses of clays and mudstones. Products that we see and use every day include toothpaste and face powder, many varieties of white ware from toilets to fine china, kitty litter and adsorbents for water and oil, tiles ranging from common every day drain tile to elegant tiles for floors and walls plus many, many more. Most of these and other uses have special requirements of the clay source if they are to fulfill their function. As a result, clay mineralogy, particle size, shape and surface area, viscosity, abrasion, cation exchange, trace element purity, etc. are critical to successful use. To insure performance, study by the Scanning Electron Microscope, Transmission Electron Microscopy, Infrared Adsorption and Mossbauer Spectroscopy is often used in addition to standard X-ray analysis to determine what minerals are present and in what proportions. It is only the use of such high technology instrumentation that creates a market for these clays and makes modern life so much better.

Three families of clay minerals dominate all these uses (common clay excepted) – the kaolin, smectite, and palygorskite families (please see Appendix A.3). Size, shape, surface area, viscosity and cation exchange capacity of the clay minerals are all important in applications so not just "kaolin" or "palygorskite" suffice, but rather a kaolin or palygorskite with very specific properties. In almost all of these uses, plasticity of the clay minerals is fundamental so the less quartz, feldspar, silt, and heavy minerals (collectively called *grit*), the better. These clay mineral requirements and the absence of mineral grit commonly demand very careful exploration and testing of specific beds in a deposit, be it residual, sedimentary or hydrothermal. So here again understanding the local depositional system is key to minimizing exploration costs, expanding market share, and adding value to the final product. The dependence of settling velocity on slight differences in size, shape, density, and flocculation plus the presence or absence of organic matter cause one clay bed to differ from another and thus to have different industrial uses.

The origin of these industrial clays is most diverse. Bentonites, altered ash falls in marine basins, are most common along convergent, continental-oceanic plate boundaries, but also are carried far downwind into nearby foreland and cratonic basins

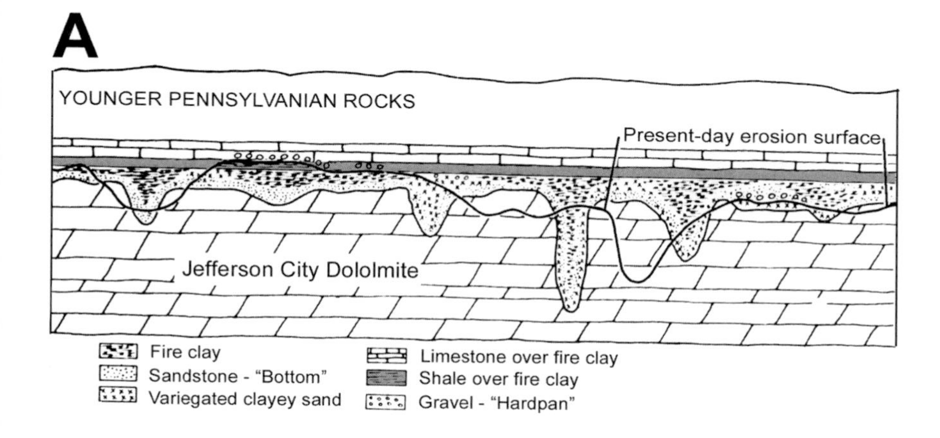

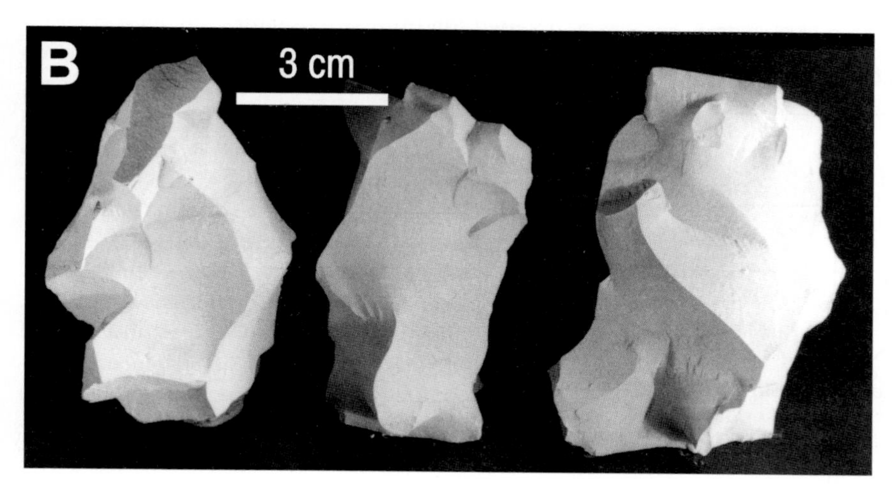

Fig. 9.8A,B. Residual flint clays of Missouri most of which occur in multiple paleosoils developed along the margin of the Ozark Dome, where sedimentation was slow (Keller 1984); idealized cross section (**A**) and samples with beautiful conchoidal fracture (**B**). Deepest sinkhole is about 10 m. Published by permission of Southern Illinois University Press

and occur in some rift basins as well. Both the residual and sedimentary kaolinitic clays, on the other hand, require a warm, wet climate and tectonic stability. The well known flint clays of Missouri are one example of such residual clays (Fig. 9.8). More diverse are the smectitic and palygorskite clays. These form in environments as varied as early alteration of volcanic detritus on the sea floor to the subaerial weathering of volcanics to deposition in hypersaline lakes or seas. Ball clay, a general field term for plastic clay used chiefly for ceramic white ware, represents still another example. These clays consist mostly of kaolinite and most occur as elongate to subcircular lenses in fluvial deposits such as those in western Kentucky, where their dimensions typically range from 250 to 600 m long by 3 to 10 meters thick (Fig. 9.9). Many of these lenses have gradational bases, sharp erosional tops, and abrupt lateral terminations (but additional information on their geometry and underlying and overlying environments – where these clays occur in a sequence stratigraphy cycle – would be helpful). Even more variable are the origins of common clay and mudstone. In all of these, *the processes are far more important than the place.*

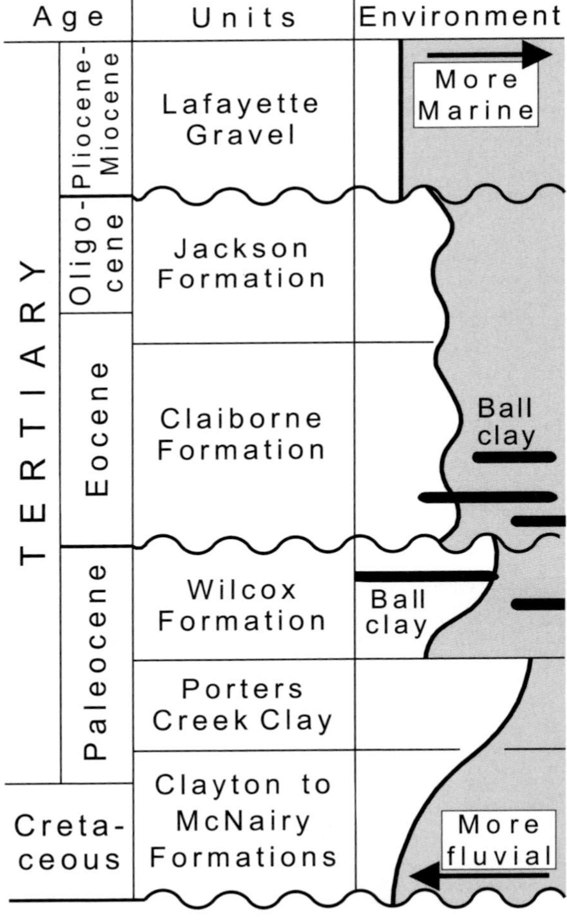

Fig. 9.9. Stratigraphic position of ball clays of Tertiary age in the Jackson Purchase of far western Kentucky, USA. These clays occur in a paralic sequence associated with some peats and much plant debris

9.5. Hydrocarbons

Estimates of the world's hydrocarbon reserves have and remain a moving target (McCabe 1998), but currently are estimated to be the world's principal energy source for many decades in the future. The expression, "moving target," is a particularly appropriate for hydrocarbons, because what constitutes a resource of any kind depends both on its price and on technology – our increasing ability to reach and extract inaccessible and leaner deposits once found. Thus there is much disagreement about their size in relation to current demand, but everyone agrees that organic-rich shales are by far the chief source of the world's petroleum. In addition, some organic-rich shales are both the source and *the reservoir* (interbedded, thin, but fractured sandstones, carbonates and even fractured mudstones). These source beds have been widely studied by geologists, micropaleontologists, chemists, and chemical engineers, who have produced a truly vast literature. The summary below is taken from Klemme and Ulmishek (1991), who identified seven key generalizations.

1. The majority of hydrocarbon reserves are very young – about 60% are believed to have been generated in the last 80 to 90 Ma.
2. Organic-rich shales with more than 0.5% TOC are the principal source of most petroleum reserves, distantly followed by coal beds.
3. Needed conditions for organic-rich mudstones include the initial preservation of organic matter on the bottom (stratified water column or overloading), warm climates (about 60% of all source rocks lie between $\pm 45^\circ$ of the equator) and tectonics (creates both deep basins with restricted circulation followed by deep burial so organic maturation occurs).
4. More than 90% of the world's petroleum reserves occur in six restricted time intervals – Silurian, Upper Devonian-Tournasian, Upper Jurassic and Middle Cretaceous (mostly deposited as transgressive units), Pennsylvanian-Lower Permian and Oligocene-Miocene (mostly deposited as regressive units). Two points are notable. The above

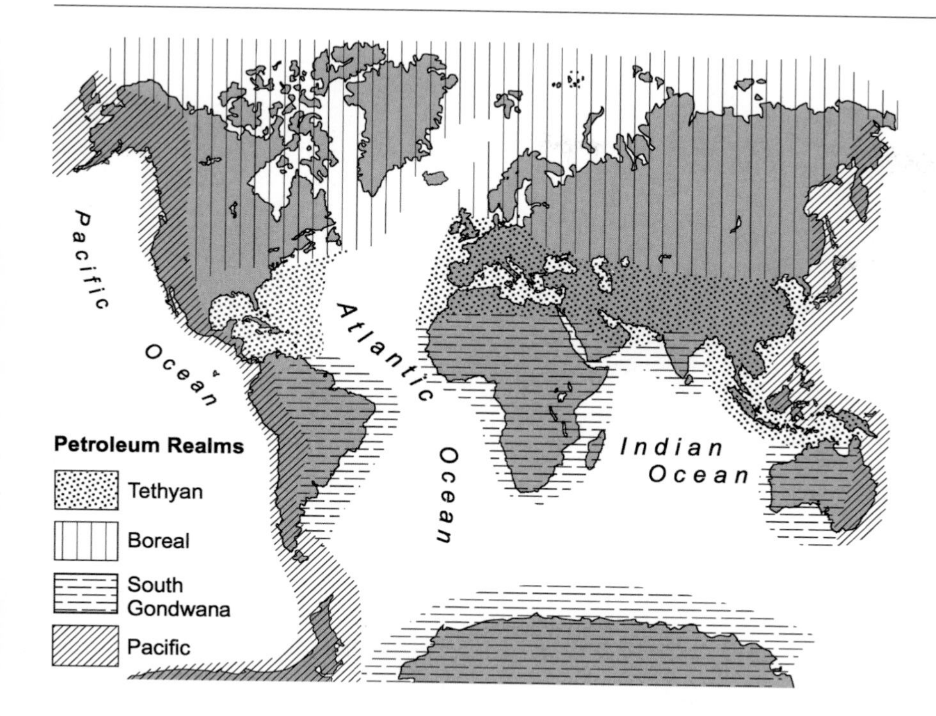

Fig. 9.10. The four great petroleum realms of the world are the result of four different geologic histories (Klemme and Ulmishek 1991, Fig. 29). Published by permission of the authors and the American Association of Petroleum Geologists

time intervals occupy less than 30% of Phanerozoic time and virtually all of these source beds are marine, distantly followed by lacustrine.

5. The sites and kinds of organic deposition have changed with time and are largely, but not entirely controlled by the evolution of both plants and animals and thus possibly linked to changing abundance of atmospheric oxygen. Open shelves, first shallow and later deep, were sites of deposition for organic-rich shales from the Upper Proterozoic through the Devonian, partially isolated deeps prevailed in the Mesozoic, and largely isolated deeps related to the Alpine and Andean Orogenies occur in the Tertiary and Quaternary. Quite possibly, evolutionary changes in the kinds of producers and consumers of organics may be responsible for these contrasts. Type II kerogen is dominant, but Type III, typical of land plants, appears in the Carboniferous, as a result of their progressively wider invasion of land. Kerogen types also appear to be related to latitude, when viewed globally. In low latitudes, source beds are likely to have principally kerogen Types I and II whereas at high latitudes Type III kerogen is more abundant (with the exception of the Oligocene-Miocene).
6. Four great realms exist (Fig. 9.10). The Tethyan with about 68% of the world's reserves, but only 17% of its area; the Boreal with 23% reserves versus 28% of area; Pacific borderlands with 5% and 17% of the area; and southern Gondwana with only 4% of the world's reserves, but 38% of its area.
7. Oil shales, although present from the Neoproterozoic through the Tertiary, do not loom large in current world oil and gas production, because of higher production and environmental costs (what to do with the great volumes of ash). As source rocks, however, they are most significant.

An oil shale is a kerogen-rich, brown to black, well laminated shale that yields appreciable oil and gas upon distillation. Many, but not all, are lacustrine in origin and, although called oil shales, many are really argillaceous limestones (marlstones) or dolostones rich in exotic carbonate and saline minerals. Some are also metal rich. The largest and richest is the lacustrine Green River Formation with 1.5×10^{13} barrels of shale oil-equivalents (Figs. Fig. 9.11A and B). This lake existed for about 45 Ma in the Cretaceous and early Tertiary in the western United States. See Smith (1990) and Russell (1990) for a summary of other lacustrine oil shales. Lacustrine oil shales appear to have mostly developed at low latitudes and elevations, which jointly favored high organic productivity (Katz 1990).

A number of organic-rich shales also directly yield natural gas – the Devonian gas shales of the Appalachian Basin are one of the best known and productive; another is the Barnett Shale of the Fort Worth Basin in Texas. Free natural gas occurs in

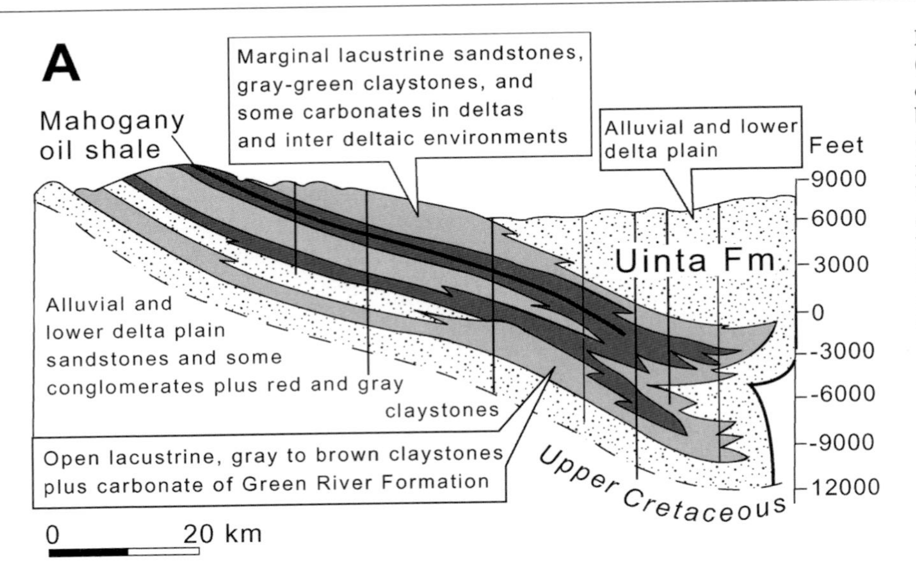

Fig. 9.11A,B. Oil shales: (**A**) down dip cross section of a classic intermontane basin (Fouch 1975, Fig. 2), and (**B**) Green River Shale as seen from Hell's Hole Overlook, Uinta County, Utah (Fouch et al. 1976, Fig. 16). Published by permission of the authors, the Rocky Mountain Geologists Association, and the Colorado School of Mines

pores and fractures and there is also adsorbed gas on organic material (Fig. 9.12). Although flow rates are far less than in sandstones or carbonates, gas shale wells have long lives and, as the price of fossil energy climbs long term and technology improves, gas from organic-rich shales should become more important.

9.6. Construction and the Environment

Fine-grained debris such as soils, clays, loess, mudstones and related clay-rich materials cause more problems in construction than other earth materials for several reasons: they are the most common lithology at the Earth's surface and they erode, weather and fail easily through change of volume, creep, liquefaction, sliding and faulting. And even a small amount of mud in sediments or rocks commonly has a significant effect on their uses as is reflected in their classifications – minor amounts of clay or mudstone invite a different name (see Figs. A.13 and A.14 in the Appendix.) This is so because most muds, clays, and mudstones are very reactive to water, and in addition, mudstones are weak structural members.

A vast geotechnical engineering literature exists on clays and mudstones (see Digging Deeper). Many, if not most, of the problems of these deposits are best resolved by multidisciplinary teams

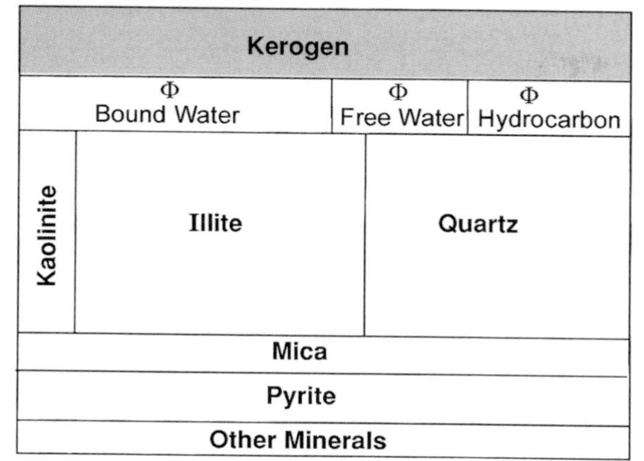

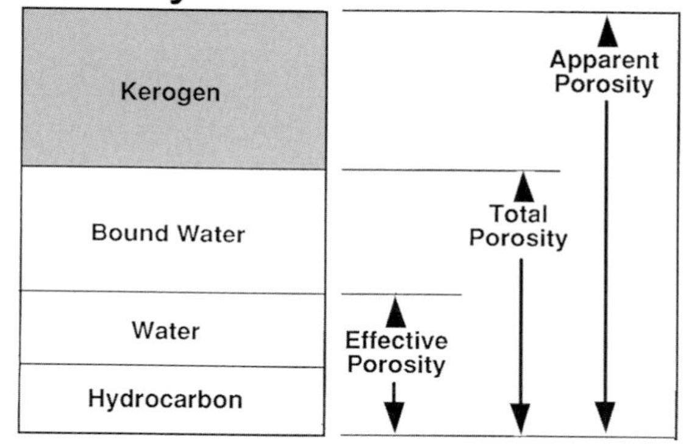

Fig. 9.12. Conceptual models of the proportions of rock and porosity (by volume) in gas producing Devonian shales of the Appalachian Basin. These probably apply to many other gas shales with appropriate local modifications (Campbell 1987, Figs. 2, 4 and 5)

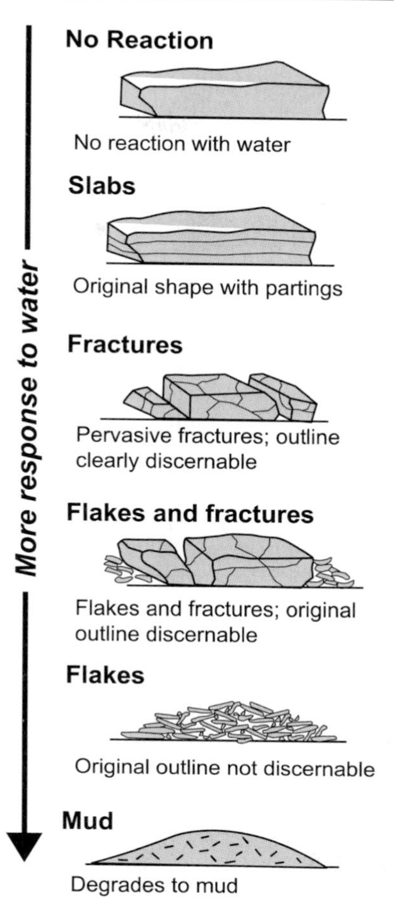

Fig. 9.13. Differential responses of mudstone to water and weathering – from total disintegration to mud, on the one hand, to no reaction on the other (adapted from Santi 1998, Fig. 1). This test shows that a "quick and dirty" way to evaluate a mudstone is simply submerge it in water. Published by permission of the author and the Geological Society of America

of geotechnical engineers, geologists, and soil scientists working with designers and architects. Here, parts of this literature are summarized with emphasis on the consolidated fine-grained deposits, especially the claystones, which cause the most problems. Our point of view is largely geological and we focus on slope stability and erosion, volume changes resulting from both physical and chemical processes, and the need to give even thin mudstones in deep cuts and mines attention.

Consolidated or not, the stability or durability of fine-grained materials, especially the argillaceous ones, in foundations, fills and cuts and in mines and on slopes depends largely on *how they react with water*. For example, when a mudstone weathers will it become a paste, flake into small pieces, expand many times its initial volume or perhaps will only some of its selected laminations expand (Fig. 9.13)? Such possible changes range mudstone – does it have many or few expandable clays, is it laminated, slickensided, massive, cemented by carbonate or silica, rich or poor in pyrite or gypsum, does it have little or much silt, and will it be subject to much or little water? California slopes illustrate how shear strength of colluvium decreases as abundance of smectite increases (Fig. 9.14). It is always well to remember that in comparison to almost all other lithologies, clays and mudstones alter very rapidly, especially where water is readily available. The several *slake durability* (Santi 1998) and *swelling tests* (Shakoor and Sarman 1992) are widely used for evaluating this property. These test alternately wet and dry mudstone under somewhat different conditions. *Freeze-thaw* tests are also important in cold climates (Lienhart and Stransky 1981). The standard procedures of soil mechanics – determination of the plastic and liquid limits and

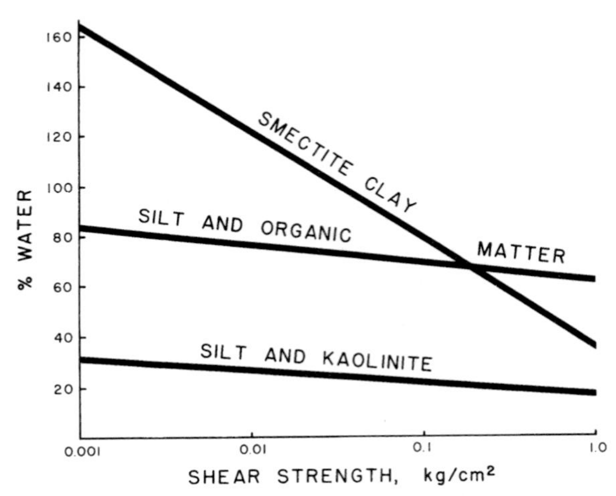

Fig. 9.14. On many California slopes the more smectite, the lower is the shear strength of the colluvium (Borchardt 1977, Fig. 2). In contrast, neither the abundance of organic matter nor kaolinite have great effect as shown by these generalized regression lines

tests to determine plasticity index and shear strength – are always valuable and commonly essential. See Table 9.2 to recognize in the field special soils.

Mudstone type is key to understanding failure. Claystones are the most susceptible to failure of all the argillaceous rocks, because they have the least silt and are thus both the weakest and most readily altered. The presence of expandable clay minerals greatly lessens stability by change in volume. The more organic matter, the greater the compaction of muds and the less their shear strength. Organic matter in a silt or mud also creates methane gas leading to overpressure. Methane gas also causes *whiteouts* (loss of detail) in shallow high-resolution seismic profiles made for construction of off shore drilling platforms and pipelines and bridges in large estuaries or lakes. Passing storm waves or seismic shocks alter near surface pore water pressures of unconsolidated sediment (dynamic cyclic loading) and reduce effective pore water pressure (see Transport and Deposition in Chap. 3 and Liquefaction below) leading to possible failure. The exsolution of methane gas in the pore waters of organic-rich mud at the bottom of the pressure pulse accentuates this process. Conversely, in Paleozoic and older black shales, organic matter, now mature kerogen, acts as a cement and binder and makes them hard, resistant, and brittle. However, even these black shales are likely to be fissile when weathered, because their finely interlaminated gray mudstones easily alter whereas the black, organic-rich laminations do not. An exception are argillites, which do not become fissile, because they are so well indurated.

Table 9.2. Recognizing expansive, saline, and pyritic soils. Adapted from Costa and Baker (1981, Table 8.7)

Expansive Soils
- *Dry*: Hard dry and cracked; glazed when cut
- *Wet*: Great stickiness makes walking difficult and impedes heavy machinery; easily rolls into a ball (hands powdery when dry)
- *Dry or Wet*: Cracked, rigid structures inside or out (but also from other causes)

Saline soils
- Surface of soil powdery and loose when dry
- "Frosted or lightly snow covered" appearance when cool near sunrise or sunset
- Wooden posts enlarge at base possibly with salt crystals
- Cracking of light structures, but not heavy ones, in semi arid and arid climates

Pyritic soils and fills
- Sulfur dioxide and hydrogen sulfide smell in fresh cuts
- Smoking or steaming on cold days
- Fissility in laminated slabs
- Brownish, iron-stained surfaces and water
- Secondary gypsum

9.6.1. Slopes

Undercutting or overloading a hillside (placing a new fill or building on its soil or colluvium), steepening its slope, or cutting its vegetation (reducing transpiration by plants adds water and weight to the slope) all promote failure (Fig. 9.15). Even where thin, resistant sandstone and limestone ledges are present in

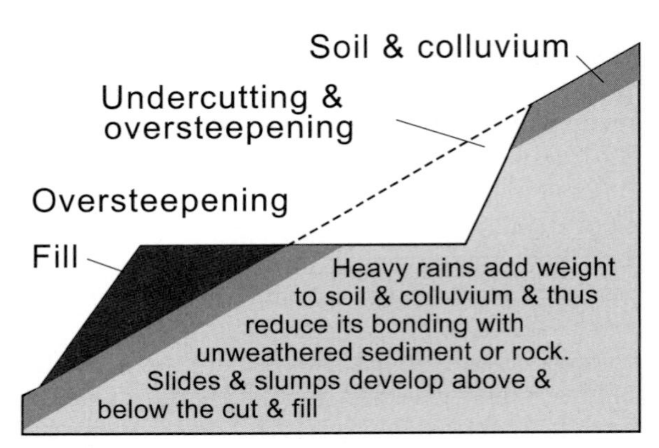

Fig. 9.15. Benching a slope for a building site needs special attention to hillside drainage, especially for hillsides underlain by mudstones, which commonly have thick colluvium. But everywhere, fill on slopes deposited directly over soil and colluvium invites trouble, because the soil and colluvium both have low shear strengths and transmit water

Fig. 9.16. "Landslide collapse" of State Route 32 east of Peebles in Adams County of southern Ohio caused by expandable clay minerals in the Olentangy Shale of Devonian age

cuts, they can be quickly undermined by the recession of the more easily weathered mudstone so that rock falls and other failures are common (Shakoor 1995). Along active continental margins, mudstones with much original fine volcanic glass – now mostly smectite and zeolite – are especially prone to failure as demonstrated by the famous deep Culebra section of the Gaillard Cut of the Panama Canal (Legget and Karrow 1983, pp 29–5 to 29–7).

Where a mudstone forms even a single thin bed in a carbonate or sandstone cut – perhaps a thin marine transgressive mudstone, an underclay, or a bentonite, it deserves attention, especially if it contains expandable clay minerals and dips toward the cut. Such a thin mudstone is a potential surface of failure much as a water-saturated siltstone or sensitive clay is in a muddy or sandy section. Such thin beds can liquefy during an earthquake and become a slide plane with disastrous results for overlying structures (see Hansen in Digging Deeper). Slope failures also occur where unconsolidated materials (informally called, "soil", by engineers) are used as cover for landfills, for embankments along highways, and for bridge abutments (Fig. 9.16).

9.6.2. Volume Changes

Changes in volume, commonly known as *heaving*, refer to the expansion of a soil or mudstone and may have either a physical or chemical origin. Heaving occurs in soils, under roads and basements, in deep cuts and underground and, while its causes vary widely, water is usually a common denominator. In cold climates, freezing is a major cause of heaving. Expandable clay minerals, pyrite, and gypsum (Fig. 9.17) all favor heaving, which produces irregular, small-scale open folds, slickensides, and open porous textures. Many additional criteria can be used to identify heave-prone, gilgaí soils (Table 9.3). These are likely to develop over mudstones rich in either pyrite or gypsum, especially in the low areas of semi-arid regions, where the water table is close to the surface (Burkart et al. 1999).

On the other hand, the swelling of smectitic clays and mudstones is used to great advantage for the construction of permeability barriers for landfills, underground tanks and for the burial of hazardous waste. Here expandable clays block and seal fractures and they flow into irregularly shaped voids. Bentonites are widely used for this purpose. Swelling clays as permeability barriers can, however, develop microfractures through wetting and drying or, because of low shear strength, such barriers may fail. Thus careful design is necessary. See Sethi and Schieber (1998, pp 245–248) for a good summary of the uses of swelling clays.

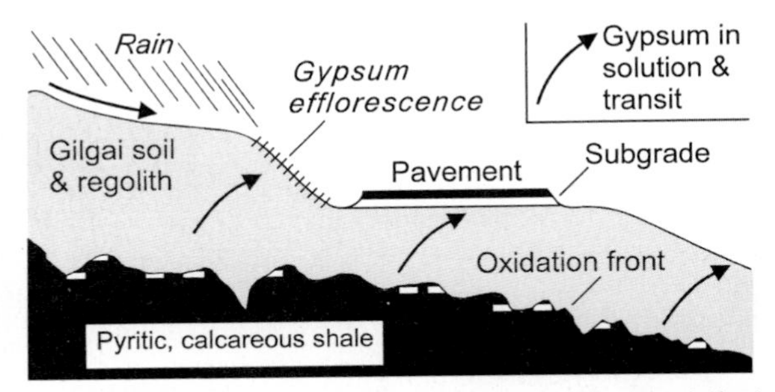

Fig. 9.17. Broken highway concrete and heaving (gilgai soils) are the consequence of the weathering of pyrite in the presence of carbonate (after Burkart et al. 1999, Fig. 3.4). Published by permission of the author and the Association of Engineering Geologists

Reduction of volume of otherwise stable unconsolidated clays can occur after long dry periods, which reduce pore water pressures above the water table. In many tropical soils, an open fabric of quartz grains

Table 9.3. Pennsylvanian shale facies, environment and roof quality (after Breyer 1992, Table 1)

Facies	Rock Types	Structures and Fossils	Environment/Roof Quality
Carbonaceous shale	Dark gray claystone; some light gray, graded mudstone	Thicker dark gray laminations alternate with 1–5 mm light gray laminations; bioturbation; plant debris and brachiopod valves	Probable shallow anoxic lake rich in carbonaceous debris; *good roof*
Gray shale	Medium to light gray shale	Massive to finely laminated with some deformational structures and abraded plant debris throughout, but lacks bioturbation; rare joints and small faults	Crevasse splay deposits-proximal facies, has sand streaks that are absent in distal facies; generally *good roof*
Gray shale with sand	Medium to light gray shale	Most sand laminations less than 1 mm, but some as thick as five and cross laminated; abraded plant debris but no bioturbation; small pyrite nodules common. A few joints and small faults	Quiet water deposition, possibly marine or brackish?
Greenish-gray mudrock	Dark to light green mottled claystone	Massive to weakly laminated with much burrow mottling. Isolated pyrite framboids common as are some calcite concretions; lacks fossils	
Mudrock with slickensides	Greenish to dark gray shale	Massively bedded possibly by rootlets or soil formation. Pedogenic slickensides	Slickensided shale is underclay; *poor roof*
Laminated shale	Dark gray claystone interlaminated with light gray mudstone	Laminations typically 2–4 mm, fallen trees, slumps, lags, and some brachiopod valves; little bioturbation and soft sediment deformation; disseminated siderite in coarser layers; joints and small faults common	Probable shallow lake with gray laminations representing flood phases; *poor roof*

Fig. 9.18. Settling of a house wall on collapsible, tropical oxisoil in Rio Claro, Sao Paulo, Brazil. Note cracks in wall and wooden supports in windows

and kaolinitic clay is responsible for a similar effect – excess water passing downward through the soil causes a reduction in volume of its clay matrix and thus compaction of the soil (Fig. 9.18). Here clay coats on framework grains as well as aggregates of clay or silt "wash out" when subject to excess water. A large void ratio and a moisture content less than complete saturation are needed for this type of compaction.

9.6.3. Liquefaction

Muds, silts, and fine sands are susceptible to liquefaction, which transforms a water-saturated solid into a liquid. Liquefaction occurs at the surface in unconsolidated muds, silts and fine sands or in the shallow subsurface. There are several common ways this happens – seismic shock, passing storm waves, and the deflocculation and compaction of marine clays.

Seismic shock produces liquefaction in fine silts and sands by rapidly oscillating pore water pressures, much as a vibrocore liquefies mud at the bottom of its core barrel (see Box 3.1). Shaking also causes compaction and frees pore water until its pressure equals overburden pressure and rupture occurs. Such overpressures will fluidize a bed, may fracture and fault overlying beds and inject dikes, cause some layers to founder, and form elongate or spherical pillows. Where the bed has a low dip and is hydraulically confined, liquefaction may cause downslope movement of overlying deposits. Silty muds, silts and fine sands are susceptible to liquefaction, because they have high porosities, but relatively low permeabilities so pore water can escape only slowly.

Storm waves impinging muddy bottoms also create rapid oscillations of pore pressure, which reduces shear strength leading to mass wasting, remolding, liquefaction and flowage (Suhayda et al. 1976; Prior and Coleman 1982). This is sufficient to destabilize large offshore drilling rigs, especially those on the foreslopes of modern deltas. (Some deformed beds on cratons called "seismites" may also have such an origin). Methane gas, generated from organic matter and released from gas hydrates in bottom mud, also significantly contributes to loss of shear strength in deltaic muds. Such gas causes seismic whiteouts and, where vented on the sea bottom, produces pockmarks, sediment and gas plumes and even mud volcanoes, some with carbonate caps (Roberts 2001).

Liquefaction also can occur where marine muds and silts are uplifted and flushed by fresh water as is well documented in the quick clays of Canada and Scandinavia. A well-known example is the Leda Clay of Canada (Torrance 1988), which consists predominantly of finely ground rock flour – about 50% feldspar and quartz followed by illite and chlorite, all minerals with low cation exchange capacity. This rock flour was deposited about 10,000 years ago on the widespread Champlain Sea floor with an open, flocculated fabric loosely tied together by Na ions. Post-glacial isostatic uplift has exposed these still-unconsolidated soft marine clays to fresh water flushing, which makes the fabric susceptible to failure by seismic shock or by either overloading or undercutting slopes, leading to regressive and rotational landslides (Lefebvre 1986). In general, liquefaction is favored by an open, flocculated structure of framework grains with low cation exchange capacity deposited in marine or brackish environments. As salinity of pore water decreases through fresh water invasion, the bonding of these open fabrics by sodium ions weakens and failure can occur.

9.6.4. Pyrite

A mudstone rich in pyrite has great potential for trouble in construction when it oxidizes. The volume and mineralogy of the host mudstone rapidly change, released ferric iron stains and possibly alters nearby surfaces and solids, proximal streams are acidified and discolored, and vegetation is inhibited from growing on outcrops and artificial fills with abundant pyrite because of acid pore water. These adverse consequences, which can occur rapidly in weeks to months, result from the formation of sulfuric acid, free sulfur, and sulfates such as jarosite and alum. Where carbonates are present, the soils tend not to be acidic, and gypsum is a prominent product. As pyrite oxidizes, a well-indurated Paleozoic mudstone quickly converts into a soft deformable mass via leaching and mineral transformations well summarized by Pye and Miller (1990). Four key chemical equations are the following:

$$\underbrace{FeS_2}_{\text{Pyrite}} + H_2O + 3.5\,O_2 \rightarrow \underbrace{FeSO_4}_{\text{Ferrous sulfate}} + \underbrace{H_2SO_4}_{\text{Sulfuric acid}}$$

$$2\,FeSO_4 + 0.5\,O_2 + H_2SO_4 \rightarrow \underbrace{Fe_2(SO_4)_3}_{\text{Ferric sulfate}} + H_2O$$

$$Fe_2(SO_4)_3 + FeS_2 \rightarrow \underbrace{3\,FeSO_4}_{\text{Ferrous sulfate}} + 2\,S$$

$$\underbrace{CaCO_3}_{\text{Calcite}} + H_2SO_4 + H_2O \rightarrow \underbrace{CaSO_4 \cdot 2\,H_2O}_{\text{Gypsum}} + CO_2\,.$$

These reactions proceed most rapidly where the pyrite is present as small crystals rather than large masses and in the presence of ample water and oxygen. The bacterium *Thiobacillus* acts as a catalyst for several of these reactions. Taken to completion, micas, chlorites, smectites and mixed layer clays plus feldspars can all be ultimately transformed into kaolinite, which itself can finally be dissolved.

The adverse consequences of these reactions are several – volume changes with cracking of overlying rigid slabs, loss of shear strength as primary diagenetic fabric is destroyed, staining of surfaces, acidification of effluents, and inhibition of plant growth at the outcrop or on the fill. Even in the high Arctic of Lapland and Finland, black, pyrite-bearing schist has a poisonous effect on vegetation. This is easily recognized by its reddish weathering crusts on blocks from recent construction (Rapp 1960, p 95). In Finland, waters of post glacial lakes with outcrops of black schists have greater concentrations of heavy metals than lakes without schists (Loukola-Ruskeeniemi et al. 1998). Another example is a recently-emergent Holocene, pyrite-bearing, marine, organic-rich mud, which quickly turns red upon oxidation and inhibits plant growth. Where the mudstone was originally calcareous and contained pyrite, gypsum will develop as it weathers, which expands the volume of the host as it precipitates causing gilgai soils (Fig. 9.17). Where embankments and fill must be constructed of pyrite-rich materials, immediate covering by soil minimizes oxidation and is essential.

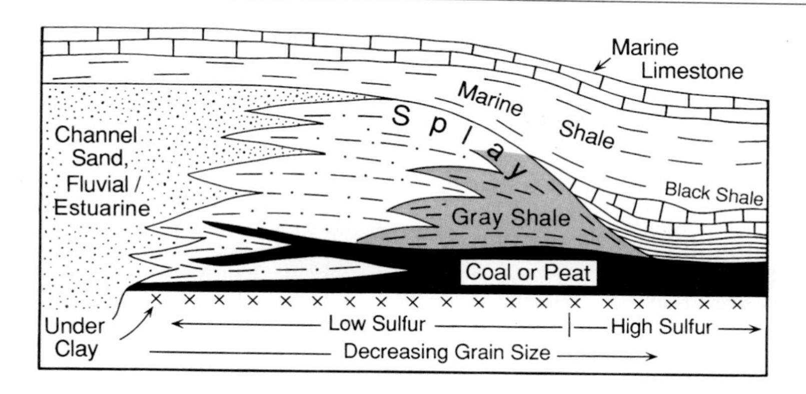

Fig. 9.19. Diagrammatic cross section showing how sulfur content of Illinois coals varies with the overlying depositional environment of the coal (Treworgy and Jacobson 1986, Fig. 6). Where black shale directly overlies coal, the coal is high in sulfur (marine waters soaked into the coal), but where intervening fresh water mud was first deposited above the coal, this was not possible. Published by permission of the authors and the Southern Illinois University Press

9.6.5. Coal Quality and Mudstones

The sulfur content of coal, perhaps its key quality in a world concerned with atmospheric pollution from fossil fuels, is closely related to the environment and water chemistry of the swamp and of the beds immediately overlying it (Fig. 9.19). Was the coal bed a brackish water swamp along a coast, was it a fresh water swamp buried by fresh water mud low in SO_4^{-2}, or was it a fresh water swamp buried by sediment deposited from marine waters high in SO_4^{-2} shortly after deposition? The coastal coals of western Indiana illustrate well these possibilities and also show the great economic importance of distinguishing marine from non-marine mudstones and their importance as seals.

The Danville Coal of the Pennsylvanian Upper Des Moinesian Series of the Illinois Basin has both low and medium high sulfur contents relatively close together. Where there are three or more meters of gray silty and sandy mudstones above it, the Danville has less sulfur than where the overlying Pusseron Sandstone is close to the coal and served as a conduit for marine waters (Mastalerz and Padgett 2002). The second example from the western Indiana area is the Block Coals (Mastalerz et al. 1999), which locally are covered by two coastal tidal flats – one more offshore and sandy (with tidal rhythmites) and the other more inshore and muddy (remember that on a tidal coast mud is carried farther inland in suspension by tidal currents than is sand and is finally deposited at "dead water"). Coal covered by the up-dip mud flat has high C/S values (low sulfur coal) and $\delta^{13}C$ values more negative than −21‰ whereas the coal covered by the marine waters on the sand flat has low C/S values (high sulfur coal) and less negative $\delta^{13}C$ values of −20 to −18‰. Both examples illustrate the importance of combining sedimentology, stratigraphy and geochemistry for best results.

9.6.6. Importance of Defining Mudstone Lithologies

Careful definition of the different lithologies of mudstones (defined by texture, stratification, the presence of pyrite or gypsum, expandable clay minerals, etc) and how they react with ground water or mineralizing fluids is the key to working successfully with them. Above all, we need to remember that mudstones are not "just shale" and that laminations of different compositions, different silt and organic contents, different kinds of stratification, and different forms, sizes, and abundances of pyrite cause great contrasts in the response of a mudstone to water and air or to a new stress environment. A detailed lithologic classification is likely to be useful here (Macquaker and Adams 2003). Differential response can occur in a deep cut, tunnel or underground mine where residual stresses are likely to be released abruptly and thus alter significantly the engineering properties of some mudstones more than others (Moon and Beattie 1995). Mudstone types in mine roofs of underground coal mines provide a good example (Table 9.3). Prediction of roof quality – mudstone stability and its response to moisture laden air – is enhanced by understanding the entire depositional system: *where and why different mudstones within it vary, and their role as barriers to flow.*

9.7. Summary

The practical applications of mudstones to the modern world – unappreciated by all too many of us – are vast and range from making talcum powder easier and more uniform to apply, to paint, toothpaste, fertilizer, kitty litter, catalysts, ceramics, and impermeable barriers for landfills and toxic and nuclear waste. Most importantly, organic-rich mudstones are the source of almost all our petroleum. For all these diverse uses, finding and developing a deposit is always enhanced by recognition of its correct depositional system. Even more critical is establishing the geometry of its internal sub-facies and their properties. Collectively, these determine the architecture of the deposit and guide one to its most valuable parts.

References

Almon WR, Dawson WC, Sutton SJ, Ethridge FG, Castelblanco B (2002) Sequence stratigraphy, facies variation and petrophysical properties in deepwater shales, Upper Cretaceous Lewis Shale, south-central Wyoming. Gulf Coast Association of Geological Societies Transactions 52:1041–1053

Ampian SG (1985) Clays. In: Mineral Facts and Problems, 1985 Edition. US Bureau Mines Bulletin 675:1–13

Arkimsa H, Hyvönen E, Lerssi J, Loukola-Ruskeeniemi K, Vanne J (1999) Compilation of maps of black shales in Finland: applications for exploration and environmental studies. In: Autio S (ed) Current Research, 1997–1998. Geological Survey Finland (Special Paper) 27:11–114

Aughenbaugh NB (1990) The geotechnical importance of clay fabric. In: Bennett RH, Bryant WR, and Hubert ME (eds) Microstructure of Fine-grained Sediments. Springer, Berlin Heidelberg New York, pp 515–518

Borchardt GA (1977) Clay mineralogy and slope stability. California Division of Mines and Geology (Special Report) 133:15

Breyer JA (1992) Shale facies and mine roof stability: a case study from the Illinois Basin. Geol Soc Am 104:1024–1030

Burkart B, Goss, Glenn C, Kern JP (1999) The role of gypsum in production of sulfide-induced deformation of lime-stabilized soils. Environ Eng Geosci 5:173–187

Campbell RL (1987) A Devonian Shale-specific formation evaluation log. Eastern Gas Shale Technol Rev 4:19–29

Costa JE, Baker VR (1981) Building with the Earth. Wiley, New York, 498 p

Dawson WC, Almon WR (1999) Top seal character and sequence stratigraphy of selected marine shales in Gulf Coast style basins. Gulf Coast Association of Geological Societies Transactions 49:190–197

Dawson WC, Almon AR (2002) Top seal potential of Tertiary deep-water Gulf of Mexico shales. Gulf Coast Association of Geological Societies Transactions 52:167–176

Fouch TD (1975) Lithofacies and related hydrocarbon accumulations in Tertiary strata of the western and central Uinta Basin. In: Bolyard DW (ed) Deep Drilling Frontiers of the Central Rocky Mountains. Rocky Mountain Association Geologists, Denver CO, pp 163–173

Fouch TD, Cashion WB, Ryder RT et al. (1976) Field guide to the lacustrine and related non-marine depositional environments in Tertiary rocks, Unita Basin, Utah. In: Epis RC et al. (eds) Studies in Colorado Field Geology. Colorado School Mines Professional Contribution 8:358–385

Franca AB, Araújo LM, Maynard JB, Potter PE (2003) Secondary porosity formed by deep meteoric leaching: southern South America. American Association Petroleum Geologists Bulletin 87:1–10

Garven G (1992) Continental-scale groundwater flow and geologic processes. Annu Rev Earth Planet Sci 23:89–117

Geehan GW, Lawton TF, Sakurat S, Klob H, Clifton TR, Iman KF, Nitzberg KE (1986) Geologic prediction of shale continuity Prudhoe Bay Field. In: Lake LW, Carroll HB Jr (eds) Reservoir Characterization. Academic Press, Orlando, pp 63–82

Goodfellow WD, Lydon JW, Turner RJW (1993) Geology and genesis of stratiform sediment-hosted (SEDEX) zinc-lead sulphide deposits. In: Kirkham RV, Sinclair WD, Thrope RI, Duke JM (eds) Mineral Deposits Modeling. Geological Association Canada (Special Paper) 40:201–251

Grainger P (1985) The micro fabric of deformed mud rocks of the Cracking Formation, southwest England. Proceedings Geologists Association 96:143–152

Hembley JJ, Hunt JP (1992) Hydrothermal ore-forming processes in the light of studies in rock-buffered systems: II Some geologic applications. Economic Geology 87:23–41

Huyck HLO (1990) When is a metalliferous black shale not a black shale? In: Grauch RI, Huyck HLO (eds) Metalliferous Black Shales and Related Ore Deposits – Proceedings (1989). United States Working Group Meeting, International Geological Correlation Program Project 254. US Geological Survey Circular 1058:42–56

Jewell PW (2000) Bedded barite in the geologic record. In: Marine Authigenesis from Global to Microbial. Society Economic Paleontologists Mineralogists Special Publication 66:147–161

Katz BJ (1990) Controls on distribution of lacustrine source rocks through time and space. In: Katz BJ (ed) Lacustrine Basin Exploration, Case Studies and Modern Analogs. American Association Petroleum Geologists Memoir 50:61–76

Keller WD (1984) Flint-clay facies in Pennsylvanian-age rock of Missouri. In: Belt ES (ed) Compte Rendu. Neuviène Congres International de Stratigraphie et de Géologie du Carbonifere, Washington/Champaign-Urbana, May 17–26, 1979, vol. 3, pp 619–625

Klemme HD, Ulmishek GF (1991) Effective petroleum source rocks of the world: stratigraphic distribution and controlling depositional factors. American Association Petroleum Geologists Bulletin 75:1809–1851

Lefebvre G (1986) Slope instability and valley formation in Canadian soft clay deposits. Canadian Geotechnical Journal 20:261–270

Legget RF, Karrow PI (1983) Handbook of Geology in Engineering: McGraw Hill, New York, various paging

Leventhal JS (1998) Metal-rich black shales; Formation, economic geology and environmental consideration. In: Schieber J, Zimmerle W, Sethi P (eds) Shales and Mudstones, vol II. E. Schweizerbart'sche, Stuttgart, pp 255–282

Lienhart DA, Stransky TE (1981) Evaluation of potential sources of riprap and armor stone – methods and considerations. Bulletin Association Engineering Geologists 18:323–332

Loukola-Ruskeeniemi K, Uutela A, Tenhola M, Paukola T (1998) Environmental impact of metalliferous black shales at Talvivarra in Finland, with indication of lake acidification 9000 years ago. Journal Geochemical Exploration 64:395–407

MacQuaker JHS, Adams AE (2003) Maximizing information from fine-grained sedimentary rocks: An inclusive nomenclature for mudstones. Journal Sedimentary Research 7b:735–744

Mastalerz M, Kvale EP, Stankiewicz BA, Porte K (1999) Organic chemistry in Pennsylvanian tidally influenced sediments from SW Indiana. Organic Geochemistry 30:57–73

Mastalerz M, Padgett PL (2002) Coal quality controls of the Danville Coal in Indiana (Illinois Basin, central USA). International Journal of Coal Geology 48:217–231

Maynard JB (1991) Shale-hosted deposits of Pb, Zn, and Ba: syngenetic deposition from exhaled brines in deep marine basins. In: Force ER, Eidel JJ, Maynard JB (eds) Sedimentary and Diagenetic Mineral Deposits: A Basin Analysis Approach to Exploration. Review in Economic Geology 5:177–185

Maynard JB, Morton J, Valdes-Nodarse EL, Diaz-Carmona A (1995) Sr isotopes of bedded barites: a guide to tectonic setting with implications for Pb-Zn mineralization. Economic Geology 90:2058–2064

McCabe PJ (1998) Energy resources – cornucopia or empty barrel? American Association Petroleum Geologists Bulletin 82:2110–2134

Moon VG, Beattie AG (1995) Textural and micro structural influences on the durability of Waikato Coal measures mudrocks. Journal Engineering Geology 28:303–312

Moore DW, Young LE, Modene JS, Plahuta JT (1986) Geologic setting and genesis of the Red Dog zinc-lead-silver deposit

western Brooks Range, Alaska. Economic Geology 81:1696–1727
Murray HH (2000) Traditional and new applications for kaolin, smectite and palygorskite: a general review. Applied Clay Science 17:207–221
Neuzil CE (1994) How permeable are clays and shales? Water Resources Research 30:145–150
Neuzil CE, Belitz K (1998) Fracture control of the hydrology of the North American mid-continental shales. In: Fluid Flow Through Faults and Fractures in Argillaceous Formations. Proceedings of the 1996 OECD International Workshop, Berne, Switzerland, pp 157–162
OECD Nuclear Energy Agency (1998) Fluid Flow through Faults and Fractures in Argillaceous Formations. Proceedings Joint NEA/EC Workshop, Berne, Switzerland, 10–12 June 1996 (Nuclear Energy Agency/European Commission). Organization, Economic Co-operation and Development, 399 p
Pasăva J (1996) A group of papers devoted to the metallogeny of black shales, Preface. Economic Geology 91:1–3
Prior DB, Coleman JM (1982) Active slides and flows in unconsolidated marine sediments on slopes of the Mississippi delta. In: Saxov S, Nieuwenhuis JK (eds) Marine Slides and other Mass Movements. Plenum Press, New York, pp 21–48
Pye K, Miller JA (1990) Chemical and biochemical weathering of pyritic mudrocks in a shale embankment. Quarterly Journal Engineering Geology 23:365–381
Rapp A (1960) Recent developments of mountain slopes in Kärkevagge and surroundings northern Scandinavia. Geografiska Annaler 42:71–200
Roberts HH (2001) Fluid and gas expulsion on the northern Gulf of Mexico continental slope: Mud-prone to mineral-prone responses. In: Paull CK, Dillon WP (eds) Geophysical Monograph 124. American Geophysical Union, Washington DC, pp 143–161
Russell PL (1990) Oil Shales of the World: Their Origin, Occurrence, and Exploration. Pergamon Press, New York, 753 p
Santi PM (1994) Classification of weathered rock materials. Association Engineering Geologists News 37:35–40
Santi PM (1998) Improving the jar slake, slake index and slake durability tests for shales. Environmental and Engineering Geoscience 4:385–396
Sethi PS, Schieber J (1998) Economic aspects of shales and clays: an overview. In: Schieber, Jürgen, Zimmerle W, Sethi PS Shales and Mudstones. E. Schweizerbart'sche, Stuttgart, 2, pp 237–253
Shakoor A (1995) Slope stability considerations in differently weathered mudrocks. In: Haneberg WC, Anderson SC (eds) Clay and Shale Slope Instability. Geological Society America, Reviews in Engineering Geology 10:131–138
Shakoor A, Sarman R (1992) Swelling potential classification of mudrocks based on geotechnical properties of selected shales from northeast Ohio. Bulletin Association Engineering Geologists 24:363–379
Smith MA (1990) Lacustrine oil in the geologic record. In: Katz BJ (ed) Lacustrine Basin Exploration, Case Studies and Modern Analogs. American Association Petroleum Geologists Memoir 50, pp 43–60
Stow DAV, Piper DJW (eds) (1984) Deep-water, fine-grained sediments; history methodology and terminology. In: Stow DAV, Piper DJW (eds) Fine-Grained Sediments: Deep-Water Processes and Facies. Geological Society (Special Paper) 15:3–14
Suhayda JN, Whelan T, III, Coleman JM, Booth JS, Garrison LE (1976) Marine sediment in stability: interaction of hydrodynamic forces and bottom sediments. Proceedings Offshore Technology Conference, Houston, TX, Paper 2426, pp 29–40
Torrance JK (1998) Mineralogy, pore-water chemistry, and geotechnical behavior of Champlain Sea and related sediments. In: Gadd NR (ed) The late Quaternary Development of the Champlain Sea Basin. Geol Assoc Can (Special Paper) 35:259–275
Treworgy CG, Jacobson RJ (1985) Paleo environments and the distribution of low-sulfur coal in Illinois. In: Cross AT (ed) Neuvième Congrès International de Stratigraphie et de Geologie du Carbonifere, Compte Rendu, May17 to 26 (1979) Washington/Champaign and Urbana. Southern Illinois University Press, Carbondale, 4, pp 349–359
Virta RL (1998) Clay and Shale. In: Mineral Industry Surveys. US Geological Survey, pp R1–R27

Digging Deeper

Bjerrum L, Laken T, Heiberg S (1979) A field study of the factors responsible for quick clay slides. Norges Geotekniske Institutt, Pub 85, pp 17–25

Classic, easy-to-understand summary of quick clay slides is a must for all those who have to live or work in such areas. Excellent.

Das BM (2000) Fundamentals of Geotechnical Engineering. Brooks/Cole, Pacific Grove, CA, 593 p

Good reference for geologists to begin learning the language and tests of geotechnical engineering.

Haneberg WC, Anderson SA (eds) (1995) Clay and Shale Slope Instability. Geological Society of America, Reviews Engineering Geology 10, 160p

Empirical and theoretical aspects of slope stability in fine-grained materials addressed in ten papers some of which require background.

Hansen WR (1965) The Alaska Earthquake March 27 (1965) effects on communities, Anchorage. US Geological Survey Professional Paper 542-A, 68 p

Full details of geology and engineering properties of the Bootlegger Cove Clay that underlies much of Anchorage and how it failed during a severe 8.5 earthquake. Important reading.

Harris GM, Beardow AP (1995) The destruction of Sodom and Gomorrah: a geotechnical perspective. Quarterly Journal Engineering Geology 28:349–362

A fascinating detective story based on the Book of Genesis, local geology, seismology, Strabo's historical account, archeology, and perhaps most important, the response of silt to cyclic shear stress.

Hunt JM (1996) Petroleum Geochemistry and Geology, 2nd edn. Freeman, New York, 741 p

This in depth treatise on all aspects of the origin and geology of petroleum by a life long researcher has four parts: (Introduction, Origin and Migration, Habitat, and Applications) plus 32 excellent colored photographs. A very well written, referenced and illustrated book that shows well the many interactions between organic carbon and shales. A must, if you want to dig deeper into this important topic.

Kingery WD (1984) Ancient technology to modern science, Vol. 1. American Ceramic Society, Columbus, 341 p

Ceramics are probably the most obvious use of clays in society. This book is the first volume in a Ceramics and Civilization series, and has 3 parts and 17 articles. Titles range from brick making in Babylon to ceramic making in Medieval Spain to the history of ceramic patents.

Large DE (1981) Sediment-hosted submarine exhalative lead-zinc deposits – a review of their geological characteristics and genesis. In: Wolf KH (ed) Handbook of Strata-bound and Stratiform Ore Deposits. Elsevier, Amsterdam, 9, pp 469–507

The first comprehensive work on these deposits. Still worth your attention.

Manning DAC (1995) Introduction to Industrial Minerals. Chapman and Hall, London, 276 p

A good starting point, this paperback has 11 chapters and 3 appendices (phase diagrams, tables of quality variation for understanding assessments of mineral deposits, and computer ware both hard-and soft), many illustrations and tables and even a section on landfills. Chapter 3, Industrial Clays, most relevant.

McClellan GH, Eades JL, Johnson NA (1998) Simple method to detect sulfide sulfur in rocks. Environmental Engineering Geoscience 4:115–116

A semiquantitative, fast, and inexpensive method to determine the presence of this important element as a sulfide.

Meyers D, Howard JJ (eds) (1983) Evaluation of Clays and Clay ! Minerals for Application to Repository Sealing. Battle Memorial Institute, Technical Report Columbus, OH (prepared for the Office of Nuclear Waste Isolation), 180 p

This report provides in six chapters a clear account of shale properties and their application for repository sealing. Good starting place.

Santi PM (1994) Classification of weathered rock materials. Association Engineering Geologists News 37:35–40

Tables 1, 2, and 3 compile almost all common tests and fully reference their sources – essential information for mudstones and other fine-grained rocks.

Seedsman RW (1993) Characterizing clay shales. In: Hudson JA (ed) Comprehensive Rock Engineering, Vol. 3. Pergamon Press, Oxford, pp 151–165

An in-depth review of why clay shales behave as they do and a summary of tests. Fifty references. Useful paper.

Środoñ J (1999) Use of clay minerals in reconstructing geological processes: recent advances and some perspectives. Clay Minerals 34:27–37

This short paper reviews the literature of the years 1987 to 1997 in terms of weathering and clay minerals, sedimentation and early and late diagenesis concluding with a brief summary. Over 100 references, whose perusal we found to be most useful.

Taylor RK, Smith TJ (1986) The engineering geology of clay minerals: swelling shrinking and mudrock breakdown. Clay Minerals 21:235–260

Strong emphasis on clay mineral properties makes this paper valuable along with its geological perspective. Read also Seedsman (1993).

Torrance JK, Ohtsubo M (1995) Ariake Bay quick clays: a comparison with the general model. Soils and Foundations 35:11–19

Exceptionally clear writing combined with good soil mechanics and over 40 references.

Way DS (1973) Terrain Analysis. Dowden Hutchinson Ross, Stroudsburg, 392 p

The section on shale, pp 90–97, provides one of the best short overviews of how to recognize a shale from an aerial photograph or map and also discusses site development, topsoil, groundwater, foundations, highways, and ponds. Still one of the best sources for insights to the recognition of mudstone terrains, their characteristics and significance for construction.

Appendices

The study of mudstones, like the study of medicine, draws on a wide range of techniques

Introduction

Many different methodologies (Table A.1) are available to study mudstones from diverse field observations to both stable and radioactive isotopes. These methodologies range from a variety of coring methods of muds to X-ray micro analysis (Figs. A.1A and B). In reality, descriptions of mudstones are the same as those of any sedimentary rocks, although some would say that we only need to look a little harder... Thus the variables are color, stratification, sedimentary structures, grain size (most easily approximated by the proportion of clay to silt to sand), bioturbation and fossils, secondary minerals and structures (concretions and nodules and their composition), and most important of all, *recognition of cycles defined by fining or coarsening upward plus low relief discontinuities (parasequence boundaries)*. Here we focus on only seven – field description, nomenclature, clay mineralogy, soils, petrology, organics, and isotopes. Each appendix includes references to facilitate additional understanding. The Encyclopedia of

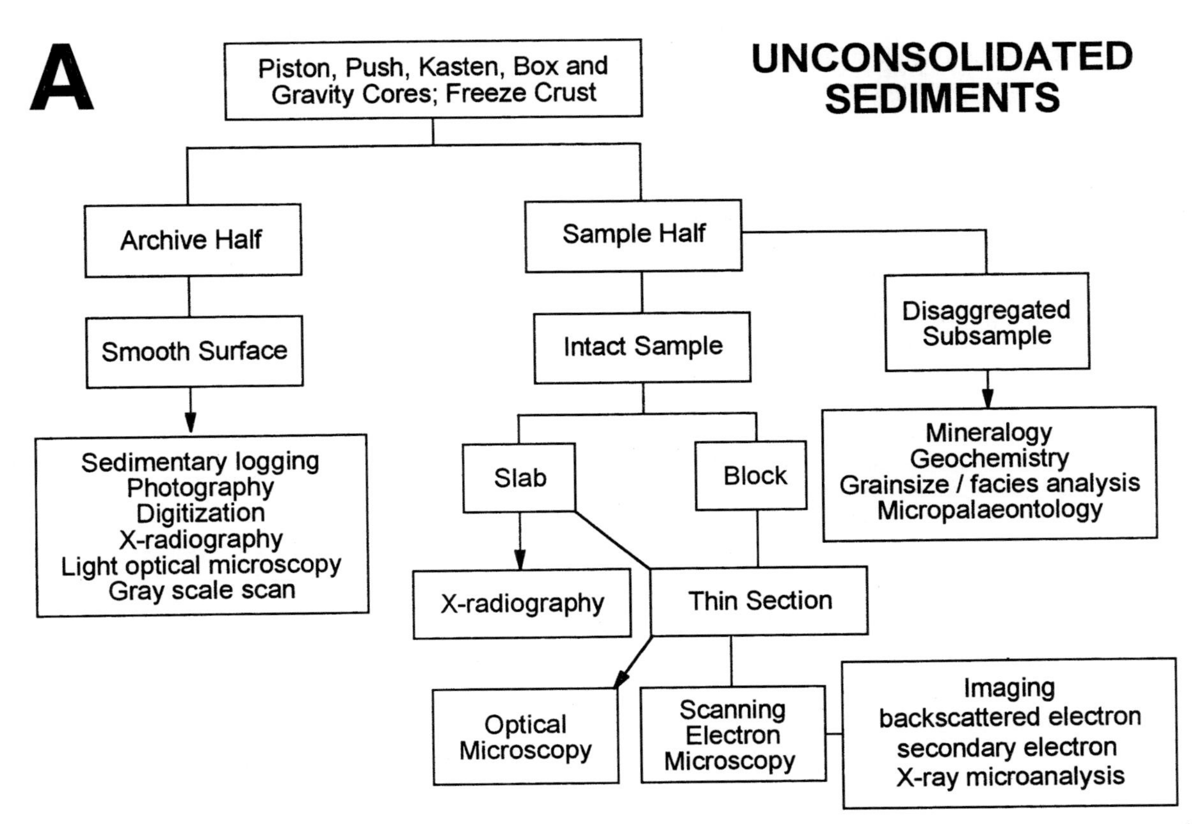

Fig. A.1A,B. Flow charts for the standard analysis of mudstones: (**A**) unconsolidated, and (**B**) consolidated (Pike and Kemp 1996, Figs. 1 and 2) (*continued on next page*). Published by permission of the authors and the Geological Society of London

Table A.1. Techniques to study mudstones

Field	Laboratory
Color	Petrography/mineralogy – Thin section, SEM, backscatter, TEM, X-ray diffraction
Texture/field name	Texture – Pipette or Coulter counter, SEM
Stratification and structures	Sedimentary structures – X-radiography, polished slabs
Macro/trace fossils	Micropaleontology/Palynology – Macerals; vitrinite reflectance
Sequences/cycles	Statistical Analysis – Milankovitch and tidal cycles
Total gamma radiation	Geochemistry: Organic – TOC, H/C, Rock-Eval, biomarkers
Spectral gamma ray – U, Th, K	Geochemistry: Inorganic – Major and trace elements, REE, stable isotopes, radioactive isotopes
Weathering features	Engineering properties – Plastic and liquid limits, undrained shear strength, unconfined compressive strength, slake durability

Sediments and Sedimentary Rocks (Middleton 2003) provides information about many of the topics below, and we also recommend the Glossary of Geology (Jackson 1997) to help with terminology.

References

Jackson JA (1997) Glossary of Geology, 4th edn. American Geological Institute, Alexandria, VA, 769 p
Middleton GV (ed) (2003) Encyclopedia of Sediments and Sedimentary Rocks. Kluwer Academic, Dordrecht, 821 p

A.1. Field and Core Description

Describing mud, clay and mudstone in outcrops and cores has long challenged many of us over the years, because it is all too easy to think, "Gosh, what is here that I can usefully describe besides color?" Today, fortunately, we have a good start on what to look for and how to interpret what we see (Chaps. 3, 4, and 5). A standard set of symbols is always helpful (Fig. A.2) as are standard definitions of stratification, thickness and weathering terms (Fig. A.3). For some unconsolidated muddy sediments, *relief peels* provide much additional information about their stratification and structures (Hattingh and Zawadam 1996) and polished slabs are helpful for some mudstones.

In outcrop how mudstone weathers provides much helpful information about its stratification and small-scale fabric (Fig. A.4), and to some degree, its clay mineral composition. Key terms are *massive, fissile (papery), blocky or concoidal, and popcorn weathering* and their equivalents that represent

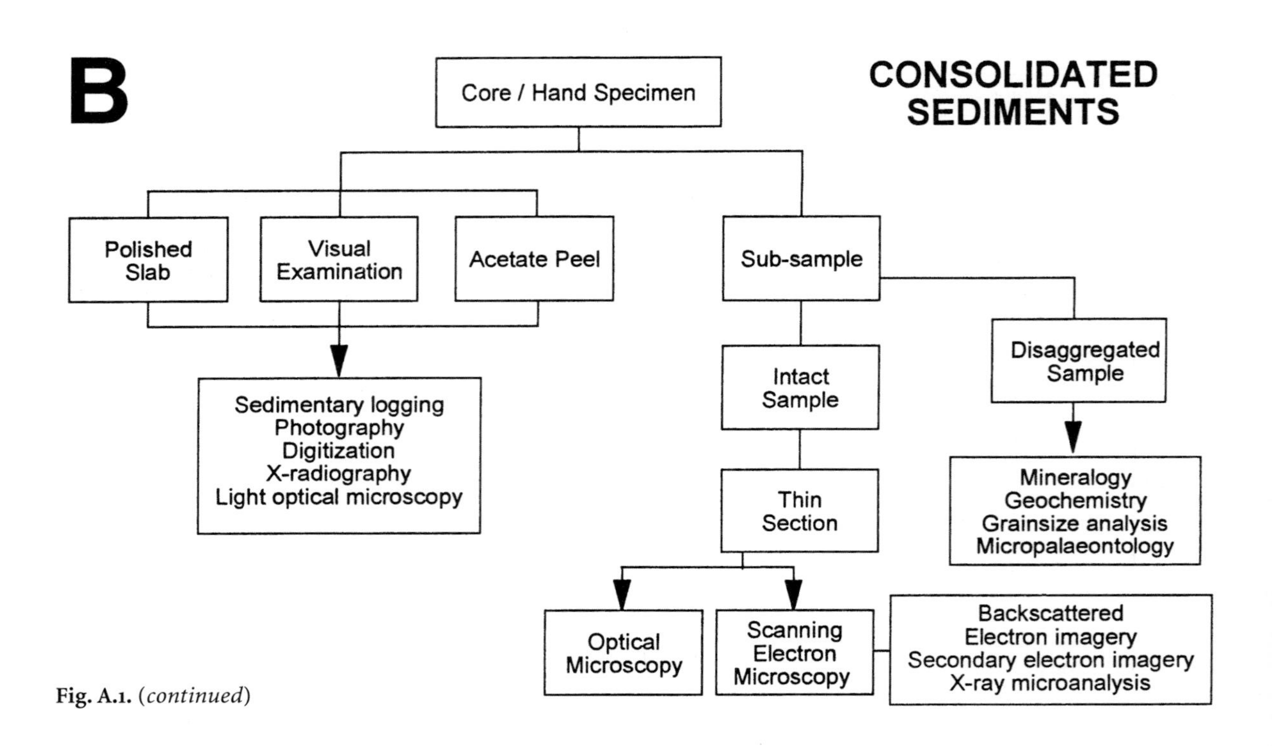

Fig. A.1. (*continued*)

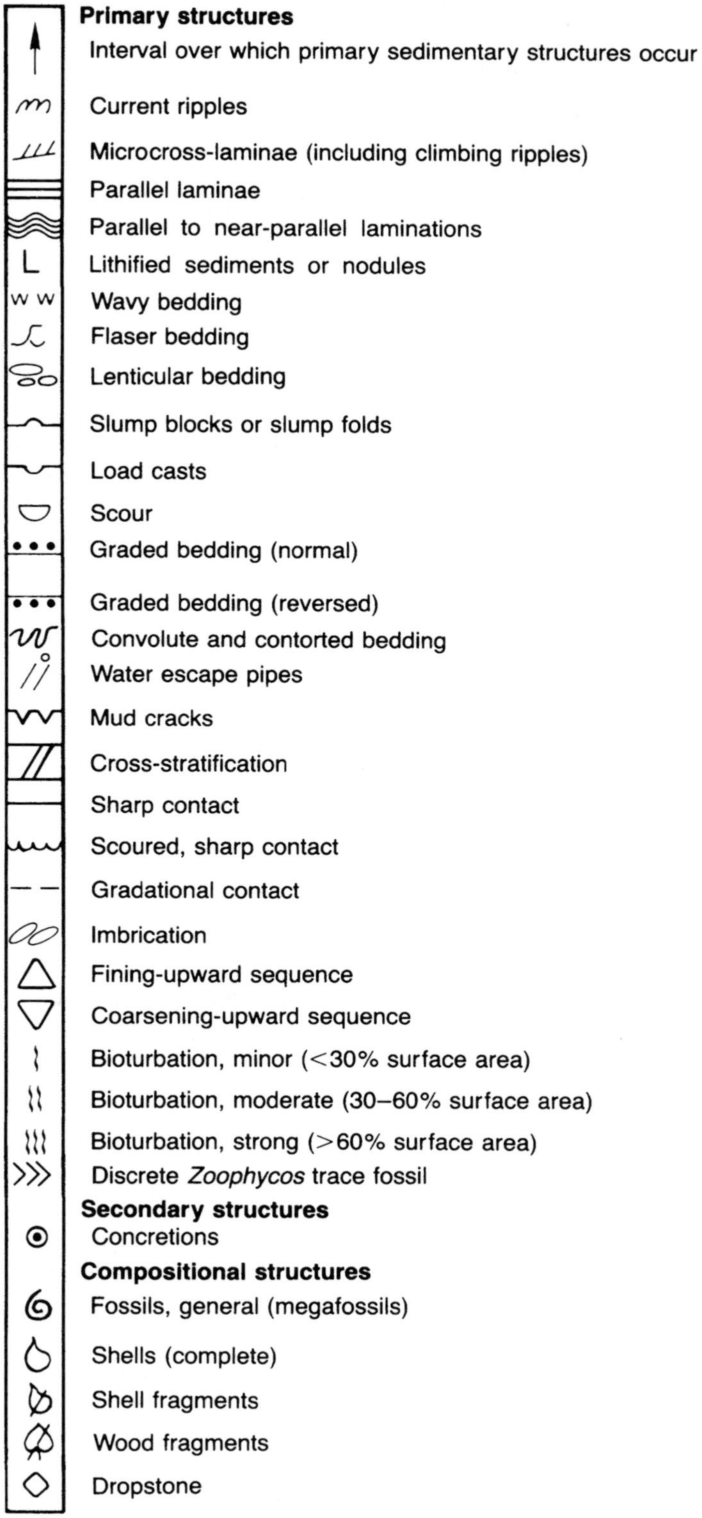

Primary structures
Interval over which primary sedimentary structures occur
Current ripples
Microcross-laminae (including climbing ripples)
Parallel laminae
Parallel to near-parallel laminations
Lithified sediments or nodules
Wavy bedding
Flaser bedding
Lenticular bedding
Slump blocks or slump folds
Load casts
Scour
Graded bedding (normal)
Graded bedding (reversed)
Convolute and contorted bedding
Water escape pipes
Mud cracks
Cross-stratification
Sharp contact
Scoured, sharp contact
Gradational contact
Imbrication
Fining-upward sequence
Coarsening-upward sequence
Bioturbation, minor (<30% surface area)
Bioturbation, moderate (30–60% surface area)
Bioturbation, strong (>60% surface area)
Discrete *Zoophycos* trace fossil
Secondary structures
Concretions
Compositional structures
Fossils, general (megafossils)
Shells (complete)
Shell fragments
Wood fragments
Dropstone

Fig. A.2. Symbols used by Scientific Shipboard Party (Ludwig et al. 1983, Fig. 6) to log DSDP cores

Fig. A.3. Terms to describe stratification and its weathering expression of fine-grained sedimentary rocks (Potter et al., their Table 1-3)

Thickness	Stratification		Parting	Composition
30 cm	Thin	Bedding	Slabby	Clay and organic content (increasing downward ↓); sand, silt and carbonate content (increasing upward ↑)
3 cm	Very thin	Bedding	Slabby	
10 mm	Thick	Lamination	Flaggy	
5 mm	Medium	Lamination	Platy	
1 mm	Thin	Lamination	Fissile	
0.5 mm	Very thin	Lamination	Papery	

end members (Fig. A.5): fissility and papery describe well-defined, fine laminations that separate into thin sheets upon weathering, whereas massive is just the opposite. The term *blocky*, like massive, also indicates an absence of fissility while its opposite is a very porous, mud cracked, flaky surface texture called *popcorn weathering*, which forms when clays rich in smectites or mixed lattice minerals are exposed to water and expand. Color also is a useful guide to some compositions, especially pyrite content; for example, in wet climates a strong, rusty reddish brown color indicates pyrite whereas some dark, organic-rich shales acquire a silvery sheen on their fissile fragments. Fortunately, a rather small number of basic descriptors suffice for most mudstones even though there are many additional terms for minor components (Table A.2). Use of a color chart (Goddard et al. undated) is always recommended, but color description can also be usefully automated, especially for rhythmically-stratified mudstones reducing several colors to a *single color density log* (Algeo et al. 1994; Schaaf and Thurow 1994). Such quantitative, continuous scans facilitate study by time series (see Box 8.2) to identify the different frequencies of cycles.

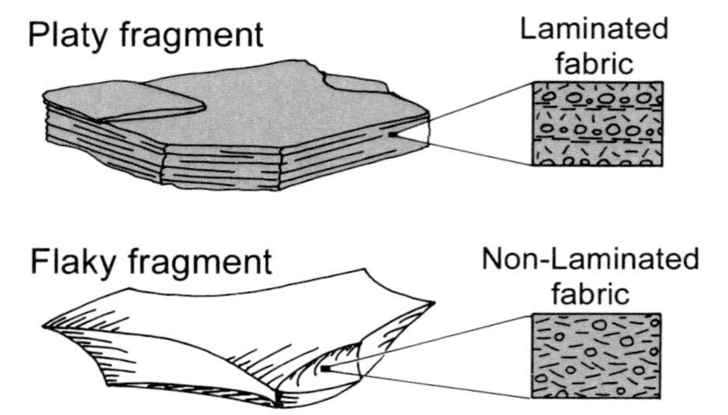

Fig. A.4. Platy versus flaky fragmentation of a shale and contrasts in fabric. After Grainger (1985, Fig. 2). Published by permission of the authors and the Geologists Association

Another technique that is widely used is X-radiography of either individual pieces of thin slabs of cores or continuous images of a vibrocore or sliced core using a traveling stage (Algeo et al. 1994).

Guides to field and core description are given by Potter et al. (1980, pp 87–119), and Ludwig et al. (1983)

Fig. A.5A–D. Four end members of weathering: (**A**) massive mudrock of the Corumbatai Formation (Permian) on SP 191 near Rio Claro, Sao Paulo, Brazil; (**B**) *total fissility* of the Devonian-Mississippian New Albany Shale on the Bluegrass Parkway in Nelson Co., KY (Photograph courtesy of Brandon Nuttal, Kentucky Geological Survey); (**C**) blocky (concoidal) weathering of the Eocene Porters Creek Clay rich in smectite in Marshal County, KY; and (**D**) "Popcorn" weathering developed on smectite-rich bed of Cretaceous Mowry Shale along Cottonwood Creek, sec. 31 T. 39 N, R 85 W, Natrona Co., WY. Photograph courtesy of Kevin Bohacs (turn to page 250)

plus several atlases and core books (see Digging Deeper). Full attention to the trace and body fossils of mudstones and their interbedded lithologies is also always recommended. See Brett and Allison (1998) in Digging Deeper for a thorough discussion.

A vibrocore (Fig. A.7) is essential to study most modern muds and is widely used; radiography and study by the scanning electron microscope of the resultant core is commonly helpful to observe many of its primary and second features (Fig. A.6). Polished sections of indurated samples also yield much information.

A hand held gamma ray scintillometer (Fig. A.8) is useful in the field (Ettensohn et al. 1979) and readily indicates the proportion of mudstone in an outcrop or core (Fig. A.9); it also gives an idea of silt content – the more silt, the lower the response of the scintillometer. Keep in mind, however, that three common sedimentary minerals, illite, the potash feldspars, and detrital muscovite all have high radioactivity. Thus, an exceptionally high gamma ray count may *not* always indicate a clay-rich (illitic rich) mudstone. More information is provided by a hand held spectrometer, which indicates the proportions of

potassium, thorium and uranium. In mudstones clay minerals (chiefly illite and 2 M Micas) and feldspar (especially albite) both contain potassium, uranium is associated with organic matter and phosphate, whereas volcanic ash and heavy minerals contain thorium. Consequently, the spectrometer provides a better stratigraphic tool than the Geiger counter and at the same time gives some indication about the provenance of the mudstone. Both instruments facilitate quick correlation to nearby geophysical logs obtained from wells, and, with experience, much environmental information can be obtained from downhole geophysical logging (Fig. A.10). See Cole and Crowley (2001) for applications and discussion of best use. Both the spectrometer and the Geiger counter should be routinely used in the field and for core description. Recognition of the end member forms of a log is always helpful (Fig. A.11). In the subsurface spectral gamma logs and amplitude induction logs (permit almost centimeter-by-centimeter discrimination between thinly inter bedded sandstones and mudstones) are becoming more common in some basins.

Spectral gamma ray logging in wells (Albertin, Darling et al. 1996; Kumar and Kear 2003) in combination with an imaging device (Thompson 1999

Table A.2. Terminology for mudstones (modified from Stow and Piper 1984, Table 3) [Published by permission of the Geological Society of London]

Basic Terms			
Unlithified	**Lithified,** non-fissile	**Lithified,** fissile/laminated	**Proportions and** grain size
Silt	Siltstone	Silt-shale	> 2/3 silt-sized (> 63 µm)
Mud	Mudrock	Shale	Silt-clay mixture (4–62 µm)
Clay	Claystone	Clay-shale	> 2/3 clay-sized (< 4 µm)
Textural Descriptors			
Silty	10% silt-size		
Muddy	10% silt- or clay-size (applied to non-mudstones)		
Clayey	> 10% clay-size		
Sandy, pebbly, etc.	> 10% sand-size, pebble-size, etc.		
Compositional Descriptors			
Calcareous	> 10% $CaCO_3$ (foraminiferal, nannofossil, etc.)		Pyritiferous, ferruginous, micaneous phosphatic, etc. typically 1 to 5%
Siliceous	> 10% SiO_2 (diatomaceous, radiolarian, etc.)		
Carbonaceous	> 1% organic carbon		
Metamorphic Terms			
Argillite	Slightly metamorphosed, non-fissile		Silt and clay mixture
Slate	Metamorphosed, fissile		Silt and clay mixture

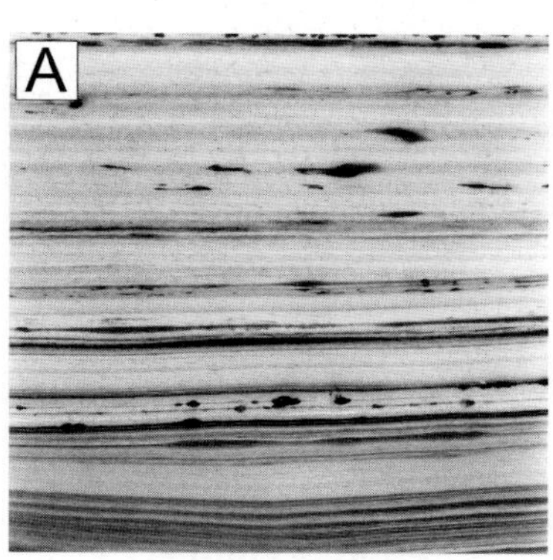

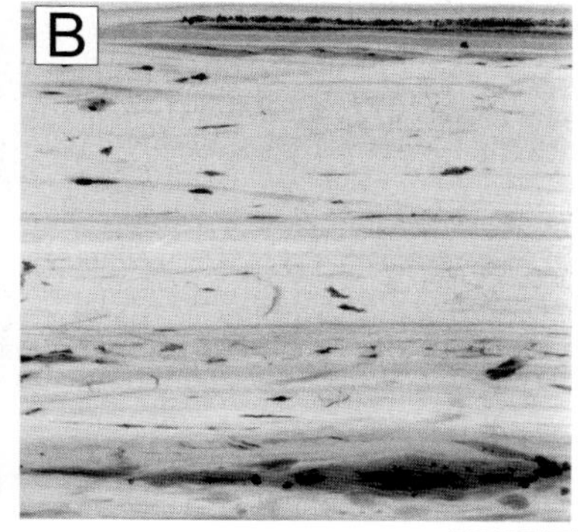

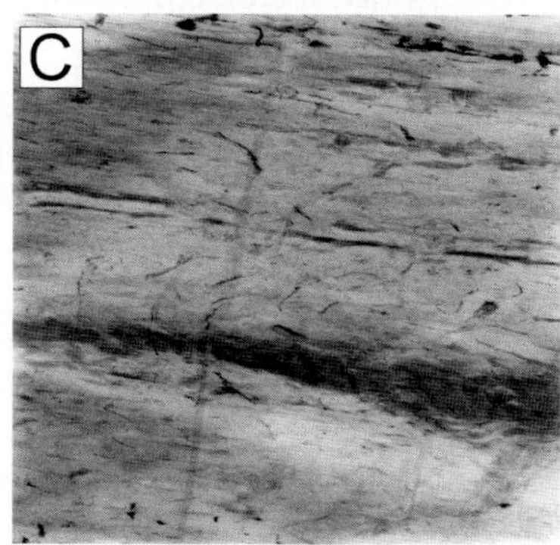

Fig. A.6A–C. Insights to mudstones from radiography: (**A**) thinly and "perfectly" laminated Devonian Blocher Shale with small scattered pyrite nodules along bedding planes (strongly euxinic) from Christian County, KY; (**B**) discontinuous, widely spaced laminations in Devonian Selmier Shale with small pyrite nodules and some thin, curved subvertical, pyrite-filled burrows (dysaerobic?) from Christian County, KY; and, (**C**) indistinctly laminated and bioturbated Selmier Shale with many thin, pyrite-filled burrows (dysaerobic-oxic?). Scale is one centimeter (Cluff et al. 1981, Figs. 8, 14, and 18)

Fig. A.7. Vibrocoring is a must for the study of modern muddy sediments as shown by a field exercise of students from the University of Cincinnati near Reelfoot Lake, Lake County, TN, USA

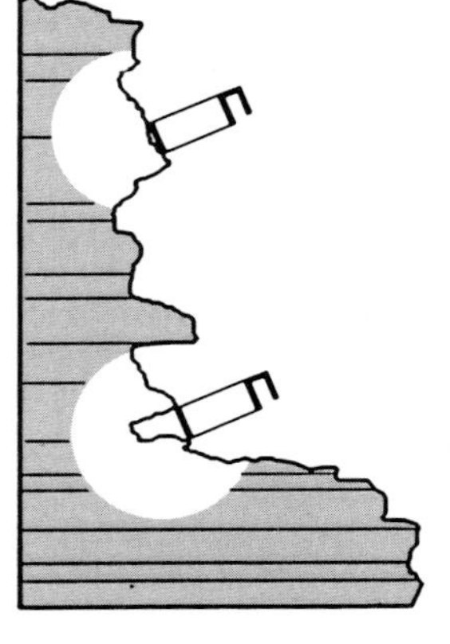

Fig. A.8. Using a scintillometer in the field (After Schwalbach and Bohacs 1992, Fig. 14). Because the reading depends on both the radioactivity of the mudstone and the amount of mass, the geometry of each sample point should be held as constant as possible from point to point. Published by permission of the authors and the Pacific Section of the American Association of Petroleum Geologists

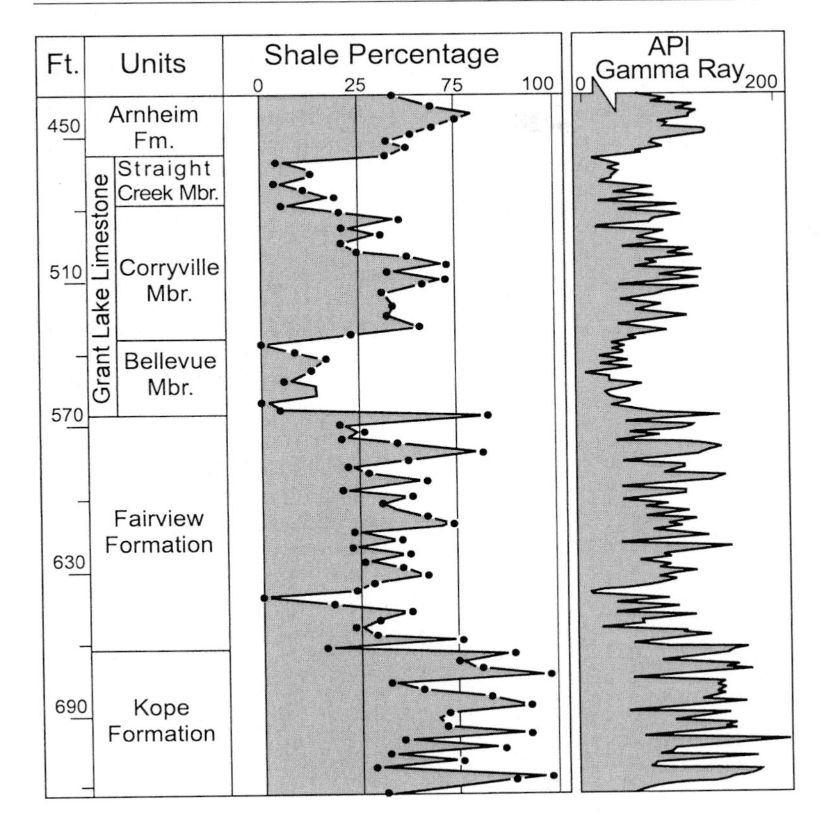

Fig. A.9. Handheld scintillometer profile of interbedded limestones and mudstones of an Upper Ordovician core from southwestern Ohio faithfully reflects proportions of mudstone present in core (after Schumacher et al. 1991, Fig. 5A). Published by permission of the author and the Ohio Academy of Science

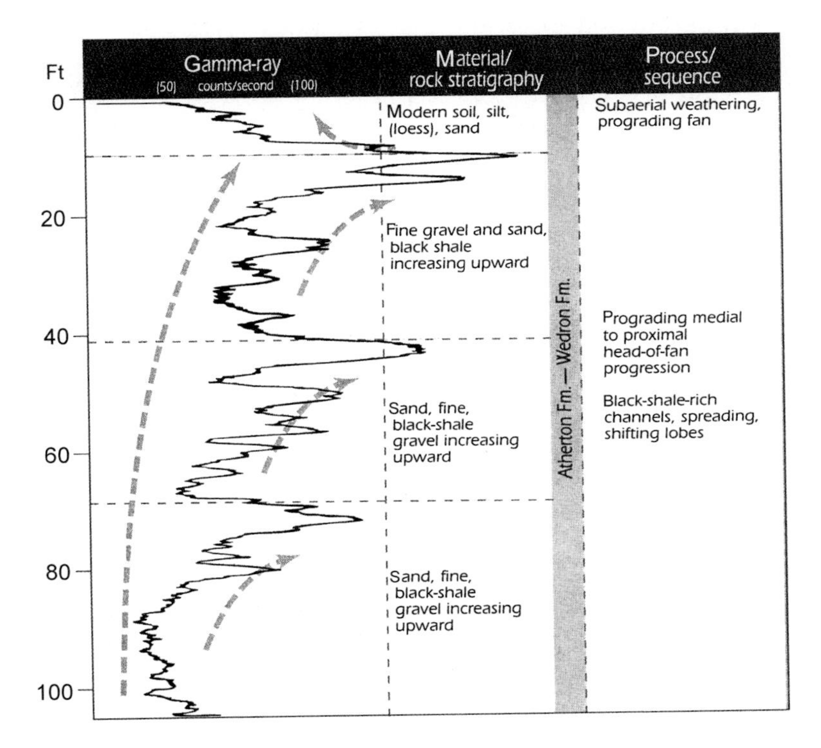

Fig. A.10. Portable downhole gamma ray profile of Wisconsin glacial deposits in northern Indiana demonstrates its value for routine mapping of unconsolidated materials (Bleuer 2004, Fig. 14)

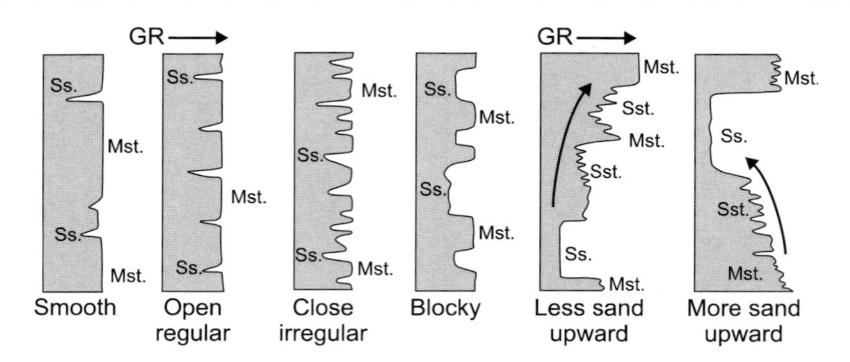

Fig. A.11. Typical shapes of gamma ray log profiles when clay alone determines their radiation response. After Serra and Sulpice (1975, Fig. 2.9) and Krassay (1998, Fig. 3). Recognition of these end member forms on a log such as shown in Fig. A-10 helps segregate it into segments and thus identify more easily environments and cycles. The same method is also applied to self-potential, sonic and other downhole logs used in deep drilling. *Arrows* indicate fining upward (*left*) and coarsening upward (*right*)

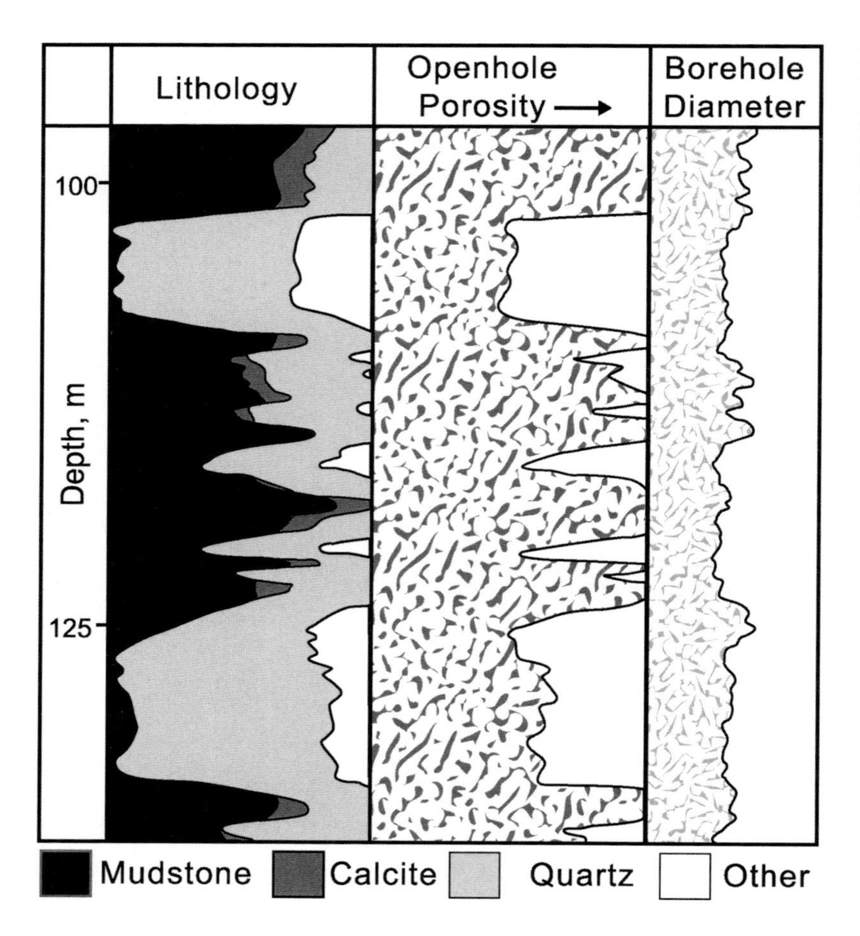

Fig. A.12. Lithologic proportions estimated from the combined use of a spectra gamma ray and a borehole image log (after Albertin et al. 1996, p 38). Published by permission of the authors and Schlumberger Limited

in Digging Deeper) provides continuous down-hole identification of lithologies (Fig. A.12). In addition, most of the sedimentary structures of the interval can be identified to cm scales. Consequently, much essential information for a sedimentological interpretation of the mudstones is in hand. Recognition of end member log forms from geophysical logging (self potential, resistivity, sonic, etc.) goes far to identify muddy environments.

References

Albertin I, Darling H et al. (1996) The many facets of pulsed casehole logging. Oilfield Review 3:28–41

Algeo TJ, Phillips M, Jaminski J, Fenwick M (1994) High-resolution X-radiography of laminated sediment cores. J Sediment Res A64:665–703

Bleuer N (2004) Slow logging, subtle sequences. Indiana Geological Survey Special Report 65:39

Clare AP, Crowley AJ (2001) Qualitative analysis of spectral gamma ray data as a tool for field wide and regional stratigraphic cor-

relation, Enderby Terrace, Carnavon Basin, Western Australia. The APPEA Journal 41:449–462
Cluff RM, Reinbold ML, Lineback JA (1981) The New Albany Shale Group of Illinois. Illinois Geological Survey, Circular 518:81
Dunham RJ (1962) Classification of carbonate rocks according to depositional texture. In: Ham WE (ed) Classification of Carbonate Rocks. American Association Petroleum Geologists Memoir 1:108–121
Ettensohn FR, Fulton LD, Kepferle RC (1979) Use of scintillometer and gamma ray logs for correlation and stratigraphy in homogeneous black shales. Geol Soc America Bulletin 90:828–849
Folk RL (1968) Petrology of Sedimentary Rocks. Hemphill's Bookstore, Austin, Texas, 170 p
Goddard EN, Trask PD, de Ford RK, Rove ON, Singlewald JT, Overbeck RM (Undated) Rock-Color Chart. Geol Soc America, Boulder, unpaged
Grainger P (1985) The microfabric of deformed mudrocks of the Cracking Formation, southwest England. Proceedings Geologists Association 96:143–152
Hattingh J, Zawada PK (1996) Relief peels in the study of paleoflood slack-writer sediments. Geomorphology 16:121–126
Krassay AA (1998) Outcrop and drill hole gamma-ray logging integrated with sequence stratigraphy: example from Proterozoic sedimentary successions of Northern Australia. AGSO Journal Geology, Geophysics 17:285–299
Kumar A, Kear GR (2003) Lithofacies classification based on spectral yields and borehole micro-resistivity images. Gulf Coast Association Geological Societies/Gulf Coast Section SEPM Transactions, 53rd Annual Convention Baton Rouge, pp 434–442
Leighton MW, Pendexter C (1962) Carbonate rock types. In: Ham WE (ed) Classification of Carbonate Rocks. American Association Petroleum Geologists Memoir 1:33–61
Ludwig WJ, Krasheninnikov VA, Wise SA (1983) Introduction and explanatory notes. In: Initial Reports of the Deep-sea Drilling Project, Vol. 71. US Government Printing Office, Washington, DC, pp 1–19
Pike J, Kemp AES (1996) Preparation and techniques for studies of laminated sediments. In: Kemp AES (ed) Paleoclimatology and Paleoceanography from Laminated Sediments. Geol Soc (Special Publication) 116:37–48
Potter PE, Maynard JB, Pryor WA (1980) Sedimentology of Shale. Springer, Berlin Heidelberg New York, 303 p
Schaaf M, Thurow J (1994) A fast and easy method to derive high-resolution time series data sets from drill cores and rock samples. Sediment Geol 94:1–10
Schieber J, Zimmerle W (1998) Petrography of shales: a survey of techniques. In: Schieber J, Zimmerle W, Sethi P (eds) Shales and Mudstones, Vol. II. E. Schweizerbart'sche, Stuttgart, pp 3–12
Schumacher G, Swinford EM, Shrake DL (1991) Lithostratigraphy of the Grant Lake Limestone and Grant Lake Formation (Upper Ordovician) in southwestern Ohio. Ohio Journal Science 91:56–68
Schwalbach JR, Bohacs KM (1992) Chapter III-Field investigation techniques for analysis of the Monterey Formation. In: Schwalbach JR, Bohacs KM (eds) Sequence Stratigraphy in Fine-Grained Rocks: Examples from the Monterey Formation. SEPM Pacific Section (Society Sediment Geol), 7 November 1992, pp 21–30
Serra J, Sulpice L (1975) Apports des diagraphics aux études sedimentologiques de séries argilo-sableuses traverses en sondage. In: Proceedings International Sedimentological Congress No. 9, Vol. 3, pp 85–94
Shipboard Scientific Party (1984) Introduction and explanatory notes. In: Hay WW, Sibuet J-C et al. Initial Reports of the Deep-sea Drilling Project. US Government Printing Office, Washington, DC, 75:3–25
Shipboard Scientific Party (1989) 2 Explanatory Notes. In: Prell WL, Niitisuma N et al. (eds) Proceedings Ocean Drilling Project. Ocean Drilling Program, College Station Texas 117:11–42

Digging Deeper

Barnhill ML, Zhou H (1996) Corebook of Pennsylvanian Rocks in the Illinois Basin. Illinois Basin Studies 3 (Published by the Indiana Geological Survey, Bloomington, IN, for the Illinois Basin Consortium), various paging.

Paired colored photographs of polished cores, many of which are mudstones and siltstones, plus a brief description and the gamma ray of the cored interval. Outstanding opportunity to take a trip to the core barn while at your desk.

Brett CW, Allison PA (1998) Paleontological approaches to the environmental interpretation of marine mudrocks. In: Schieber J, Zimmerle W, Sethi P (eds) Shales and Mudstones. E. Schweizerbart'sche, Stuttgart, pp 301–349.

An extended, detailed discussion by two well-known experts of an important, all-to-often understudied aspect of mudstones – their fossils and the many insights they provide. Body and trace fossils provide insights to water depth and light, current systems, temperature, salinity and oxygen content of the water column, firmness of the bottom plus sedimentation rates and turbidity. Their Fig. 14, a ternary classification of three major types of life/feeding habitants of fossils, is most useful. Well referenced.

Ferm JC, Weisenfluh G (1981) Cored Rocks of the Southern Appalachian Coal Fields. University of Kentucky Department Geology, Lexington, 93 p.

Many mudstones and fire clays illustrated by excellent colored plates each with a simple name.

Ferm JC, Smith GC (Undated) A Guide to the Cored Rocks in the Pittsburg Basin. University of Kentucky Department of Geology, Lexington, 109 p.

Colored plates show fireclays and many mudstone types – green to red mudstones and every color in between! Most plates are vertical sections of cores, but some show broken surfaces parallel to bedding.

Ferm JC, Melton RA (1977) A Guide to Cored Rocks in the Pocahontas Basin. University South Carolina, Department Geology, Columbia, 90 p.

Short but effective descriptions of many diverse mudstones and fireclays. Beautiful colored plates.

Stow DAV, Piper DJW (1984) Deep-water fine-grained sediments; history, methodology, and terminology. In: Stow DAV, Piper DJW (eds) Fine-grained sediments: deep-water processes and facies. Geol Soc (Special Paper) 15:3–14.

Based on the ratio of information (much) to length (11 pages), this is one of the very best articles to start your study of fine-grained sediments.

Syvitski JPM (ed) (1991) Principles, Methods, and Application of Particle Size Analysis. Cambridge University Press, Cambridge, 368 p.

An in-depth, technical book for the advanced specialist, has 24 articles arranged in five parts, is well referenced, and is the first place to look for the methodology of size.

Thompson LB (ed) (1999) Atlas of Borehole Imagery, Disks 1and 2. American Association Petroleum Geologists, Tulsa, CD ROMs.

Disk 1 has nine parts consisting principally of Tools (77 p.), Images (48 p.) and a Bibliography (34 p.). The electronically imaged downhole structures include faults, folds, fractures, fault-clay smears, stratification of different types, and bioturbation. Paper copies of these images are useful to have when logging core (to relate the images to rocks), and conversely, photos of structures go far to help understand the images. The wave of the future or too expensive for common use? Disk 2 is entitled "Imagery Software Demonstration and Data".

A.2. Terminology

Different definitions of the terms clay, loam, shale, mudstone, mudrock, claystone, clay rock etc., have created confusion for over 50 years and more (Figs. A.13 and A.14 and Table A.3). Why? Because many terms have multiple uses and definitions, which have changed with time. For example, the term *shale* is used as a formation name such as Bright Angel Shale; the term shale is also widely used as the generic, field term for *any* fine-grained argillaceous rock (think red shale, black shale or, "nothing but shale on the other side of the creek", etc.), and it is used for a specific type of fine-grained argillaceous rock, a *laminated* argillaceous rock as defined by Hooson (1747). "Untidy" is a mild word to use here.

If one is a specialist studying ancient mudstones, it is fairly easy to live with this lack of standardization simply by consulting Fig. A.14, which summarizes four of the most widely used classifications. But rather than continue with this untidiness, we propose that *mudstone be the generic term* for all fine-grained argillaceous rocks and that shale be restricted to laminated fine-grained argillaceous rocks, following its original definition by Hooson. We offer two supporting arguments. First, mudstone would then take its place along with sandstone, siltstone, limestone, micstone, dolostone, wackestone, etc., so that all the important sedimentary rock sub families would all be "stones". Secondly, we note that mudstone is becoming increasingly popular as the generic name. And finally, we note that our choice of mudstone as the general term fine-grained for argillaceous rocks conflicts with its use by Dunham (1962), who used it for fine-grained limestone. In the same year Leighton and Pendexter (1962, p 62) used *micritic limestone* for fine-grained limestone. Shorten micritic limestone to *micstone* and much of the nomenclatural confusion of sedimentary rocks disappears. Thus, although we grew up with and like shale (only one syllable is needed for pronunciation) as the general term for argillaceous rocks, here we restrict

Table A.3. Common terms used in the study of fine-grained sediments and sedimentary rocks

Part A Terrigenous

Mud, mudstone, mudshale and mudrock – Mud is a field term for any soft plastic, silt-clay mixture with more than 50% of its size fraction smaller than four micrometers. While there is little disagreement here, it is the naming of the "hard rock" equivalents of mud that have generated most of the nomenclature untidiness of fine-grained argillaceous rocks. Both mudstone and mudrock have been proposed to replace shale as the general generic term for fine-grained argillaceous rocks. These terms arose because shale was originally defined as being laminated, so the problem became, "What shall we call non-laminated, argillaceous rocks?"

Clay, claystone, and clayshale – The term clay has at least three uses: A field term for a clay-rich, fine-grained, plastic sediment with little or no silt; a size fraction, all of which is finer than 4 micrometers; and a formation name; e. g., London Clay and Yazoo Clay formations. Three "hardrock" derivative terms are clayrock (broad field term proposed to replace shale), claystone (massive or thickly bedded) and clayshale (fissile or thinly bedded)

Shale – A word of teutonic origin originally applied to an indurated, laminated, fine-grained argillaceous rock, but also widely used as a broad term for the entire class of such rocks, laminated or not, and as a formation name; e. g., the Nolichucky, Exshaw, La Luna, Pilot, and Monterey Shales

Lutite and pelite – Both are old, general field terms for argillaceous, fine- grained rocks (lutites for sedimentary rocks and pelites for their low grade metamorphic equivalents); the term lutite is not widely used today whereas metamorphic geologists use pelite

Silt and siltstone – Both need at least 50% silt-sized material which typically can be determined easily in the field; silts are non-cohesive and non-plastic, but liquefy easily

Part B Non-terrigenous

Marl and marlstone, ooze, sapropel, diatomite, micrite and micstone – All are fine grained sediments and rocks composed of 50% or more carbonate, biogenic silica or phosphate that either may be gradational to argillaceous, fine-grained rocks or interbedded with them. Consequently, they may be present, where terrigenous sedimentation was restricted, so we need knowledge of them

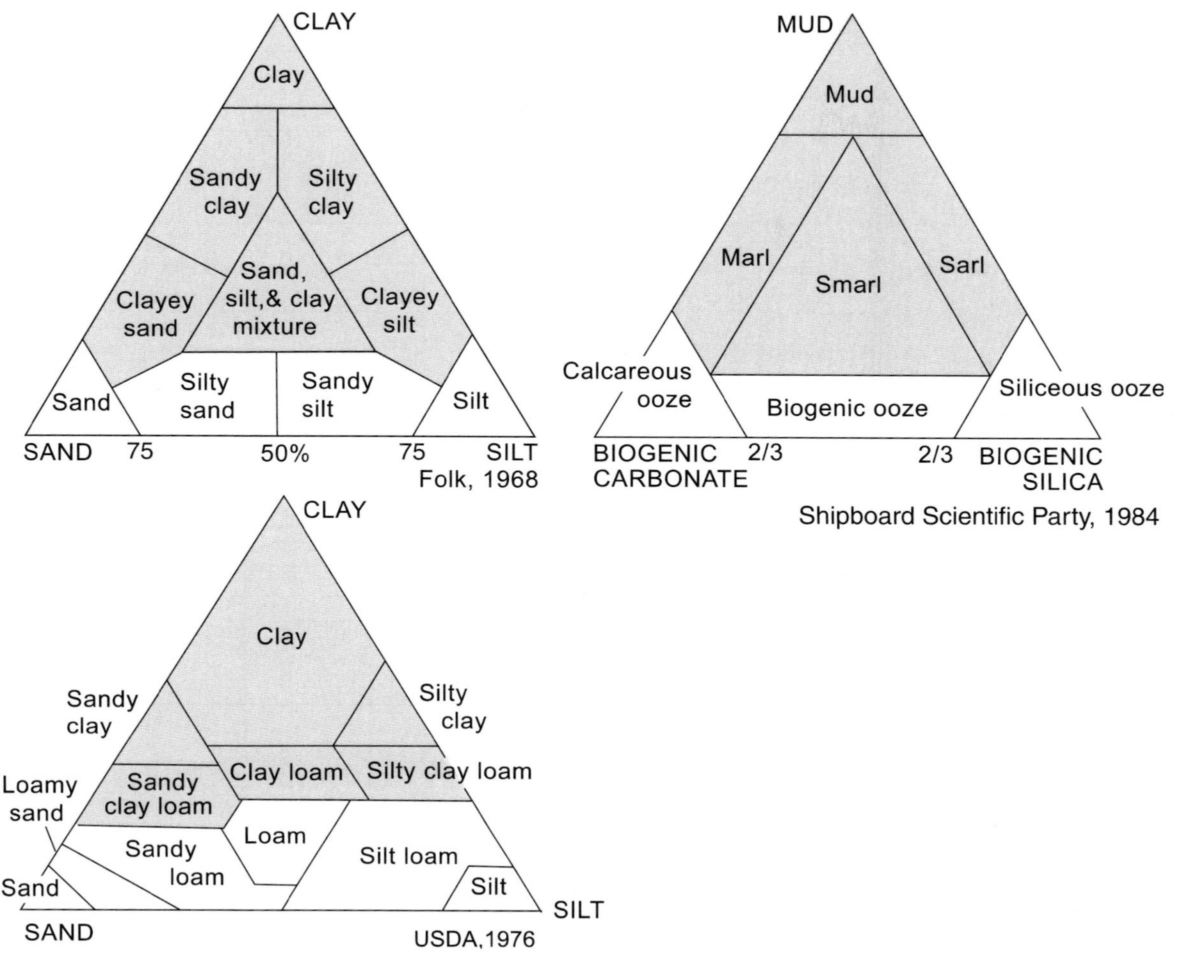

Fig. A.13. Two triangular diagrams describe mixtures of clay, silt, and sand in unconsolidated sediments (Folk 1968 and US Department of Agriculture, 1975) and an adaptation for admixtures of biogenic carbonate and silica found in marine sediments (Shipboard Scientific Party 1984, Fig. 9). Notice how even small quantities of clay are recognized in these triangles. This shows its great importance

it to its original sense of a laminated, argillaceous rock.

Therefore we recommend the terminology shown in Table A.2, which mostly follows Stow and Piper (1984). Box 1.1 and Table A.3 give equivalents and additional definitions of commonly interbedded fine-grained rocks found in mudstones. Where we report on the work of others, we try to use their terminology and where we write, we use mudstone as the general term, and shale to refer to obviously fissile varieties.

See Macquaker and Adams (2003) for a classification that emphasizes the many variations of mudstones as proportions of silt clay, carbonate, organics, pyrite, etc., vary.

References

Blatt H, Middleton G, Murray R (1980) Origin of Sedimentary Rocks, 2nd edn. Prentice Hall, Englewood Cliffs, NJ, 782 p

Folk RL (1968) Petrology of Sedimentary Rocks. Hemphill's Bookstore, Austin, TX, 170 p

Hooson W (1747) The Miners Dictionary. Wrexhem, England, unpaged

Ingram RL (1953) Fissility of mudrocks. Geol Soc America Bulletin 65:869–878

Macquaker JHS, Adams AE (2003) Maximizing information from fine-grained sedimentary rocks: an inclusive nomenclature for mudstones. J Sediment Res 73:735–744

Potter PE, Maynard JB, Pryor WA (1980) Sedimentology of Shale. Springer, Berlin Heidelberg New York, 310 p

Shipboard Scientific Party (1984) 1. Introduction and explanatory notes. In: Hay WW, Sibuet J-C etc. Initial Reports Deep Sea Drilling Project Leg 75, Pt 1, pp 3–25

SEDIMENTARY ROCK CONTAINING MORE THAN 50% SILT AND/OR CLAY			
	NO CONNOTATIONS AS TO BREAKING CHARACTERISTICS	MASSIVE	FISSILE
NO CONNOTATIONS AS TO RELATIVE AMOUNTS OF SILT & CLAY	MUDROCK	MUDSTONE	MUDSHALE
SILT > CLAY	SILTROCK	SILTSTONE	SILTSHALE
CLAY > SILT	CLAYROCK	CLAYSTONE	CLAYSHALE

Ingram (1953, Table 1)

CLASSIFICATION OF MUDROCKS			
IDEAL SIZE DEFINITION	FIELD CRITERIA	FISSILE MUDROCK	NONFISSILE MUDROCK
> 2/3 SILT	ABUNDANT SILT VISIBLE WITH HAND LENS	SILT - SHALE	SILTSTONE
>1/3 < 2/3 SILT	FEELS GRITTY WHEN CHEWED	MUD - SHALE	MUDSTONE
> 2/3 CLAY	FEELS SMOOTH WHEN CHEWED	CLAY - SHALE	CLAYSTONE

Blatt, Middelton & Murray (1980, Table 11-1)

MUDROCK DIVISION BASED UPON TEXTURE AND STRUCTURE			
GRAIN SIZE OF MUD FRACTION	SOFT	INDURATED, NONFISSILE	INDURATED, FISSILE
OVER 2/3 SILT	SILT	SILTSTONE	SILT - SHALE
SUBEQUAL SILT AND CLAY	MUD	MUDSTONE	MUD - SHALE
OVER 2/3 CLAY	CLAY	CLAYSTONE	CLAY - SHALE

Folk (1965, p. 130)

PERCENTAGE CLAY - SIZE CONSTITUENTS			0 - 32	33 - 65	66 - 100
FIELD ADJECTIVE			GRITTY	LOAMY	FAT OR SLICK
INDURATED	BEDS	GREATER THAN 10mm	BEDDED SILTSTONE	MUDSTONE	CLAYSTONE
	LAMINAE	LESS THAN 10mm	LAMINATED SILTSTONE	MUDSHALE	CLAYSHALE

Modified from Potter et al. (1980, Table 1.2)

Fig. A.14. Petrology: Four widely used classifications of mudstones and their related definitions. Each uses somewhat different terminology; Ingram, 1953 (*upper left*), Blatt et al. 1980 (*upper right*), Folk; 1965 (*lower left*), and Potter et al. 1980 (*lower right*)

Stow DAV, Piper DJW (1984) Deep-water fine-grained sediments; history, methodology, and terminology. In: Stow DAV, Piper DJW (eds) Fine-grained sediments: deep-water processes and facies: Geol Soc (Special Paper) 15:3–14

Digging Deeper

Tourtelot HA (1960) The use of the word "shale." American Journal Science, Bradley 258-A:335–343.

An in depth, scholarly examination of this primordial term that deserves your attention. Does such a paper exist in French, Italian, German, or Russian?

A.3. Clay Minerals

Clay minerals are a major product of weathering processes and the most characteristic components of mud and mudstones.

It is the clay minerals that make muds cohesive and plastic in addition to supplying many diverse industrial and technological applications. A small volume of clays also comes from hydrothermal processes, mostly from water-rock interactions at temperatures of 100–250 °C. The term clay is also used to denote a grain size of less than 4 μm. This size-class may also contain appreciable quartz, feldspar, iron oxides, and carbonates, although most clay-sized materials in nature are clay minerals. In addition, some clay minerals, particularly the kaolinites and chlorites, when detrital, can be larger than 2 μm and a variable fraction of clay particles are even colloidal in nature (< 0.2 μm). Whereas cave deposits, deep oceanic sediments, many lacustrine muds, and claystones consist of particles dominantly smaller than 4 μm, most of the fluvial and tidal muds and many deltaic muds contain much silt and are noticeably gritty to the touch (try using the front teeth).

How are clay minerals used in the study of mudstones? Without hesitation, we can say that their uses are vast – from pills to provenance to basins – and, of

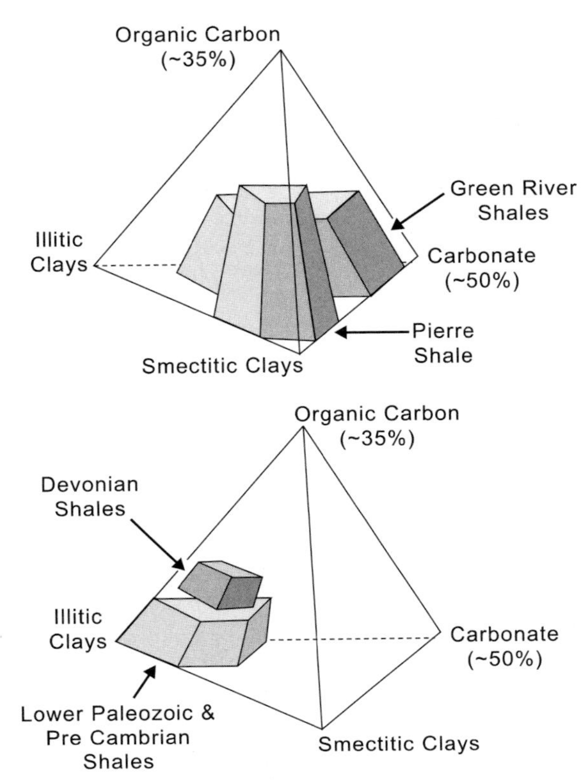

Fig. A.15. Generalized composition of four contrasting mudstones (redrawn from Brookins and Stow 1986)

all instrumental methods, X-ray identification of the clays is the most common. Why is this true?

Foremost is the fact that the clay minerals are always either first or second in volume in every mudstone and, as we will see, many bulk mudstone properties such as shear strength, cation exchange capacity, slaking or swelling in water, etc., are directly related to clay mineral composition. Thus for landfills, slope stability, absorbents, construction and a plethora of industrial uses knowledge of the types of clay minerals in a mudstone is essential. Another reason for their popularity is the ease with which they can be determined today. And what about geological problems? The clay mineralogy of different mudstones is quite different and demands explanation (Fig. A.15). In addition, clay mineralogy can change up section and can contribute significantly to the interpretation of stratigraphic history (Fig. A.16).

Clay mineral composition in Mesozoic and Tertiary mudstones reflects mostly some combination of climate, relief, and source rocks (think of soil types) – just as does the composition of the sandstones, although there are also three additional factors to consider. The first is differential transport in suspension while the second is differential sensitivity of clay minerals to a change of fresh to saline water. The smectites are finest and travel farthest in suspension, whereas the kaolinites and micas are coarser and settle sooner to the bottom. Thus there is mineral segregation by Stokes Law during transport, especially by weak currents (see Box 3.3 in Chap. 3). Sodium and other cations in seawater and saline lakes cause flocculation of clay, which also produces segregation (because these ions attract other clay flakes to form aggregates which settle faster). And finally, the clay minerals change more during burial than the sandstones. Thus original mineralogy changes with time. Nonetheless, once these factors are understood, there are many processes and problems in the geology of mudstones to which clay mineral composition contributes importantly.

Structurally, clay minerals are almost identical to the mica group, but are generally less than 2 micrometers (2 μm, 1 μm = 10^{-6} m) in grain size. Clays are hydrous phyllosilicates with continuous sheet structures (Figs. A.17 and A.18), and they are classified on the basis of the number of sheets making up their basic structure or unit cell (Bailey 1980; Moore and Reynolds 1989; Velde 1992, pp 42–68). There are two types of sheets, *tetrahedral* and *octahedral* sheets. The tetrahedral sheets consist of linked silica tetrahedra similar to the mica group of minerals, whereas octahedral sheets are composed of linked octahedra, each consisting of oxygen and hydroxide ions around an aluminum or magnesium ion. When aluminum ions are in the centers of the octahedra, only two-thirds of the octahedral centers are filled (gibbsite sheet), but when magnesium ions are in the centers, all the octahedra centers are filled (brucite sheet). Different clay minerals result from the different kinds of cations and amount of OH that tie the sheets together. Water is also held on both surfaces and between layers and plays an important role in the physical properties of clays such as swelling and dehydration. Cation substitution in the tetrahedral and octahedral sheets may produce a residual charge deficiency that is compensated by the incorporation of mono- and divalent cations in the interlayer position.

The most common clay mineral groups in muds and mudstones are kaolinite, illite, chlorite, smectite, and sepiolite-palygorskite. A further division into subgroups and species is based on the octahedral sheet type, chemical composition, and the geometry of layer and interlayer superposition. All in all, there are over 35 different clay mineral species. The smectite group tends to have the finest grain-size distribution of all the clay mineral family.

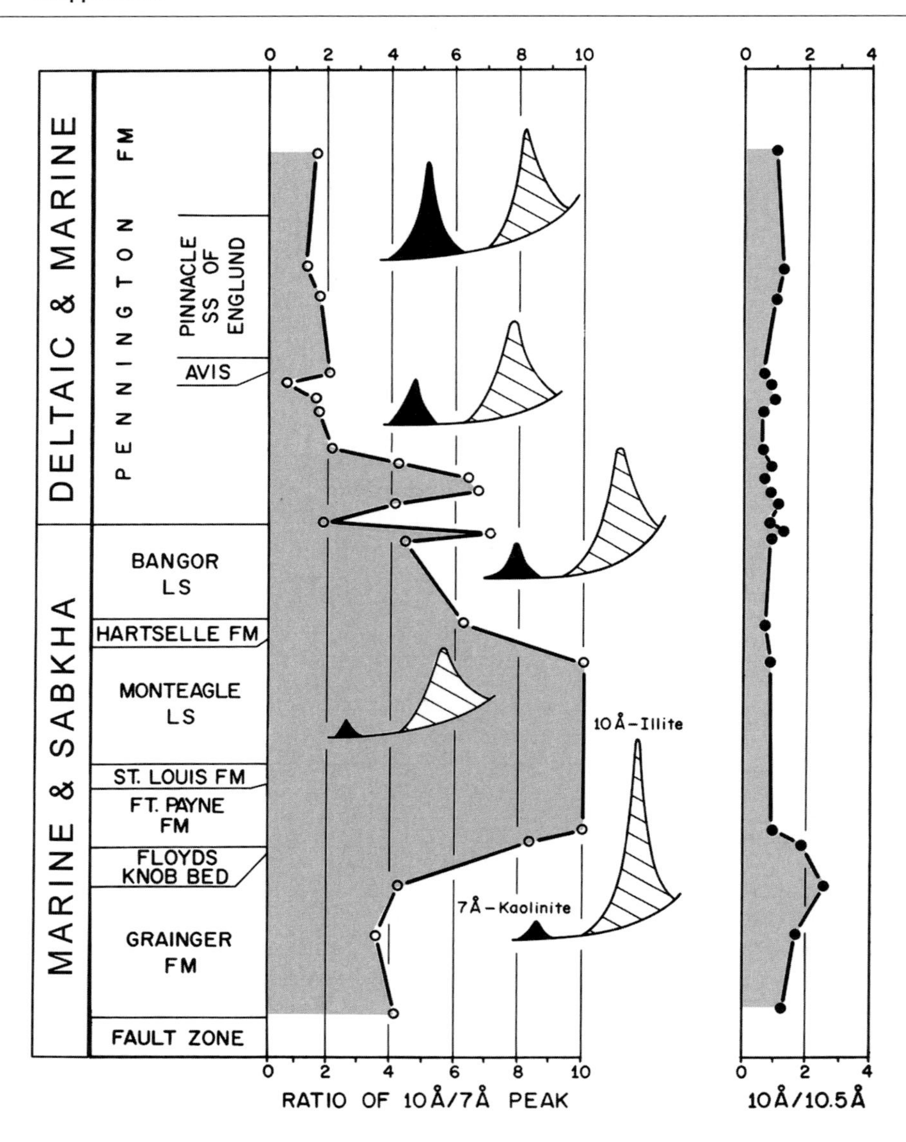

Fig. A.16. Clay mineral profile of Lower to Middle Mississippian rocks exposed along Interstate 75 as it climbs the Pine Mountain Overthrust in Campbell Co., Tennessee (Sedimentation Seminar 1981, Fig. 20). The lower part of the section consists mostly of marine mudstones deposited beyond the shelf edge (Fort Payne Formation) followed by shelfal and tidal flat carbonates and some sabkhas which are all illite dominated. Higher in the section in the Pennington Formation, there are inner shelf and deltaic terrigenous environments and kaolinite becomes more prominent (shoreline segregation). Clearly, clay minerals help strengthen the environmental interpretation based on evidence from sandstones and carbonates alone. Illite crystallinity changes but little, however, because the depth of burial range is too small. Total length of section is slightly more than 300 m. Is it not time to resample this section for isotopic study of its carbonate concretions as in Fig. 6.20?

Sometimes, particularly in older literature, the term montmorillonite – which in fact is a species of the smectite group – is used as a synonym of smectite (Table A.4). Therefore, suspensions dominated by smectite can be expected to travel farther than coarser illite or kaolinite in the same suspension (Gibbs 1997). This characteristic also has a practical implication: smectite suspensions are much slower to filter through very fine porous media (i. e., less than 0.2 μm).

The term *mixed-layer* is used for clays that consist of random or regularly alternating layers of two or more minerals. Because practically any combination of minerals can occur, mixed-layer clays may have almost any composition between the end members represented by the individual clay minerals. The more common mixed layer clay minerals, however, are illite/smectite (also known as *rectorite*, when the I/S ratio is 1:1) and smectite/chlorite (or *corrensite*, when the ratio S/C is also 1:1), both mostly of diagenetic origin.

Because clay minerals are very small, their identification by means of the optical petrographic microscope is rarely effective. Powder X-ray diffraction (XRD) is, instead, the method almost always used, because it allows a rapid and precise determination of the clay minerals present in a sample as well as some other crystallographic attributes such as the crystallinity of illite. Among several kinds of applications of XRD, the determination of the crystallinity of illite (as measured by the sharpness ratio of the 10-Å diffraction peak) is most important because it increases with depth and temperature, and is a proxy for maximum

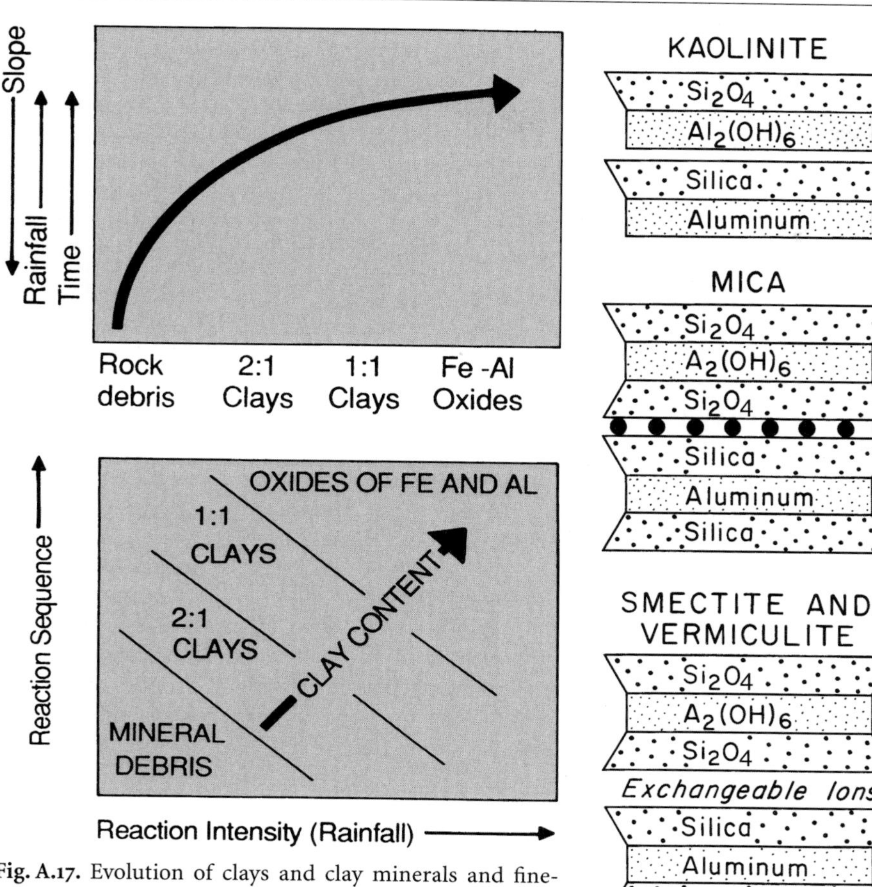

Fig. A.17. Evolution of clays and clay minerals and fine-grained materials from mineral debris. After Velde (1992, Figs. 4.4 and 4.5)

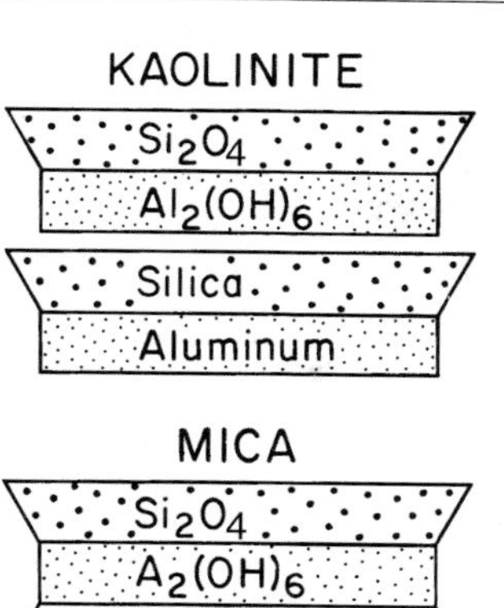

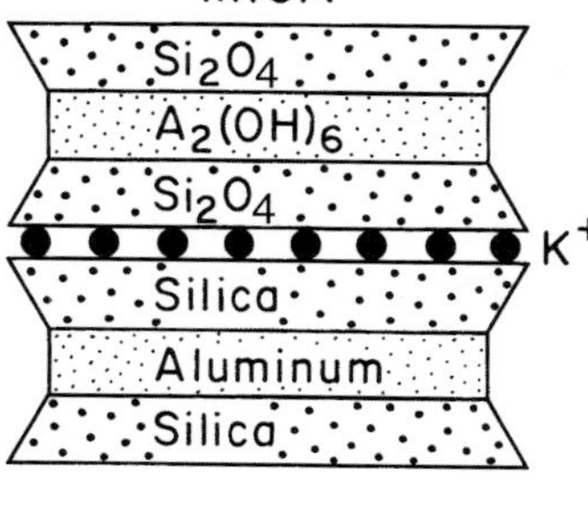

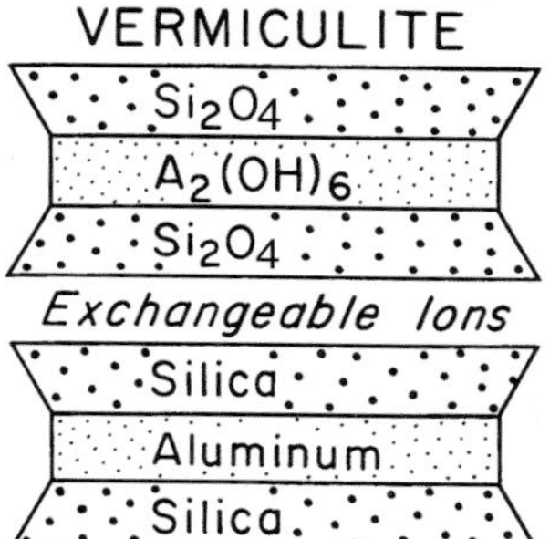

Fig. A.18. Basic model of sheet-like clay mineral structures (Potter et al., Fig. 1.22)

Table A.4. Principal clay-mineral groups: structure, chemical composition and cation exchange capacity (CEC)

Sheets/Group	Examples	Ideal formula	CEC, (meq/100 g)
Two			
Kaolin	Kaolinite	$Al_2Si_2O_5(OH)_4$	1–18
Serpentine	Chrysotile	$Mg_3Si_2O_5(OH)_4$	0
Three			
Smectite	Montmorillonite	$Al_2Si_4O_{10}(OH)_2 \cdot n\,H_2O$	60–150
	Saponite	$Mg_3Si_4O_{10}(OH)_2 \cdot n\,H_2O$	60–150
Three			
Vermiculite	*di*-Vermiculite	$Al_2Si_4O_{10}(OH)_2 \cdot n\,H_2O$	120–200
	tri-Vermiculite	$Mg_3Si_4O_{10}(OH)_2 \cdot n\,H_2O$	120–200
Three			
Mica	Illite	$KAl_2(AlSi_3)O_{10}(OH)_2$	10–40
	Phlogopite	$KMg_3(AlSi_3)O_{10}(OH)_2$	0
Three plus one			
Chlorite	Donbassite	$Al_2Si_4O_{10}(OH)_2 \cdot Mg_3(OH)_6$	1–5
	Clinochlore	$Mg_3Si_4O_{10}(OH)_2 \cdot Mg_3(OH)_6$	1–5

di, dioctahedral; tri, trioctahedral

burial temperature. See Frey and Robinson (1999) for a complete treatment of the many facets of how the mineralogy of mudstones changes with low temperature metamorphism.

Clay minerals are also classified as *expandable* and *non-expandable*, according to their capacity to change volume by absorbing water molecules or other polar ions into their structure. Expandable clays swell when wet (as water is adsorbed within or on their outer surfaces) or decrease in volume as water is removed, perhaps during a long drought. This change in volume is unique to the clay mineral world and among them, smectites and vermiculite are the only minerals that exhibit this peculiar property.

Rarely do two natural clay samples have identical composition. This chemical diversity results from either isomorphous substitution or ion exchange. Substitution may take place within either the octahedral or the tetrahedral layers or in interlayers by adding ions on the surface of the layer sheet of the tetrahedral-octahedral network. In contrast, ion exchange proceeds at the solid-liquid interface between anions and cations held in unbalanced charges at or near the surface of the solid material and ions in the aqueous medium. In addition to substitution of ions in their structural units, most of the clay minerals have the ability to exchange ions adsorbed on their surfaces, particularly cations, as for example, H^+, K^+, Ca^{2+}, Mg^{2+}, and Al^{3+}. This exchange of ions is called the *cation-exchange capacity*, CEC, and varies greatly among the clay minerals (Table A.4). Exchange of adsorbed ions occurs when water flows past stationary clay particles as in the passage of water through soil or when a clay particle moves into a new chemical environment – when river-borne clays enter seawater, for example. Cations that are adsorbed on the charged surfaces of small particles in contact with dilute electrolyte solutions are exchangeable for ions in the solution, as shown below

$$[\text{COLLOID}]\text{Ca}^{2+} + 2\,\text{H}^+(\text{aq.}) \rightarrow [\text{COLLOID}]\,2\,\text{H}^+ + \text{Ca}^{2+}(\text{aq.})\,.$$

In this example, protons in acidic water replace exchangeable Ca^{2+}, releasing Ca^{2+} ions into the water and attaching two H^+ ions to the clay surface. Such exchange reactions can achieve a state of equilibrium that causes water-particle mixtures to respond to changes in water compositions in accordance with Le Châtelier's principle. It is the clays with a high cation exchange capacity – highly sorbent clays, such as palygorskite and sepiolite – that make a good containment barrier for electrolytes, organics, or metals leaking from a point source such as a land fill, tank, or low level nuclear waste facility. Smectites are also good sorbents but are most valued for their sealing properties – expansion in grouting or by collapse of their structure with the addition of sodium to perhaps seal a sludge pond.

Because of electrical charges on their surfaces, clay minerals and other small particles form colloidal suspensions in dilute electrolyte solutions. Suspensions with these characteristics may be stable for long periods of time but flocculate or coagulate as the concentration of electrolytes increases (high ionic strength), as happens when clay suspended in fresh water enters supersaline water. The polarity of the electrical charges of colloidal particles of different compositions depends on the pH of the solution. Clay minerals are negatively charged except in acidic environments with pH less than 2.

Clay minerals in modern muds, as in most freshwater or saline lakes, alluvial deposits and in estuaries, are virtually all inherited (detrital) and result entirely from the erosion of upstream drainage basins. Exceptions to this generalization are the clay minerals formed at or very near the interface and in concentrated brines in playas, such as sepiolite-palygorskite. Conversely, outcropping mudstones and those of the deep subsurface, which have been subjected to deep burial, have a mixture of minerals that had been produced in situ (diagenetic) or had been inherited from their source rocks. Diagenesis or very low-grade metamorphism is revealed by the transformation of detrital minerals, such as smectites, which change to illite at depths on the order of 3,000 m (90 °C).

The stable isotopic composition of oxygen ($\delta^{18}O$) and hydrogen (δD) in clay minerals depends primarily on the temperature and the composition of water during their formation (Faure 1998, p 311). Hence, the stable isotopic signature inherited during clay minerals diagenesis, particularly the hydrogen isotope composition, has been related to the isotopic composition and source of the coexisting formation waters (Masuda et al. 1992). In general, the $\delta^{18}O$ and δD values of clays formed on the continents in contact with meteoric water (weathering or diagenetic environments), are generally different from those where oceanic water is involved in their formation.

Certain authigenic K-bearing clay minerals are dateable by the K-Ar and Rb-Sr methods (Clauer and Chaudhuri 1992, 1995) and therefore provide potential to determine the ages of mudstones. The resulting dates, however, tend to reflect a mixture of ages that are inherited from both detrital and authigenic clays, and thus may depend on burial temperatures, time, and circulating solutions and possibly even structural deformation as well as age of the detritus. It is clear, then, that the isotopic approach to determining provenance of muds and mudstones will be most definitive when multiple isotopic and mineralogical signatures are employed.

Clay minerals play a key role in many parts of today's industrial world (Murray 1991), especially the kaolinite and smectite families. The principal uses of the stable and largely inert kaolinite family are in the paper, paint, plastic, rubber, and catalytic cracking industries (Murray et al. 1993; Murray 1999), while those of the expandable and reactive smectite family are in drilling muds, as bonds for foundry sands and iron pellets, as sealants, and as absorbents for pet litter, agricultural chemicals, and water, oil, and gasoline. From toothpaste to high-pressure complex chemical processes, to thermally resistant ceramics in space, clay minerals serve well the many diverse technological needs of modern society.

References

Bailey SW (1980) Summary and recommendations nomenclature committee. Clays and Clay Minerals 22:73–78

Brookins DG, Stow SS (1986) Proceedings of the First Geochemical Workshop on Shale. Oak Ridge National Laboratory, Environmental Sciences Division, Publication 2845, 61 p

Clauer N, Chaudhuri S (1992) Isotopic Signatures and Sedimentary Records. Lecture Notes in Earth Sciences 43, Springer, Berlin Heidelberg New York, 529 p

Clauer N, Chaudhuri S (1995) Clays in Crustal Environments: Isotopic Dating and Tracing. Springer, Berlin Heidelberg New York, 359 p

Faure G (1998) Principles and Applications of Geochemistry, 2nd edn. Prentice Hall, Upper Saddle River, 600 p

Frey M, Robinson D (eds) (1999) Low-Grade metamorphism. Blackwell Science, Oxford, 313 p

Gibbs RJ (1977) Clay mineral segregation in the marine environment. J Sediment Petrol 47:237–243

Hughes RE, DeMaris PJ, White WA, Cowin DK (1987) Origin of clay minerals in Pennsylvanian strata of the Illinois Basin. In: Proceedings of the International Clay Conference, Denver, 1985 Clay Minerals Society, Bloomington, Indiana, pp 97–104

Masuda H, Kusakabe M, Sakai H (1992) Hydrogen and oxygen isotope ratios of shale and characteristics of formation waters in sedimentary complexes accreted at different times, Kinki District, Southwest Japan. Geochim Cosmochim Acta 56:3505–3511

Moore, D, Reynolds R (1989) X-ray diffraction and the identification and analysis of clay minerals. Oxford University Press, Oxford, 332 p

Murray HH (1991) Overview – Clay mineral applications. Appl Clay Sci 5:379–395

Murray HH (1999) Applied clay mineralogy today and tomorrow. Clay Minerals 34:39–49

Murray HH, Bundy W, Harvey C (eds) (1993) Kaolin Genesis and Utilization. The Clay Minerals Society, Boulder, 341 p

Sedimentation Seminar (1981) Mississippian and Pennsylvanian section on Interstate 75 south of Jellico, Campbell County, Tennessee. Tennessee Division Geology, Report Investigations 38, 41 p

Terry BO, LaGarry HE, Wells WB (1995) The White River Group revisited; vertebrate track ways, ecosystems, and lithostratigraphic revisions. In: Diffendal RF, Flowerday C (eds) Geologic Field Trips in Nebraska and adjacent parts of Kansas and South Dakota, Guidebook #10. Conservation Survey Division/Institute Agriculture and Natural Resources/University Nebraska (29th Annual Meetings North and South Central Sections, Geol Soc America), pp 43–57

Velde B (1992) Introduction to Clay Minerals. Chapman and Hall, London, 195 p

Digging Deeper

Bouchet A, Meunier A, Sardini P (2000) Minéraux argileux: structure crystalline, identification par diffraction de rayons X. Bulletin Centre Recherche Elf, Exploration Production (Pau), Mémoir 23:136.

This is the reference for anyone who wants to learn, step-by-step, about the structure of clay minerals and how that structure controls their many properties. To paraphrase the authors, "This book offers simplified do-it-yourself exercises on clay mineral identification." Four chapters – structures, compositions and chemical properties, X-ray diffraction and identification plus five appendices and many elegant colored illustrations and a CD-ROM. There is an introduction in English and the rest of the text is in easy-to-read French. This is the most beautifully illustrated book on clay mineral identification that we have seen.

A.4. Soils and Paleosols

There are countless reasons to study modern soils and many reasons to study their ancient equivalents. Modern soils show us how the factors of soil formation change with variation of relief, slope, parent material, climate and vegetation and thus provide a model for interpreting ancient soils. In addition, both modern and ancient soils are the source of almost all clay minerals as weathering converts rock fragments and debris into secondary minerals (Fig. A.17). Paleosols are of interest for two major reasons: the insights they give us to paleoclimate and because paleosols provide important stratigraphic markers both in Quaternary and older deposits (Figs. A.19 and A.20).

Paleosols are abundant and widely studied in Quaternary and Tertiary sediments, but are less commonly recognized in Mesozoic and Paleozoic deposits with the important exception of underclays in Coal Measures. In most of these studies, it is the climate record of the soil that is sought – did the paleosol develop under a humid or dry climate, warm or cold (e.g. Koch et al. 2003)? In addition, paleosols also have great application in coastal deposits for sequence stratigraphy. Lower sea levels cause lower water tables and thus better oxidized paleosols, whereas rising water tables

Fig. A.19. Alluvial soils: Multiple Holocene alluvial soils along Haines Branch in Pioneers Park near Lincoln, Lancaster Co., Nebraska (Photograph by Brian Nicklen)

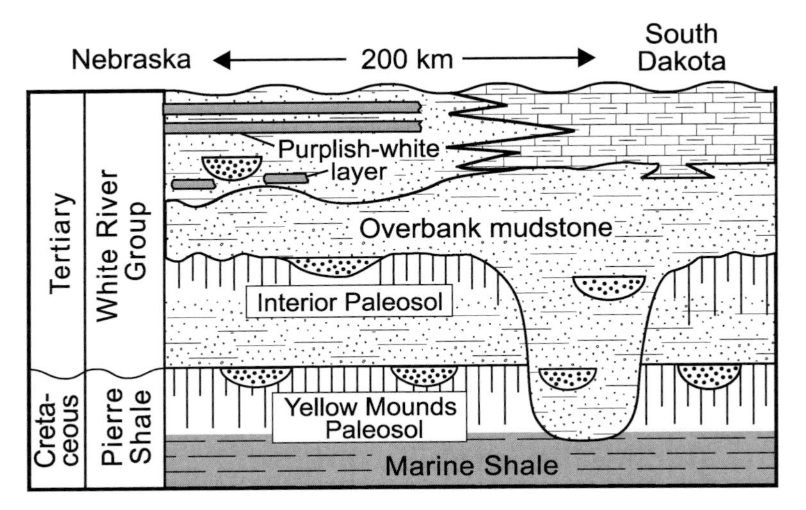

Fig. A.20. Paleosols as time lines (event beds) in the Miocene molasse sediments of eastern Nebraska and southern South Dakota (after Terry et al. 1995, Fig. 5)

cause water logged (gley) soils possibly followed by swamps (peats) and finally transgressive bay fills and open shelf deposits (Chap. 8). Thus recognition of paleosols is important for an understanding of basin fill. Precambrian paleosols also have received much attention chiefly as witnesses to the changing oxygen content of the atmosphere in the Proterozoic (e. g., Rye and Holland 1998). However, the interpretation of these paleosols has been quite controversial, and more convincing evidence for a low oxygen atmosphere before 2.2 to 2.3 Ga comes from sulfur isotope studies (e. g. Mojzsis et al. 2003). Good examples of paleoclimate interpretations of paleosols are provided by Joeckel (1999) and Mora and Driese (1999).

An outstanding study of soil and climate, using a single, widespread Pennsylvanian underclay is Cecil et al. (2003). See especially their figures 22 and 23. They correlated a Desmoinesian underclay from the Appalachian Basin across the United States as far west as Arizona and Nevada. The characteristics of the underclay (and its equivalent beds) systematically change from wet and waterlogged in the east to dry and arid in the west, parallel to present climate. Recognition of the changing character of the underclay as linked to climate is central to their argument. Worldwide fall in sea level (lowstand systems tract position in the cycle) is responsible for the great extent of this bed. In the same volume, Cecil and DuLong, (2003) also emphasize the greater influence of

Table A.5. Soil nomenclature and weathering (after Steila and Pond 1989, Table 5.3)

Soil Order	Weathering Intensity	Earlier Marbut Names
Entisols	Very low	Azonal soils, some Low Humic Soils
Vertisols	Low	Grumusols
Inceptisols	Low	Andosols, Sol Brun Acide, some Brown Forest Soils, Humic Gley Soils, and Low Humic Gley Soils
Aridisols	Low-moderate	Desert and Reddish Desert Soils, Serozem, Solonchak, Solonetz, plus some Brown and Reddish Brown Soils
Mollisols	Moderate	Chestnut Soils, Chernozem, Brunizem, Rendzina, plus some Brown, Brown Forest, Humic Gley and Solonetz Soils
Spodosols	Moderate	Podzols, Brown Podzolic soils, and Groundwater Podzols
Alfisols	Moderate-strong	Gray-Brown Podzolic, Gray-Wooded, and Noncalcic Brown Soils, Degraded Chernozem, and some Planosols and some Half-Bog Soils
Ultisols	Strong	Red-Yellow Podzolic, Reddish-Brown Lateritic, plus some Planosols and Half-Bog Soils
Oxisols	Very Strong	Laterites, Latosols
Histosols		Bog Soils

climate compared to other factors in the generation of coals and underclays.

Useful observations to make about soils and paleosols include color, soil structure, root and animal traces, destruction and solution of bedrock minerals and the precipitation of minerals such as calcite, silica, iron and dolomite to produce caliches, silica crusts and laterites. In all these studies recognition of zoning is essential.

Soils have been extensively studied, have a vast literature, and a diverse nomenclature. We found Table A.5 helpful to understand much of the terminology found in the literature.

References

Cecil CB, Dulong FT, West RR, Edgar NT (2003) Climate controls on the stratigraphy of a Middle Pennsylvanian cyclothem in North America. In: Cecil CB, Edgar NT (eds) Climate Controls on Stratigraphy. Tulsa, OK, Society of Economic Paleontologists and Mineralogists (Special Publication) 77:151–180

Cecil CB, Dulong FT (2003) Precipitation models for sediment supply in warm climates. In: Cecil CB, Edgar NT (eds) Climate Controls on Stratigraphy. Tulsa, OK, Society of Economic Paleontologists and Mineralogists, (Special Publication) 77:21–27

Joeckel RM (1999) Paleosol in Galesburg Formation (Kansas City Group, Upper Pennsylvanian), northern Midcontinent, USA; evidence for climate change and mechanisms of marine transgression. J Sediment Res 69:720–737

Koch PL, Clyde WC, Hepple RP, Fogel ML, Wing SL, Zachos JC (2003) Carbon and oxygen isotope records from paleosols spanning the Paleocene-Eocene boundary, Bighorn Basin, Wyoming. In: Wing S L, Gingerich P D, Schmitz B, Thomas E (eds) Causes and Consequences of Globally Warm Climates in the Early Paleogene. Geol Soc of America (Special Paper) 369:49–64

Mora CI, Driese SG (1999) Palaeoenvironment, palaeoclimate and stable carbon isotopes of Palaeozoic red-bed palaeosols, Appalachian Basin, USA and Canada. In: Thiry, Médard, Simon-Coinçon R (eds) Palaeoweathering, Palaeosurfaces, and Related Continental Deposits. Blackwell Science, Malden, MA, pp 61–86

Mojzsis SJ, Coath CD, Greenwood JP, Harrison TM, McKeegan KD (2003) Mass-independent isotope effects in Archean (2.5 to 3.8 Ga) sedimentary sulfides determined by ion microprobe analysis. Geochim Cosmochim Acta 67:1635–1658

Rye R, Holland HD (1998) Paleosols and the evolution of atmospheric oxygen: A critical review. Am J Sci 298:621–672

Steila D, Pond TE (1989) The Geography of Soils, 2nd ed. Rowman and Littleford, Savage, MD, 239 p

Terry DO, LaGarry HE, Wells WB (1995) The White River Group: Vertebrate trackways, ecosystems, and lithostratigraphic revision, redefinition, and redescription. In: Diffendal RF Jr, Flowerday CA (eds) Geologic Field Trips in Nebraska and Adjacent Parts of Kansas and South Dakota, Guidebook 10. 29th Annual Meeting North Central – South Central Sections, Geol Soc of America (Conservation and Survey Division, Institute of Agriculture and Natural Resources, University of Nebraska, Lincoln) pp 43–57

US Department Agriculture (1975) Soil Taxonomy: A Basic System of Soil, Classification for Making and Interpreting Soil Surveys. US Department Agriculture, Washington, DC, Handbook 436, 750 pp

Digging Deeper

Fitzpatrick EA (1984) Micromorphology of Soils. Chapman and Hall, London, 433 p.

A useful source of photomicrographs of soil thin sections plus methods, detailed descriptions arranged by the major USDA soil families and a helpful glossary. An excellent "how to" and "what it means" book.

Gardiner DT, Miller RW (2004) Soils in Our Environment, 10th edn. Pearson Education, Upper Saddle River, NJ, 641 p.

Twenty well-illustrated chapters (including 89 small colored plates) many tables, and a helpful glossary. Strong emphasis throughout on practical uses and problems. Ten editions prove the lasting value of this book.

Meyer R (1997) Paleolaterites and Paleosoils. Balkemia, Rotterdam, 151 p.

Subtitled, "Imprints of Terrestrial Processes in Sedimentary Rocks", this short book has six chapters, clear line drawings, and is unusual in that it has a special chapter devoted to the knowledge provided by paleosols and another to general conclusions. References provide a good introduction to the French literature.

Retallack GJ (2001) Soils of the Past, 2nd edn. Blackwell Science, Cambridge, 404 p.

Three parts – "Soils and Paleosoils, Soil Formation, and the Fossil Record of Soils" – plus 21 short chapters, a glossary, and over 1,325 references make this book an excellent one to understand paleopedology. Start here.

Summer ME (ed) (1999) Handbook of Soil Science. CRC Press, Boca Raton, Various paging.

See "Geomorphology of Soil Landscapes, Pedogenic Processes, Soil Taxonomy, other Systems of Soil Classification and Alteration and Formation of Soil Minerals in Weathering." In depth reviews, background required.

A.5. Petrographic Study

Petrographic study of mudstones has been based on thin sections since Sorby in the 19th century and is basic to their understanding, because it provides essential information about texture, lamination and many other structures, and with skill, permits the identification of the types of silt-sized grain. Collectively, this information gives us key insights to the depositional environment of the mudstone and, to a degree, its provenance. Texture and micro stratification are especially important, when mudstones are studied as seals for hydrocarbons – relating the pore size distribution of a mudstone to its petrographic and lithologic type. The drawings (Fig. A.21) by Schieber (1990 and 1999) provide a good starting point for microscopic study, as do the microphotographs of Macquaker and Gawthorpe (1993). Matching your thin sections against the drawings of Fig. A.21 (think of this as the *microfacies* approach to shale petrology) is a very good way to start your study of thin sections of mudstones. To help you, Zimmerle (1992) provides unique photomicrographs of individual grains in mudstones and Schieber and Zimmerle (1998) an introduction to shale petrology. A good initial overview of the petrography and origin of fine-grained argillaceous rocks is also found in Carozzi (1993). Most comprehen-

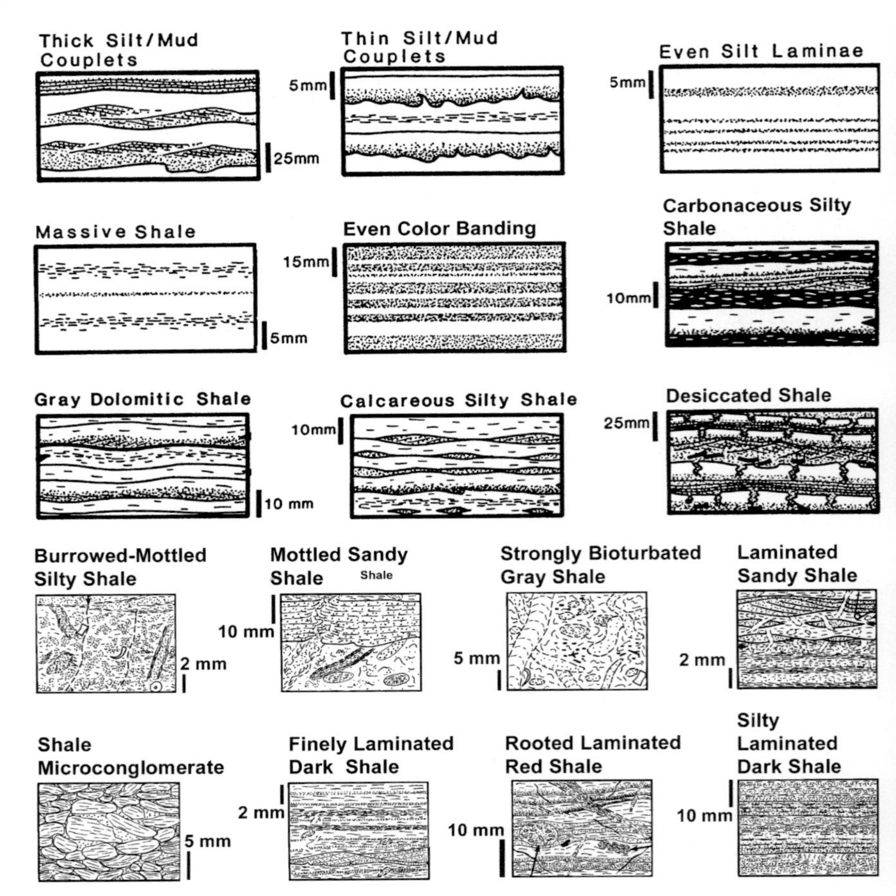

Fig. A.21. Informative drawings of thin sections of many of the mudstone types you are likely to encounter (Schieber et al. 1990 and 1999). Published by permission of the author, the Society for Sedimentary Geology and Elsevier Science

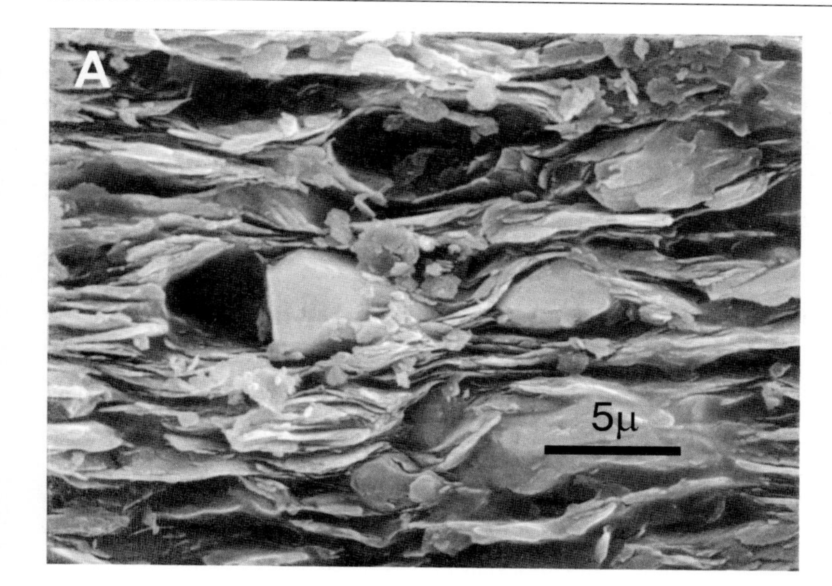

Fig. A.22A,B. Electronic imaging: (**A**) Clay fabric deformed about very fine, scattered quartz silt grains in Devonian Shale at Palmer Glenn near Rochester, NY (Image courtesy of Neal O'Brien) and (**B**) back-scattered electron images (BSEI) from a Cretaceous shale from offshore Brazil (Courtesy of PETROBRAS)

sive, however, and very well illustrated, is the atlas of O'Brien and Slatt (1990) with its many colored illustrations each accompanied by an extended interpretation. X-radiography of the rocks from the sampled interval always goes far to place the thin section in its correct depositional setting. Ultra thin, wedge-shaped and polished sections of mudstones aid study of thin sections with the petrographic microscope (Newtwich and Yale 1991) and with many of the techniques of Fig. A.1.

Other techniques build upon insights obtained from optical petrographic study. These include the Scanning Electron Microscope (SEM) to define fine-scale clay fabrics (Fig. A.22A) and Backscattered Electron Microscopy (BSE), which is becoming more and more routine to identify the mineralogy of individual particles. Back Scatter techniques also image the form, orientation, texture, and internal structure of single grains (Fig. A.22B).

Special mention needs to be made of heavy minerals in mudstones. Rarely studied, they may be the best provenance indicator of a mudstone, because the low permeability of mudstones shields heavy minerals from much, perhaps even most, solution after burial. See Totten et al. (1998) for a comprehensive study and Commeau et al. (1992) for the methodology of separation.

Bulk techniques are commonly used to supplement information gained from the above petrographic ones, which focus on particles. These include bulk mineralogy determined by powder X-rays, diffraction, clay mineralogy (using oriented slides coated with <2 micrometers clay), major element chemistry from X-ray fluorescence, and Rock-Eval analysis to determine total organic carbon (TOC). Mineral composition from bulk chemical analyses is also possible. See Tucker (1988) for a summary of the above techniques.

References

Carozzi AV (1993) Sedimentary Petrography. PTR Prentice Hall, Englewood Cliffs, NJ, 263 p

Commeau JA, Poppe LJ, Commeau RF (1992) Separation and identification of the silt-sized heavy mineral fraction in sediments. US Geological Survey Circular 1071:13

Macquaker JHS, Gawthorpe RL (1993) Mudstone lithofacies in the Kimmeridge Clay Formation, Wessex Basin, Southern England: Implications for the origin and controls of the distribution of mudstones. J Sediment Petrol 63:1129–1143

O'Brien NR, Slatt RM (1990) Argillaceous Rock Atlas. Springer, Berlin Heidelberg New York, 141 p

Pike J, Kemp AES (1996) Preparation and analysis techniques for studies of laminated sediments. In: Kemp AES (ed) Paleoclimatology and Paleoceanography from Laminated Sediments. Geol Soc (Special Publication) 116:37–48

Schieber J (1990) Significance of styles of epicontinental shale sedimentation in the Belt Basin, Mid-Proterozoic of Montana, USA. Sediment Geol 69:297–312

Schieber J (1999) Distribution and deposition of mudstone facies in the Upper Devonian Sonyea Group of New York. J Sediment Res 69:909–925

Totten MW, Hanan MA (1998) The accessory-mineral fraction of mudrocks and its significance for whole-rock geochemistry. In: Schieber J, Zimmerle W, Sethi P (eds) Shales and Mudstones, Part II. E. Schweizerbart'sche, Stuttgart, pp 35–53

Tucker M (1988) Techniques in Sedimentology. Blackwell, Oxford, 394 p

Zimmerle W (1992) Thin-section petrography of argillaceous rocks. Zentralablatt Geologie Paläontologie Part 1 5:365–390

Digging Deeper

Dawson WC (2000) Shale microfacies: Eagle Ford Group (Cenomanian-Turonian) North Central Texas outcrops and subsurface equivalents. Gulf Coast Association Geological Societies Transactions 50:607–621.

Good example of what is needed for understanding seals. See also Schieber (1990 and 1999) in References Cited.

O'Brien, Neal, Slatt, Roger M (1990) Argillaceous Rock Atlas. Springer, Berlin Heidelberg New York, 141 p.

All who are going to describe mudstones petrographically and relate their results to the geologic setting of the mudstone need to have this atlas. Why? Because its 10 chapters provide many case histories with photomicrographs as well as methodology. Included are 40 colored photomicrographs, many black and white ones, plus SEM photomicrographs of mudstone fabrics and compositions of the studied units.

A.6. Organic Matter

The organic content of mudstones ranges from less than 0.2 to 0.5% (all maroon and red ones plus many gray ones) to more than 15 to 25% for some dark gray to black shales. This organic matter has been studied extensively both chemically and optically to learn about the initial conditions of its deposition and its later burial history (Table A.6). The terminology used in the study of organic matter is complex, in part because of the nature of the material, but also because terminologies differ in different fields – geochemists use the term *kerogen*, organic petrologists using mostly reflected light employ the term *maceral* (Latin, *macerare*), and palynologists use biological terminology. Preparation methods also vary among these groups. Here we follow Tyson (1995, pp 15–16; see Digging Deeper) and use the term kerogen for all the organic matter in sedimentary rocks that is insoluble in organic solvents such as chloroform, carbon disulfide, etc. Kerogen is also defined chemically as all the high-oxygen pyrobitumens other than peat and coal that occurs in sediments and sedimentary rocks (Killops and Killops 1993, pp 106–118).

In more detail, the types of organic matter, called kerogen I, II, III or IV (Fig. A.23), provide information about its source: was the organic matter derived from land plants or from the marine realm. The sedimentology of this debris – its size, sorting, and abrasion – help us reconstruct its abrasion history and thus better understand the final depositional environment (Tyson 1995; see Digging Deeper). In addition, both the color and reflectance of the particulate organic matter provide information about burial history, a key factor in petroleum exploration: will the organic matter yield mostly CO_2 and H_2O, gas or oil or is it already depleted in these volatiles? The abundance of particulate organic matter in a mudstone is also the prime determinant of its color. In sum, the preserved organic matter of a mudstone permits us to fine-tune its environment of deposition (paleoxygen levels and source and sedimentology of its organic particles). But even more useful to us, the study of kerogen provides insights to temperature history during its burial and thus better appraise

Table A.6. Importance of organic matter in muds (Adapted from Tyson, 1995, Table 1.1)

- Biochemical reactivity critical for bacterially controlled (affected) early diagenesis
- Major influence on Eh of both water and mud
- Affects food quality for deposit feeders
- Controls amount and type of potential hydrocarbons and some sulfide deposits
- Helps assess original depositional environment including separation of marine from nonmarine mudstones

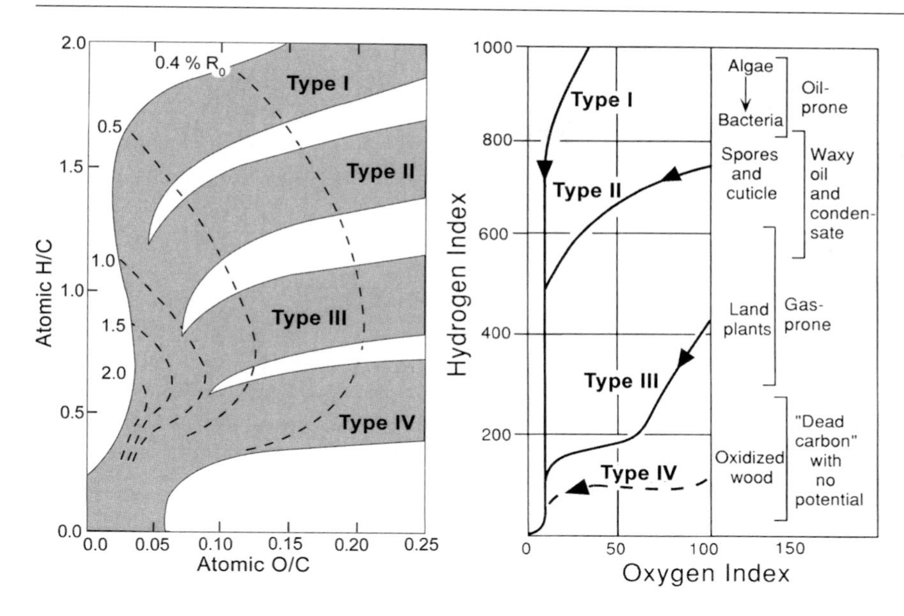

Fig. A.23. Kerogen types defined by hydrogen index, H/C, and oxygen index O/C, (after Hunt 1996, Fig. 10–5) and as defined by contents of hydrogen and oxygen (after Cornford 1990, Fig 11.7). Published by permission of the authors, W.H. Freeman and Blackwell Scientific

a mudstone for its value as a source rock. Below we present the needed background to initially understand and appreciate these techniques, because today the comprehensive study of a mud or mudstone nearly always includes at least some of them.

Table A.7A. Types of kerogen (Waples 1985, Table 4.1). [Published by permission of the author and IHRDC]

Maceral	Kerogen Type	Source
Alginite	I	Fresh-water algae
Exinite	II	Pollen, spores
Cutinite	II	Land-plant cuticle
Resinite	II	Land-plant resins
Liptinite	II	Land-plant lipids; marine algae
Vitrinite	III	Land-plants wood and cellulose
Inertinite	IV	Charcoal; oxidized material

Background: All organic matter is mainly composed of carbon, oxygen, and hydrogen. This material is divided into five groups: carbohydrates, proteins, lipids, lignins, and pigments.

In continental water bodies and in marginal seas, organic matter covers a continuous size spectrum ranging from free small molecules on one extreme to macromolecules, aggregates, and living organisms on the other. By arbitrary convention, organic matter is separated into particulate (organic material retained on a 0.5–1.0 µm pore-size filter) and dissolved (organic material passing on to the filtrate) fractions.

Particulate organic matter, produced either in the water column or on land, is divided into living (i. e., bacteria, plankton) and non-living, detrital fractions. Most of the living fraction is rapidly recycled along with some of the detritus that has not been subjected to significant bacterial reworking. Carbohydrates, lipids, proteins, etc. make up this fraction, generally identified as *labile* organic matter in opposition to

Table A.7B. Kerogen maturation (after Hunt 1996, Fig. 10.38). Published by permission of W.H. Freeman

T_{max} (°C)	Vitrinite reflectance	Kerogen color	Wt.% C	Coal Rank
50	0.2–0.5	Yellow	65–75	Lignite-Subbituminous
50–100	0.5–1.0	Orange	75–83	High-volatile bituminous
100–140	1.0–1.5	Light brown	83–87	Medium-volatile bituminous
140–170	1.5–1.9	Brown	87–88	Low-volatile Bituminous
170–200	1.5–2.5	Dark rown	88–90	Semianthracite
200–250	2.5–4.0	Black	90–95	Anthracite
> 250	> 4.0	Black	> 95	Meta-anthracite

the *refractory* fraction, which resists chemical break down. Accordingly, particulate organic matter in a mud or mudstone can originate from either in situ biological production or be derived from a source area.

Techniques: Organic solvents are used to separate the several kinds of organic matter present in muds and mudstones. The part that dissolves is called *bitumen*, whereas the insoluble part is called *pyrobitumen* or *kerogen*. Kerogen is concentrated in coals and oil shales, but is also present in virtually all fine-grained sedimentary rocks. It is the most common organic material on Earth and is present in most mudstones, most abundantly in oil shales. Kerogen has been divided into four types that are defined by both chemical and microscopic study (Table A.7).

The total organic matter of a mudstone, TOC for short, is determined by LECO combustion or by ROCK-EVAL pyrolysis in the laboratory and also is approximated by a gamma ray, density, or sonic log (if more than 4%) and very, very roughly by mudstone color (with much local experience!). The Rock-Eval procedure also is used to determine the kerogen types in terms of the hydrogen index, HI (mg hydrocarbons/g organic C) and the oxygen index (mg CO_2/g organic C). Determining total organic matter is the first step in the study of the organic matter of a mudstone followed by its kerogen types, reflectance, and the sedimentology of its particles.

Processes: The composition and physical properties of kerogen change in a systematic way with time and temperature during burial. During this process, known as *kerogen maturation*, lighter molecules are first driven off as gas and later, as heavier molecules are liberated, to form oil and finally dry gas. Because the maturation process is not reversible, geologists can map and study the thermal evolution of a mudstone in a basin by using the downhole changes in the chemical and physical properties of kerogen. In general, as kerogen matures with temperature, it becomes darker just like a piece of toast (Table A.7). This evolution is made quantitative by a microscopic technique called *vitrinite reflectance* (Waples 1985, pp 87–88; Taylor et al. 1998, pp 372–391). Reflected light and polished sections are standard to identify and count the different proportions of particles, the macerals, in organic-rich shales and coals. Organic matter is also profitably studied with the ordinary petrographic microscope by noting its concentration, size distribution, abrasion, biological affinities, and color in muds and mudstones, much as sedimentologists study and analyze individual particles of terrigenous and carbonate silts and sands (See Tyson 1995, Table 25-2 in Digging Deeper). Collectively, all of the above petrographic techniques are called *organic petrology* and rare is the mud or mudstone whose understanding is not expanded by their use along with determination of total organic carbon and the hydrogen and oxygen indices.

The variations of both total organic carbon and its kinds in muds and mudstones are closely linked to relative sea level (sequence stratigraphy), bottom oxygen levels, and rate of sedimentation in muddy basins (Creaney and Passey 1993). Porosity and resistivity logs provide initial information about the organic content of mudstones, when cores are not available (Passey et al. 1990). See Wüst and Bustin (2003) for more details on maturation of organic matter with burial.

References

Cornford C (1990) Source rocks and hydrocarbons of the North Sea. In: Glennie KW (ed) Introduction to Petroleum Geology of the North Sea, 3rd edn. Blackwell, Oxford, pp 291–361

Creaney S, Passey QR (1993) Recurring patterns of total organic carbon and source rock quality within a sequence stratigraphic framework. Am Assoc Petrol Geol Bull 77:386–401

Hunt JM (1996) Petroleum Geochemistry and Geology, 2nd edn. Freeman, San Francisco, 743 p

Killops SD, Killops VJ (1993) An Introduction to Organic Geochemistry. Longman Scientific and Technical, Singapore, 265 p

Passey QR, Creaney S, Kulla JB, Moretti FJ, Stroud JD (1990) A practical model of organic richness from porosity and resistivity logs. Am Assoc Petrol Geol Bull 74:1777–1794

Taylor GH, Teichmüller M, Davis A, Diessel CFK, Littke R, Robert P (1998) Organic Petrology. Gebrüder Borntraeger, Berlin, 704 p

Wûst RAJ, Bustin RM (2003) Kerogen. In: Middleton GV (ed) Encyclopedia of Sediments and Sedimentary Rocks. Kluwer Academic, Dordrecht, pp 400–403

Waples DW (1985) Geochemistry in Petroleum Exploration. IHRDC, Boston, 232 p

Digging Deeper

Peters KE, Moldowan JM (1993) The Biomarker Guide. Prentice Hall Englewood Cliffs, NJ, 363 p.

We welcomed our discovery of this book, because it does just what we needed – it gave us insight to what biomarkers can do (learning mode) and provides the already knowledgeable with detailed, technical information (selected access mode). To achieve these goals the book has four parts: an introduction, procedures, guidelines to interpretations, and future problems.

Tyson RV (1995) Sedimentary Organic Matter. Chapman and Hall, London, 615 p.

Aided by a clearly written text plus many informative tables and figures, the 25 chapters of this book provide one of the best insights to organic matter in sediments available. Especially valuable are the Introduction with its many informative tables, the thorough integration of biology and sedimentology, and the fine set of color plates to guide identification of organic particles. The chapter on carbon isotopes is particularly helpful.

A.7. Isotopes: The Stable Isotopes of C, O, H, S, and N

Isotopes are one, two or more species of a chemical element that have slightly different chemical and physical properties, because of small differences in mass related to different numbers of neutrons. Isotopes may be stable (non-radioactive) or unstable (radioactive).

Five light elements with stable isotopes are widely used for the study of muds and mudstones: C, O, H, S, and N. All of these isotopes vary sufficiently and have been studied enough to provide fairly standard patterns of interpretation. Hence, stable isotopes are one of the foremost geochemical methods of study of mud and mudstones, particularly for their minor carbonates. Two additional reasons for their study are that they are relatively inexpensive to determine and that only a few grams are needed. The study of stable isotopes in sedimentary geology only started after World War II.

In nature, chemical, physical, or biological processes cause some isotopes of C, O, H, S, and N to change in their relative abundance thus causing small differences in their isotopic ratios. Although evaporation and diffusion are important in specific situations, most significant fractionations are controlled by biological or chemical reactions. Typically, because of differences in mass, two isotopes of an element will be fractionated during an exchange reaction between two phases as for example, air and water or solid and water. Hence, their isotopic ratios will be different in each phase. Another example is the change of the carbonate $^{18}O/^{16}O$ ratio with temperature. This allows the determination of oceanic paleotemperature differences of less than 1 °C in carbonate material formed by inorganic or biological precipitation. This was the first extended geologic use of stable isotopes (Urey 1947).

Table A.8. $\delta^{13}C$ in various reservoirs (adapted from Degens 1989, Fig. 11.19)

Reservoir/Material	$\delta^{13}C$ (‰)
Atmospheric CO_2	−6 to −10
Freshwater H_2CO_3	−4 to −14
Kerogen	−22 to −27
Marine carbonates (inorganic)	+2 to +7
Marine carbonates (biogenic)	−4 to +3.5
Fresh water carbonates	−2 to −12.5
Methane (Mesozoic)	−32 to −54
Methane (Quaternary)	−56 to −82

Ratios of stable isotopes are expressed as δ values, in parts per thousand (‰) and are related to specific standards:

$$\delta(‰) = \left[\frac{\text{isotopic ratio of sample}}{\text{isotopic ratio of standard}} - 1\right] \times 10^3 .$$

Ratios are expressed as the heavier to the lighter isotopes; e.g., $^{18}O/^{16}O$, and the δ value is the relative difference between a sample and an arbitrary standard: a higher ratio in the sample compared to the standard yields a positive (heavier) number, whereas a lower ratio yields a negative (lighter) number (Fig. A.24). See Longstaffe (2003) for a recent concise overview.

Carbon: Carbon is the major element in the biosphere, but it occurs also in the atmosphere, the lithosphere, and the hydrosphere of the Earth. In mud and mudstones, carbon is the major component of organic matter and of carbonate minerals. Two stable isotopes of carbon, ^{12}C and ^{13}C are widely employed to differentiate carbon sources in mud and mudstones (Table A.8). The fractionation of carbon isotopes separates carbon with a biogenic origin (depleted in ^{13}C) from carbon in equilibrium with the atmosphere (enriched in ^{13}C). During photosynthesis, for example, plants utilize more of the lighter ^{12}C than the heavier ^{13}C; in other words, they discriminate against ^{13}C in favor of ^{12}C. As a result, the $^{13}C/^{12}C$ ratio of biogenic material is lower than that of atmospheric CO_2. Consistent with their derivation from biogenic matter, fossil fuels are depleted in ^{13}C.

Natural fractionation processes determine a variation of almost 150‰ in the stable isotopic ratio, heaviest being a carbonate from a meteorite ($\delta^{13}C = +60$‰), and lightest a bacterial methane ($\delta^{13}C = -90$‰). Numerous authors have documented the partitioning of carbon along its biogeochemical pathways. The ranges of $\delta^{13}C$ ranges for various carbon-bearing materials have been extensively summarized (e.g., Degens 1989; Faure 1986).

Stable carbon isotopes are currently used to study both ancient and modern organic matter in varied biogeochemical systems (e.g., Gearing et al. 1984; Mook and Tan 1991; Altabet 1996). Such data is helpful in evaluating attempts to use stable isotopes to trace carbon in systems having two or more isotopically distinguishable sources. An example is the use of $\delta^{13}C$ variations to track the proportion of marine and terrestrial carbon or to track carbon from C4 plants like grasses.

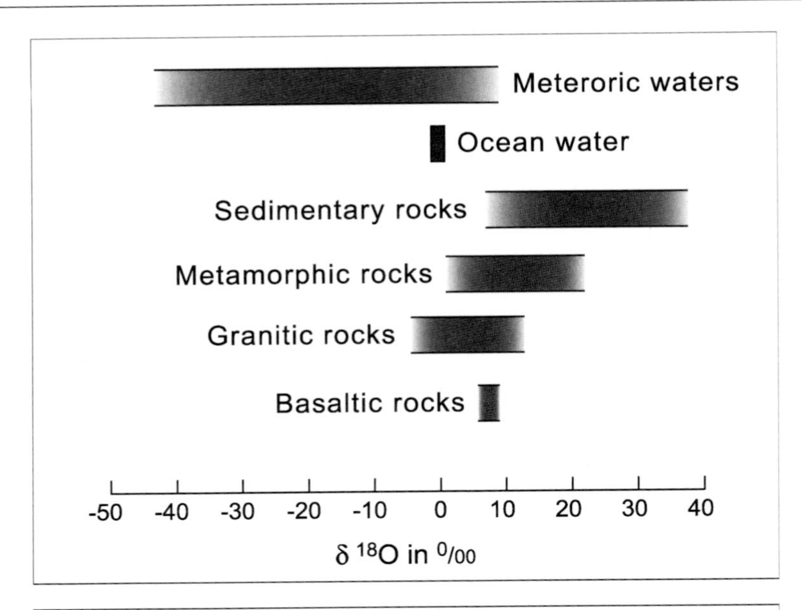

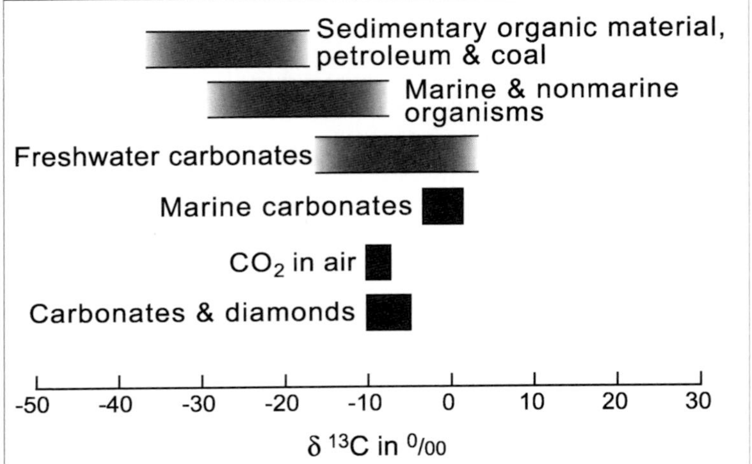

Fig. A.24. The variability $\delta^{18}O$ of $\delta^{13}C$ in various materials spans a wide range (after Hoefs, Figs. 20 and 25). Isotopes of O are useful as tracers of sources of water and carbonate minerals; isotopes of C help identify biologic influences for carbonate minerals and organic matter

Table A.9. Examples of variations of $\delta^{18}O$ in minerals and rocks (adapted from Brownlow 1996, Fig. 2.17)

Material	$\delta^{18}O$ (‰)
Quartz	+89 to +103
Biotite	+44 to +66
K-feldspar	+70 to +91
Plagioclase	+65 to +91
Marine limes to ne	+22 to +30
Freshwater limes to ne	+18 to +25
Mudstones	+14 to +19
Ocean water	−05 to +05
Temperate fresh water	−10 to −4
Snow and ice	−60 to −20

Oxygen and Hydrogen: The important stable isotopes of oxygen are ^{16}O and ^{18}O (Table A.9), and those of hydrogen are ^{1}H and ^{2}H, ordinarily identified as D, for Deuterium. The two systems are almost always determined together for water samples. The ratios D/H and $^{18}O/^{16}O$ exhibit measurable variations in natural waters ($\delta^{18}O$ of −0.5 to +0.5 in ocean waters, −0 to −4 in temperate fresh waters) and can advantageously be used to trace their source and history (Fig. A.24).

A major use of oxygen and hydrogen isotopes is the study of the complex interactions existing between minerals and the water cycle. Measurements of oxygen isotopes are also used to obtain insights to the sources of waters that precipitated minerals, their temperature of formation, and the degree of equilibrium attained (e. g., Faure 1998, pp 308–313). A good

Table A.10. Sulfur isotopes and sedimentation rates

Relationship from modern sediments: $\log \omega = 1.33 - 0.042 \Delta\delta^{34}S$
where ω = sedimentation rate in cm/yr and $\Delta\delta^{34}S = \delta^{34}S_{sulfate} - \delta^{34}S_{pyrite}$

	$\delta^{34}S$, permil	$\Delta\delta^{34}S$, permil	ω, cm/yr	Ratio
Devonian-Mississippian mudstones of Appalachian Basin[1]				
Seawater sulfate	+17			
Pyrite in highstand gray mudstones	+11.6	5.4	13	
Pyrite in transgressive black shales	−8.9	26	1.8	
Relative rate				7.3
Nearshore and offshore black shales of the Illinois Basin[2]				
Seawater sulfate	+15			
Pyrite in nearshore black shales	−10.0	25	1.9	
Pyrite in offshore black shales	−14.5	30	1.2	
Relative rate				1.5
Cretaceous of western Canada[3]				
Seawater sulfate	+17			
Pyrite in highstand mudstones	−16	0 33	0.88	
Pyrite in transgressive shales	−27	0 44	0.30	
Relative rate				2.9

Sources: [1]Maynard, 1980; [2]Coveney and Shaffer, 1988; [3]Bloch et al. 1999; Seawater sulfate values from Holser et al. 1988, Fig. 4.7

example is the already mentioned development by Harold C. Urey (1947) of a *paleo-thermometer* based on the isotope composition of oxygen in marine calcite and aragonite. Another application is the reaction between feldspar and acidified meteoric water to form clay minerals that takes place in the presence of an excess of water. The water/rock ratio is large, and so clay minerals and hydroxides formed during weathering are normally in isotopic equilibrium with local meteoric water (Faure 1986, pp 477–481). Clay minerals formed on land and redeposited in the oceans – like other detrital minerals associated with clays, for example, in marine mud – do not equilibrate isotopically at the time of deposition. Instead they preserve the isotopic signature acquired at their formation, although their oxygen isotope composition appears to be reset at 300 °C or more (O'Neil and Kharaka 1976). On the other hand, hydrogen may undergo retrograde exchange with meteoric water at lower temperatures (Kyser and Kerrich 1991).

The original standard for oxygen and hydrogen isotopes (Craig 1961) was a hypothetical water sample known as SMOW (Standard Mean Ocean Water), but new standards, such as V-SMOW and SLAP (Standard Light Antarctic Precipitation) are currently being used.

There is a linear relationship between δD and $\delta^{18}O$ for present-day meteoric waters

$$\delta D = 8\delta^{18}O + 10$$

which defines a line, known as the *meteoric water line* (MWL). The model that describes the condensation processes that defines the MWL is known as *Rayleigh isotopic fractionation*. Because the isotopic compositions of clay minerals formed at surface temperatures are controlled by the isotopic composition of meteoric water, it follows that the isotopic compositions of clay minerals formed at a particular temperature should be related by equations similar to the one for the MWL.

Sulfur: There are four stable isotopes of sulfur, ^{32}S, ^{33}S, ^{34}S, and ^{36}S. The most abundant isotopic species, with 95.02% and 4.21% respectively, are ^{32}S and ^{34}S. The $^{32}S/^{34}S$ ratio is usually measured, and the isotopic composition of sulfur is commonly expressed as $\delta^{34}S$, in a manner similar to other stable isotopes. The standard value ($\delta^{34}S = 0‰$) of $^{32}S/^{34}S = 22.22$ has been measured in the Cañón Diablo troilite. The troilite phase of other meteorites show only slight variations from this standard value. Igneous rocks also tend to be close to this value, indicating derivation from chondritic-type material with little fractionation by processes within the Earth's mantle.

Sulfur isotopic fractionation occurs primarily in oxidation-reduction reactions, which can be either entirely inorganic or complex, multi-step biologically-mediated processes. Isotopic values of sulfides in sedimentary rocks range from $-45‰ < \delta^{34}S < +42‰$ so there is ample variation of isotopic composition. This probably reflects a wide variety of processes and conditions. Sulfides in igneous rocks show a much narrower range of values, $-2‰ < \delta^{34}S < +10‰$ (Brownlow 1996, pp 101–109).

Because of bacterial sulfate reduction, hydrogen sulfide and iron sulfide minerals in Recent marine muds are depleted in ^{33}S when compared to the heavier values of contemporary sulfate of overlying waters. Thus the sulfur that has passed through the sedimentary cycle shows increased variability in $\delta^{34}S$ because of this separation into light sulfide and heavy sulfate minerals. The degree of separation is a function of the sedimentation rate of the mud: the faster the sedimentation rate, the faster the bacteria work, and the less they fractionate sulfur isotopically. By applying this relationship to ancient mudstones, the relative sedimentation rates of different facies can be calculated (Table A.10).

Nitrogen: Nitrogen plays an important role in the biological world and has two stable isotopes, ^{14}N and ^{15}N. These, as with carbon isotopes, have been used to trace the source of organic matter in modern muds and in coal and petroleum. The atmospheric relative abundance of these isotopes is 99.64% and 0.36%, respectively. The isotopic composition of samples is usually expressed as $\delta^{15}N$. The N_2 of the atmosphere is used as a standard with a value of 0‰.

Marine organisms have $\delta^{15}N$ between +4‰ and +20‰; whereas terrestrial organisms have $\delta^{15}N$ between −8‰ and +15‰. These differences have been used, for instance, to determine the relative contributions of terrestrial and marine organic nitrogen to marine mud (e.g., Schäfer and Ittekkot 1993). Likewise, the isotopic composition of N in coal varies from −2.5‰ to +6.3‰, depending on the origin of the organic material and on the rank of the coal. Coal derived from terrestrial plants has lower $\delta^{15}N$ values than coal formed from zooplankton and phytoplankton. Petroleum $\delta^{15}N$ values range from +0.7‰ to +8.3‰ indicating that N isotopes are not appreciably fractionated by denitrification, and that they carry the signature of the biogenic source of oil (Faure 1986, pp 519).

References

Altabet MA (1996) Nitrogen and carbon isotopic tracers of the source and transformation of particles in the deep sea. In: Ittekkot V, Schäfer P, Honjo S, Depetris PJ (eds) Particle Flux in the Ocean, SCOPE 57. Wiley, New York, pp 155–184

Bloch JD, Schröder-Adams CJ, Leckie DA, Craig J, McIntyre DJ (1999) Sedimentology, micropaleontology, geochemistry, and hydrocarbon potential of shale from the Cretaceous Lower Colorado Group in western Canada. Geological Survey of Canada Bulletin 531:185

Brownlow AH (1996) Geochemistry, 2nd edn. Prentice-Hall, Englewood Cliffs, New Jersey, 580 p

Coveney RM, Shaffer NR (1988) Sulfur-isotope variations in Pennsylvanian shales of the midwestern United States. Geology 16:18–21

Craig H (1961) Standard for reporting concentrations of deuterium and oxygen-18 in natural waters. Science 133:1833–1834

Degens ET (1989) Perspectives on Biogeochemistry. Springer, Berlin Heidelberg New York, 423 p

Drever JI (1997) The Geochemistry of Natural Waters, Surface and Groundwater Environments, 3rd edn. Prentice-Hall, Englewood Cliffs, New Jersey, 436 p

Faure G (1986) Principles of Isotope Geology, 2nd edn. Wiley, New York, 589 p

Faure G (1998) Principles and Applications of Geochemistry, 2nd edn. Prentice Hall, Englewood Cliffs, New Jersey, 600 p

Gearing JN, Gearing PJ, Rudnick DT, Requejo AG, Hutchings MJ (1984) Isotopic variability of organic carbon in a phytoplankton-based, temperate estuary. Geochim Cosmochim Acta 48:1089–1098

Hoefs J (1997) Stable Isotope Geochemistry, 4th Edn. Springer, Berlin Heidelberg New York, 201 p

Holser WT, Schidlowski M, MacKenzie FT, Maynard JB (1988) Biogeochemical cycles of carbon and sulfur. In: Gregor CB, Garrels RM, MacKenzie FT, Maynard JB (eds) Chemical Cycles in the History of the Earth. Wiley, New York, pp 105-174

Kyser TK, Kerrich R (1991) Retrograde exchange of hydrogen isotopes between hydrous minerals and water at low temperatures. In: Taylor HP, O'Neil JR, Kaplan IR (eds) Stable Isotope Geochemistry. A Tribute to Samuel Epstein. The Geochemical Society (Special Publication) 3:409–422

Longstaffe F (2003) Isotopic Methods in Sedimentology. In: Middleton GV (ed) Encyclopedia of Sediments and Sedimentary Rocks. Kluwer Academic, Dordrecht, pp 385–397

Maynard JB (1980) Sulfur isotopes of iron sulfides in Devonian-Mississippian shales of the Appalachian Basin: Control by rate of sedimentation. Am J Sci 280:772–786

Mook WG, Tan FC (1991) Stable carbon isotopes in rivers and estuaries. In: Degens ET, Kempe S, Richey JE (eds) Biogeochemistry of Major World Rivers, SCOPE 42. Wiley, New York, pp 245–264

O'Neil JR, Kharaka YK (1976) Hydrogen and oxygen isotope exchange reactions between clay minerals and water. Geochim Cosmochim Acta 40:241–246

Schäfer P, Ittekkot V (1993) Seasonal variability of $\delta^{15}N$ in settling particles in the Arabian Sea and its palaeogeochemical significance. Naturwissenschaften 80:511–513

Urey HC (1947) The thermodynamic properties of isotopic substances. Journal Chemical Society 1947:562–581

Glossary

Vocabulary is half the battle

Below are many terms useful in the study of muddy sediments and rocks. Some terms are related to the theory and study of sedimentary and burial processes, others are used for field, petrographic, or engineering descriptions and some are used in industrial applications. For more terms and additional insights consult Jackson (1997) and Middleton (2003).

A-horizon Uppermost, organic-rich layer of soil. May have subdivisions. Transition to underlying B-horizon may be abrupt, transitional, planar or irregular.

Abandonment facies Refers to the fill of an abandoned chute, channel, or canyon, principally by fines deposited from suspension.

Absorption Uptake of a chemical substance into the interior of a solid. Especially characteristic of 2:1 clay minerals and zeolites. Dominantly a physical process. Compare adsorption.

Abyssal Water depths in the ocean greater than 1,000 m; depths with temperatures of 4 °C or less.

Accommodation Space available for deposition; depends on interaction of tectonic subsidence, terrigenous supply, compaction, submarine topography, and sea or lake level.

Adobe Widely used term for sun-dried brick made of mud and silt.

Adsorption Trapping or adherence of ions or organic molecules in solution to the broken bonds of a surface with which they are in contact. Dominantly a chemical process.

Advection The lateral flow of a fluid.

Aerobic Biofacies representing normally oxygenated bottom water – no oxygen restriction on fauna (Tyson and Pearson 1991, Table 2). The corresponding term for the environment is oxic.

Alum shale An old name for a fine-grained, carbonaceous rock rich in alum, $KAl(SO_4)_2 \cdot 12\,H_2O$, formed by the alteration of iron sulfides in organic-rich shales. Also alum schist and slate. Cambrian Alum Shale Formation of Sweden is a famous example.

Anaerobic Lack of free oxygen. Also a biologic community (mostly bacterial) that can live and grow under these conditions (Tyson and Pearson 1991, Table 2).

Anoxia Absence of dissolved oxygen in pore water or bottom water.

Aquifer A permeable bed or formation that transmits groundwater to wells or springs in economically significant quantities.

Aquitard A bed or layer that retards, but does not totally block, the flow of water. Most mudstones act as aquitards. Same as confining bed.

Argille scagliose An Italian term for a thick, slumped deposit of deformed, chaotic plastic clay or mudstone produced by a gravity slide.

Argillite Field term for a totally indurated mudstone that no longer fractures or splits along stratification.

Ash Volcanic debris finer than 2 mm and thus an important airborne contributor to many mudstones along active margins.

Attapulgite Industrial (trade name) for palygorskite.

B-horizon Intermediate zone of accumulation of fine clays and colloids derived from downward transport from the A-horizon. Blocky, prismatic structure is typical, as is a dense, tight texture. Commonly leached. Many subdivisions. See fragipan.

Backswamp Refers to the floodplain of a river that is low-lying, swampy, frequently flooded and commonly poorly drained. Major site of alluvial muds and silts. Peats possible.

Ball clay A plastic clay mostly composed of kaolinite; used for its bonding properties in ceramic materials.

Basin floor deposits Distal fines deposited mostly from suspension in the topographically lowest part of a basin. Either subaerial or subaqueous.

Bathyal Water depths between 200 and 1,000 m.

Bearing capacity The weight a soil can support before failure. The ultimate bearing capacity is the average load per unit area required to produce failure by rupture of the soil mass. The allowable bearing capacity is, on the other hand, the maximum pressure that can be reasonably permitted on a soil.

Bed load Silt, sand and larger clasts that are pushed or pulled (rolled, slid, or saltated) along the bottom at a much lower velocity than the main body of the current. This detritus comes to rest as flat lamination, ripples, dunes, and bars. Also called contact load.

Beidellite A high aluminum smectite.

Bentonite An altered ash fall. A fine-grained, soft argillaceous sediment that is greasy and soap-like and absorbs much water. Its dominant clay mineral is smectite. Swelling bentonites are high in Na, whereas non-swelling varieties are high in Ca or K. Bentonites are deposited in the marine realm from volcanic ash clouds over very wide areas and thus commonly make good stratigraphic markers. Easily recognized on gamma-ray logs in carbonate rocks by their high radioactivity.

Biomarker A term applied to an organic molecule or compound for which a biological origin can be inferred, a molecular fossil. Biomarkers can be used as fingerprints for depositional environments (much as microfossils) and are routinely used for typing oils and making oil-to-source rock correlations. Also used as thermal stress markers. Synonyms include geochemical fossils, biological markers and molecular fossils. See Philp (1985).

Biotite A trioctahedral mica whose ideal composition is K(Mg, Fe)3 AlSi3O10(OH)2.

Bitumen Organic matter soluble in organic solvents. Principal chemical constituents are C and H, whereas O and S are minor.

Bituminous shale See oil shale.

Black shale A field term for any argillaceous, fine-grained (silt size or finer), dark-colored (black or dark gray), laminated rock rich in organic matter; e. g., an organic-rich (more than 0.5%) claystone, clay shale, etc. (See Huyck 1990). But also used more loosely to include fine-grained, organic-rich argillaceous rocks that are not laminated; i. e., dark, organic-rich mudstones. Many diverse origins in both deep and shallow water, but all involve anoxia. The most studied of all mudstones.

Boulder clay Old name for a clay-rich glacial till containing striated boulders.

Brown clay A fine-grained, oxidized, deep-ocean pelagic sediment with less than 30% microfossils. Totally lacks calcareous skeletals, because they have been dissolved, but may contain siliceous ones.

Brucite A Mg hydroxide that is a structural component of some trioctahedral micas and clays.

Burial curve See depth-subsidence curve.

C-horizon Lowermost soil horizon consisting of altered oxidized bedrock, somewhat leached.

CIA index The Chemical Index of Alteration, an index of chemical weathering based on the formula $[Al_2O_3/(Al_2O_3 + CaO^* + Na_2O + K_2O)] \times 100$, where CaO^* equals total CaO minus that in calcite, dolomite, and apatite. Scale ranges from 100 (complete weathering) to about 50 (depending on the source rock). This formula emphasizes the conversion of feldspar to clays. See also the ICV index, Index of Compositional Variability.

Cannel coal An oil shale, dull black to brownish black, that contains abundant liptinite derived from spores and pollen.

Catagenesis The intermediate stage of alteration of organic matter that thermally degrades kerogen to produce oil and wet gas (Tissot and Welte 1984, pp 71–73).

Cation exchange capacity, CEC Number of chemical equivalents per unit weight of dry clay (milliequivalents per 100 grams). Measures the concentration of unfixed cations in interior and surface layers of a clay mineral. These unfixed ions can exchange with a solution and thus CEC is a measure of the "reactivity" of clay. A key parameter for many environmental applications.

Chemical fossil See biomarker.

China clay Industrial term for a residual Fe-poor clay composed mostly of kaolin, is very inert, and gives a white color to both fired and non-fired materials.

Chlorite group A mica-like mineral with an additional octahedral layer in the interlayer position and therefore a somewhat variable chemistry.

Clay Three common usages. (1) A textural term for a fine-grained sediment with appreciable material smaller than 4 microns (sedimentology) in which clay minerals and silt-sized quartz and feldspar predominate. May or not be organic- or carbonate rich. (2) A fine-grained soil that exhibits plasticity within a range of water contents, but when dried, exhibits considerable strength (soil mechanics). (3) A fine-grained deposit consisting predominantly of clay minerals such as illite, smectite, kaolinite, etc. (mineralogical viewpoint).

Clay belt A linear to ovate area, where clay-rich sediment predominates at the surface. Diverse origins glacial lake bottoms formed between a topographic barrier and an ice sheet, coastal deposits of tidal origin, or the flood deposits of a low-gradient stream.

Clay breccia A breccia with a clay matrix. Commonly results from overpressuring.

Clay curtain A thin sheet of clay draped over depositional topography. Common in fluvial and tidal environments. Also clay or mud drape.

Clay dike A dike consisting of clay or mudstone; commonly associated with mud or mudstone diapirs.

Clay dune An eolian dune formed by grains or pellets of clay (Coffey 1909).

Clay factory Analogous to the carbonate factory of carbonate sedimentology, the term clay factory refers to the ultimate source of clay minerals, chemical weathering in soils.

Clay ironstone An iron-impregnated mudstone, argillaceous siltstone, or fine-grained sandstone; commonly consists of siderite when fresh, but hematitic when weathered. Typical as nodules, concretions, or thin beds.

Clay pebble conglomerate A conglomerate composed essentially of clasts of subrounded clay or mudstone locally eroded; common in fluvial deposits and many proximal turbidites.

Clay plug An abandoned channel, straight or curved, filled largely with clay or mudstone. Common in meandering stream deposits.

Clay skins Coats of clay-sized material that occur on grains, aggregates or peds mostly in soils, see also cutans.

Clayshale An indurated, fine-grained rock composed of more than 66% clay-sized constituents with stratification less than 10 mm. See also Folk (1954, Table 20) and Lewan (1978, Fig. 1).

Claystone An indurated, fine-grained rock composed of more than 66% clay-sized constituents. See also Folk (1954, Table 2).

Clinoform Subaqueous deposit with consistent, internal, inclined beds formed by deposition prograding into deeper water. This deposit, first proposed by Rich (1951, Fig. 1), defines a paleotopography and thus the direction of infilling of a basin. Recognition of clinoforms is essential for understanding the architecture and history of filling of muddy basins and for the study of seismic stratigraphy. Many varieties recognized in seismic stratigraphy such as sigmoid, oblique, sigmoid-oblique, complex, shingled, and hummocky (Mitchum et al. 1977, Fig. 6).

Cohesion Technically defined as the shear strength of soil or sediment in excess of that produced by interparticle friction. Less technically, the tendency of any fine-grained material to stick together.

Condensed section Thin, time-rich, distal stratigraphic units deposited during slow sedimentation at maximum landward migration of the shoreline (maximum transgression/maximum relative sea level rise). Equivalent beds are thicker and closer to basin margin. Commonly identified by high radioactivity and diverse pelagic fauna, but may also have hard grounds, fossil lags, pyrite, etc. Key element when using sequence stratigraphy to study mudstones.

Confining bed (layer) Term used to describe a layer of low permeability sediment, commonly argillaceous, that restricts groundwater flow. Same as aquitard.

Conodont alteration index, CAI Color changes of conodonts from light yellow to dark brown and black reflect increasing burial temperatures. Most useful in Paleozoic sediments that have passed the peak of oil generation (Epstein et al. 1977).

Consolidation (compaction) The gradual reduction of the volume of a sediment resulting from an increase in compressive stress and the expulsion of fluid from its pores and rotation of framework grains into more stable configuration.

Corg content Shorthand for total organic carbon in a sediment.

Corrensite A regularly interstratified mixed-layer clay mineral containing chlorite and smectite. Named after a famous German petrologist.

Cutans Soil term for thin films of clay deposited on free surfaces within a soil.

Cyclopel A name proposed by Mackiewicz et al. (1984, p. 129) for diurnal interlaminations of fine sand or silt followed by clay produced by glacial discharge. Base of silt or sand is sharp and overlying mud poorly sorted. Lithified equivalent is cyclopelite.

Degree of pyritization, DOP A simple quantitative measure of pyrite abundance whose variation is relatable to the oxygen level of the sediment; high values correspond to restricted and inhospitable bottoms (Berner 1970; Raiswell et al. 1988).

Depth-subsidence curve Plot of the burial history of a formation, now an almost standard feature of mudstone study, because it tracks burial through time. With this curve one can relate changes in mudstone properties to absolute ages and thus to far distant events. Also called a geohistory curve.

Detached turbidity current A weak, low-density turbidity current separated from the bottom by a denser body of water. The detached current travels along this interface to deposit a gentle rain of fine silt or mud along the bottom. The resulting laminations are believed to lack grading, to be very continuous, may be as thin as a few grains of silt, and are commonly interlaminated with organic-rich facies (Stanley 1983; O'Brien 1989). See also suspension cascading.

Diagenesis All the secondary chemical and physical changes from deposition up to metamorphism of both inorganic and organic constituents; Tissot and Welte (1984, p. 73), however, restricted the term to those changes in organic matter that occur at shallow depths that transform organic matter into methane and kerogen. See also catagenesis.

Diamicton A widely used descriptive term for an unconsolidated, poorly sorted sediment commonly with appreciable fine-grained matrix; examples included clay-rich till, many landslide deposits, colluvium, etc. If lithified, diamictite.

Diatomite A fine-grained, siliceous rock composed dominantly of diatoms with some admixed clays. Typical of many deep marine basins.

Dickite A member of the kaolinite group, but with a distinctive stacking (polytype) of units.

Digital sediment color analysis, DSCA A quantitative measure of color variation from black to gray to white and thus very good to quantitatively define rhythmically stratified, fine-grained sediments and rocks. Also termed "gray value analysis". See Milankovitch cycles.

Dispersive pressure Term used for fluid pressure in a grainflow or mudflow that support its solids while they are in motion. High dispersive pressures facilitate transport over long distances on gentle slopes and support large clasts and boulders in mud and debris flows.

Downlap The convergence (pinch out) of the prograding front of a clinoform with an underlying surface composed of older sediments; also the "turn around" between transgressive and regressive deposits.

Dropstones Isolated or clustered clasts of diverse composition that fell vertically or obliquely into a fine-grained matrix; clasts rafted into deeper water and thus dropped into fine-grained sediment. The best indicator of a glacial marine or glacio-lacustrine origin of a mudstone.

Dysaerobic Impoverished biofacies with some oxygen restriction. Develops between 0.2 to 2.0 ml/l of oxygen in bottom waters (Tyson and Pearson 1991, Table 2).

Dysoxic Water low in dissolved oxygen between 0.2 to 2.0 ml/l. (Tyson and Pearson 1991, Table 2)

Effective stress Difference between normal burial pressure and pore water pressure; when this pressure difference is zero, the sediment liquefies. Effective stress is a key parameter for understanding soft sediment deformation and slope failures.

Eh The oxidation potential, Eh provides a quantitative measure of the oxidation state of an environment by determining the potential difference between platinum and hydrogen electrodes; thus a quick and important field measurement. Also referred to as oxidation-reduction potential or ORP.

Euxinic An environment or a basin with no measurable dissolved oxygen and free H_2S.

Expandable clay or soil Any clay or soil that expands in the presence of water. See Gilgai microtopography.

Exaerobic A biofacies at the boundary between anaerobic and dysaerobic. Characterized by a few opportunistic species proliferating in great abundance on bedding planes in otherwise barren strata or at the transition from laminated to bioturbated strata.

Facies Sum total of all observable properties of a rock unit. Expressed another way, facies can be thought as rock units distinguished in the field by their origin, lithology, organic components, and structure; e.g., the channel-fill mud facies, the overbank facies, organic-rich facies, greenschist facies, etc.

Fall-out or settling structures (deposits) Sedimentary structures or deposits made by deposition from suspension rather than traction; see also clay drape.

Fat clay A field term earlier much used for a clay that is plastic when wet (the CH clays of the Soil Classification System).

Fireclay Synonym for underclay.

Fissility A secondary weathering property of mudstones that causes them to have many thin, paper-like laminations.

Flint clay A smooth, hard kaolinitic clay that breaks with conchoidal fracture and slakes but little (See Fig. 9.8B).

Floc Short for floccule.

Flocculation A general term for the loose bonding of individual flakes of clay and grains of silt in suspension to form loosely connected aggregates or flocs. Flocs of clay-sized particles settle much faster than individual ones.

Floccule A cluster of loosely bound clay minerals in association with fine silt and organic debris held together by electrostatic forces or organic sheaths. Probably the dominant particle that is deposited to form most marine clays. Informally called a floc.

Fluff mud Soupy mud that absorbs wave energy (See fluid mud).

Fluidization Loss of strength sufficient to convert semi-consolidated deposits of mud or silt into thick, soupy masses. Loss of strength results from increase of pore water pressure or presence of methane gas. Responsible for soft sediment deformation and mud flows (See effective pressure).

Fluid mud A fine-grained, sediment-water mixture with a sediment concentration of more than 104 mg/l. A bottom with fluid mud effectively dampens wave energy and thus becomes a type of energy absorbing mud trap. Also known as sling mud (Surinam), fluff mud, and crème de vase.

Fragipan The hard, dense horizon of a soil with fine colloidal clays, clay films (cutans), blocky structure, and little organic matter. Commonly oxidized and leached. Poor permeability impedes flow of water, root development, etc. Many paleosols preserve only fragipans or C-horizons.

Fuller's earth An industrial term for a fine-grained, non-plastic natural clay with high adsorbing capacity (high surface area) commonly composed of calcium smectite as well as colloidal silica and opal. Palygorskite may also be a component. Many industrial uses and much appreciated by pet owners and those who work with grease.

Geohistory curve See depth-subsidence curve.

Gibbsite An aluminum hydroxide that represents the penultimate stage of weathering. A major component of bauxite. Also a structural component of dioctahedral micas and clay minerals.

Gilgai microtopography A characteristic weathering feature of expansive, smectitic clays caused by their high shrink-swell potential, which produces disruptive cracks and microtopography. See Paton (1974).

Glauconite Fine grained, iron-rich dioctahedral clay mineral. Forms penecontemporaneously on the sea bottom during slow sedimentation; distinct, greenish color is typical. Many morphologies.

Grayness Informal expression for digital sediment color analysis.

Gyttja A Swedish term for a freshwater, organic-rich mud.

Halocline A rapid downward increase in salinity in a body of water. See thermocline.

Halloysite A poorly ordered 1:1 clay mineral of the kaolinite group similar to kaolinite, but with a different stacking pattern; has tubular morphology in hydrated form.

Hemipelagic Fine terrigenous material carried in suspension by oceanic currents and deposited around the continents. For a deposit to be termed hemipelagic, it should have more than 25% terrigenous material smaller than 5 microns (Jackson 1997, p. 295).

Highstand The period of time when relative sea level and shoreline are farthest cratonward. Maximum retreat of the shoreline with terrigenous deposition far inshore from the shelf break. See related concepts of maximum flooding surface and condensed section.

Hydraulic conductivity Term closely allied to permeability, used by groundwater geologists to quantitatively describe the ability of a layer to transmit water.

Hydrogen index, HI The amount of hydrocarbons that may be generated from a rock or sediment (total hydrocarbons/organic carbon); thus used to give some indication of petroleum potential. Commonly plotted against the oxygen index. A high hydrogen index and low oxygen index indicate a good source rock for oil. One of the important parameters determined by Rock-Eval analysis.

Hypoxia Dissolved oxygen less than 2 ml/l (ppm).

ICV index The Index of Compositional Variability is similar in purpose to the CIA index except that weights of oxides are used rather than moles; also, Fe and Mg are included.

Illite/muscovite group A clay mineral group with a 2:1 structure and a 10 Å spacing. Members include muscovite, which is dioctahedral and has the formula $K\,Al_2\,(Al\,Si_3)O_{10}(OH)_2$. When the muscovite mineral is very small, 4 microns or less, and has less than 1 Al in the tetrahedral sheet it is called illite. Glauconite is a greenish, iron-rich variety. Illite comes in 1 Md (diagenetic) and 2 M (greenschist metamorphic) varieties.

Illite crystallinity The sharpness of the illite peak at 10 Å. The better the crystallinity, for a given original value at deposition, the greater the degree of burial metamorphism. The preferred way to measure is peak width at half height in degrees 2θ (Kubler 2000).

Illite-smectite The most common of the mixed layer clay minerals; the proportion of smectite interlayers decreases with increasing depth of burial.

Illitization Increase of illite with temperature and depth, generally at the expense of smectite and feldspar.

Kaolinite group A clay mineral group with uncharged dioctahedral layers with a 1:1 structure and 7 Å spacing; includes kaolinite, nacrite, dickite, and halloysite. The kaolinite group is chemically stable at the Earth's surface, but not at depth.

Kerogen Chemically altered sedimentary organic matter that is insoluble in organic solvents and acids such as HF and HCl. Dead organisms, both terrestrial and marine, are the source of kerogen. Three major types recognized. Kerogens derived from higher plants (Type II) tend to yield gas whereas kerogens derived from phytoplankton and algae (Type I) tend to yield oil.

Kukersite Ordovician oil shale (up to 30% TOC) of the Baltic region with abundant benthic fossils.

Lama Portuguese word used as a general field term for mud; limo is the Spanish equivalent.

Lamination Stratification thinner than one centimeter. Many different kinds.

Lamosite An olive-gray brown or brownish black oil shale that consists mostly of lamalginite derived from lacustrine and planktonic algae (Hutton 1986, p. 86).

Lean clay A clay with a high silt content and only somewhat plastic, a CL type clay of the Soil Classification System. Now less used than in earlier years.

Lehm German field term for loam, clay, or mud.

Level of organic metamorphism, LOM Defined by Hood et al. (1975, p. 987–989) to establish the equivalence of different maturity indicators such as vitrinite reflectance, spore carbonization, thermal alteration index, etc.

Liquefaction Transformation of semi-consolidated mud, silt, or sand into a slurry, which may then flow down slope or be injected along a fracture upward into overlying beds as a dike. When injected onto the sea floor or at the surface, small mud or sand volcanoes form. Overpressured pore water, commonly methane rich is responsible.

Liquid limit, LL The water content of a sediment corresponding to the arbitrary limit between the liquid and plastic states of consistency of a soil. More technically, it is the water content at which a pot of soil, cut by a groove of standard dimensions, will flow together for a distance of one-halfinch under the impact of 25 blows in a standard liquid limit apparatus (Das 2000, p. 27–28). One of the Atterberg limits.

Limo Spanish term commonly used for a silty mud; textural term for silt. Limolita is used for siltstone.

Littoral Sea bottom between high and low tides.

Loess A fine-grained, dominantly wind-blown deposit composed chiefly of silt-sized quartz and some feldspar and calcite that covers preexisting topography. Has two origins – glacial (dust blown from valley trains) and nonglacial (dry plains, deserts). The parent material of many rich soils. Widespread in Quaternary deposits.

Loop structure A group of laminae that pinch out abruptly, possibly by flow of intervening mud or by dewatering.

Lowstand Maximum lowering of relative sea level, possibly even below the shelf break, with good development of shelf channels, deep sea fans and much hemipelagic mud carried to the deep sea.

Lutite An older name for a fine-grained, consolidated argillaceous rock.

Maceral Name originally applied to the components of a coal, but now also used for visually different organic components in fine-grained sediments and rocks.

Marine snow Fine-grained, diverse aggregates such as fecal pellets, organic debris, and particles bound by organic sheaths, surface tension, and Van der Wals forces. Collectively, these form loose, fragile, macroscopic aggregates that settle slowly to the bottom.

Marinite A grayish brown to dark brown oil shale containing lamalginite derived mostly from marine algae and bituminite (Hutton l986, p. 86).

Maximum flooding surface Term used in sequence stratigraphy to describe the surface at the transition between transgressive and regressive deposits – we think of this as the "turn around" surface. Highstand deposits downlap onto this surface. Recognized by maximum fossil diversity, high clay content and high gamma radiation. A key stratigraphic horizon for the study of mudstones.

Metabentonite A somewhat older term for a bentonite that has been buried and has had time to alter mostly to illite and illite-smectite. Hence "metabentonite". See potassium bentonite.

Metagenesis Final stage of alteration of organic matter at great depth (vitrinite reflectance of about 2%) that occurs prior to mineral alteration (Tissot and Welte 1984, p. 73). Methane generated.

Metalliferous shale A shale rich in metals. Such shales are nearly always black, are variously defined but one approach is the following: For a shale to qualify as metalliferous, any given metal in it should be enriched at least two times over the values reported in the U.S. Geological Survey Standard Reference Material SDO-1, except for Be, Co, and U for which only equivalence is needed (Huyck 1990). Also known as metal-rich black shales (MRBS).

Metashale Field term for a deformed and/or severely metamorphosed shale.

Mica See Illite/muscovite group and Biotite.

Microfabric Refers to the fine-scale arrangement of clay-sized particles in a mud or mudstone as seen with a scanning or transmission electron microscope. Fabric includes the arrangement, orientation, and particle-to-particle contacts of the components. Variation of fabric types affects physical properties of mud or mudstone and may reflect initial conditions as well as bioturbation.

Micstone Short for micritic limestone (Leighton and Pendexter 1962, p. 62). A fine-grained carbonate rock consisting of calcite or dolomite (dolomicstone). A lithified carbonate mud, approximately equivalent to a mudstone, especially one with little or no silt.

Milankovitch cycles These cycles are increasingly recognized in argillaceous and other fine-grained rocks deposited in lakes and in both deep and shallow (but protected) marine basins. Caused by variations in solar energy with periodicities of 19,000 to 403,000 years (see Box 8.2).

Mixed-layer (interstratified) clays Two different clay minerals regularly or randomly interstratified, usually with an expandable clay mineral; the most common examples are illite-smectite and chlorite-smectite.

Molecular fossil See biomarker.

Montmorillonite Old group name for swelling clays, now replaced by smectite, with montmorillonite reserved for dioctahedral varieties.

Mud Field term for an unconsolidated, fine-grained, sticky or slimy sediment of any origin. May be clay- or carbonate rich and normally contains some silt-sized components and organic material. Plastic when wet. The Oxford English Dictionary lists 57 other popular terms for mud!

Mud blanket A distinctive, relatively uniform and easily mapped, widespread deposit of mud. Useful as local markers and seals.

Mud coastal fringe Mud deposit along shore commonly downstream from a muddy river (or may be deposited by tides).

Mud delta Informal term for a low-energy, Mississippi River-type delta rich in mud, because either inshore wave energy is low or tidal range small. Sand much subordinate to mud, which accumulates chiefly as prodelta marine muds and in interdistributary bays.

Mud drape (curtain) A relatively thin and laterally extensive sheet of mud deposited at slack water over a surface formed by ripple marks, sand waves, the side of a riverine or turbidite bar in a channel, etc.

Mud flood A flowing mass of mud. Such floods occur at the lower end of hyper-concentrated flows, have concentrations of mud between about 25 and 45%, and have vertical size gradients (Liow 1989, p. 111–112). Common on alluvial and subsea fans.

Mudflow A flow intermediate between a mud flood and a landside; mud concentrations greater than 45% by volume (Liow 1989, p. 112). Loosely speaking, a thick, viscous, muddy mass.

Mudflow gully Term used for regressive subsea failure on a delta front consisting mostly of mud and silt. Linear gullies terminate downdip in muddy lobes. Once formed, these gullies are conduits for turbidity currents.

Mudline The line on a shelf that separates mud from silt, sand, or carbonate; i. e., the limit of mud deposition (Stanley and Wear 1978) and thus useful to circumscribe the limits of a "muddy basin".

Mud lump An extrusion of soft plastic mud that reaches the surface to create a temporary obstacle or island. Essentially the same as a mudstone diapir.

Mudrock An indurated, terrigenous, fine-grained rock composed of 33 to 65% clay-size constituents with stratification greater than 10 mm. See Folk (1954, Table 2) and Lewan (1978, Fig. 1). May be siliceous, dolomitic or phosphatic. See also micstone, its carbonate equivalent.

Mudshale An indurated, fine-grained sedimentary rock composed of 33 to 65% clay-sized constituents with stratification less than 10 mm. See also Folk (1954, Table 2).

Mudshale prism A wedge of mud or mudstone that thickens downdip.

Mudstone Used by us as the general family name for all fine-grained argillaceous rocks; parallel to sandstone, siltstone, limestone, and dolostone.

Mudstone diapir Intrusion of mudstone or clay into overlying sediment. See overpressured mudstone.

Mudstream A plume of river-derived, muddy water along a coast; also where a muddy river joins a clear river.

Mud wave A large, low ridge or wave-like body of low viscosity mud on the sea floor or along the shore.

Mud withdrawal Refers to overpressured mud or mudstone that migrates laterally into a vertically rising mud or mudstone diapir.

Muscovite See Illite/muscovite group.

Nepheloid layer A turbid layer in the ocean, usually hundreds of meters thick, that contains appreciable, fine suspended matter and thus has a density contrast with water above and below. Intermediate nepheloid layers (INL) occur at the shelf edge and on the slope. Bottom nepheloid layers (BNL) are widespread and thick, up to 2,000 m, in deep trenches. Nepheloid layers have been suggested as the origin of even, "single grain" laminations observed in mudstones. See overview by McCave (1986).

Neritic The bottom of the ocean from low tide to 200 m, commonly to the shelf break.

Nontronite An iron-rich dioctahedral smectite, usually green.

Normally consolidated soil A soil that has never been subjected to an effective pressure greater than the existing overburden pressure and one that is also completely consolidated by the existing overburden.

Obrution bed A mudstone bed formed by a slurry that smothers and kills most benthic organisms in its path. Commonly the fossils of such organisms exhibit exceptional preservation, because they escape both bottom scavengers and oxidation.

Oil Shale A widely accepted definition is that of Gavin (1922, p. 26), "Oil-shale is a compact, laminated rock of sedimentary origin, yielding over 33 percent of ash and containing organic matter that yields oil when distilled, but not appreciably when extracted with the ordinary solvents for petroleum".

Ooze A mixture of fine, terrigenous mud plus at least 30% siliceous or calcareous pelagic organisms. Many modifiers and mixtures possible; e. g., calcareous, siliceous ooze, diatomaceous ooze, calcareous nannofossil ooze, etc.

Organic facies A sediment or rock, usually fine-grained, consisting of more than 5% organic matter such as peats and coals, some lacustrine shales and many black shales.

Organic overloading Informal term used to describe anoxia caused by more organic material than can be oxidized by available oxygen.

Overbanking Refers to the deposition of mostly fine sediment during high flows outside of or beyond normal channel limits. Can occur on floodplains, in interdistributary bays, along the margins of tidal channels or on subsea fans between their distributary channels. Muds and silts predominate.

Overconsolidated soil A soil that has been subjected to an effective pressure greater than the present overburden pressure.

Overloading Oxygen demand greater than oxygen supply – a key condition for anoxia.

Overpressured mud or mudstone A mechanically unstable mud or mudstone whose internal hydrostatic pressure exceeds lithostatic pressure and thus is unstable and tends to move upward in the section. See also mudstone diapir and mud lump.

Overturning The disruption by wind, storms or exceptional currents of a stratified body of water; poorly oxygenated bottom waters come to the surface and may kill life in the photic zone.

Oxic Water with more than 2.0 ml/l of dissolved oxygen. Supports a diverse, aerobic fauna.

Oxygen-minimum zone, OMZ A mass of water that is lower in oxygen than the water above or below it. Also known as oxygen minimum layer.

Oxygen Index A measure of the amount of oxygen (as carbon dioxide) released by the pyrolysis of kerogen. Used with the hydrogen index to determine kerogen types.

Palygorskite A 2:1 clay mineral with an elongate, fibrous shape (commonly called attapulgite in industry) is prized for its sorptive and bleaching qualities. A principal component of Fullers earth.

Palynofacies The detailed integration of palynology with depositional environments and sequence stratigraphy (changing proportions of continental and marine palynomorphs, phytoclasts and their sorting, degree of abrasion, weathering, etc.) as guides to the depositional environment.

Palynomorph Cellular acid-insoluble microfossils such as spores, acritarchs, and pollen that are present in many mudstones. Useful for correlation, interpretation of the depositional environment (via their sedimentology), burial history (coloration and vitrinite reflectance), and establishing proximity to a shoreline.

Parasequence Described as the "key building block of stratigraphy," a parasequence is formally defined as "a relatively conformable succession of genetically related beds bounded by marine flooding surfaces (commonly mudstone zones) and their correlative beds." Parasequences shoal or coarsen upwards; thickness varies greatly from tens of meters to decimeters or less in deep, muddy basins. Recognizable on logs, in cores, and in outcrops. From a genetic point of view, parasequences are a response to fourth or fifth order oscillations of relative sea level.

Parasequence sets Two or more stacked parasequences; the sets may prograde seaward, regress, or be static (stable shoreline position).

Parna An Australian term for loess.

Pebbly mudstone An argillaceous sediment that contains floating pebbles or cobbles; also diamictite.

Ped A soil term for an individual soil aggregate separated from others by natural voids, planes of weakness or cutans. Peds may be hard, soft, irregular or blocky.

Pelagic deposits Fine-grained sediments deposited from a dilute suspension of pelagic fossil debris in the deep ocean. Skeletal debris consists mostly of pelagic organisms such as diatoms, foraminifera, radiolarians, etc.

Pelite A fine-grained, argillaceous rock now metamorphosed, but commonly with much original stratification still recognizable. Used mostly by metamorphic petrologists.

Pelletal sand A sand-sized sediment composed mostly of clay or carbonate pellets.

Permeability Broadly defined as the ability of a rock to transmit a fluid (water, oil, natural gas) through its pores and fractures. Petroleum geologists favor permeability, whereas groundwater geologists use the related term hydraulic conductivity. These parameters vary over many orders of magnitude and are commonly the most sensitive term in equations of flow; hence the great emphasis placed on their measurement or estimation.

Phytoclast Term for a detrital clast of plant debris. Size, abrasion, and preservation of the clast give clues to transport history.

Plastic index, PI A numerical index defined by the difference between liquid and plastic limits; i. e., $PI = LL - PL$.

Plastic limit, PL The water content corresponding to an arbitrary limit between the plastic and semisolid states of consistency of a soil. The plastic limit is determined by the water content at which a soil will just begin to crumble, when rolled into a thread approximately 1/8 inch in diameter (Das 2000, p. 29)

Polytypism A special case of polymorphism found in clay minerals that refers to C-axis stacking of structural units.

Porcellanite A siliceous fine-grained rock with a dull luster that may be somewhat argillaceous. Commonly less dense, less vitreous and softer than chert.

Porosity The ratio of void space to total volume of a substance. Compare with void ratio. See also solidity.

Potassium bentonite Current usage for a bentonite that has been diagenetically altered to an illite-rich, fine-grained sedimentary deposit, which may still contain some volcanogenic minerals such as biotite, zircon, plagioclase, etc. Volcanic ash beds provide good correlation markers and are well known in the marine sequences of many sedimentary basins. Also see tonstein.

Pozzolan Siliceous volcanic ash that, when mixed with calcium hydroxide, hardens underwater. First used by the Romans from a locality at Puzzuoli, Italy.

Primary migration Movement of hydrocarbons through and out of a maturing source rock.

Pycnocline An abrupt density discontinuity between two water layers. May result from differences in salinity (halocline) or temperature (thermocline). The presence of a pycnocline and its relation to the bottom is a key control on the properties of muds and mudstones.

Quick clay A clay that suddenly loses strength when disturbed, because of excess pore-water pressure or fresh water invasion. A clay that liquefies readily.

Rare-earth elements Trace elements from lanthanum to lutetium on the periodic table. Their chemical properties are very similar, and they are not generally fractionated by sedimentary processes. Thus they make good tracers of original igneous rocks.

Red clay A fine-grained, deep-sea deposit commonly found below 3,500 m that was deposited slowly and thus oxidized either in transit to or at the sea bottom. May be bright red, brownish red, or maroon.

Redox front Term used to describe the contact between oxidizing and reducing water; front may be above the sediment, at or below the sediment-water interface or localized within a mud around a decaying fossil or plant. See Eh.

Remolded clay A clay that, while still soft, has been extruded in a landslide, mud lump or mudstone diapir and thus lacks lamination. Commonly contains poorly sorted clasts of the country rock. A very special type of diamicton.

Residual organic matter A measure of the non-volatile organic matter in a rock.

Runout Distance from point of failure that a debris flow travels downslope (the height-distance ratio is a rough measure of the potential distance a debris flow may be a hazard). As a first approximation, we found runout to be a useful concept, when thinking about the caliber of load on an alluvial fan, large river, or subsea fan.

Rock-Eval A rapid, widely used screening technique based on pyrolysis for the characterization and evaluation of source rocks (Espitalié et al. 1977). Now an oil industry standard.

Saponite A magnesium-rich, trioctahedral smectite.

Saprolite A weathered, soft, rotten rock, commonly red or brown in color and clay rich, in which relic structures such as veins and even bedding can be seen as well as scattered rounded clasts of country rock. Typically, saprolites are relics from Tertiary and Mesozoic weathering.

Sapropel An unconsolidated, organic-rich sludge or ooze composed of higher plant or algal material formed in an anaerobic environment.

Sarl Proposed by Shipboard Scientific Party (1984, Fig. 9) for sediment consisting chiefly of clay and siliceous ooze and having less than 15% calcareous ooze.

Scanning Electron Microscopy (SEM) Imaging of a surface using an electron beam scanned at a high rate. Provides lower resolution than TEM, but shows surface morphology better. Used to study mudstone fabric.

Schist A metamorphosed, fine-grained rock, normally derived from a mudstone, with abundant micas and strong foliation or cleavage.

Seat-earth See underclay.

Secondary migration Movement of oil and gas once it has been expelled from a source rock. Movement of oil and gas from one reservoir to another is referred to as tertiary migration.

SEDEX deposits See shale-hosted massive sulfides below.

Sensitivity A term used by engineers to describe the decrease in shear strength of a clay, at constant moisture content, following remolding. A clay that remolds or liquefies easily is called a sensitive clay and one that does not, non-sensitive.

Sequence A generally conformable succession of genetically related sedimentary rocks limited at their top and bottom by unconformities or correlative conformities. One or more systems tracts form a sequence.

Shale A general field term for a fine-grained indurated laminated sedimentary rock, which mostly consists of clay minerals and appreciable quartz and feldspar silt. Commonly considered to have at least 50% particles smaller than 0.062 mm. Many varieties such as calcareous, organic-rich, etc.; also used as a field term for any consolidated terrigenous fine-grained rock, laminated or not.

Shale arenite A sandstone composed mostly of shale fragments (Folk 1968, p. 124). Rare, but see clay dunes and pelletal sand.

Shale baseline The reference line on a geophysical log corresponding to shale. The exact value of the baseline varies with the type of mudstone and the conditions under which the log was run.

Shale break A thin shale in a thick limestone or sandstone. Easily identifiable with gamma ray or sonic logs or when drilling.

Shale heave Field term for the swelling of a shale or mudstone rich in expandable clays or gypsum into low mounds, folds, or buckles when exposed to water (or reduced confining pressure in a mine). Common problem in the floors of mine galleries.

Shale-hosted massive sulfides Stratiform galena and sphalerite mineralization mostly in dark, organic-rich, basinal shales, commonly with turbidites. Massive pyrite and barite are usually associated. Also called by the more genetic name SEDEX, sedimentary exhalative deposits. The parallel deposit type in volcanic rocks is referred to as volcanic massive sulfide or VMS.

Shale out Informal but useful term to describe the change from limestone or sandstone into mudstone or the abrupt termination of beds against a mudstone-filled channel. Significant in blocking fluid flow.

Shale ridge An elongate intrusion of mudstone or mud that forms a ridge. A special type of mudstone diapir that forms elongate, sub-parallel ridges.

Shale smear A mudstone that occurs along a fault plane. As a fault moves, the intercepted mudstone bed may be stretched and distributed along the fault plane. In this way, it may become a barrier to horizontal fluid flow. Also a technique to make a rough size analysis of mud and semi-consolidated mudstones.

Sharpness index See illite crystallinity.

Shear strength The maximum resistance of a soil or rock to applied shearing forces.

Shelf-slope break On many modern continental shelves this occurs at about 200 m of water depth. Beyond the break mass movements have a much greater role in sediment transport whereas between the break and the shoreline storms, intruding oceanic currents, and tidal currents are more important. Shelf-slope breaks are widely recognized in ancient basins and, as in the modern, have a strong influence on sedimentary facies.

Shell pavement Concentrations of shells on bedding planes; can provide estimates of paleocurrent direction and added insight to environmental conditions. Important for the interpretation of many mudstones. See exaerobic environment.

Siliceous mudstone A mudstone cemented by silica usually the result of diagenesis of original silica-rich debris such as volcanic glass, diatoms, sponge spicules, or radiolarians.

Sill The lowest point in a basin through which water can enter or leave. Large reductions of inflow and outflow favor clay deposition, evaporates or anoxia.

Slaking The crumbling or disintegration of a fine-grained sediment or rock, when exposed to air or moisture. A most important engineering property of muds and mudstones. There are several closely related slake durability tests used in civil engineering.

Slate A metamorphosed mudstone with well-defined cleavage cutting stratification.

Slickensides Formed by ductile deformation in expansive soils as well as in underclays and many mudrocks of varied clay mineralogy. Recognized by shiny, commonly dipping, slightly undulating surfaces with many fine ridges and grooves. Brittle slickensides form after lithification.

Sling mud Local term used on the Surinam coast for a gel-like marine mud. See fluid mud.

Smarl Proposed by Shipboard Scientific Party (1984, Fig. 9) for sediment consisting of about equal parts terrigenous mud, siliceous ooze, and calcareous ooze.

Smectite group Fine-grained clays with a 2:1 structure and with variable spacing from 10 to 15 Å depending upon interlayer cations. Widespread in many soils, in volcanics and in modern sediments; used in drilling fluids, as waterproofing agents, kitty litter, etc. Have high cation exchange capacity. Common group members are montmorillonite, beidellite, nontronite and saponite.

Soil A three-dimensional layer at the Earth's surface capable of supporting plants. Formed by a combination of physical and chemical weathering. Soil types vary with parent material, time, slope, and climate. As used by engineers, soil is any solid material that weakens with water.

Soil horizon A layer of soil, parallel to the surface, with distinct characteristics such as a fragipan, A-horizon, C-horizon, etc.

Solum The combined A and B-horizons of soil terminology. The A- and B-horizons are where the processes of soil formation are most active.

Solidity The volume percent of solids in a mud or mudstone; the complement of porosity.

Spectral gamma ray log Made by a sonde lowered into a well or by hand in the field, this log subdivides total gamma radiation into contributions by uranium (organic matter, some heavy minerals), potassium (K-feldspar, illite, glauconite, micas) and thorium (some heavy minerals, especially monazite and zircon, and phosphates). Ratios such as Th/K or U/K help distinguish a black shale from a bentonite, help identify maximum flooding surfaces, etc.

Spore Small, heavy-walled reproductive part of a plant. Outer wall very resistant and thus a "good survivor" during transport and weathering.

Steinkern Mud infill of a fossil shell or mold.

Stratified water body A water mass consisting of two or more layers differing in density or temperature. Stratification inhibits or prevents mixing and thus insulates bottom sediment from oxygenated surface waters. Stratification of a water body is far more important to the formation of muds and mudstones than it is to sandstones. See pycnocline.

Structural clay Industrial term used for any clay suitable for brick or tile making.

Suboxic Between 0.0 and 0.2 ml/l of dissolved oxygen in bottom waters (Tyson and Pearson 1991, Table 2). Fauna inhabiting this zone are referred to as dysaerobic.

Suspended load Clay, silt, and fine sand supported by turbulent eddies within the flow, which move down current with the same velocity as the current. Upon deposition these particles form "fall-out structures".

Suspension cascading Progressive separation of different density portions of a turbidity current as it encounters successively denser water masses flowing downslope. A "peeling off" of its different water densities on a submarine slope.

Swelling clays 2:1 clays such as the smectite family and bentonites that absorb much water and thus crack when subsequently dry.

System tract A linkage of genetically related depositional systems. High stand, low stand, and transgressive system tracts commonly recognized.

Tasmanite A grayish-brown to black oil shale that contains telalginite derived from Tasmanites (Hutton 1986, p. 86).

Tempestite A bed or lamina of silt or sand with distinctive structure believed to have been formed by storm waves impinging on a sea or lake bottom. Common in the interbedded siltstones and fine sandstones of most shelf and lacustrine mudstones.

Tephra A Greek word for ash, used collectively for all pyroclastic materials.

Terrestrial organic carbon That part of the total organic carbon derived from land plants. See phytoclast.

Thermal alteration index, TAI Pollen and spore colors as seen under the microscope and calibrated to stages of kerogen maturity.

Thermocline A layer of water having a rapid decrease in temperature. See also halocline and pycnocline.

Tmax A laboratory parameter from Rock-Eval analysis. The temperature at which maximum release of hydrocarbons from pyrolysis of kerogen occurs. Used to indicate the stage of maturation of the organic matter.

Tonstein A German term for a thin, kaolinite-rich, argillaceous band or bed deposited from volcanic ash in a coal swamp. These can extend over wide areas in many coal beds. Several subtypes recognized. See Williamson (1970) for details.

Torbanite A black to greenish-black oil shale that contains abundant telalginite derived from the colonial alga Reinschia and Pila (Hutton 1986, p. 85).

Total organic carbon, TOC The weight percent of organic carbon in a rock; used to estimate organic content. As a rock matures, H, O, and N are driven off so that C becomes the predominant organic component.

Total organic matter, TOM The weight percent of organic matter in a rock; included are C, H, N, O and S. Often approximated by total organic carbon, which is easier to measure.

Transgressive surface First important flooding surface above lowstand deposits or an unconformity.

Transformation ratio A laboratory parameter derived from Rock-Eval analysis used to estimate the stage of petroleum generation.

Transmission electron microscopy, TEM An electron-beam imaging system with resolution to a few tenths of a nanometer. Permits recognition of the spacing planes of atoms in clay minerals. Compare SEM.

Tuffaceous mudstone A mudstone rich in fine volcanic debris from either air or water transport, common along active margins, and likely to be unstable in construction.

Turbidity Broadly speaking, a measure of the opaqueness of water usually produced by fine suspended matter.

Turbidity current A bottom-flowing density current caused by sediment in suspension. Many deepwater mudrocks were formed by distal turbidity currents. See unifite.

Underclay Soft, fairly plastic, lightly colored, silty clay beneath a coal bed commonly with many slickensides and plant rootlets; may be partially calcareous. Also called fireclay and seat earth. Resists high temperatures and thus good for refractory products.

Unifite A descriptive term applied to a structureless or faintly laminated, commonly thick mud layer with a fining-upward trend believed to have been deposited by a low density gravity flow (Stanley 1981).

Upwelling The rising of cold, nutrient-rich ocean water along a continent or landmass or in the open ocean far from land. Upwelling currents, because of their nutrients, favor marine life of all kinds, help precipitate carbonates and phosphates, and also produce black shales because of oxygen restriction. See also oxygen minimum and organic overloading.

Van Houten cycles Transgressive-regressive lacustrine sedimentary sequences useful in correlation, named after Van Houten (1964). A variety of Milankovitch cycles.

Varve Annual couplet, typically fine-grained, of a summer (silt-fine sand) and winter (clay) deposit, glacial or non-glacial. Commonly record climate signals.

Vermiculite group Has a 2:1 structure and 10 to 15 Å spacing depending on interlayer cations. The clay mineral vermiculite is similar to smectite, but has a higher layer charge. Strong swelling properties. The term vermiculite is used both for a coarse mica of silt- and sand size as well as a clay mineral.

Vitrinite A kerogen material derived from woody higher plants.

Vitrinite reflectance, R_o A quantitative microscopic measure of the intensity of reflectance of vitrinite. Used to assess both coal rank and hydrocarbon source rock maturity.

Void ratio Engineering term for the ratio of volume of void space in a sediment to the volume of its solids.

Wadden A Dutch word for the muddy tidal flats bordering the Wadden Sea, a shallow part of the North Sea, where important early studies of muddy tidal deposits were made. Many modern and ancient equivalents.

Weathering front The contact between weathered and unweathered rock, which can be either regular or irregular following fractures, burrows, roots, etc. Engineering properties vary greatly above and below such fronts, hence their recognition is all-important.

References

Berner RA (1970) Sedimentary pyrite formation. Am J Sci 266:1–23

Coffey GN (1909) Clay dunes. J Geol 17:754–755

Das BM (2000) Principles of geotechnical engineering. PWS Publishing, Boston, 571 p

Epstein AG, Epstein JB, Harris LD (1977) Conodont color alteration-an index to organic metamorphism. US Geology Survey (Professional Paper) 995:27

Espitalié J and six others (1977) Méthode rapide de caractérisation des roches méres, de leur potential pétrolier et de leur degré d'évolution. Revue Institut Francais de Pétrole 312:23–42

Folk RL (1954) The distinction between grain size and mineral composition in sedimentary rock nomenclature. J Geol 62:344–359

Folk RL (1968) The petrology of sedimentary rocks. Hemphill's Book Store, Austin, Texas, p 170

Gavin MJ (1992) Oil-shale. An historical, technical and economic study. US Department Interior Bureau Mines Bulletin 210:201

Hood A, Gutjahr CCM, Heacock RL (1975) Organic metamorphism and the generation of petroleum. Am Assoc Petrol Geol Bull 59:986–996

Hutton AC (1986) Classification of Australian oil shales: Energy Exploration and Exploitation 4:61–93

Huyck HLO (1990) When is a metalliferous black shale not a black shale? In: Grauch R, Huyck HLO (eds) Metalliferous black shales and related ore deposits: Proceedings 1989 US Working Group Meeting, International Geology Correlation Program Project 254, US Geological Survey (Washington, D.C.), pp 42–56

Jackson JA (ed) (1997) Glossary of Geology, 4th edn. American Geological Institute, Alexandria, Virginia, 769 p

Kubler B, Jaboyedoff M (2000) Illite crystallinity: Compte Rendu Académie Sciences Paris, Sciences de la Terre et des Planètes 331:75–89

Leighton MW Jr, Pendexter C (1962) Carbonate rock types, in Ham, W.E (ed) Classification of Carbonate Rocks. Am Assoc Petrol Geol Memoir 1:33–61

Lewan MD (1978) Laboratory classification of very fine-grained sedimentary rocks: Geology 6:745–748

Liow J (1989) Mud flood and mudflow mapping in Davis County, Utah, in Proceedings of the Conference on Arid West Flood Plain Management Issues, Las Vegas, Nevada, October 19–21, 1988: Association of Flood Plain Managers, Madison, Wisconsin, pp 111–146

Mackiewicz NE, Powell RD, Carlson PR, Moling BF (1984) Interlaminated ice-proximal glaciomarine sediments in Muir Inlet, Alaska. Marine Geol 57:113–147

McCave IN (1986) Local and global aspects of the bottom nepheloid layers in the world ocean: Nederlands J Sea Res 20:167–181

Middleton GV (ed) (2003) Encyclopedia of Sediments and Sedimentary Rocks. Kluwer Academic, Dordrecht, 821 p

Mitchum RM Jr, Vail PR, Sangree JB (1977) Seismic stratigraphy and global changes of sea level, Part 6: Stratigraphic interpretation of seismic reflection patterns in depositional sequences. In: Payton CE (ed) Seismic stratigraphy-applications to hydrocarbon exploration. Am Assoc Petrol Geol Memoir 26:117–134

O'Brien N (1989) Origin of lamination in Middle and Upper Devonian black shales, New York State. Northeastern Geology 11:159–165

Paton TR (1974) Origin and terminology of gilgai in Australia. Geoderma 11:221–242

Philp RP (1985) Fossil fuel biomarkers: applications and spectra. Elsevier, Amsterdam, 294 p

Raiswell R, Buckley F, Berner RA, Anderson TF (1988) Degree of pyritization as a paleoenvironmental indicator of bottom-water oxygenation. J Sediment Petrol 58:812–819

Rich JL (1951) Three critical environments of deposition, and criteria for recognition of rocks deposited in each of them. Geol Soc Am Bull 62:1–20

Shipboard Scientific Party (1984) 1. Introduction and explanatory notes, Deep Sea Drilling Project Leg 75 in Amidei R, ED, Initial Reports of the Deep Sea Drilling Project, v. LXXV, Part 1, US National Science Foundation, Washington pp 3–25

Stanley DJ (1981) Unifites: structureless muds of gravity-flow origin in Mediterranean basins. Geo-Marine Lett 1:77–83

Stanley DJ (1983) Parallel laminated deep-sea muds and coupled gravity flow hemipelagic settling in the Mediterranean. Smithsonian Contributions to Marine Science 19:19

Stanley DJ, Wear CM (1978) The "mud-line": an erosion-deposition boundary on the upper continental slope. Marine Geol 28:M19-M29

Tissot BP, Welte DH (1984) Petroleum formation and occurrence, 2nd edn. Springer, Berlin Heidelberg New York, p 699

Tyson RV, Pearson TH (1991) Modern and ancient shelf anoxia: an overview in Tyson RV. and Pearson TH., Eds., Modern and ancient shelf anoxia shelf anoxia. Geol Soc (Special Publication) 58:1–26

Van Houten FB (1964) Cyclic lacustrine sedimentation, Upper Triassic Lockatong Formation, central New Jersey and adjacent Pennsylvania. Kansas Geology Survey Bulletin 169:497–531

Williamson IA (1970) Tonsteins: their nature, origin and uses. Mining Magazine 122:119–125 and 203–211

Subject Index